DRAGONFLY

ALSO BY BRYAN BURROUGH

Vendetta
Barbarians at the Gate
(with John Helyar)

DRAGONFLY

NASA and the Crisis Aboard Mir

BRYAN BURROUGH

HarperCollinsPublishers

Page 1: Photo courtesy of NASA; Photo courtesy of Jo Tyler/Corbis/Bettman. Page 2: Photo courtesy of Pierre du Charme/Reuters/archive photo; Photo courtesy of NASA. Page 3: Photo courtesy of AP/Wide World; Photo courtesy of NASA; Photo courtesy of Joe Skipper/Reuters/archive photo. Page 4: Photo courtesy of Sipa Press; Photo courtesy of NASA. Page 5: Photo courtesy of T.Globa; Photo courtesy of Bert Vis. Page 6: Photos courtesy of NASA. Page 7: Photo courtesy of Slim Films. Page 8: Photos courtesy of NASA; Photo courtesy of AP/Wide World. Page 9: Photos courtesy of NASA. Page 10: Photos courtesy of NASA. Page 11: Photos courtesy of NASA. Page 12: Photo courtesy of NASA; Photo courtesy of John McBrine; Photo courtesy of AP/Wide World. Page 13: Photo courtesy of NASA; Photo courtesy of Sergei Karpukhin/Reuters/archive photo; Photo courtesy of AP/Wide World. Page 14: Photo courtesy of D. Kirkland/Sygma; Photo courtesy of Vladimir Sichov/Sipa Press. Page 15: Photos courtesy of NASA; Photo courtesy of AP/Wide World. Page 16: Photo courtesy of AP/Wide World.

HarperCollins books may be purchased for educational, business, or sales promotional use. For information please write: Special Markets Department, HarperCollins Publishers, Inc., 10 East 53rd Street, New York, NY 10022.

FIRST EDITION

Designed by Stanley S. Drate / Folio Graphics Co. Inc.

Library of Congress Cataloging-in-Publication Data

Burrough, Bryan, 1961–
Dragonfly : NASA and the crisis aboard the MIR / Bryan Burrough.—
1st ed.
p. cm.
Includes index.
ISBN 0-88730-783-3
1. Mir (Space station)—Accidents. 2. Space vehicle accidents.
3. United States. National Aeronautics and Space Administration.
4. Astronautics—Government policy—United States. I. Title.
TL867.B87 1998
363.12'492—dc21 98-43309

98 99 00 01 02 10 9 8 7 6 5 4 3 2 1

This book is dedicated to the men and women who risk their lives in the exploration of outer space, and to the families and ground teams who love and support them.

CONTENTS

PART ONE

January 10 to May 18, 1997

1

PART TWO

1992 to 1996

237

PART THREE

May 18 to September 25, 1997

339

EPILOGUE

503

ACKNOWLEDGMENTS

515

INDEX

521

AUTHOR'S NOTE

The information contained in this book is largely derived from one-on-one interviews with the Russian and American participants.

Transcripts and logs of air-to-ground communications with the Mir space station during 1997 were obtained exclusively from the Russian space program and, through the use of requests under the Freedom of Information Act, from the National Aeronautics & Space Administration. These communications are italicized where they appear in the text.

While voluminous, neither government's transcripts are complete.

Russian-language transcripts, translated by the author, were made available for only select days, including February 23–24, April 1–10, and June 25–27. NASA tape-recorded daily conversations between Mir and Russian Mission Control only from June 25 to mid-September. Even these transcripts are incomplete. In some cases NASA provided transcripts with omissions; in others, recorded sessions were inaudible or never transcribed. For other days, including most of the period from January until June 1997, the only source of information on Mir communications are handwritten summaries transcribed by NASA interpreters in Moscow. In a handful of cases I have relied upon participants' memories of air-to-ground conversations not indicated in transcripts.

Copies of NASA e-mail traffic and other internal correspondence and reports were obtained using Freedom of Information Act requests. NASA made significant reductions in materials it made available. In general the agency declined to release transcripts of astronaut debriefings. Where debriefings are cited in the text, they were obtained from the astronauts involved or from individuals within NASA.

CAST OF CHARACTERS

The Mir Astronauts

Norman E. Thagard—March to July 1995
Shannon W. Lucid—March to September 1996
John E. Blaha—September 1996 to January 1997
Jerry Linenger—January to May 1997
C. Michael Foale—May to October 1997

The Mir Cosmonauts

Valery Korzun and Aleksandr "Sasha" Kaleri—August 1996 to March 1997
Vasily Tsibliyev and Aleksandr "Sasha" Lazutkin—February to August 1997
Anatoli Solovyov and Pavel Vinogradov—August 1997 to January 1998

At Russian Mission Control (TsUP)

Vladimir Solovyov, flight director
Viktor Blagov, deputy flight director
Valery Ryumin, director, Phase One program

At NASA

Daniel S. Goldin, administrator
George Abbey, director, Johnson Space Center
Frank Culbertson, director, Phase One program
Jim Van Laak, assistant director, Phase One program
Tony Sang and Tom Marshburn, members of Linenger's ground team
Keith Zimmerman, Mark Severance, and Terry Taddeo, members of Foale's ground team
Blaine Hammond, safety chief, Astronaut Office
David Leestma, director of Flight Crew Operations

MILESTONES OF RUSSIAN-AMERICAN SPACE COOPERATION IN THE 1990s

June 1992 ✯ U.S. President George Bush and Russian President Boris Yeltsin agree to a pioneering space-cooperation agreement. One American astronaut will fly aboard the Mir space station; two Russian cosmonauts will fly aboard the U.S. Space Shuttle.

September 1993 ✯ U.S. Vice President Al Gore and Russian Prime Minister Viktor Chernomyrdin announce plans for Russia to help the U.S. build a new International Space Station. As part of this agreement, NASA agrees to pay the Russian Space Agency $400 million to send five (later seven) astronauts to live aboard the Mir space station.

February 1994 ✯ The first American astronauts begin training in Russia.

March to July 1995 ✯ Astronaut Norman E. Thagard becomes the first American to live aboard Mir.

March to September 1996 ✯ The second Mir astronaut, Shannon W. Lucid, breaks the world record for time in space by an American.

September 1996 to January 1997 ✯ John E. Blaha becomes the third American to live aboard Mir.

January 1997 ✯ Jerry Linenger begins the fourth American mission to Mir.

PART ONE

January 10 to May 18, 1997

1

Friday, January 10, 1997
5 A.M.
Kennedy Space Center, Fla.

The boys could lease any type of Chevrolet they wanted for practically nothing per year. Eventually Gus and Gordo had Corvettes like Al Shepard's; Wally moved up from an Austin-Healy to a Maserati; and Scott Carpenter got a Shelby Cobra, a true racing vehicle. Al was continally coming by Rathmann's to have his gear ratios changed. Gus wanted flared fenders and magnesium wheels. The fever gripped them all, but Gus and Gordo especially. . . .

Gus would go out rat-racing at night at the Cape, racing full-bore for the next curve, dealing with the oncoming headlights by psychokinesis, spinning off the shoulders and then scrambling back up on the highway for more of it. It made you cover up your eyes and chuckle at the same time. The boys were fearless in an automobile, they were determined to hang their hides right out over the edge—and they had no idea what mediocre drivers they actually were, at least by the standards of professional racing. Which is to say, they were like every group of pilot trainees at every base in America who ever reached that crazed hour of the night when it came time to prove that the right stuff works in all areas of life.

—Tom Wolfe, *The Right Stuff*

The little red convertible rockets through the predawn blackness, its sole occupant pressing the accelerator past sixty miles per hour. Down the narrow, two-lane blacktop it speeds, shooting past darkened ditches where a lonely alligator could be spied lying low in the grassy water, a pair of reptilian eyes gleaming yellow in the still night. The car drives erratically, spitting up gravel at the side of the road, accelerating to sixty-five miles an hour. Out toward the beach it runs, hitting seventy on the straightaways.

Behind the wheel, the cold wind blasting his close-cropped brown hair, Jerry Linenger takes some of the last free gulps of real air he will get for four months. In forty-eight hours, after almost two years of training, two years of countless meetings and planning sessions and studying, two years of dunking in swimming pools, lakes, and hydrolabs, two years of spinning in centrifuges, endless hours of simulations and constant poking by legions of faceless doctors, he will return to space. But this, Linenger's second mission for NASA, will be far different from his first, back in 1994, when he had served as a mission specialist on an unremarkable run aboard the Space Shuttle Discovery. This time he is going to live aboard the Russian space station Mir. He will be the fourth, and at forty-one by far the youngest, American astronaut to do so.

Linenger keeps the accelerator floored as he hurtles through the inky Florida night. There is no one on the road; even the raccoons and shorebirds who populate this end of the peninsula stay out of his headlights, as if they too want to give an astronaut some breathing room on his last earthly adventure. He drinks in the salty ocean breeze, savoring the scent of boxwood and morning glory in the air. Far off to the left Linenger can see the orbiter itself on launchpad 39B. Bathed in spotlights, Atlantis clings to the side of its rocket, a fragile bird strapped to a hundred tons of explosives. Every time he glances over at it he gets a thrill.

Linenger revels in the moment. It is, after all, an image dripping with American myth, the astronaut streaking through a tropical night on the eve of launch, or gunning his TR-3 down some lonely California highway, lost in profound reverie: It conjures up memories of Dennis Quaid as Gordo Cooper in *The Right Stuff,* the last of the original Mercury astronauts, beaming that crazy smile as he runs his convertible out toward Edwards Air Force Base. Of Jack Nicholson whooping out loud while hot-rodding *his* convertible down a gorgeous Texas beach as the over-the-hill astronaut Garrett Breedlove in *Terms of Endearment.* But Linenger's last private moments on Earth evoke nothing so much as those of Sam Shepard as the legendary test pilot Chuck Yeager in *The Right Stuff.* There is Shepard on horseback, in an imagined scene, sauntering out of the desert to eye his new rocket plane, which is wreathed in white steam. The plane appears to be a living, breathing animal, the toughest bronco Yeager will ever bust.

The irony, of course, is that Jerry Linenger is none of those men, nor will he ever be. The Mercury astronauts went down in myth as hot dogs, flyboys careening around The Cape in their beloved Maseratis and Corvettes and Shelby Cobras like a pack of drunken teenagers. Linenger's little convertible is a rental. He isn't even a pilot. When shuttle astronauts climb into their sleek T-38 jet trainers at Houston's Ellington Air Force Base for the three-hour flights to The Cape, Linenger rides in the backseat as a passenger. He is a thin, sinewy navy doctor, a marathon runner and triathlete with half as many college degrees as the first seven Mercury astronauts combined.

Linenger's life as an astronaut in the final years of the twentieth century bears little resemblance to those of the first men in space. His predecessors had blasted off in balls of orange flame to chart the unknown. Across the country schoolchildren gathered in front of television sets in gyms and auditoriums to watch their missions. Today, swathed in the smothering layers of NASA's safety bureaucracy, shuttle flights pack all the suspense of a crosstown bus. They are routine. No one other than science teachers, *Star Trek* fans, and documentary filmmakers much cares what the astronauts actually do up in space. "Looking at stars, pissing in jars" is the snide catchphrase for astronaut work you hear at Kennedy.

Underlying the humdrum attitude is the fact that the shuttles never actually *shuttle* anywhere; in the 1990s, other than the flights to Mir, there are no moons or planets or space stations for astronauts to visit. They go nowhere man hasn't gone before. Every month or two another group of anonymous men and women simply shuffles onto another anonymous flight to poke and prod each other, draw syringes of blood, grow slender plant seedlings, or baby-sit vomiting monkeys; a week or two later they return to Earth to start the cycle all over again. Sometimes their missions make the nightly news, sometimes they don't. To the frustration of NASA's hardworking engineers and scientists, the only time anyone other than CNN or the Discovery Channel pays much attention is when something goes wrong: a lost satellite, usually, or a busted fuel line. No one but their families and hometown newspapers even knows who the astronauts are anymore. Yeager, Shepard, Armstrong, Glenn—those men were American heroes. Jerry Linenger is a space worker.

Linenger pushes the car past seventy, past seventy-five, as the Florida scrub hurtles by unseen. Then, just as he reaches eighty, he taps the brakes, once, softly, then harder, then harder still, abruptly slowing to crawl along as if entering a school zone. As he slows the car, he cranes his neck in the brisk night air, looking for anyone who could stop him for speeding. There is no one in sight. He has to grin at his own caution.

What could they do, anyway? he thinks. *Arrest me?*

He can envision the scene.

Uh, I'm sorry, officer, you can't take me to jail. I'm going into orbit tomorrow.

Linenger eases off the brake and starts driving at a prudent speed. He is not an especially popular astronaut among his peers. A loner, he belongs to none of the standard astronaut cliques, the spacewalkers, the Christians, the sports enthusiasts, the commanders. A member of the astronaut class of 1992, Linenger is still considered a rookie by the older astronauts from the first shuttle classes in 1978 and 1980, yet he doesn't do a good job of disguising his ambition, nor the fact that it lies outside NASA. He is allergic to chitchat. The only time most astronauts see Linenger open up is around the press; so outgoing was he around reporters and photographers that his 1992 classmates nicknamed him Hollywood. Mike Foale, the British-born astronaut who is to succeed Linenger aboard Mir in May, noticed one other group that could draw him out: children. Linenger's whole demeanor changes when confronted by innocence, especially that of his one-year-old son, John. His face jumps to life, he actually relaxes, kids around. But when children aren't near—and that's most of the time—he can come across as tense and humorless, almost scarily focused.

There is about Linenger a hint of calculation, a sense, however slight, that he is perfectly aware of the potential drama he is walking into aboard the Russian space station, and of the possibilities it opens for him upon his return to Earth. He is a man with big plans. Congress. Or lieutenant governor, maybe in his home state of Michigan. "Jerry always seems to have his own agenda that no one else knows," observes one fellow astronaut. "He just seems like he's passing through," says another. Ply some of those who trained him in Russia with a few beers, and you hear the same observation repeated over and over: "Jerry's in it for Jerry."

On paper Linenger is cut in the classic NASA mold, a lean, white, clean-shaven Annapolis graduate who grew up in a granite-solid family of five children in the blue-collar suburb of East Detroit. Everything about the Linenger household bespoke a life of constancy and normality. Don Linenger kept his job as a lineman at Michigan Bell for forty-one years, retiring shortly before his death in 1990. Every single workday he came home for lunch, and if the kids were out of school, they were told to be quiet while he caught a quick nap before returning to his job. Don and his wife, Fran, were the kind of involved parents who volunteered to chaperone the ninth-grade prom. On weekends Don coached Little League baseball.

As a child Jerry was the only one of the five Linenger children who enjoyed playing alone. He would sit out for hours in the backyard sandbox by himself, commanding battalions of Army men. He was never one to join cliques or crowds. At thirteen he watched Neil Armstrong step on the Moon and decided he wanted to be an astronaut, an idea Don and Fran encouraged even if they didn't exactly take it seriously. A head injury he suffered in ninth-grade football knocked him out of contact sports, so he became captain of the tennis and swim teams. Sitting outside a counselor's office during his senior year, he picked up a Naval Academy brochure that asked the question: Are you good enough? The idea intrigued him, he applied, and was stunned when he was ultimately accepted. It was a good thing, his father always said, because there wasn't enough money to send him to any college that wasn't free.

At Annapolis Linenger thrived, graduating fourth in his class. His eyesight wasn't sharp enough to be a pilot, so, with the idea of NASA always in the back of his mind, he entered the Navy flight surgeon program, a career track that would have made him head doctor on an aircraft carrier or an admiral's personal staff. For the next ten years Linenger ricocheted around the United States, gobbling up degrees like popcorn: First there were four years of medical school at Wayne State in Detroit; then a yearlong surgical internship in San Diego; flight training in Pensacola; a tour with a carrier group in the Philippines; back to San Diego on an admiral's staff; then two years at North Carolina State, where he completed his doctorate in epidemiology, at the same time finishing his master's in public health and completing his residency; then back to San Diego to work at a Navy clinic. Along the way he met and married his wife, Kathryn, a student at San Diego State he met at a beach party.

Linenger first applied to NASA to be accepted into the astronaut class of 1988. He couldn't even get an interview. In 1990 he tried again; this time the Navy declined to include him in its batch of nominees. Not until the spring of 1992 did he finally manage to get to Houston for an interview, and after returning to San Diego he felt he had nailed it. When the Astronaut Office called him at the hospital and told him he had been accepted, he let out a whoop of joy. But in the next breath he was asked to tell no one; the official announcement wouldn't be made till the next day.

"What's going on?" a friend asked when he heard the commotion.

"Oh, uh, nothing," Linenger said. "Results of a test I needed."

For twenty-four hours Linenger told no one, not even his wife. The next day he and Kathryn boarded a flight to Colorado, where they planned to join friends on a ski vacation. Linenger slipped the NASA press release to the pilot and whispered a few words. Not long after the plane reached cruising altitude, the pilot's voice came over the intercom.

"Would Kathryn Linenger please identify herself," he said.

A puzzled look crossed Kathryn's face. She popped her call button.

"Would you please stand up," the pilot said. As she looked at Jerry, he shrugged.

"We have a press release here," the pilot went on. "Kathryn, you better like Tex-Mex food, because you are going to Houston. Captain Jerry Linenger, U.S. Navy, has been selected to be a part of the 1992 NASA astronaut class."

There is a parking lot at the beach at the end of the road, and when he reaches it Linenger brakes and spins the convertible in tight little circles, tires spraying gravel and sand across the empty asphalt. Doing donuts is an exhilaratingly childish way to celebrate his last hours of freedom, and Linenger knows it. In his mind he has it coming; for the next four months scientists and bureaucrats will map out his days to the last five-minute increment. By and by, Linenger brings the little convertible to a stop and parks. Unfolding himself from the front seat, he spies a rusty old bicycle someone has cast aside. He rides it in wobbly circles for a few minutes, keeping his eyes glued to the pavement. The last thing he needs is to hit a pothole, twist an ankle, and have his mission scrubbed.

But then Linenger knows nothing less than a heart attack will prevent NASA from sending him aloft. The dirty little secret of the Mir program is that virtually none of the astronauts want anything to do with it. Of the 125 or so astronauts in the corps, you could count the number of volunteers for Mir missions on two hands.* Linenger had read about the psychology of isolation in the Navy, pored over studies of how men survived in the Antarctic, and was fascinated by the challenges of long-duration space flight, something NASA knew very little about. Finding Kathryn enthusiastic about an overseas adventure, he had volunteered for the program right out of training, and had briefly been under consideration to be Norm Thagard's backup on the very first Mir mission, in March 1995.

The Russians, however, demanded that every American who stayed on

*As of August 1998, there were 119 American astronauts—160 counting the international astronauts from Japan and Europe and other countries who fly on the shuttle.

Mir already have experience in space. So Robert L. "Hoot" Gibson, the popular head of the Astronaut Office at the time, assigned Linenger to a later Mir increment and rushed him into the crew of STS-64, scheduled for a September 1994 launch, for the sole purpose of qualifying him for Mir. Linenger's last-minute appointment prompted grumbling among the STS-64 crew, and his performance in orbit did little to endear him to his crewmates. Blaine Hammond, STS-64's pilot, recalls Linenger as a rookie "in the worst sense of the word. He was so raw. But he was one of those guys you couldn't offer up helpful hints about how to live and work in space. He wouldn't listen to anybody. He had to learn it himself. He always knew best. He didn't exhibit the kind of humility and the ability to learn from others that I would've expected from a first-time flier."

By his own admission, Linenger is headstrong. Inside the Astronaut Office the charitable view is that he is a perfectionist, and like most perfectionists, he has a reputation as a complainer, and a loud one at that. His outbursts during the long months of training at the famed Russian military base known as Star City are renowned. He would get on the phone with Frank Culbertson, the astronaut who heads the Shuttle-Mir exchange program back in Houston—NASA calls it the Phase One office—and rant about how stupid the Russians' rote-memorization techniques were, how ridiculous their little ink-blot psychological tests were, how their yoga and breathing exercises were wasting his time. The Russian instructors would lecture Linenger three times on how to use a piece of equipment before letting him put his hands on it, which drove Linenger crazy. "It just doesn't make sense the way we do things," he fumed.

If there was anything Linenger hated, it was having his time wasted. At Star City he rarely engaged in the little social niceties and cultural exchanges that astronauts like Mike Foale enjoyed. Foale and his wife, Rhonda, overcame the Russians' ingrained resistance to mingling by befriending everyone, from cosmonauts and generals down to the lowly soldiers who manned the Star City gates. At first few of the Russians would invite the Americans into their cramped little flats, and the Americans could never understand why, although they gradually deduced that the Russians were simply embarrassed by their primitive living conditions. The Foales barged their way into Star City's coldwater flats with an endearing mix of goodwill and laughter, and were beloved because of it. Not Linenger. He had little time to press the flesh. Though Kathryn was charming and popular, Linenger had days when he couldn't be bothered to look up and smile when passing a Russian on Star City's forested walkways. "The best thing about Jerry," the Americans took to saying, "is Kathryn."

Nor did Linenger socialize much with his own ground team, which will follow his flight from a suite of offices at the Russian Mission Control center in the northern Moscow suburbs. The man who will lead that team, Tony Sang, a beefy Chinese-American from Mississippi known for his easygoing manner, trained with Linenger at Star City for six weeks that fall and was frustrated in his attempts to bond with the man he was to spend the next four months supporting. Sang loved to cook and often served up his special barbecue or General Tso's Chicken in informal dinner parties for the American contingent. Linenger never bothered to show. "Where's Jerry?" Sang was forever asking Kathryn, who was left to make excuses. Jerry was in the gym. Jerry was doing homework. Jerry was resting.

Tom Marshburn, the genial young North Carolinian who served as Linenger's flight surgeon, spent nine months training with Linenger at Star City and saw a side of the astronaut that Sang wouldn't discover for months. Like his immediate predecessor aboard Mir, the shuttle commander John Blaha, Linenger expected a level of support at Star City comparable with what astronauts received in Houston. It simply wasn't possible; there weren't enough flight docs and trainers to do it. Linenger spent hours working out at the worn Star City gymnasium and was incensed that the Russians didn't have water to drink there. He complained to Marshburn, who took it upon himself to begin lugging large bottles of water out to the gym for him. If Linenger ever said thank you, Marshburn didn't remember it. Linenger would ask Marshburn to put together reports on his progress, and the flight doc would stay up all night writing one up, only to discover the next morning that Linenger had forgotten what he asked for. "I did a lot of things for Jerry that I don't think he knows," Marshburn remembers quietly. "And I don't think he would care if he did know." Yet there were moments when Linenger could surprise him with a gesture of genuine warmth. When Marshburn asked to borrow a book, Linenger inscribed in the flyleaf a heartfelt thanks for all his hard work. Marshburn was so startled he didn't know how to respond. And so he didn't.

But that was Linenger. The Americans cut him slack; he was an astronaut, and, well, that was just how some astronauts were. The Russians were a different matter. Linenger's three predecessors aboard Mir—stubborn old Norm Thagard, grandmotherly Shannon Lucid, and Blaha—had each been veteran space fliers over fifty years of age and had been careful to remain properly deferential to their Russian commanders. Not so Linenger, who was younger, quick to anger, and fiercely independent. To an unstudied observer, of course, everything appeared fine. Linenger and his Russian commander, Vasily Tsibliyev, were cordial enough. But Matt Muller, the NASA contract worker who followed Linenger closely at Star City, watched the two men intently and worried how they would interact in space. During training exercises, Muller remembers, "Jerry would be in the middle of something, and Vasily would say, 'Jerry, it's time to do this [with me].' And Jerry would continue doing what he was doing. Vasily would say it a second time, Jerry would then acknowledge what Vasily was saying, then would finish what he was doing before he went to help Vasily."

The Russians, whose collectivist emphasis on teamwork and sacrifice was legend, took a strong dislike to Linenger. Among themselves they sniped about everything from his selfishness to his performance at cosmonaut banquets. "Jerry's toasts were pathetic," sniffs one Russian colonel. Until July 1996—six months before his flight—Linenger's willful ways remained the stuff of hallway gossip and late-night gripe sessions. Then came the day when Linenger, along with Marshburn and two other NASA flight surgeons, walked into Building Three at Star City to stand before the Grand Medical Commission, the periodic gathering of the top doctors in Russia, which reviewed medical reports on each of the astronauts and cosmonauts and laid down the final word on who would fly and who wouldn't. It was the day John Blaha was to be medically certified for his flight; Linenger was obliged to undergo the same process as a member of Blaha's backup crew.

Linenger was expecting trouble even before entering the auditorium that day. For two years the Russian and American doctors had quarreled about Star City's right to rule on an American astronaut's fitness for flight; if NASA said its fliers were good to go, the Americans argued, then damn it, they were good to go. The Russians countered by bringing up an endless series of niggling complaints about the American astronauts' health, as if to say, "We know better." The disputes were aggravated by the Russians' insecurity about their medical expertise. Linenger, a doctor himself, thought Russian medicine had only advanced to the level Western medicine reached during the 1950s; he had already endured repeated run-ins with the Russian medical establishment.

He had a slow heartbeat, for instance, a not uncommon condition among long-distance runners. Underwater, in the cavernous Russian hydrolab where crews trained for spacewalks, his heart sometimes skipped a beat. The Russian doctors had pulled him out of the tank at one point and demanded to run tests. Linenger, in no uncertain terms, refused. "The Russians did things that didn't make any sense, they were just silly," Linenger recalls. "And yeah, I would object to things. I was very outspoken. I was not meek. They felt they had your life in their hands, and I didn't acknowledge it. I'm not afraid of not being sent to Mir."

Another time Linenger mentioned to trainers at the hydrolab that he might wear glasses or contact lenses during the spacewalk planned for his increment; his vision wasn't quite twenty-twenty. The next day the doctors demanded he see a Russian optometrist, who asked him to read an eye chart.

"I'm not going to read that chart," Linenger said.

"What do you mean?" the surprised doctor asked.

"I've read it before. You know there's no problems with my eyes."

The Russian doctor, unaccustomed to defiance from a cosmonaut, stammered that he *had* to read the chart.

"Why? What's the problem?" Linenger asked.

"I can't discuss it," the doctor said. "But we can't allow you to wear glasses during your spacewalk."

"Fine, I won't. I don't need them anyway. But don't waste my time with these silly charts."

Only later, during a walk in the woods with one of his Russian trainers, did Linenger fathom why the doctors didn't want him to wear glasses in his space suit. The visors in the new Russian Orlan space suits had a habit of fogging up: A guest cosmonaut during an earlier Mir mission had nearly been blinded by fog on his glasses. "They had a deficiency in the suit," Linenger recalls, "and they didn't want to say it, because they wanted to sell it to the Americans for use on the International Space Station. They didn't want to tell anyone that, of course."

On that July morning the Grand Medical Commission convened, Marshburn and another young NASA flight surgeon, David Ward, escorted Linenger through a series of examination rooms, each filled with a dozen Russian specialists who pored over every muscle and almost every orifice in his body—NASA had explicitly forbidden any Russian doctor to administer rectal exams. The astronaut Dave Wolf, a notorious cutup, once strode into a roomful of Russian optometrists, held out his hand in greeting, and walked straight into a wall, just for laughs; Linenger, who saw nothing funny about the whole exercise, trudged through the examination rooms wearing his characteristic look of stoic determination. Everything went smoothly until the trio entered the surgeons' room. Suddenly a Russian doctor, a man none of the NASA contingent had ever seen, raised an objection to an old injury Linenger had once suffered, a nerve palsy behind his right shoulder. It was something NASA doctors had signed off on when Linenger was first named an astronaut. He had never given it a second thought.

Ward, who had the most experience with the Russians and thought he knew how to handle them, immediately lost his temper. Just weeks before, after months of rancorous negotiations, NASA and the Russian doctors had reached an awkward compromise: The Russians would be allowed to pass medical judgment on NASA astronauts, but only within forty-five days of their flight. Linenger was still six months from his actual flight; as far as the NASA doctors were concerned, they were enduring the commission's review that day purely as a courtesy.

"Look, we've got an agreement here!" Ward told the commission's cochairman, an Air Force general named Yevgeny Berezhnov. "You know Jerry is not under your system yet. You cannot do this! For all intents and purposes he is still our guy to train. You know this is true. Your people negotiated this. You signed this yourself. And even if he was yours, this is not a disqualifying abnormality."

Berezhnov listened patiently but said only, "Well, I'm not going to make any guarantees. We'll see what happens in the commission."

Later that day, when it came time to face the commission, Marshburn, Ward, and another NASA doctor, all still angry, filed into the auditorium and took seats in the front row. Linenger walked alone onto the stage, where he took a seat in a straight-backed chair beside Berezhnov. The setting was reminiscent of a military court, with Linenger sitting alone in the dock beside his judge. "It is just so incredibly demeaning," says Ward. "It makes my toes curl when I think about it. You have this whole auditorium full of people who have poked and prodded you, they are holding the keys to your future, and they pronounce you fit or unfit."

Linenger looked out on an auditorium filled with the fifty best doctors in all Russia. Most were elderly and had no regular contact with the space program. Almost all were hearing details of his health for the first time. As Linenger sat quietly, the head of each medical group stood and spoke. The doctor in charge of the internal medicine group pronounced him fit to fly. The doctor in charge of ear, nose, and throat pronounced him fit to fly. Dentistry pronounced him fit to fly. Opthalmology did the same.

Then the snowy-haired head of the surgical group stood. He briefly described the nerve palsy situation. "We find that this is a disqualifying condition," the doctor said. "Pending further testing, including ultrasound and an MRI, we will reserve judgment."

"Is there any dissenting vote?" Berezhnov asked.

Ward leaped to his feet. "This is a normal anatomic variant!" he shouted. "This is not a disqualifying condition!"

"Well, I understand this," Berezhnov said, "but I must take the advice of the surgeons. I find this is the prudent move." Turning to Linenger, who sat on the stage, stunned, he said, "Jerry, you are qualified only pending the normal results of these tests."

"No, Jerry, you're not going to do that," Ward said.

"Jerry, we have not agreed to any of this," Tom Marshburn said.

"Nyet, nyet, nyet," Linenger said, growing visibly angry. "I'm not going to do that. This has never been a problem before. I'm fine. I'm fine."

"Well, it's your choice," Berezhnov said. "Then you are disqualified."

As the Americans protested, murmurs of surprise were trilling through the crowd. These were the finest minds in Russian medicine, and they weren't accustomed to a bunch of wet-behind-the-ears thirtysomethings—from America, of all places—questioning their judgment. They were used to pliant, servile cosmonauts who treated their word as gospel. "What surprised them," Linenger recalls, "was we were not some meek little third-world country, going in there prostrating ourselves before them, begging to go on their Mir. We were not the Afghans [who had sent a cosmonaut to Mir in 1988]."

Despite the American protests, the meeting ended with the commission's decision unchanged. "We left going, 'Holy shit,' they just disqualified our crew member. What do we do now?" remembers Ward.

The commission's decision was considered top secret. Months later, many of the Americans involved in the Russian program still had no idea that anything unusual had happened to Linenger. "None of these details were *ever* discussed," remembers Matt Muller. "But living in a Russian dorm, you hear things." Still, even those few Americans who learned of Linenger's struggle with the Russian doctors never discovered that it had all been a charade. There was nothing wrong with Linenger's shoulder, and the Russians knew it. The problem, though Russian doctors were loathe to admit this to their NASA counterparts, was Linenger's attitude.

"There were some peculiarities in the relationship between the crew and Jerry; it worried the psychologists" is how Valery Morgun, chief of Star City's medical division, puts it. Morgun, a languid-eyed doctor with a large, blocky head, sighs at the naïveté of the young NASA flight surgeons. "We Russians, we know far more about psychological issues than the Americans; even the Americans admit this," he says. "Jerry, you see, is not a team player. He can't work as part of some collective family. He was just not capable of working as a member of a team. The psychologists, they saw this before the flight and they were very, very concerned about it. They put a report together and put it on my desk."

The report was authored by a senior doctor in Star City's psychological group, a chatty, ebullient fellow named Rostislav Bogdashevsky, who had been analyzing cosmonaut crews—and had met every cosmonaut—since joining the Russian space program in 1962. Bogdashevsky, whom all the Americans called Steve, spent weeks watching Linenger in training before making a recommendation he says he had made only two or three times in the previous thirty-four years. "I did not recommend that Jerry be included in that crew," Bogdashevsky acknowledges. "This decision was the subject of consultations with all the directors of all the medical programs. In my report I wrote that Jerry, he has an egocentric personality. He is an egotist. His communication skills are very low. He is not a good team player. He views people as a means. I wrote that that could become a big problem [in orbit]."

The commission's nerve palsy excuse was a typical Russian ploy, the kind of silly scheme the military bureaucrats at Star City would hatch when they were too embarrassed or too uncomfortable to confront NASA with painful truths. After the meeting Linenger stormed back to his apartment. Part of him wanted to leave Star City that very minute and go back to Houston. *Screw the Russians.* But he wasn't a quitter. He stayed, and as the American and Russian doctors debated diagnoses behind the scenes, he returned to his classwork. But when after a week his flight status remained in limbo, Linenger decided to try and force the Russians' hand. He announced that he was boycotting his classes. Only a call from someone in the Phase One office in Houston forced him to change his mind. Within days, in fact, the storm passed. None of the doctors involved, citing medical privacy, will discuss the incident in detail, but Bogdashevsky admits that the Russians knew their nerve palsy excuse would never stand up.

"I tried to explain the real problem [to NASA]," he recalls. "They wouldn't listen. It was because Jerry was a professional, a good doctor, they liked him as a scientist. And he was eager to fly. They believed the crew would manage all the psychological problems."

They were wrong.

2

Tossing the bike aside, Linenger trots out the path behind the Beachhouse, the old astronaut party shack NASA has transformed into a conference center. He jogs through gentle dunes crowned with thickets of palmetto, then steps down the two wooden steps onto the beach, a six-mile stretch of pristine coastline the agency reserves for use by astronauts, their families, and other agency personnel.

This morning the beach lies wild and empty. Tiny ghost crabs scuttle by unseen as Linenger sucks in the cool sea air and begins his jog down the sand. He loves running. It is as much a part of his daily routine as brushing his teeth; if he misses his run he tends to feel sluggish and out of sorts. As he heads down the beach in the gathering dawn, the first rays of sunlight begin to shoot out over the Atlantic, shafts of pink and orange and magenta. It is a gorgeous sunrise, and for the first time Linenger realizes he won't be able to do this again for four months. *I'm going to miss this.*

He runs a mile and a half down the beach, and then back, and by the time he returns to the Beachhouse the sun is a shimmering yellow ball rising out of the ocean. There is a shovel thrust into the sand where the crew of STS-81 hopes to dig a barbecue pit for their families' farewell party that evening. Linenger falls to his knees in the sand and begins digging. It takes a while to dig the pit, and when he is done he shapes the resulting sandpile into a circular hummock, then decorates the top of the little mound with seaweed to make a face. Then he strips off his sweaty running clothes and plunges naked into the surf for a swim. The water is freezing, and he doesn't last long. After several minutes he emerges, shivering, and goes to put his clothes back on.

Up in the parking lot Linenger spies Mike Baker and his wife, Deidra, walking down toward the beach. He hurries to finish dressing. Baker, whom everyone calls Bakes, is his commander.

"Hey, man," Baker says.

"Hey," Linenger replies.

After a bit Deidra walks off to make a phone call. For several minutes the two astronauts sit on the back porch of the Beachhouse in silence, dangling their toes in the sand and gazing out at the rising sun.

"Man," Linenger finally says, "it doesn't get any better than this."

This remains one of the very best parts of being an astronaut, even in the workaday Shuttle era: the delicious anticipation of flight, the long hours of training behind, hanging out at dawn on a private beach pondering the meaning of life. And the worst part? The worst part is something neither Baker nor Linenger, nor for that matter any astronaut, will ever discuss publicly. The worst part is the fear. Not the fear of dying in some fiery launchpad fire. Not the fear of a Challenger-like explosion as the orbiter climbs "up the hill" toward space. Not the fear of dying frozen out in the cosmos, lost in some Apollo 13–like disaster. It is the fear of saying the wrong thing. The fear of having one too many Lone Stars at the Outpost—the ramshackle astronaut tavern just down from the NASA front gates in Houston—and being reported drunk and disorderly. Fear of the rumor of an extramarital affair. Fear of making waves. Fear of being seen as disloyal to the program. Fear of not flying.

For Linenger and Baker and every other NASA astronaut of the last twenty years, this fear has had a name: George Abbey. It is difficult to convey to outsiders the mixture of awe and power and omniscience that the mere mention of the name conjures for an astronaut. *George Abbey.* To a generation of astronauts Abbey has been God, the all-seeing, all-knowing judge of their fates. Every astronaut of the Shuttle era knows that to find one's way into Abbey's good graces means a long and distinguished career in space. To earn Abbey's disfavor means a descent into purgatory, a period marked by lengthy delays between missions and painful self-examination, an endless unanswerable series of "What-did-I-do-wrong?" questions, followed by a gradual alienation from NASA and, eventually, an ignominious retirement. Mention Abbey's name to people at the Johnson Space Center in Houston, and the first thing they do is close the door. As he shambles through the institutional-white corridors of his headquarters building, NASA employees have been known to hum the foreboding theme music that accompanied Darth Vader's entrances in the movie *Star Wars.*

In his new posting as director of the Johnson Space Center, a job he has held only since January 1996, and for much of the previous two decades as the man in charge of the shuttle program's crucial flight assignments, Abbey is regarded by many as the J. Edgar Hoover of NASA, a mysterious figure shrouded in myth and legend. Astronauts whisper about the files he is said to keep on every center employee. A thick green binder he totes to meetings is regarded as a source of secrets on a par with Pandora's box. Among the astronauts and the hundreds who support them, it is axiomatic that it is Abbey who actually runs NASA, not the bubbly Administrator Daniel S. Goldin, who spends his days in far-off Washington glad-handing politicians and flattering the poor, hapless Russians. In fact, the two men run the agency as a team: Goldin is the visionary, cowboy-boot-wearing Jewish kid from the Bronx who functions as NASA's Mr. Outside; Abbey is the complex, Machiavellian Mr. Inside. "Dan Goldin and George Abbey," Goldin himself likes to say, "are connected at the hip."

Yet for all his power, the single most notable thing about Abbey is his invisibility. In twenty years no one could remember a single major magazine or newspaper profile of him. Old-timers at NASA recall a rare newspaper photo of Abbey with a pair of shuttle astronauts in the early 1980s; the caption called Abbey "an unidentified man."

"The real book about the manned space program would be a book about George Abbey," says Vice Admiral Richard Truly, the decorated astronaut brought in to head NASA following the Challenger disaster in 1986. "You would have to interview hundreds of people to write that book. He is the most mysterious man I've ever met. Someone once told me working with George was like playing chess covered by a blanket. You could play for years and years, thinking you were winning, then one day you come out from under that blanket and find George smiling, because you've made all the moves he wanted you to make."

"If one man owned and controlled the supply of rice in Asia, he'd be very powerful; George Abbey controls the rice of space," says Drew Gaffney, a Vanderbilt University researcher who supervised several NASA experiments and flew as a payload specialist on a shuttle flight in 1991. Adds Patricia A. Santy, the former head of psychology at Johnson: "George Abbey has dominated the space program for the last twenty years, and all behind the scenes. Forget Goldin. He's a figurehead. Whoever controls the astronauts controls NASA. Period. The astronauts *are* NASA. And Abbey controls the astronauts."

The man who inspires such hyperbole is a jowly, Buddha-like career bureaucrat who mumbles so softly at times he is almost incoherent. He rarely makes eye contact and seems ill at ease around strangers. Those who underestimate Abbey, however, do so at their own risk. He is the kind of man who prizes loyalty, and who tends to characterize NASA's internal critics as "traitors." His memory is said to be both encyclopedic and photographic. In thirty-four years at NASA, Abbey has built an unrivaled network of contacts throughout the aerospace industry and in Washington. Longtime friends say slavish devotion to his job cost Abbey his marriage, which only led him to plunge even deeper into the minutiae of manned spaceflight. Yet Abbey is entirely unprepossessing. Some nights he can be seen dining alone at one of his favorite restaurants, an Italian place called Frenchie's, in a shopping center a mile down NASA Road One; if you meet him coming out after finishing his lasagna, you might guess he is a butcher or a bus driver.

Today Abbey's Johnson Space Center, which all the locals call JSC, sprawls in the wilting Texas heat much as it has for the last thirty-six years, a vast campuslike conglomeration of low-lying buildings plunked down in a former cattle pasture at the suburban southeast edge of the city of Houston. The surrounding streets are lined with Seven-Elevens and fast-food joints and space-themed strip malls with names like Challenger Plaza. Inside the security gates, long, snakelike trams creep through the center's sunbaked streets packed with tourists who lean out snapping pictures. The employee parking lots, which begin filling up by seven and emptying by five, are lined with the cars of middle-class Middle America, minivans and Ford pickup trucks and rusting Supras bearing "Starfleet Academy" and "Space *Is* Our Future" bumper stickers. The JSC workforce is largely white, polite, and well fed, the kind of badge-wearing government workers who nod and smile when you pass them on the sidewalk. The engineers, with their earnest tucked-in shirts and tie tacks, are no longer Harvard, Princeton, and Stanford men; because each of the NASA centers recruit regionally, they tend to hale from Rice and Louisiana Tech and especially Texas A&M. Half the phones at JSC are answered with thick southern drawls.

Abbey's headquarters is Building One, a white concrete-and-glass stack whose exterior resembles a giant ticktacktoe grid. On its roof, at the extreme northwest corner directly above Abbey's ninth-floor corner suite, is a single surveillance camera; the symbolism is lost on no one. Abbey first came here in 1964. The Seattle-raised son of a Welsh immigrant mother, he attended the Naval Academy, after which he joined the Air

Force and was detailed to what was then called the Manned Spacecraft Center. He made his mark in the wake of the notorious launchpad fire that killed astronauts Gus Grissom, Ed White, and Roger Chaffee in 1967. After the fire hundreds of design changes were proposed, and all had to be approved by a NASA panel called the Change Control Board. The board's meetings would convene at seven in the morning and run till midnight, and as its secretary, Abbey was responsible for keeping track of everything that had been accomplished or promised. A board meeting might go till one or two in the morning, but NASA engineers would arrive in their offices at dawn and find on their desks a detailed summation of the day's deliberations and "action items" from Abbey. Engineers whispered that Abbey knew everything and forgot nothing. He was the consummate example of the anonymous NASA staffer the astronauts liked to deride as a "note taker" and a "horse holder." He was, in fact, a born bureaucrat, the kind of glue that held NASA together.

His prowess was recognized in December 1969 when the center director, Christopher Kraft, named Abbey as his technical assistant, a post that made him the de facto No. 3 man in Houston. He functioned as Kraft's consigliere, silently hunching beside him in meetings but offering counsel only in private. For seven years nothing went in or out of Kraft's office—not a single financial statement, not a single crew roster—that Abbey didn't approve. "Remember in the *Green Hornet,* the good and faithful Kato? That was George," says his longtime friend Tom Tate. "He was always there, for everybody. He always anticipated every move. He was always one or two steps ahead of anyone working a problem, and he always had a backup plan if someone screwed up. He basically ran the place."

Still, Abbey's elevation in January 1976 to the post of director of flight operations stunned many inside NASA. It placed him in control of all the agency's astronauts and flight controllers, even though he had never worked in either job. His promotion created a visceral rivalry with Gene Kranz, the legendary flight director of the Mercury, Apollo, and Skylab missions. One of the space program's true icons, Kranz is best known to movie viewers as the steely NASA ground controller portrayed in the movie *Apollo 13* by Ed Harris, whose directive that "failure is not an option" passed into American lore. To Kranz, who as assistant director of flight operations was passed over for the job and now found himself Abbey's deputy, Abbey had been a nonentity, "a career horse holder."

The difference in style between the two men was the difference between Abbey and almost every other NASA administrator. Kranz, like so many of the engineers who rose through Apollo, was a straight talker who tended to begin sentences with the phrase "To be blunt." Abbey, in turn, was a master of secrecy and misdirection; he veiled himself in an aura of mystery and oracular wisdom. He seldom said what he thought. "You have the feeling he's a Doctor Strangelove, exactly that kind of crazy, scheming character," Kranz told one associate. The two men coexisted for nine years by striking a Solomonic compromise: Kranz supervised the flight controllers in what became the Missions Operations Directorate, called MOD; Abbey oversaw the astronauts. For the most part neither interfered in the other's business.

Then, as now, the crux of Abbey's power was his control of the astronaut crew-selection process. The chance to fly into space, after all, is the grail that all astronauts seek; in most cases, it is a dream they have pursued since childhood. But while there are 125 or so astronauts, there are only five or six slots on each shuttle flight. How those slots will be filled is a source of endless speculation in the closely knit astronaut community. Theoretically, crews are chosen by the astronaut who heads the Astronaut Office—since late 1994 a slim, mannered Marine colonel named Bob Cabana, for the three years before that Hoot Gibson. But everyone knows it is Abbey who makes the decisions.

"The process is really unknown," says Michael R. "Rich" Clifford, an astronaut who retired in 1997 after three shuttle missions, including the first American spacewalk outside the Mir station. "You're always being watched, but you're never sure who's doing the watching. And you will never know if you do something wrong, because they won't tell you. We assume the chief of the office makes the new selections and feeds them to Mr. Abbey for approval. What happens after that, no one knows." Adds Dick Truly: "George had a huge amount of power in crew selection and crew assignment, and he never showed anybody how he wielded that, or what process he used. I've seen hour-long meetings where he would not say a word."

Currying favor with Abbey has been an obsession for Shuttle-era astronauts, the subject of endless late-night bull sessions. Even today, many take to hanging out at Abbey's favorite watering hole, the Outpost, hoping for a chance to drink with him. Frenchie's, just across the intersection from the Outpost, is a magnet for astronauts for the same reason. Because Abbey backs the center's annual chili cookoff every May, no astronaut dares miss it. Half the class of 1984 begged off a group scuba excursion to the Caribbean at the mere suggestion that the trip might displease Abbey.

"For astronauts the first rule is 'If you suck up, you go up,' " says Drew Gaffney. "If you don't, you're cooked. If you look at some of the people who flew over and over [on the shuttle], their biggest quality is their closeness with Abbey, and the willingness to do what they're told. It borders on something out of *The Stepford Wives.*"

By the mid-1980s Abbey's favorite astronauts, including a number who had served as his technical assistant, men like Bob Crippen and Jim Wetherbee and Sonny Carter, had earned a nickname: the Bubbas. Friday nights Abbey could be found at the Outpost tossing back Budweisers with the Bubbas, a crowd of Bubba wannabes hanging on the edge of the group drinking in every word of the old Apollo war stories Abbey loved to tell. More than one astronaut remembers nights in which Abbey would fall asleep in his chair, or at least they thought he had; the power of Abbey's presence was such that others swore he was just faking it. "You're always at your peril if you carry on a conversation about him in his presence and he's sleeping," chuckles Tom Tate. "You're liable to hear about it later."

Evenings at the Outpost sometimes ended in a ritual known as "the Abbey Watch." The Abbey Watch fell to the unfortunate astronaut who was obliged to drive Abbey home after a long night of drinking. Hoot Gibson recalls how many of the Bubbas would sneak out of these drinking bouts early rather than be caught on Abbey Watch. "I wound up with it one night, a Friday night after happy hour," Gibson recalls. "I loaded George into my car. So there was George, quasi asleep-slash-drunk in the right seat of my car, and he couldn't tell me where he lived. We drove all around. 'George, is this your house?' He would wake up, look kind of sleepily, [and say] 'No.' 'George, is *this* your house?' 'No.' It was awful."

While Abbey's secrecy draws comparisons to Hoover, a more accurate comparison might be to famed Chicago mayor Richard Daley. By tradition JSC is a collection of quarreling fiefdoms, but Abbey discovered that his leverage over the astronauts afforded him a unique way to control the system. He began placing astronauts, many still on active flight status, in key management positions; while moving astronauts into decision-making posts had been a recommendation of post-Challenger review commission, it also gave Abbey considerable control over them. Before Frank Culbertson, who ran the Shuttle-Mir program, or John Casper, the astronaut in charge of JSC's safety division, did anything contrary to Abbey's wishes, they had to think twice; both men intended to fly again aboard the shuttle—that is, if Abbey let them. It is an arrangement that even those who admire Abbey are uncomfortable with.

"It's a real conflict of interest and absolutely the wrong thing to do," argues Bryan D. O'Connor, a former shuttle commander and NASA administrator who worked closely with Abbey for years. "You can't really be independent and step to the side and feel like you don't have any pressure from George to do what he wants." Norm Thagard, the veteran mission specialist who flew on the shuttle four times before becoming the first astronaut to live aboard Mir in 1995, is more blunt. "Abbey," he says, "is a godfather type who runs a patronage system."

Abbey has long been plagued by allegations of favoritism and cronyism. Astronauts who failed to thrive under his system were forever griping about how often the Bubbas flew. When a Bubba ran into trouble—as Hoot Gibson once did by involving himself in a fatal midair collision in an airplane race he had been forbidden to enter—he always seemed to work himself back into Abbey's good graces and return to space. For years murmurs of discontent rumbled through the astronaut corps, especially during the cathartic self-analysis NASA endured following the Challenger disaster in 1986, but no astronaut was willing to risk his career by criticizing Abbey publicly. Finally Drew Gaffney, who had friends in Congress, decided it was time to tell several legislators of the astronauts' concerns. He went up to the Hill, Gaffney recalls, and said: "Here's what's happening, and here is why it's so damaging. You have this guy, a masterful manipulator who was using his total power to select astronaut and flight crews for his own empire building. You wound up in a situation where people weren't willing to say, 'This is wrong, this is bad, this is unsafe.' "

At one point Abbey did lose his job. He was done in, it appears, not by congressional pressure, not by any dramatic astronaut coup, but by something as simple as an opinion poll. In the wake of the Challenger disaster, NASA endured a collective bout of soul-searching, and one part of that bout was what agency officials called a cultural survey. In one survey, the JSC personnel office polled about 20 percent of the workers, asking them to rank their bosses in a wide range of categories. According to former NASA officials, Abbey's Astronaut Office was ranked dead last in categories ranging from competency to trust to overall morale. "It absolutely shocked center management that the relationship between Abbey and the astronauts was on that shaky ground," recalls Gene Kranz. "It was clear that there was no way they could keep Abbey after that." After consulting with headquarters, center director Aaron Cohen removed Abbey from his duties.

There is a saying at NASA that no one is ever fired, and Abbey is sometimes offered up as exhibit A. In 1988 he was detailed to Washington as a deputy associate administrator for space flight. But Abbey's exile to Washington only set the stage for his amazing resurrection. In early 1991, at an impromptu Washington wake for the astronaut Sonny Carter, who died in the crash of a commuter plane, Abbey was introduced to an aide to Vice President Dan Quayle named Mark Albrecht, who was staff director at the National Space Council, a White House advisory committee that was trying to focus the nation's space efforts in the wake of the Challenger accident. Albrecht was immediately impressed by Abbey's knowledge of NASA's inner workings and asked the agency to detail him to the Space Council.

Overnight Abbey was transformed into the Space Council's in-house wise man, a post that gave him considerable influence over space policy. It was Abbey who many at NASA felt—with some evidence—was behind the Space Council–inspired resignation of Truly a year later. When Albrecht in 1992 nominated a little-known outsider named Daniel S. Goldin to replace Truly at NASA's helm, it was Abbey whom the Space Council installed as Goldin's Rasputin, guiding and prodding him through the agency's tangled internal politics; the two became so inseparable, aides referred to them as "Chief Dan George." And it was Abbey who, just weeks into Goldin's tenure, hatched the ideas that led to the merging of the Russian and American space-station programs in 1993.

Abbey's return to Houston in 1994, and his elevation to the center directorship two years later, prompted groans from many in the astronaut corps, who feared he would once again politicize the crew-selection process. The politics of crew selection under Abbey can be seen vividly in the Byzantine intrigues leading up to the naming of who would command the most prestigious mission of the 1990s, STS-71 in 1995, the first shuttle mission to dock with Mir. During Abbey's exile, Hoot Gibson had spearheaded the selection process. Gibson took months assembling crews with the proper mix of skills and experience for each mission. "Until Abbey took over, every single name that I ever submitted was approved and passed, and they should have been. I did my homework," Gibson recalls.

"I worked on planning for six months leading up to the assignment of a crew. When George came in, things went downhill just about immediately. George had no business jerking with all those things. However, he felt that he needed to hold up just about everything."

Gibson's problems began immediately after Abbey returned to Houston in early 1994, when Gibson sent up his candidates for the crew of STS-63, the so-called fly-around mission, which was scheduled to take the Russian cosmonaut Vladimir Titov on a fly-by of Mir in February 1995. Gibson was nervous about the mission for two reasons: It was a rehearsal for the coveted STS-71 docking mission, and it included NASA's first woman pilot, a rookie Air Force flier named Eileen Collins, for whom Gibson wanted extra training time. For some reason Abbey refused to approve Gibson's selections. The crew was scheduled to enter training one year before the flight, but the twelve-month mark came and went with no word from Abbey. Administrators in the training division began pelting Gibson with panicky calls, but still Abbey didn't move. It was a process that Gibson would come to know well in coming months.

"George would hem and haw and say, 'No, that's not the right name,' but he wouldn't make a decision, he would never say anything specific," says Gibson. "He wanted you to go back to your office, scratch off [the] name, and put in a new one. That's the way George operates. Then you did it, not him. Then his fingerprints aren't on it."

Finally, Gibson figured out what Abbey wanted. It had to do with one of Abbey's favorite Bubbas, the commander of STS-63, Jim Wetherbee.

"George wanted to cancel STS-63, because he wanted to give his boy Jim Wetherbee STS-71," Gibson recalls. "If we canceled 63, we'd have to do something with that crew, and the logical choice would be [to give them] 71. That was totally against what I wanted to do, and I was head of the astronaut corps. I was not going to knuckle under. So I worked behind the scenes to sabotage him." Gibson went to one of NASA's many outside advisory commissions, this one chaired by the former astronaut General Thomas Stafford, and in his words "outflanked" Abbey by coaxing from Stafford a recommendation that they go ahead and fly STS-63. "We shoved that one right down ole George's throat," Gibson recalls with a note of triumph.

Also drawn into the STS-71 intrigues was the veteran shuttle commander Bill Readdy. In September 1993 Readdy had surprised everyone in the corps by volunteering for the thankless job of nonflying backup on Norm Thagard's first stay aboard Mir, an assignment that would require him to train in Russia for a year. What the other astronauts didn't know was that Readdy had taken the job only after Abbey promised him a mission he called Mir 18B. As Readdy understood it, he would fly to Mir aboard a shuttle and replace Thagard aboard the station. A month later he would return to Earth with two cosmonauts aboard a Soyuz capsule, something no American had done before. It would be an excellent way for NASA to check out the usefulness of the Soyuz, which was then under consideration as the emergency evacuation vehicle for the International Space Station.

But when Readdy visited Star City on an inspection tour, he received a rude surprise: The Russians told him they had never signed off on any 18B mission. Readdy returned to the United States and phoned Gibson in Houston. Gibson said he didn't know anything about an 18B mission either. "Hey, look, that was part of the deal here," Readdy told Dave Leestma, the head of Flight Crew Operations in Houston, the man who served as Gibson's immediate boss. Before he knew it, Readdy found himself summoned to Abbey's office. In that meeting Abbey offered Readdy his choice of three jobs. He could take the newly created position of director of operations in Russia, the senior NASA official at Star City. He could remain Thagard's backup; Abbey hinted strongly that if Readdy accepted the assignment, Thagard would be eased aside and Readdy could fly the mission himself. Or, Abbey said, Readdy could be commander of STS-71.

This was the reason so many of the astronauts loathed Abbey. Hoot Gibson had already promised STS-71 to Steven R. Nagel, a Navy flier who had commanded two shuttle missions; Readdy would not betray Nagel by taking the STS-71 command. Nor would he betray Thagard, who had been in training for the Mir mission for eighteen months. And he had no interest in the Russian management job. After leaving Abbey's office, Readdy walked out onto the sidewalk and thought about the price he would pay for defying the most powerful man at NASA. "There was no doubt in my mind," he recalls, "that I was probably done with the shuttle program and the astronauting business."

Readdy was lucky; after a brief period in purgatory, he returned to command the fourth Mir-docking mission, in September 1996. But Gibson's problems were just beginning. In early 1995 he sent Abbey his proposed crew roster for STS-71 with Nagel as the commander. Three weeks later Gibson received a summons to the office of Dave Leestma.

"Hoot, he liked your crew except for one," Leestma said. "Nagel."

Before Gibson could say anything, Leestma delivered the punchline. "They want you to command it."

This was Gibson's worst fear. His predecessor, Dan Brandenstein, had earned the ire of many in the astronaut corps by taking for himself two of the best missions of the early 1990s, including the first flight of the new Shuttle Endeavour in 1992. After Gibson took over from Brandenstein in the fall of 1992, a *Space News* reporter had asked him straight out whether he intended to take the first Mir docking mission. "Absolutely not," Gibson had said.

Now Abbey was offering it to him. Gibson was certain he knew why: If he took STS-71, he would have to step down as head of the Astronaut Office; Abbey would be free to appoint a more pliable astronaut in his place. "I don't want it," Gibson told Leestma.

Leestma promised to take his message back to Abbey, but Gibson had little confidence he would be able to avoid the assignment. There was no way Leestma could stand up to George Abbey. "George would just annihilate Dave," remembers Gibson. "George can be extremely disagreeable, and David is just too nice a guy. George would just control him completely."

Within days Leestma returned to Gibson. "Hoot, you gotta do me a favor," he said. "Go do [STS-]71. Do it for me."

Gibson had seen this coming. "Okay, Dave, if that's what we've got to do," he said. "But you have to do one thing. Will you come over and tell the corps that I'm not doing this myself, that it's what you want me to do?" Leestma agreed. Gibson took over STS-71, and Nagel took an administrative job. Abbey had won, as he always seemed to do.

No one wants to fall out of George Abbey's favor for one simple reason. You could end up like Blaine Hammond.

Hammond, the forty-five-year-old pilot on Linenger's shuttle flight in September 1994, was a St. Louis kid, the happy-go-lucky son of a General Electric sales manager who grew up watching McDonnell Douglas F-4s swooping in and out of Lambert Field. After the Air Force Academy he had gone on to be a test pilot, and was teaching in the Air Force's test flight program at Edwards Air Force Base when he learned his application to become an astronaut had been approved. Named by Abbey as one of seven military pilots among the seventeen members of the star-studded astronaut class of 1984, Hammond joined a group that included Bill Shepherd, chosen in 1996 to be the first commander of the International Space Station; Jim Wetherbee, who later emerged as Abbey's right-hand man; and Frank Culbertson, the straight-arrow Navy pilot now running the Shuttle-Mir program.

Even with the introduction of women in 1978, the astronaut corps remains a macho, male-dominated fraternity, and like fraternities everywhere, it has its own hazing rituals. For reasons he never fully understood, Hammond emerged as the focus of hazing in the class of 1984, whose members dubbed themselves "the Maggots." For Hammond it began when he was still a rookie, on a commercial flight the class was taking back to Houston from The Cape. Settling into the plane, Hammond and Wetherbee struck up a conversation with a young woman, who flashed a skeptical look when informed the two were astronauts. To convince her, Hammond stood, removed his NASA badge from his luggage, and showed it to her.

It was a fateful mistake. Changing planes in Atlanta, Culbertson and an astronaut named Ken Cameron approached the copilot, an old friend of Culbertson's, and asked his help in a practical joke. The plane had barely left the gate when one of the flight attendants announced over the intercom that the crew was happy to have aboard "the world-famous astronaut Blaine Hammond," and invited the passengers to approach Hammond for autographs; while his fellow astronauts snickered and rolled their eyes, Hammond spent the rest of the flight signing his name to everything handed him.

His new monicker stuck. Before Hammond knew what was happening, the other astronauts began taking copies of his official NASA photo on trips, signing them, and handing them out to people. *Hey, here's an autographed picture of Blaine Hammond, the world-famous astronaut! Want one?* Someone printed up dozens of Blaine Hammond nametags and began handing those out as well. At conferences and meetings all over the world, astronauts began wearing Blaine Hammond nametags and introducing themselves as *Blaine Hammond, World-Famous Astronaut*. Whenever someone got drunk or said something off-color, he always seemed to be wearing a Blaine Hammond nametag. Hammond would get the blame until he explained the situation to his superiors in the Astronaut Office. The hazing reached its zenith on STS-26 in October 1988, when Rick Hauck and Dick Covey—two astronauts not even in Hammond's class—had a photo taken in the shuttle middeck. In their hands, brandished for the camera, was an autographed photo of *Blaine Hammond, World-Famous Astronaut.*

Hammond was a good sport. He tried to laugh it all off, but it was demeaning. He was a goat. No one seemed to take him seriously. As the years went by, he began to suspect the rampant teasing was having an effect on his career. He flew as a shuttle pilot once, on STS-39 in April 1991, without incident. But as 1991 stretched into 1992, he received no word about his next flight. By early 1993 he had still heard nothing and was beginning to think something was seriously wrong. By then every pilot in his class but one had flown twice, the second time usually as a commander. Even pilots in the class of 1985 were flying their second missions, some as commanders.

He searched his mind for anything he had done to embarrass the office. Had he offended someone? Did Abbey have it out for him? Hammond pestered Hoot Gibson and Bob Cabana and everyone else he could think of. *What have I done?* No one knew. But then no one ever knew. That was the way the Abbey regime worked. One day you were in Abbey's good graces, flying one mission and scheduled for another. The next you were in purgatory, a Wandering Astronaut, cursed to walk the halls of Building Four's astronaut offices for years with no explanation why you weren't flying.

Hammond suspected his divorce hadn't helped him in Abbey's eyes. Before he and his wife split up, Hammond had carried on an affair with a woman who worked at The Cape. "Blaine's divorce was very messy, and very public, at least in the Astronaut Office," says his friend Rich Clifford. "He wasn't very subtle about the affair, although he thought he was. Blaine's big problem with the Astronaut Office [was] he doesn't perceive that others see what he thinks he's getting away with." But when Hammond asked, Mike Coats, at one point the acting chief of the Astronaut Office, assured him that his divorce wouldn't prevent him from flying.

As far as Hammond could tell, the only black mark on his record was something that happened at The Cape in October 1989, an episode that Hammond ruefully came to call the Maggot Stamp Incident. During each shuttle flight a small group of nonflying astronauts are designated Astronaut Support Personnel, or ASPs. Their job is to watch the orbiter before launch, help prepare flight documentation, and generally make sure the mission astronauts have everything they need. The ASP point man, the astronaut who straps the crew into their seats and wishes them farewell, is called the Caped Crusader or C-squared.

Hammond was the Caped Crusader for STS-34. On launch day, as he crawled through the flight deck checking and double-checking the crew's checklists, he thought to take out his Maggot Stamp. This was a rubber stamp astronaut Marsha Ivins had designed for each of the members of the class of 1984; it featured the Maggot logo, a space shuttle whose payload bay had been replaced with a garbage can swimming with little maggots in space suits. The Maggots had the design emblazoned on T-shirts and playfully decorated their notes to each other with the stamp. It was a silly thing, but it gave the Maggots a sense of identity, a dash of esprit de corps.

The pilot on STS-34 that day was Hammond's fellow Maggot Mike McCulley. As a little morale booster, Hammond took out his Maggot Stamp there in the shuttle and stamped the Maggot design on several of McCulley's checklists; Hammond thought McCulley would get a kick out of it. Then, to make sure the commander, Don Williams, a member of the class of 1978, didn't feel left out, Hammond stamped a few of Williams's checklists too. He was careful not to obscure any writing. When he finished, he grinned. This was the kind of fun things astronauts did for each other. The only problem, as Rich Clifford later pointed out, was that Hammond had a little bit too much fun. By one count, he used the Maggot Stamp over fifty times.

Later, when Hammond was helping Williams strap into his seat on the flight deck, the commander held up his checklists.

"What's this?" he asked, pointing to the Maggot Stamp.

"Oh yeah," said Hammond. "What's that?" He figured Williams knew exactly what it was and who had done it.

"Who did this?"

"I wonder," Hammond said, smiling.

There was bad weather on The Cape that day, and after sitting on the pad for five hours the mission was aborted and the crew was forced to crawl back out of the shuttle. By the time Williams returned to crew quarters he was furious. Someone, he fumed, had defaced his data files, and he wanted to know who.

Hammond didn't learn of the commander's anger until he returned to crew quarters after tidying up the flight deck. A young NASA engineer met him at the door with a frightened look on his face. In his hand he held a Xerox copy of one of Williams's Maggot-stamped checklists.

"Eric, what's the matter?" Hammond asked.

"Do you know anything about this?" the engineer asked, showing him the checklist. "Williams thinks we did it. He wants to fire us."

"Sure," said Hammond. "I did it."

Worried, Hammond sought out his friend Mike Coats and explained the situation. "Mike, you know, I did this, I did it just in good humor, a spirit thing."

Coats rolled his eyes and told him not to worry. He would take care of the problem. But Don Williams would not be mollified. When he learned it was Hammond, he went straight to the Astronaut Office and tried to have Hammond thrown out of the corps, a tempest that blew past only after Mike McCulley intervened on Hammond's behalf.

Finally, in late 1993, Hammond got a flight assignment, for STS-64 in September 1994. It was as a pilot, not a commander, but Hammond was happy just to fly again. The mission went smoothly, and Hammond returned to Earth in high spirits, almost giddy, certain that his next flight would be as a commander. While still in orbit his commander, Dick Richards, asked him if, after their landing, he would like to do the final "walk-around," the largely ceremonial inspection of the orbiter as it sat on the runway. It was an honor, a pilot thing, a badge of having finally arrived after so many years in limbo. Hammond was so excited, he told his family to watch for him doing the walk-around on television.

But once the shuttle came to a stop, the first ASP into the orbiter, the astronaut Scott Horowitz, climbed up to the flight deck and said, "Welcome home, guys, and Blaine, you're not to do the walk-around."

Furious, Hammond stared at Richards. He was on the verge of exploding.

"Blaine, don't do it," Richards cautioned. "You've had a great flight. Don't do anything to jeopardize the next one."

"I know, I know," said Hammond, barely managing to swallow his anger and hurt.

"I'm sorry, Blaine," Horowitz said. "The orders came directly from Abbey."

"I'm sorry, Blaine," Dave Leestma repeated the next day when the crew arrived back in Houston. "I think you know this wasn't my decision." The official explanation was that it wouldn't look good for Hammond to do the walk-around alone. People might wonder where the rest of the crew was. "It was a lame excuse," recalls Hammond. "They just wanted to screw me."

Hammond, disguising his disappointment as best he could, returned to the warren of astronaut offices in Building Four determined to be a good soldier, to do everything asked of him in order to make sure his next flight was as a commander. The Russian program was ramping up by then, and the Astronaut Office was desperate for seasoned pilots to rotate to Star City as director of operations, in effect the den mother to the American astronauts in training there. Hammond, though fluent in Russian, wasn't asked at first. Dick Richards was and declined, but suggested Hammond instead. Hammond leaped at the chance, even though it meant relocating his family to Russia for several months. Every pilot who had served as Russian ops officer, he knew, had subsequently been selected to command one of the Mir rendezvous missions.

"Well, I can't *promise* you a mission" if you go, Bob Cabana told him. But Hammond volunteered anyway, sensing that it was his fastest way back into space. He immediately plunged into Russian-language refresher courses, eight hours a day, five days a week. If all went well, he would be in Star City by the spring of 1995.

By the end of February, after five months of language training, Hammond was shopping for winter clothes and making plans to get a Russian visa. But when he returned to Cabana's office to get his starting date, he got a rude disappointment.

"Stand by," Cabana said, hurrying out of his office on an errand. "You're not confirmed for Russia."

"What do you mean, 'I'm not confirmed for Russia'?" Hammond asked plaintively. "I've been taking all these classes."

"Come back in a week," Cabana suggested, and he did. But Cabana had no new details. All he could say was Abbey hadn't yet confirmed him for the Russian post. Instead, word came from Abbey's office that Hammond had been selected to be NASA's liaison to the Air Force, a job whose duties no one in the Astronaut Office seemed able to explain. In frustration, but still determined to behave himself, Hammond made an appointment with Abbey himself, who in a polite little meeting told the astronaut to fly out to Colorado Springs and see what he could do to help the Air Force in a series of joint NASA–Air Force projects. It was make-work, a nonjob that required him to shuttle back and forth to Colorado every few days, but Hammond did it without complaint. He strongly suspected Abbey hoped someone out in Colorado Springs would try and hire him back into the Air Force. "They were trying to put me out to pasture," Hammond recalls. "I was spitting nails, I was so mad, but I was going to keep my nose clean."

The Air Force assignment went well, and by that fall of 1995 word had spread through JSC that the next crew announcement, for STS-78 in July 1996, would be Hammond's to command. Hammond allowed himself to get excited. The mission was perfect for him. He considered several of the astronauts rumored for the flight to be friends. STS-78 was a Mir docking mission, and Hammond figured his Russian skills would be needed. People began sticking their heads into his office, flashing thumbs-up, and wishing him good luck. Surely, he felt, whatever sins he had committed in the past—the Maggot Stamp, the divorce—had been forgotten. This was it. This was his flight.

At a quarter to five one afternoon that September, Bob Cabana popped his head around the corner of Hammond's office and said, "Come down and see me." Hammond wanted to shout, *Yes!* This was how it always happened. Cabana called you down to his big corner office on the sixth floor at the end of a day and gave you the good news. A few minutes later, Hammond fairly floated out of his office and down the corridor to see Cabana. This was it. This was his mission.

"Well, Blaine," Cabana began. "I'm sorry to tell you this, but I haven't had any luck getting you assigned to a flight."

For a moment Hammond didn't understand. *This is my flight. This is my mission.* "What do you mean?" he said, trying to remain calm.

"I put your name in for [STS-]78, and it didn't go through." Put it in to Abbey.

Hammond's mind reeled. This couldn't be happening. "Why?" he finally asked. "Did he give a reason?"

"No," Cabana said. "It just didn't go through."

Hammond tried to get his thoughts together. "Does this mean I won't fly again?"

"I don't know."

Hammond returned to his office, destroyed. In the ensuing days he asked everyone—Cabana, Leestma, Jim Wetherbee—what he had done to be frozen out. No one knew. He thought of appealing to Abbey himself but decided against it. That would be suicide. In the end, it really made no difference. Hammond's career had already been dead for months; he just didn't know it. Crushed, he disappeared into a series of technical assignments that culminated, strangely, in his July 1996 appointment as the Astronaut Office's safety officer. It was a job that until Abbey's resurrection had been held by a series of strong-willed astronauts including Dan Brandenstein, the former head of the corps, and Fred Gregory, who had gone on to become the safety chief at NASA headquarters in Washington. The appointment of Hammond, a laughingstock to many in the corps, signaled to a number of astronauts that the safety position had lost all its clout. "I don't know why they put Blaine in that job in the first place, when they didn't trust him," says Rich Clifford. "It may be they were setting him up to fail, so they could get rid of him."

By early 1997 Hammond had given up his hopes of returning to space and was preparing to retire. In an agency where many astronauts were frightened to speak out for fear of damaging their careers, this made Blaine Hammond a dangerous man: an astronaut with nothing to lose.

Today the hot-button question at JSC is whether the widespread fear of Abbey inhibits the kind of open, frank discussions that are crucial to safety reviews. A number of recently retired astronauts believe it does. "People don't dare speak out, because George makes pretty short work of you if you do," says Hoot Gibson. "No one in the program will ever criticize what we're doing in space," seconds Story Musgrave, who retired in 1997. "They're all too afraid of George." Others don't want to believe that anyone would put their jobs ahead of safety concerns. Abbey, his allies point out, has devoted his career to safe spaceflight. And indeed, the corridors at JSC are lined with posters promoting safety consciousness.

Still, the fear persists. "People are afraid to confront center management," says Clifford. "That's just a fact. People are just plain afraid of Mr. Abbey. If you get on his bad side, you won't get a flight assignment. The smart ones, which is 95 percent of us, know to confine their protests to safety issues. If you're talking about safety, there's never been a compromise on that. But there are other parts, issues dealing with training and the selection of crews, that you don't dare speak up about."

Rarely do these concerns creep into public view, and when they do they have received little or no attention in the press. The NASA inspector general, Roberta Gross, whose office reviewed the safety of the Shuttle-Mir program, noted in a September 1997 report repeated complaints from JSC employees about a "chilling impact [on] free discussion and criticism. . . . Some of those employees have also said they they feel it would jeopardize their careers to be frank in their opinions, observations, and assessments of the Mir program. These remarks were made by even those employees who support the mission and characterize it as being safe. In a human space program, free and open communication is an essential component."

No less a figure than Gene Kranz echoes this concern. In a private letter to the Inspector General's Office in 1997, Kranz wrote that "NASA is no longer capable of the freewheeling and open discussions that ultimately result in safe and effective space system design and operations." In a 1998 interview for this book, Kranz put the blame squarely on Abbey. "You can't point to anything specific; if you could, he would have been nailed long ago," says Kranz, who retired from NASA in 1994. "It's a feeling. You have to contrast the feeling [today] with the way we did business ten, twenty years ago. Then you had people in a management position who came up through the ranks, who had the respect of people who were working for them. As such, you felt capable of addressing issues that need change, debating topics where maybe you were pressing the edge too much. Here today you have a group of handpicked people who did not come up through the ranks, who do not have that healthy respect that engenders full communication. Full and open communication is the key to the quality and safety of the program. . . . Those things that are obvious to everybody, like a fire, they are comfortable to discuss. Those things that are more subtle don't even see the light of day."

Whatever its implications, the subject of Abbey's power and its effect on safety debates at NASA was to be thrown into sharp relief by the events aboard the Mir space station in 1997. The NASA flights to Mir were part of an ambitious plan conceived by the Clinton administration in 1993 to merge the American and Russian space station programs and launch a single jointly run International Space Station by 2003. As politicians in both countries saw it, the new program solved a host of problems. NASA, which had squandered $8 billion over the course of a decade in a vain effort to launch its own station, would finally get one. By building the new station in partnership with the Russians—and with the Japanese, Europeans, and Canadians—Boris Yeltsin's unsteady regime would be drawn farther into the community of Western nations. Russian engineers and missile experts would be given productive work to keep them from sharing their secrets with the Irans and Libyas of the world. And the United States would gain the leverage it needed to prevent Russia from selling nuclear technology to longtime trading partners like India. From Russia's standpoint, the $400 million NASA agreed to pay for the Mir flights kept its space program alive. The International Space Station, or ISS, is a cornerstone of Bill Clinton's Russian strategy.

The Russian program, with Abbey and Dan Goldin as its principal architects, comprises three distinct phases. The final phase, Phase Three, will be the actual operation of the new ISS, which is to be the size of a city block, and by some estimates the brightest object in the night sky after Venus and the Moon. The second phase, Phase Two, to begin in late 1998, consists of forty-three separate Russian and American launches, during which all the various modules and station components are to be lifted into space and assembled there by spacewalking astronauts. Phase One, also called the Shuttle-Mir program, serves as the dress rehearsal, seven four-and-a-half-month missions by American astronauts aboard Russia's aging Mir station between 1995 and the spring of 1998. The Americans spend their days aboard Mir performing scientific experiments, studying things like the effects of long-term spaceflight on the human body, but the main goal of Phase One is to give the Russian and American space programs the chance to learn how to work together. Until the opening weeks of 1997, Phase One, while marked by rushed and sloppy administration behind the scenes, had gone smoothly enough. All that is about to change.

3

Friday night
The Beachhouse

Friday evening the families of the STS-81 astronauts gather at the Beachhouse to say their good-byes. Linenger's mother, Fran, down from Detroit for the launch, is there with his three sisters, all four women having held their breaths through the physical examinations NASA doctors had given them that afternoon. It was necessary, Jerry explained, since the astronauts were already in quarantine and couldn't risk coming into contact with anyone with even a slight cold. Fran had been petrified that the doctors would use her high blood pressure to keep her out, but she passed the exam easily. Jerry's brother, Ken, had gotten a scare back before his first mission in 1994, when one of the NASA doctors told him "something doesn't look right" and asked for a colonoscope. Even as Ken's mind began to reel with images of a tumor in his colon, the doctor had broken out in a grin; Jerry had put him up to it.

This year nearly 1,200 of the Linengers' far-flung friends and family make the trip for Jerry's launch. Don Linenger came from a family of nine children, and this weekend Cocoa Beach is practically swimming in Linenger uncles, aunts, and cousins. Fran has a huge contingent of family who drove down from Chicago in a snowstorm and barely made the launch in time. There are local schoolkids from the Linengers' old neighborhood and Detroit newspaper reporters and even a priest the Linengers had known when Jerry was a boy. All of them descended on the Day's Inn in Cocoa Beach and took over the place, launching a three-day tailgate party that had people strewn all around the pool and out onto the beach. For two nights, Jerry's sister Karen and her husband, Jack, took groups out to karaoke bars.

Once at the Beachhouse, Fran and her three daughters wait with the other families for about twenty minutes until Jerry and the rest of the crew arrive. It's too cold for a barbecue, so they pluck food from a nice government buffet. Everyone gives Jerry birthday presents and sings "Happy Birthday" to him; his birthday is the next week. Fran presents him with a coffee mug embossed with a picture of Jerry and his brother Ken she had taken in Russia the year before; she also slips him his birthday check. It is for twenty-five dollars. One of his sisters gives him a book about Michigan. Some of the other guests rib Jerry that he has brought along his "harem." There are video cameras humming and lots of people to meet, and before they know it, it is time for good-byes. Before they leave, Jerry insists on taking all four women down to the beach to show them the silly face he had made in the sand that morning. Karen and her sisters spend the next few minutes pulling burrs out of their stockings.

Afterward, as all four women step back onto the NASA bus, there are tears and hugs and a few memories of Jerry's late father. Three years earlier Jerry's flight had been delayed, causing everyone in the family some anxious moments. Then, just as the launch window was set to close, a brilliant rainbow had burst over the Cape and the countdown had resumed. Fran and the whole family had taken the rainbow as a sign that Don was watching over Jerry.

"We're going to be praying for you," Fran whispers to Jerry.

And then they are gone, leaving Jerry and Kathryn alone to say their own private good-byes. Kathryn is pregnant and is due to deliver in June; she and Jerry are hoping he returns on time in May.

"I love you," she says.

"Take care of our boy," Jerry says.

Saturday night/Sunday morning January 11–12

The launch of STS-81 is set for the hour before dawn Sunday morning, the liftoff precisely timed at 4:27 A.M. The astronauts have been sleepshifting for a week now, bedding down during the days and rising, vampirelike, to work at night. Mission specialist John Grunsfeld, a member of Linenger's class of 1992, rises around 5:30 P.M. After a quick breakfast he slips into a pair of gym shorts and goes out for a jog in the cool evening air on the backroads out near the Beachhouse.

In the darkness he spots Linenger, who is also out for a run, his last before boarding the shuttle in a few hours. They wave and pass each other in the night, a fitting symbol, Grunsfeld muses, of Linenger's place in the STS-81 crew. Because he has been in Russia, the crew has trained without him. On those few occasions when Linenger returned to Houston for a few days, Grunsfeld and Mike Baker made the effort to get him into the simulator with them, if only to foster some camaraderie, but it was always a chore logistically. It was odd: Shuttle crews had traditionally trained as a single unit, yet now, with the Mir missions, they had a passenger.

That sense of separation between Linenger and his crew is underscored at the crew's midnight lunch, when everyone shows up wearing a white STS-81 sports shirt except Linenger, who has pulled on a long-sleeved denim shirt. "That's okay, you're the Mir guy anyway," someone says. Linenger smiles and drinks his three glasses of fruit juice. He avoids solid food before launch, thinking it increases his chances of spacesickness.

After the midnight lunch Linenger returns to crew quarters. On the floor outside his room the suit techs have piled his flight diaper and his white cooling garment, a bodysuit he will wear beneath his space suit that vaguely resembles long underwear. After slipping into these, he pads down the hall to a room lined with burgundy Lazy-Boy recliners. Each of the astronauts has a recliner to rest in while he is being suited up.

"Good morning, Captain Linenger," one of the suit techs says. "Are you ready to fly?"

"You bet," Linenger says.

"All right, sir, we've got all your equipment here. Would you like to check it?"

Linenger finds the suit techs' formal routine soothing and reassuring. It reinforces the feeling that he is not alone, that he really is part of a team that will take care of him before handing him over to the Russians. He takes his time checking his equipment. Some astronauts cram all manner of things into the pockets of their space suits: Kleenex, pens and pencils, tape recorders. Linenger prefers a minimalist approach. As the other astronauts begin to file in and work with their own suit techs—over there is Brent Jett, the pilot, down from him Marsha Ivins, the payload commander—Linenger slides just a few items into the pockets of his orange space suit, which the astronauts call the Launch and Entry Suit, or LES:

an emergency radio he can use in the event they are lost at sea; a vomit bag he slides into his top thigh pocket, just in case; a knife to cut his parachute should they have an emergency evacuation; and a palm-sized mirror he can use aboard the shuttle to check his connections.

To one side, Tom Marshburn watches his astronaut closely. He likes what he sees. Linenger is so excited he is almost giddy; Marshburn notices his feet are constantly tapping the floor with nervous energy. Linenger's mood assuages some of Marshburn's nagging worries about the mission. In their last weeks together at Star City, Linenger had been approaching mental and physical exhaustion. Eighteen months of six-days-a-week training with no rest or vacation was not a recipe for launching a fresh, rested astronaut. The Russians knew how to prepare for long-duration spaceflight; before launch they sent the cosmonauts off to a spa for a few days of relaxation. The Americans, to Marshburn's consternation, still had a lot to learn. In the weeks before Christmas Marshburn had tried to get Linenger excused from some of the Russian training, especially things like suit checks and centrifuge training, which he had already done many times. The Russians wouldn't listen. Training was training. No exceptions.

There are two other men standing by as Linenger wriggles into the LES and goes through his suit checks. One is Bob Cabana, the head of the Astronaut Office. The other is Frank Culbertson.

Culbertson is a small, compact man, five-foot-seven with close-cropped blond hair and welcoming, pale-blue eyes. Most days he can be seen traipsing in and out of his fifth-floor corner office in JSC's Building One in a worn blue suit, a black backpack casually slung over his right shoulder. Serious, but with a hair-trigger smile that sends wrinkles rippling behind his aviator glasses, he is an acknowledged star in the NASA firmament and a favorite of Abbey's, the kind of clean-cut, no-bullshit southern boy who has always thrived inside the agency. Inside NASA, almost no one has a bad word for him. Culbertson didn't ask to run the Mir program, and once he got the job he continually lobbied Abbey to let him out of it so he could fly again. Mention that you are surprised, given his senior position in the program, that he has flown in space only twice, and he emits a shy chuckle. "Kinda sucks, doesn't it?" he says.

Like so many of his fellow astronauts, Culbertson had dedicated his life to becoming an astronaut at an early age, in his case thirteen, after watching the flights of John Glenn and others as a South Carolina schoolboy in the early 1960s. He had made it to NASA on a standard route: Annapolis, Navy pilot, then test pilot. His first marriage had already been in trouble when, after years of trying, he finally was named to the navy test pilot school at Patuxent River, Maryland; when his application to become an astronaut was accepted in 1984, his wife refused to go, keeping his young son behind. They divorced soon after.

In Houston Culbertson fell hard for one of his fellow astronauts, Judy Resnick, and there were rumors in the Astronaut Office that they were on the verge of being married when Culbertson flew to The Cape one morning in January 1986 to watch Resnick fly into space aboard the Space Shuttle Challenger. He was sitting with the families of the other astronauts when the orbiter exploded, killing Resnick and everyone aboard. To this day, though he eventually married another woman and sired four more children, Culbertson doesn't talk about what happened. No one at NASA questions his commitment to flight safety.

Culbertson had reluctantly taken over the Mir program in September 1995, following Thagard's inaugural mission to Mir. His predecessor, a wily Arkansas-born NASA veteran named Tommy Holloway, had recommended against his assignment and told him so. Some worried that Culbertson was too soft to joust with the Russians, and even Culbertson admitted he had been petrified at the prospect of daily battles with his Russian counterpart, the blustery, vodka-loving former cosmonaut Valery Ryumin. Holloway's opposition to Culbertson had more to do with the long-standing rivalry between astronauts and the flight controllers who sit on console directing their flights. Flight controllers, who work in NASA's vaunted Missions Operations Directorate, tend to think of astronauts as unlettered passengers who ride aboard machines that MOD designs and manages. Men like Holloway, who came out of MOD, look askance at the increasing number of astronauts like Culbertson whom George Abbey was promoting into management's senior ranks.

"Their primary job qualification is being a good stickman in an airplane," sighs Phil Engelauf, one of MOD's senior shuttle flight directors. "There's always a sense of 'My God, another astronaut in a management position.' " Holloway felt the same way. "Frank is still an astronaut at heart," Holloway says, "and probably doesn't know it."

On its face, there wasn't all that much to Culbertson's job. The U.S. astronauts, like the Europeans who had been flying to Mir for years, were simply guests aboard Mir. The Russians ran the station and were responsible for everyone's safety. All Culbertson and his dozen staffers had to do was supervise the Moscow ground team that time-lined the astronaut's daily scientific experiments and keep an eye on the astronauts in training at Star City. Tommy Holloway had conceived the Phase One office as a stripped-down, science-support operation, and by and large that was what it remained.

Complicating Culbertson's job was the humbling lack of respect and attention the Shuttle-Mir program received at NASA. For the longest time there hadn't even been a Mir flag flying out with the other flags in front of Building One; it had taken a direct appeal from Culbertson's office to have one strung up. It was as if most of NASA had no idea the Mir missions were even going on.

Tom Marshburn was constantly reminding others at NASA not to pelt him with new work.

"I'm working the mission," he always said.

"What mission?" was the typical comeback.

"The Mir."

"Oh."

Six months before Linenger's launch, during an earlier rotation through Moscow, Marshburn had worked a portion of Shannon Lucid's mission. At one point there was a concern that some of the batteries on board might be leaking sulfur dioxide, and Marshburn badly needed to find out how toxic the gas might be. He paged several of the toxicologists in Houston, but his messages were never returned.

"I've been paging you guys; where ya been?" he asked when he finally managed to convene a teleconference.

"Well, we don't turn our pagers on unless there's a mission going on," one of the doctors said, meaning a shuttle mission.

"There *is* a mission going on," a senior toxicologist named Helen Lane said. "To Mir."

For many at NASA, the enduring symbol of Phase One was David A. Wolf, the hard-partying, forty-year-old astronaut who had been armtwisted into taking the final American mission to Mir, scheduled for early 1998. At the time he accepted the assignment, in the summer of 1996, Wolf's career was a train wreck. It was as if George Abbey had compiled a list of unpardonable sins and Wolf had done his darnedest to commit every one. Among Wolf's hobbies was acrobatic flying, and neighbors near the Houston-area airstrip he used had threatened to go to the FAA to have his license revoked after he buzzed low over their neighborhood; only Hoot Gibson's intervention had saved him. Then there was the arrest for public intoxication outside a bar at three in the morning. But Wolf's crowning achievement, the incident that landed him in astronaut purgatory for three full years, was his role in a bizarre FBI sting at JSC, a bit of stupidity that earned him a segment on the television show *Dateline NBC*. After that, many in the corps bet Wolf would never fly again. So did he.

Wolf, in fact, was being shipped to Mir for one reason, and one reason only: No one else would go. He was the absolute bottom of the astronaut barrel, and he knew it. "I was their last chance," he says with a sigh. "They had no other option."

Like Blaine Hammond, Wolf seemed cursed almost from the moment he joined the astronaut corps. A muscular sparkplug of a man with gray hairs at his temples and dark Sunday-morning rings beneath his eyes, he had grown up well-to-do in Indianapolis, the son of a car-salesman-turned-doctor whose family owned the oldest used-car lot in Indiana. Until the age of eight he wanted to be a garbageman. Then he watched Ed White's spacewalk outside a Gemini capsule in 1965 and decided he would be an astronaut. As a teenager he did the things so many astronauts do: build stereos with his father, fly open-cockpit biplanes with his uncle, dabble in rocketry and photography. He got his electrical engineering degree at Purdue. His motto at Indiana University medical school was "You can't have too much fun," and Wolf never let doctoring get in the way of a good Saturday night.

He first wangled a job at JSC in 1983, working as a medical researcher out in Building 37 until he was fired following a long-simmering personality conflict with his boss. "I had more failure in life than anything," Wolf remembers. "I had no great flying career; I was a backseater and only an average acrobatic flier. I was not a great doctor; I never even had a residency. And I was not the greatest electrical engineer either." Wolf's big break came after his firing, when he managed to land a research job working on a NASA project to grow human cells in space. The device he helped design, called a bioreactor, succeeded in growing cells in a solution-filled tank, a development that theoretically could help scientists study the formation of cancer tumors. Despite the skepticism of many in the American scientific community, NASA named Wolf Inventor of the Year in 1992.

Away from work, Wolf lived the astronaut's dream life, screeching through the streets of suburban Houston on his motorcycle, downing Shiner Boks at the Outpost, and flying his biplane. He bought a low-slung ranch house on a golf course a few miles north of NASA, let all his friends know where he kept the extra key, and threw raucous pool parties every weekend. Comely young women lounged in the hot tub and drunken pilots did cannonballs and backflips off the roof while Wolf, clad in flip-flops and gym shorts, stood barbecuing salami and ribs on his Weber kettle.

His former roommate John McBrine remembers being invited by a friend to one of Wolf's Friday night blowouts on his first day in Houston in 1989. Wolf hadn't yet arrived, and when the friend ran home to change clothes, McBrine was left alone in the house of a man he had never met. By and by, Wolf and a group of friends arrived, walked up the driveway, and began drinking beer out by the pool. Noticing McBrine, a man he had never met, watching television in his living room, Wolf just waved and went back to talking. It was that kind of house.

That year, 1989, Wolf made another routine application to become an astronaut, went in unprepared for the interview, and to his amazement found himself accepted into the class of 1990, also known as "the Hairballs," a nickname Wolf contributed during a beery evening at the Outpost. It should have been the beginning of a sparkling career. It wasn't.

A year later, while Wolf was rotating through a mundane technical job in anticipation of his first flight assignment, one of his neighbors asked him out to dinner with a businessman named John Clifford, who was trying to sell NASA a new piece of research equipment. Clifford and the neighbor picked up Wolf at his house in a white limousine and whisked him to the tony Rainbow Lodge restaurant in the River Oaks section of Houston, where they drained several bottles of champagne. Afterward they all repaired to Houston's leading strip club, Rick's, where Clifford slipped several hundred dollars to two women to give the group a private show. Wolf, normally a big fan of strippers, was so tired he fell asleep. They never got around to discussing Clifford's new equipment. Afterward Clifford stayed in touch, pestering Wolf with offers of free tips to the Florida Keys if the astronaut would only put in a good word with NASA management about his new equipment. Wolf blew him off. Clifford eventually stopped calling.

On its face, Wolf's career thrived. In October 1993 he made his first trip into space as a mission specialist aboard STS-58, flying side by side with future Mir astronauts John Blaha and Shannon Lucid. But a few weeks after returning to Earth, Wolf saw something on television that caught his attention. A local news broadcast was reporting details of a massive sting investigation at JSC involving an undercover FBI agent who called himself John Clifford. Wolf realized it was the man from the strip club. The news report was short on details, but it mentioned something about an astronaut under investigation for accepting bribes. Wolf wondered who it could be.

The next morning he sought out Jerry Ross, a fellow Purdue graduate who was filling in as Hoot Gibson's No. 2 in the Astronaut Office. Wolf mentioned the television report. "Jerry, I think I met these guys," he said. "I think you should know that." He explained what had happened that night, then asked Ross if he knew the astronaut who had taken a bribe.

Ross looked at Wolf, took a deep breath, and said, "It's you they're talking about, Dave."

"What?"

It was true. Ross explained that the matter had been investigated by the FBI and NASA's inspector general. The Astronaut Office had learned of it right before Wolf's shuttle flight but hadn't said anything.

"That was the beginning," Wolf remembers, "of three years of pure hell."

Standing there with Ross, Wolf demanded to talk to someone from the inspector general's office. No one would see him. Lawyers at the NASA legal office said they didn't know anything more than what was on television. For several days Wolf hung in limbo as news about the sting slowly dribbled out. It was like water torture: Because his name hadn't yet surfaced publicly, the taint of scandal was tarring everyone in the Astronaut Office. Finally Wolf couldn't take it anymore. He went on the offensive, giving interviews to local newspapers and television stations, laying out exactly what had happened. There had been no bribes, no favors. It was one night at a strip club, and he had even paid his own tab. A week after having his life turned upside down, Wolf sat up until dawn one Saturday morning at his house pleading his case with a *Dateline NBC* correspondent, Brian Ross, from whom he finally learned the full story of what was being called Operation Lightning Strike.* Ross's subsequent televised report portrayed the FBI sting as a misguided waste of taxpayer money and characterized Wolf as an unsuspecting dupe.

But the damage was done. "No one knew what all this meant," Wolf recalls. "Was I guilty of something? Was I innocent? No one knew." Hoot Gibson told Wolf the Astronaut Office stood behind him, but admitted that he shouldn't expect to fly for a while. It would take time for every-

*Operation Lightning Strike, a nineteen-month, $2 million undercover investigation spearheaded by the FBI, eventually led to a dozen plea-bargain convictions involving two midlevel NASA employees, ten NASA contractors, and two other firms. It was widely criticized as an ambitious failure.

thing to blow over. In the meantime Gibson found Wolf a job in Mission Control, working as a crew communicator—what NASA called a capcom—on shuttle missions.

Overnight Dave Wolf was a changed man. Gone was the carefree playboy with the pool parties and the hot tub lovelies. In his place was an angry man who felt betrayed by a government he had faithfully served for fifteen years. "I stayed mad," Wolf remembers. "Mornings I woke mad, at night I went to bed mad. I didn't have fun for three years, not one minute of it. I was never going to fly again." His bitterness had cost him a promising new girlfriend, and his open, trusting nature had stuck him with a roommate who ate all his food and never paid rent. Every day Wolf felt an almost physical pain when he slipped the chain holding his NASA badge around his neck. "Wearing that badge, walking through that front gate every morning, it was actually hard for me to breathe," he remembers.

"To be honest, it was hard to be around Dave for a long time," remembers John McBrine, who moved in with Wolf in 1995—replacing the problem roommate—after a stint working for NASA in Star City. "He was a bitter, bitter guy. He wouldn't talk about girls or cars or work. The subject always turned to the FBI, how he had been screwed by the FBI. Even when I started living with him, that was still the Number One topic of discussion. Even guys that loved Dave, you just got tired of hearing about it. And that was the only topic Dave would discuss."

By early 1996, a full two years after the sting's disclosure, Wolf was still sitting in Mission Control, an angry man. And then, that January, he heard the first rumors that he might fly again. Maybe, just maybe, he thought, his time in purgatory was nearing an end.

But then that February, even as he mustered the first hopes of returning to space, Wolf flew to Indianapolis to visit family. One Saturday night he stayed out late playing pool with friends at a tavern in the city's Broad Ripple section. Around 3 A.M. Wolf walked out on the sidewalk and saw a policeman yelling at a young man in handcuffs who lay on the ground between two parked cars. Wolf thought the cop was roughing up the kid.

"Is there something I can do to help?" he asked.

The cop told him to step away. Wolf didn't step away. The cop shined his flashlight into Wolf's eyes.

"You're publicly intoxicated," the policeman said.

"Then give me a test," Wolf shot back defiantly.

"You flunk."

"That's no test! I want a real test!" Wolf protested. Before he knew it the patrolman had spun him around and clamped handcuffs onto his wrists. Wolf was taken to the Marion County Jail and charged with public intoxication. The arresting officer said Wolf smelled of alcohol and exhibited slurred speech. His arrest was big news in Indianapolis, and he returned to Houston humiliated.

"I got in a little trouble up there," he mentioned to McBrine when he walked back into the house.

"Yeah, I know."

Wolf's jaw dropped. "How do you know?"

"Dave, it's in the papers down here."

"No."

"And on TV."

"Oh . . . no."

It didn't matter that six days after his arrest the local prosecutors in Indianapolis announced they would not press charges. The arresting officer, it turned out, had a lengthy disciplinary record, including twelve written reprimands and 114 total days of suspension for violations ranging from sexual harassment to leaving the scene of an accident involving his patrol car. Later that year the officer was fired.

Wolf's NASA friends could only roll their eyes: On the verge of returning to space, Wolf had come crashing back to Earth, convincing too many people that he was nothing more than a trouble-prone party animal. Not long after that, Abbey himself walked by the console where Wolf was sitting in Mission Control. "Dave," he said, not stopping, "you're going to have to be more careful."

"Yes sir, you're right" was all Wolf could bring himself to say.

Several weeks later, Wolf was still sitting in Mission Control, angrier and more depressed than ever. He was certain he would never fly again. Then, out of the blue, he heard a new rumor: NASA couldn't find anyone for the final Mir mission. Wolf couldn't imagine that the agency was so desperate it would actually choose him. And that was fine; even if it meant his only chance to return to space, he didn't want to go to Russia. He had no facility with languages. And his years flying in the Indiana Air National Guard had left him with an ingrained distrust of the Russians.

But then the unthinkable happened: Bob Cabana summoned him down to his corner office. "Dave, you're not a capcom anymore," Cabana announced. "We want you to go to Russia."

Cabana obviously expected Wolf to leap at the assignment. Instead Wolf asked for a day or two to think about it.

"You need to decide really quick," Cabana said.

John McBrine, who had thoroughly enjoyed his two Russian tours working as a ground-support man for NASA, urged Wolf to go. "Dave, you gotta do it, man," he said. "It's great over there."

"No way," Wolf would always say. "I don't wanna fucking go. Can you even get a hamburger over there?"

For days Wolf stewed. Then one Friday night he was eating dinner at Molly's Pub, a favorite astronaut haunt, along with an old pal from his Air Guard days, John Egan. Torn between the desire to return to space and his dread of moving to Russia, Wolf explained what he knew about the Mir program, which wasn't much. Egan didn't care. He felt the Russian assignment was exactly what Wolf needed to resuscitate his career. He could wait months, even years, before getting a shuttle assignment, Egan argued. This was something real, something tangible, something he could do to turn his life around right now. As he listened to Egan, Wolf couldn't believe it. He was actually getting excited.

Across the room, eating at another table, he spied Culbertson. An idea suddenly sprang into his mind. He knew there were problems with Wendy Lawrence, the astronaut who was scheduled for the penultimate Mir mission. Lawrence, at five-foot-three, didn't meet the Russian height specifications, and the Russians had repeatedly rejected her assignment.

"Frank," he said, walking over and sitting beside Culbertson. "I'm not going to volunteer for a four-month mission. But I will go for eight." Two back-to-back missions. Wolf grinned when he saw the look of surprise on Culbertson's face. "Four's been done," Wolf said. "It's time to press the envelope."

Culbertson, who to his irritation wasn't involved in the astronaut-selection process, was less than thrilled at the prospect of sending a tainted astronaut like Wolf to Mir for a four-month mission. An eight-month stay just meant twice the chance for trouble.

"You think about that, Dave," he said.

He did, and was eventually granted NASA's final four-month mission to Mir. And so, in August 1996, Dave Wolf swallowed hard and boarded a plane for Moscow. He was looking for redemption, but he would settle for a good hamburger and a shot to return to space.

* * *

"If anything goes wrong, I want to encourage you to communicate directly with me," Culbertson is telling Linenger as the astronaut finishes suiting up. "Don't be shy. Ask for me."

"Roger that," Linenger says, smiling.

"Have fun up there," Bob Cabana chimes in. There is some concern in the Astronaut Office that Linenger may be overly fixated on his science program. Both Cabana and Culbertson want to make sure, with casual little comments, that Linenger takes time away from his experiments to fully integrate himself with his Russian crewmates. This has been a problem with his predecessors. "You realize, Jerry," Cabana says, "that getting all of your science done is not 100 percent of your goal."

Linenger nods but isn't really listening. He is ready to go. He needs no further reassurances or reminders. He is a skier at the top of the slope, a racehorse in the gate.

Culbertson and the Phase One office have a lot riding on Linenger's mission. There are problems in the program, and he hopes Linenger can help correct them. The good news is that the NASA bureaucracy's dismissive attitude toward Phase One had begun to change in the months leading up to this flight. The reason was Shannon Lucid's stay aboard Mir, which had been a glittering public relations triumph. Thanks to an engine glitch that had delayed the shuttle flight to retrieve her, Lucid had stayed on the station an extra six weeks, enabling her to set the world record for time spent in space by a woman. She had returned to Earth a hero. President Clinton personally congratulated her, and her smiling face graced the cover of *Newsweek*. It was the kind of moment NASA administrators live for, earning them more goodwill with the public—and with members of congress who control the agency budget—than a dozen run-of-the-mill shuttle flights.

Lost amid the spectacle of Lucid's return, however, was the troubling mission of the astronaut who succeeded her, John Blaha. Where Lucid had charmed the Russians and complained little, Blaha, a fifty-four-year-old shuttle commander, had emerged from eighteen months of training in Star City exhausted and overwhelmed. Feeling alone and abandoned by his superiors in far-off Houston, Blaha had focused his anger on Culbertson and had blasted him during a tense, forty-five-minute confrontation in the quarantined crew quarters at Kennedy immediately before his flight that September. Culbertson walked away from the encounter badly shaken—*devastated* was the word he used. Up to that point Culbertson had seen his job as one astronaut helping another. He took pride in the support and loyalty he gave the Mir astronauts. Now, for the first time, he realized that at least some of his charges viewed him as what he called "the management enemy." It was a common-enough sentiment around JSC, but one that until then had not infected Phase One. Blaha's performance in orbit had been just as worrisome. There had been open tensions with his ground team.

Culbertson was determined to right the program during Linenger's four-month mission. During preflight preparations at The Cape he had sought out Linenger for a long talk, emphasizing how important it was to work closely with his ground team and the Russians. A complicating factor was the fact that Culbertson didn't really know Linenger. And what he saw he didn't much like. During Culbertson's visits to Star City, he had come to dread meeting with Linenger. "Jerry always had a complaint about everything; every session was a marathon," Culbertson remembers. "And he had a lot to say before he launched: This had to be changed, that had to be changed. And he didn't like this, and he didn't like that."

The chitchat ends when word comes that it is time for the crew of STS-81 to head to the launchpad. Linenger fairly bursts down the corridor to the elevator, past all the NASA techs, who line the hallways calling out encouragement.

"Go get 'em, guys!"

"Good luck, Jerry!"

Someone says there isn't much press outside. Linenger isn't surprised. It is, after all, almost two o'clock in the morning. But the moment Linenger and his five crewmates walk out the glass doors toward their waiting bus, they are blinded by an onslaught of flashbulbs and television cameras. People are clapping and shouting out his name. Linenger is reminded of celebrities stepping into the Academy Awards show. As he walks out the door and around the bus, he manages a quick wave and then is suddenly inside the bus with the other astronauts, jostling to get a seat.

"Wow," he says. "Didn't expect that!"

That's when it hits him: All this press, the cameras, the reporters, it's all for *me. I am the mission.* For the first time in two years Linenger is struck by the enormity of what he is about to do, how many people around the world wish him well, how important his mission is to the furtherance of world peace. For a moment his mind flashes back to all those late afternoons at Star City, when night already had fallen, and he had sat in some dimly lit Russian classroom poring over the minutiae of Soyuz thruster systems. *All for this. All for this.* It is at once humbling and thrilling.

It's a ten-minute ride in the bus to launchpad 39B, where Atlantis stands, wreathed in steam. Everyone piles out of the bus and takes the elevator up to the 195-foot level, where Linenger, along with John Grunsfeld and Brent Jett, walks over to the catwalk railing, looking out over the lights of The Cape. In the distance they can see lines of headlights as hundreds of cars creep over the causeway for the launch. Three miles off they can almost make out the bleachers where Fran Linenger and the rest of the Linengers are already gathering. Down from those bleachers is the press area, where the reporters and camera crews are milling about. All the usual suspects are there, Bill Harwood from *Space News*, Jay Barbee from NBC News. The author Tom Clancy is in the crowd, following the launch for one of Microsoft's on-line magazines. There too is Seth Borenstein, the thirty-five-year-old space writer for the *Orlando Sentinel*.

Borenstein is an oddity. To some at NASA he is a vulture, a harbinger of death, the living embodiment of the symbolic, dark-suited messenger in the film version of *The Right Stuff*. To anyone who cares to ask, Borenstein freely admits he came to The Cape to see the shuttle explode. In his previous job in Fort Lauderdale he covered hurricanes, and he considers himself a disaster specialist. Borenstein prepares for every shuttle launch by writing obituaries. He's got them written already—in his computer and ready to be published at the press of a button—on Linenger and the other astronauts. Each describes how the astronaut was killed in an explosion shortly after liftoff. Days before each flight Borenstein calls up all the astronauts' families and interviews them at length, although never telling them it's for an obituary. He's also written a 140,000-word story on the explosion, with a complete rundown of possible causes.

Linenger stands there at the catwalk railing for the longest time, burning the view into his mind. By and by, the fruit juices he drank begin to move through him, and he grabs Jett and Grunsfeld to walk around the corner to a small toilet set back into a closet area. Grunsfeld helps him unzip his suit so he can empty his bladder; afterward Grunsfeld rezips him and performs his final safety checks.

"Okay, guys," someone yells over their shoulders. "Let's go!"

The Caped Crusader for the mission is an astronaut named Pam Melroy, and she helps Linenger don his helmet and scrunch himself into his seat on the shuttle's middeck beside Marsha Ivins. Above him on the flight deck, Linenger can hear Baker, Jett, and Grunsfeld going through their flight checks. There is little for him to do. As a passenger, he has no real duties unless there is an emergency, at which point he and Ivins would be responsible for opening the escape hatch. He takes out an index card outlining his emergency evacuation procedures and places it across his right knee. Several times he uses his pocket mirror to double-check his connections, the air hose, his communication lines, his shoulder harness. When he is finished Melroy leans over and gives him a quick peck on the cheek. "Good luck," she says. "See you in five months."

It is a welcome feminine touch. Linenger reaches out and gives Ivins's arm a squeeze.

"We're going to space," he says.

Linenger doesn't have a window to look out, so when the first engines ignite at 4:27 he closes his eyes and concentrates on feeling and hearing the launch. First there is the shaking and vibrating, as if he is in an earthquake, and then the loud roar of the engines as Atlantis lifts off the pad. Everyone feels the kick, slamming them back into their seats. Up on the flight deck John Grunsfeld has a far better view. Three or four seconds after ignition he takes his eyes off the brightly lit panels in front of him and uses his pocket mirror to look back through an overhead window. The launchpad quickly recedes; then there is the long trail of smoke from the solid rocket boosters. At five seconds all of Kennedy Space Center sprawls out below him, then all of northern Florida, then the rest of the state. By fifteen seconds there is nothing left to see.

Down on the bleachers Fran Linenger and her daughters hold hands as the shuttle knifes up into the black morning sky. Over and over she repeats her favorite prayer:

Remember, O most gracious Virgin Mary,
that never was it known
that any one who fled to thy protection,
implored thy help or sought thy intercession,
was left unaided.

Inspired with this confidence,
I fly unto thee,
O Virgin of virgins my Mother;
to thee do I come,
before thee I stand,
sinful and sorrowful;
O Mother of thy Word Incarnate,

despise not my petitions,
but in thy clemency hear and answer me.
Amen.

Eight minutes after liftoff, in about the time it takes most people to eat a sandwich, Linenger and the crew of the Space Shuttle Atlantis are in space.

4

Monday, February 3, 1997
Outside Moscow

An hour north of downtown Moscow, past the grimy suburban apartment blocks, the trash-strewn roadside flea markets and the lines of green gingerbread dachas, a two-lane road turns off the main highway into a snowy forest. The road runs straight as a lightning rod through the pencil-thin trees, all tall pines and birch, their black-slashed white trunks gleaming like bones in the morning sun. A rail line runs beside the road, past a quarry where freight cars huddle beneath clouds of frosty steam. There are people on the roadside, scurrying along the shoulders, fur-clad babushkas with shopping bags pulling eight-year-olds across icy patches where the snow has been cleared.

About a mile into the woods the road abruptly ends, taking a sharp left turn and coming to a dead end before a ten-foot-high brick wall and a single high, silvery gate. Beside the gate is a large two-story guardhouse, its worn facade and cracked windows suggesting it once enjoyed better times. Behind the wall is a Russian military base. Twenty years ago an unwanted visitor could never have reached this gate alive, and no one passed through without a military pass. Today you can slip one of the two greatcoated guards a ten-dollar bill and have the run of the place. This is Star City, the fabled home of Russia's cosmonaut corps.*

Past the gate the forest continues, and the road narrows as it slices through the towering pines. As in Moscow there is snow piled everywhere, but unlike in the city it has been plowed seemingly within minutes of hitting the ground. Pyotr Klimuk, Star City's base commander, is renowned for supervising the most intensive snow-clearing operations in

*The official name is the Gagarin Cosmonaut Training Center, sometimes called GCTC.

Russia. The air here is cold, clean, and fresh, a world away from the smoggy city. The scent of pine is everywhere. Down from the gate a secondary road crosses the entrance road, creating an intersection. A right turn would lead you through a security gate to base headquarters, tucked away unseen in the deep woods.

Farther down, past the intersection, the main road abruptly widens into a ragged square. To the right, its roof visible above the trees, is an enormous, humpbacked building containing one of the world's largest centrifuges; inside, the three-hundred-ton machine resembles a giant, sixty-foot-long blue hot dog, one end anchored in the middle of an oversized circular gymnasium, the other capable of whirling cosmonauts around the gym's inside walls at rates of up to 30 Gs, enough to drive your teeth into your throat. Ahead, on the far side of the square through the trees, is a jarring sight, a grouping of colorful Western-style town houses, the kind you might see beside a golf course on the northern edges of Atlanta or Dallas. So out of place do the town homes seem here in the Russian woods, so at odds with the stone-age architecture of the base's Soviet-style apartment blocks, they could be a mirage. In fact, they have been built for the American astronauts and their families, whose tastes for Russian flats lasted about as long as their interest in borscht.

To the left of the square, past a small park dominated by a statue of Yuri Gagarin, the first man in space, runs the Avenue of the Heroes. There are no statues or monuments here, just a snow-covered greenway flanked by parallel sidewalks. Blink and you could be on the grounds of an American university, maybe Stanford or Princeton. In fact, despite persistent stories of its ruination in Western scientific circles, all of Star City has the bucolic air of a campus, albeit one for an elite grouping of students. Most afternoons, while the children are at school and their parents at work, there is almost no one to be seen. A handful of cars are parked helter-skelter in front of some of the buildings. You can tell which are owned by the cosmonauts. They're the ones with government plates beginning with numbers from 1 to 100, denoting their official cosmonaut numbers, as well as the order in which they went into space.

This morning, down at the far end of the Avenue of the Heroes, Cosmonaut No. 76 can be seen leading a small procession up the stairs into the Cosmonaut Museum. The cosmonaut's name is Vasily Tsibliyev, which the Americans pronounce "Si-BLEE-ef," and he is commander of the next crew to be sent into space to run the space station Mir, the pride and only remaining joy of the once-mighty Russian space program. Tsibliyev is a squat bull of a man, with a Kirk Douglas cleft chin, eyes like bullets, and, one suspects, the will to match. There are creases in his royal-blue uniform pants, his jacket is buttoned firmly across his chest, and his black tie is cinched up snugly beneath his chin. He is a military man with a military bearing, a onetime Soviet fighter pilot who lives with his wife and two children in an apartment building here at the base.

Behind Tsibliyev walks his slim flight engineer, the likable, laconic Aleksandr Lazutkin, pronounced "La-ZOOT-kin," whom everyone calls Sasha. Unlike Tsibliyev, who commanded a six-month mission aboard Mir during the winter of 1993–94, Lazutkin will be making his first trip into space, and he is excited. Like all Russian flight engineers, he is a civilian—he actually works for Energia (sometimes spelled Energiya), the giant Russian contractor that built Mir—and lives with his wife and two young daughters in a small, dimly lit flat an hour away in Moscow. No matter what his workload, Lazutkin always looks as if he hasn't slept in days; he has unruly, mousy-brown hair, a forlorn mustache, and the eyes of a mourner. To Lazutkin, Tsibliyev is a stern older brother, in the words of a cosmonaut psychologist he trusts, a "strong stone wall that will protect me." Beside Lazutkin is the crew's third member, a mustachioed German astronaut named Reinhold Ewald. Ewald is to accompany the two Russians into space in seven days but will leave Mir after only a three-week stay, returning to Earth with the two-man crew Tsibliyev and Lazutkin are to replace. Ewald is well regarded, but what the Russians like most about him is the $60 million the Germans are rumored to have paid to send him into space.

Stomping the snow from their boots, Tsibliyev and his crew enter the dim museum lobby where, along with a group of Star City officials and photographers, they take off their heavy coats and hats and leave them with an unsmiling woman at the coat-check. While it contains mementos from thirty-six years of Russian spaceflight, the museum is in many ways a shrine to Gagarin, whose single, 108-minute orbit of the Earth on April 12, 1961, remains the high point of Russian space history. Inside the museum's front doors a visitor's first sight is a mural of Gagarin. To the right is a hall containing all manner of Gagarin memorabilia: the puffy, brownish gray space suit he wore into orbit, a vase bearing his image, gifts he received from foreign dignitaries.

Heading left out of the lobby, Tsibliyev's group ascends a flight of darkened stone stairs. As in most buildings here at Star City, few of the overhead lights are turned on, a cost-saving measure. Off the second-floor landing is the main Gagarin room, where a large plaster bust of the cosmonaut gazes over several glass cases brimming with still more of his artifacts. On one wall is a chronology of Gagarin's foreign visits and more gifts he received along the way: tiny geisha dolls in Japan, a Mexican sombrero, medals from Cuba, Egypt, Hungary, and more. There is a striking photo of Gagarin's small round orbiter lying spent on the Russian steppe moments after landing, the fuzzy brown silhouettes of rescue helicopters hovering in the background. Photos of his mother, and of his days as a metallurgical student, flank items the Russians treat almost as religious objects. Here is the watch Gagarin wore in orbit, beside newspaper headlines trumpeting his achievement. The most sacred—and most morbid—objects are those taken from Gagarin's body the morning he died, in a plane crash during a training mission in March 1968. His black-leather wallet, shredded and burned, along with his driver's license and an old photo of his mentor and protector, the legendary chief designer of the Soviet space program, Sergei Pavlovich Korolev. There is even a small open vial brimming with gray dirt from the crash site. It looks like ashes.

These last items seem included almost as a response to one of the stranger myths to have arisen in Russia since Perestroika, that Gagarin didn't die in that plane crash. The stories have been around for years, but only gained purchase in the late 1980s as the Communist monolith crumbled. Some say Gagarin was abducted by aliens. Others insist he was murdered on orders from Brezhnev for some unspecified disloyalty. Or driven insane by something he saw in outer space, and left to molder anonymously in a remote psychiatric prison. The museum guides grow irritated when schoolchildren bring up these wild theories. In death, they ruefully acknowledge, Gagarin has become the Russian Elvis.

Tsibliyev's group passes this room and enters a long hall lined with glass cases containing photos and pennants heralding all the Russian space stations, from the first Salyut station sent aloft in 1971 to each of the myriad missions to Mir. At the far end of the room is a case holding the gray jumpsuits worn by the initial three cosmonauts to live aboard that very first space station: Georgiy Dobrovolskiy, forty-three; Vladislav Volkov, thirty-five; and Viktor Patsayev, thirty-eight. Their mission to Salyut—which translates as "salute," to Gagarin of course—was an unmitigated triumph, a throwback to the glory days of Gagarin's flight following a series of shadowy disasters and setbacks in the program during the late 1960s. All through June of 1971, Russians from Kiev to Kamchatka thrilled to the trio's exploits via nightly television reports; the three men cracked jokes, turned somersaults for the cameras, and cheerily demonstrated the ways they gobbled peanuts and fresh fruit as the food drifted in midair. Their return to Earth was to climax in a massive celebration and parade—until the moment recovery forces opened their capsule moments after it landed out on the central Asian steppe. All three men were dead, suffocated when a broken valve leaked their oxygen into space during reentry. The ensuing period of national mourning has been compared to what Americans endured following the assassination of John F. Kennedy.

Emerging from this hall, Tsibliyev's crew rounds a corner and enters what appears to be an office. It is in fact Gagarin's office, an exact replica of the room as he left it on the morning of his death, March 27, 1968. After Gagarin's fatal crash the first person allowed inside his office was an official photographer, who carefully shot the room from a dozen different angles. Afterward the room and its contents were painstakingly re-created here in the museum. Over the door is Gagarin's wall clock, the hands stopped forever at the moment he died, 10:31. On his desktop, now covered with a hard plastic bubble, his daily calendar lies open exactly as he left it, his plans for the afternoon of the twenty-seventh scribbled out in his own hand. To the left of the calendar, beneath the green desk lamp, are a dozen sepia-toned letters and envelopes, neatly and precisely staggered downward. Spanning the rear wall is a giant map of the Eurasian continent, from Spain to Japan. Beside it hangs a portrait of Korolev. The one addition the museum keepers have made is a glass case holding the gray greatcoat Gagarin wore to the airfield that last morning.

Lazutkin wants to pinch himself. It isn't just that Gagarin is his hero. It is more than that. This is one of the moments he has waited his entire life for. All his adolescent parachuting and gymnastic lessons, not to mention five years of cosmonaut training—everything has been for this moment, the chance to walk into Gagarin's office and sign the book. Lazutkin is surrounded by history, but at this moment it means nothing to him. This is purely a personal triumph, the high point of his life.

For every cosmonaut crew the ritual in Gagarin's office is the same, and under Tsibliyev's gaze the crew of Mir 23 does not deviate one millimeter from custom. At Gagarin's wooden conference desk, set out from his own desk at a perpendicular angle, there are two small wooden chairs. The commander takes the right-hand chair and sits; Lazutkin follows, taking the chair on the left. Ewald stands centered behind them. Behind Ewald the backup crew spreads out in front of Gagarin's bookcase.

A thick brown book is placed before Tsibliyev, who begins to write. "At the anniversary of Korolev's birthday and the 40th anniversary of the launch of the first artificial satellite of the Earth, we have to fulfill the program of the 23rd basic expedition to the orbital complex Mir. Our Russian-German crew is ready to fulfill the program. The launch of the 23rd basic expedition on board spaceship Soyuz-TM 25 was created by the genius of S. Korolev, and underscores again and again that the history of space investigation is still going on, and our crew is doing our bit.

"The commander of the crew—V. Tsibliyev

"Engineer of the crew—A. Lazutkin

"Research cosmonaut of Germany—Ewald."

It is a moment of the utmost gravity for the departing cosmonauts. Hands are shaken, official photos are taken. Afterward the three men and their families are transported in a procession of limousines into downtown Moscow, where they lay wreaths at the graves of Gagarin and other Russian space heroes at the base of the Kremlin wall. It is a day of tradition, of honor, of contemplating the sacrifices of those who have gone before them. Gagarin. Titov. Nedelin, the general who died in a 1960 missile explosion and fire. Korolev, the legendary chief designer himself, who suffered untold cruelties in Siberian prison camps.

Amid the ceremonies, no one bothers to take a minute to acknowledge some of the differences between Gagarin and the cosmonaut trio of Mir 23 some thirty-six years later. It is hard to imagine Gagarin making the kind of entrance Tsibliyev is contractually obligated to make when he finally arrives on the world's oldest space station seven days hence: As he floats through the docking hatch, the commander will be smiling, and in his arms will be a huge bunch of bananas for a European television commercial. If all goes as planned, the first words he will speak aboard Mir are "I brought the fruit!"

That morning, as Tsibliyev and his crewmates paid their respects beneath the Kremlin walls, the vaunted Russian space program—the program that had beaten the Americans into space with Sputnik in 1957, that had launched the first man into space in 1961, that had successfully launched cosmonauts onto orbiting space stations for twenty-five years—was crumbling around them. Just two months earlier Yuri Koptev, the stoic, long-suffering bureaucrat who heads the Russian Space Agency, had walked into a Kremlin conference room and told Boris Yeltsin's top advisers that without an immediate injection of cash, the entire program would collapse.

It was humiliating: The space program, as Moscow newspapers were forever reminding the Russian public, was the last vestige of the country's status as a superpower. But then humiliation was something the Russians were growing accustomed to. For ten years, since Mikhail Gorbachev first unleashed the demons of democracy upon his unsuspecting compatriots, the citizens of the former Soviet Union had endured a dizzying downward spiral of embarrassments: the loss of eastern Europe, the loss of Ukraine and the central Asian states, the 1991 coup attempt against Gorbachev, the 1993 coup attempt against Yeltsin, the war in Chechnya, the systematic disintegration of the country's army and nuclear arsenal.

Now, as the long gray winter of 1997 crept onward, the economy was a shambles. From Vladivostok to St. Petersburg, doctors, nurses, soldiers, miners, teachers, police officers, fishermen, engineers—everyone—waited months for paychecks that never seemed to come. Factories that thrived for decades on Soviet subsidies now sat idle, windows broken and doors barred, unable to function in the new free-market environment. Though the days when gangsters machine-gunned each other in daylight assassinations were largely past, in many areas of the country the Mafia was still the most efficient operation extant. Only in Moscow could the promise of a bright future be seen. The city's ambitious mayor, his eye on one day replacing Yeltsin in the Kremlin, somehow found a way to rebuild cathedrals, pave streets, and open new museums, even lay groundwork for a vast underground mall below the Kremlin walls. Everywhere else, there was just no money.

And it was cash, not rocket fuel, that kept the country's last space station, the eleven-year-old Mir, circling the globe through its low-Earth orbit. Since 1989, as Koptev constantly reminded his superiors, the country's space budget had shrunk by 80 percent. For eight years new projects had been canceled, engineers had been laid off, facilities had been closed. The Russian space shuttle, called the Buran, was among the first casualties. It went to space once, in 1988, but never flew again; today it sat glumly in Moscow's Gorky Park, a painful reminder of the country's lost glory. The world's largest rocket, the Energia, died just as ignominious a death. Mothballed too was the space program's fleet of communications ships, which in the old days crisscrossed the world's oceans providing immediate communication between Russian spacecraft and Mission Control. (Today cosmonauts aboard the Mir are only able to speak to the ground—through a network of Russian ground stations, two scratchy American radar sites and a rented German antenna outside Munich—for fifteen minutes or so every hour and a half; NASA shuttles, by contrast, remain in near constant contact with the ground twenty-four hours a day.) In central Asia, the Russians had only barely been able to retain control of their main launch complex, sprawled across the desert steppe in what was now the independent country of Kazakhstan; Yeltsin had managed to lease it only after humbling negotiations with Kazakhstan's new leaders.

Nowhere was the space program's deterioration more evident than at the Russian Mission Control Center, a dark, hulking mausoleum down a potholed side street in a northern Moscow suburb. Mission Control was known by its Russian acronym, TsUP, which the Americans pronounced "Soup." The TsUP was a chilly, drafty place—the Russians always seemed to wear sweaters—with wide corridors, cracking linoleum floors and a cavernous break area dominated by a leftover bust of Lenin that was simply too heavy to move. Outside the building, the wind sent candy wrappers and wadded napkins pinwheeling through the parking lots; in summer the grass out front grew two feet long, and weeds poked through the sidewalks. In the dimly lit lobby—as at Star City, the lights were switched off to save money—the building's cash-strapped overseers had rented out space to a Czech lighting-fixture company; the shop's gaudy chandeliers and crystal figurines mocked workers whose paychecks, if they came on time, rarely topped six hundred dollars a month. So underpaid was the staff that many of Mir's ground controllers moonlighted as taxi drivers and interpreters. Privately they muttered that the TsUP had become a sweatshop. The aging computers down on the main floor tended to greet the unwary with stiff electrical shocks. The Americans who came over to work the first Mir mission in 1995 had wondered aloud why there were cats roaming through the consoles. "How else are you going to control the mice?" one of the Russians replied.

Sometimes it seemed the Russian space program would do almost anything to raise money. Before and after the historic partnership agreement with the Americans in 1993, its main source of hard currency came from European and other countries that paid tens of millions to fly an astronaut aboard Mir. But then you didn't have to be an astronaut to get to Mir. In one notable stunt the Russians in 1990 accepted $12 million from a Japanese television station to fly its reporter aboard Mir. A decade later, cosmonauts still shook their head at the memory; they had never seen anyone vomit that much. After the Soviet Union collapsed in December 1991—briefly marooning aboard Mir cosmonaut Sergei Krikalev, who journeyed to space a citizen of the Soviet Union and returned to Earth a citizen of Russia—there were rampant rumors that the Kremlin was preparing to sell Mir itself to the highest bidder. One expert estimated the station might fetch as much as $700 million; such talk made cosmonauts aboard the station nervous. "Is it true the Russians are going to sell the Mir space station, where we are now?" a commander actually asked the ground. "And, we are asking, together with us?"

The Mir was not sold, of course, and as the months and years wore on, Russian officials discovered endless ways to make money. At the TsUP, grasping public relations men demanded—and received—as much as $1,500 for a single interview with the center's flight director, the growling former cosmonaut Vladimir Solovyov. At Star City General Klimuk's men began baby-sitting "space tourists." For $200 a day per person, busloads of British schoolchildren were allowed free run of the base, trying on space suits, crawling through training modules—"Look, it's the toilet!"—and exploring the darkened interiors of flight simulators, occasionally to the consternation of cosmonauts who were still inside.

The Russians were especially adept at working with Western advertisers. Down in the TsUP's main control room, an auditorium where Russian ground controllers hunched over four parallel rows of consoles beneath an overhanging mezzanine for observers, a large Hewlett-Packard advertisement sat beneath the main viewing screen. In spring 1996, while Shannon Lucid sat waiting inside the station, her two Russian crewmates crawled along the outer hull of Mir and unfurled and filmed what appeared to be a man-size nylon-and-aluminum soda can. On its side was the unmistakable Pepsi name, a stunt for which Pepsi paid the Russians $1 million. Nothing, it seemed, was too far-out or too demeaning for the Russians. At one point officials floated the idea of a civilian-in-space sweepstakes; at another, they entered into talks with a Russian filmmaker who seriously wanted to make a movie—a romantic comedy—aboard the station, a notion he gave up only after Emma Thompson and other Western actresses politely declined lead roles. "The Soviet program is like a chicken," one TsUP supervisor quipped in 1992. "You cut off its head, and it runs around the yard for a while thinking it's still alive."

As 1997 dawned, the program was barely alive. Funding snafus were canceling or delaying almost every major Russian space launch. Every cosmonaut crew that made it to Mir could count on having its mission extended; this way, the program saved money on rockets. The Mir 21 crew of Yuri Onufriyenko and Yuri Usachev in early 1996 had its stay aboard the station stretched by forty-four days due to production delays in the booster rocket that would carry its replacement crew. The Mir 22 crew of Valery Korzun and Aleksandr Kaleri, which worked with John Blaha and was still aboard to greet Linenger, had its time in space extended by nearly two months. The program's unmanned Progress cargo ships, which ferried food and supplies to Mir every few months, had been reduced to three flights a year from six. On October 15, three months before Linenger's mission began, Progress-M 33 was shunted into November because the program didn't have enough money to build its booster on time; the Russian military, which had regularly lent the agency boosters to help out, this time refused. It suggested calling NASA for help.

In the weeks leading up to Linenger's shuttle flight, things had gotten so bad that NASA officials were openly admitting what critics had been muttering for months: The only thing keeping Mir aloft was the shuttle resupply missions. Atlantis, which ferried Linenger to Mir in January, carried nearly 3,600 pounds of supplies to Mir—200 pounds more than the last shuttle flight—and much of it was food and clothing. Atlantis also transferred to Mir 1,400 pounds of water, which was automatically generated by the shuttle's fuel cells during its flight.

It wasn't just machinery that suffered. To the dismay of many inside the program, the obsession with money had changed the very nature of what it meant to be a cosmonaut. In the glory days of the Soviet program, cosmonauts might receive a Volga sedan, a new apartment, and a small cash reward for a successful flight. "When we flew, we performed our tasks for the Motherland, and we didn't think about money at all," remembers Vasily Sevastyanov, a decorated former cosmonaut who is now a deputy in the Russian Duma. "Today the cosmonauts fly into space only to make money. As a result, the country looks different at cosmonauts. Before, we were heroes. Now, they do things, but the attitude is 'so what?' "

The role of the cosmonaut began to change with the space program's first budget cutbacks around 1990. Their traditional perks—free trips to the beaches of Vietnam, Black Sea resort vacations, free cars and apartments—disappeared quickly. Salaries sank so low, the Duma had to pass a law to raise them. To prevent cosmonauts from fleeing the program in droves, Energia devised a new compensation system. Beginning in about 1993, every cosmonaut who flew to Mir did so with a signed contract. The contract, negotiated with the cosmonaut's legal representative, laid out exactly what the cosmonaut would do in space: how many experiments, how many spacewalks. The contract also detailed precisely how much each failure aboard the station—failure to follow an order, failure to carry out a spacewalk, failure to perform an experiment—would dock the cosmonaut's bonus. One bureaucrat at the Institute for Biomedical Problems, the IBMP, was charged with keeping track of all the additions and subtractions and toting up the cosmonaut's bonus at the end of each flight. To the cosmonauts the bonus was all-important. One year spent working on the ground might earn a salary of $20,000; Tsibliyev and Lazutkin didn't even make that much.* But the typical flight bonus was $30,000—roughly the amount stipulated in Tsibliyev's and Lazutkin's contracts. They were paid a set amount each month, received specific fees for each of the American experiments they participated in, and earned bonuses for special tasks. For every spacewalk they tried—and the two Russians were scheduled for three during their stay aboard Mir—they would receive an extra $1,000.

The cosmonauts thus entered the vanguard of Russia's new capitalists. The new contract system pervaded, and in some cases distorted, almost every aspect of space travel. Before the man who oversaw Mir on a daily basis, Viktor Blagov, the TsUP's sixty-one-year-old deputy flight director, could order a cosmonaut to perform an extra experiment in orbit, he had to negotiate a new contractual clause, inevitably containing more money, with the cosmonaut's legal representative. "There are pluses and minuses to the system," reports Blagov, a wry, white-haired engineer who has worked in Russian ground-control facilities since 1959. "Now they always say, 'We're not going to do anything extra unless we're paid to do it.' But it somehow makes the crew more disciplined, too. You know?"

The downside of the contract system, although Blagov and the cosmonauts are loath to admit it, is that it provides an incentive for cosmonauts to play down, or even cover up, problems aboard the station, out of fear that this will lead to a reduction in their flight bonuses. Former cosmonauts suggest it gives today's cosmonauts a disincentive to complain about safety. "In my day we could say frankly what was wrong with equipment," says Sevastyanov. "Today they are more cautious. They're afraid to complain. They don't want to get any black marks."

*NASA astronauts, by contrast, earn between $48,000 and $103,000, depending on their government employment level.

* * *

For the most part, few of the Russian cutbacks generated any serious concern in the West about deterioration in the program's safety standards. Still, there were worries. That September, four months before Linenger's flight, the Russian Space Agency disclosed that it would no longer blast Soyuz and Progress ships aloft on the powerful Soyuz-U2 booster rockets, which had been used since 1982. The explanation: the Russians could no longer afford its expensive kerosene propellant, called Synthin. Other, weaker boosters would be used, which would force the program to strip some backup systems off the Soyuz and Progress ships to lighten their loads.

At about the same time, a curious article appeared in a semi-official Russian technical journal called *Videocosmos*. The article reported that at an unspecified point in the near future Soyuz and Progress spacecraft would no longer be fitted with the computer-automated docking systems called Kurs (Course) units. This marked a major but almost unnoticed shift in the way the TsUP docked its spaceships. Since 1985 all Russian spacecraft had used the Kurs computers to dock automatically with the Mir station. The computer, triangulating radar signals from various antennae on the ship and the station, mapped and executed all the complex series of booster firings and brakings that allowed the ships to glide in and effortlessly dock with Mir. All the Russian commanders had to do was sit by and watch.

The differences in the Russian and American docking systems, in fact, reflected the differences in the countries' political systems. NASA, which prided itself on the expertise and decision-making abilities of its individual commanders, gave its astronauts the freedom to manually dock spacecraft themselves, whether with Mir, Skylab, or on the 1975 Apollo-Soyuz mission. The Russians, as befitting a totalitarian regime, created a centralized docking system that took control of the spacecraft out of the hands of the cosmonauts and placed it with ground controllers at the TsUP. It was as if the Russians were worried that cosmonauts, without total control from the ground, would fly off and defect.

Now, just as Russia had shed its totalitarian garb to attempt the transformation into a Western-style democracy, so too was its space program preparing to shelve its centralized docking strategy in favor of the "freer" Western model. There were two reasons for the change. One was reliability. On several occasions the Kurs systems had inexplicably failed as a

Soyuz approached the Mir, forcing the Soyuz commander to fly in manually, as the Americans did. The repeated failures of Kurs units on the unmanned Progress cargo ships were even more dangerous. On March 23, 1991, Progress-M 7, having already failed once to dock with Mir, was on its final approach toward the station's Kvant docking port when a ground controller noticed that the ship was not properly aligned with its target. At the last minute the ground controller aborted the approach, at which point the Progress veered to the left, passing within twenty feet of the station's massive solar arrays. The Russians were lucky; had the ship been approaching the docking port at the station's other end, the TsUP admitted, it would have struck the station head on, probably killing the two cosmonauts aboard.

A month later, cosmonauts Musa Manarov and Viktor Afanasayev discovered during a spacewalk that the station's main Kurs antenna was missing its satellite dish; Soviet television reported that the dish was damaged by a cosmonaut's inadvertent kick during an earlier spacewalk. The antenna was replaced, but Kurs units continued to fail. Just six months later an incoming Progress inexplicably aborted its approach just 150 meters from the station. In September 1994 the TsUP briefly considered abandoning the station when the Kurs system on Progress-M 24 failed repeatedly, preventing the crew from receiving a load of badly needed supplies; only a last-minute manual docking saved the day.

With these kinds of failures in mind, the cosmonauts had begun training in the early 1990s with a new kind of manual-docking apparatus that the Russians called the TORU system. The system itself, integrated into all Progress spacecraft and first shipped to the Mir in 1993, looked a bit like a stationary bicycle; to use it, the cosmonaut sat on a raised seat, a Sony television monitor before him, manipulating a pair of black joysticks in front of him. As problems with the Kurs units intensified in the mid-1990s, Mir commanders regularly prepared for Soyuz and Progress dockings by assembling the TORU system in the station's command module, just in case they were forced to take over control of the incoming ships.

But unreliability was only half the equation. The real problem, as always, was money. The Kurs systems were made by a government-owned company called Radiopribor, located in Kiev. With the disintegration of the Soviet Union in 1991, Kiev had emerged as the capital of the newly independent country of Ukraine. As the 1990s wore on, Ukrainian government officials demanded higher and higher prices for new Kurs computer units. According to Alexander G. Derechin, the head of Energia's international marketing division, the cost of a single new Kurs system had mushroomed 400 percent in the last five years. When the Russians protested, the Ukrainians said, in effect, "Tough, pay up." When the Russians fell behind on their bills, the Ukrainians refused to ship new units. Those it did ship sometimes didn't have all their parts. The Russians seethed. Slowly but surely, they were running out of docking computers. Every time one of the unmanned Progress ships burned up upon reentry, they lost another one. In about 1996 the cosmonauts began dismantling the Kurs units from Progress spacecraft before reentry, and the Russian Space Agency quietly begged NASA to begin carrying the dismantled units back on the shuttle to be reused.

Calmer heads should have prevailed, but where Ukraine and Russia were concerned, there weren't a lot of calm heads. Ethnic rivalries between the two countries only exacerbated the problem. Ukrainians felt they had been colonized by Russians for centuries, and weren't inclined to lend a hand just because Star City needed it. After Ukrainian independence a host of far larger disputes, including a nasty fight over which country would control the former Soviet Union's Black Sea fleet and naval bases, drove a deeper wedge between the two countries.

Behind the scenes Energia was furiously trying to design and produce its own Kurs units, using new Western technology, but that effort would take time, maybe years. In the meantime the TsUP, as it confirmed that fall of 1996, was going to attempt to dock at least some of its spacecraft with the TORU manual-docking system. Real questions remained as to how effective the new system could be. Cosmonauts had used it successfully on several occasions, but always in the immediate vicinity of the Mir, docking ships from a distance of ten meters or so. Could cosmonauts in space use the TORU system to manually dock spacecraft from a distance of seven kilometers instead of seven meters? That was one of many daunting tasks the TsUP had planned for the Mir 23 crew of Vasily Tsibliyev and Sasha Lazutkin.

5

Wednesday, February 5
Star City

It's still early when Tsibliyev slips out of the quarantine rooms in the Prophylactorum conference center and trots toward his family's flat in Building 47. In a matter of hours he and his two crewmates, Lazutkin and Ewald, will board a Star City jet for the four-hour flight to the Russians' main launch facilities in central Asia. Tsibliyev's breath hangs a frosty white in the frigid air as he makes his way out past the frozen lake and across the grounds. He has exactly one hour to spend with his family before leaving, and he plans to use it enduring a newish, and informal, cosmonaut tradition: checking his household appliances.

The joke—and there is some truth to it—is the moment a cosmonaut blasts into space, his family's apartment begins falling into disrepair. An older cosmonaut, Viktor Afanaseyev, started the tradition of checking his appliances before departure, and it caught on. When Tsibliyev reaches Building 47, his wife, Larissa, and his seventeen-year-old son, Vasily Jr., give him plenty of room. He checks the refrigerator, the stove, the stereo, and the lamp bulbs. He checks the radiators, and he checks the television. Everything works. Pleased with his handy ways, the commander of Mir 23 steps back into the hallway to return to the Prophylactorum and smiles. "Everything is perfect," he tells Larissa. And then he closes the door—and hears the coat rack on the wall behind it fall to the floor.

This, some will say, is the essence of Vasily Tsibliyev. Responsible. Hardworking. Caring. Always the best intentions. And yet somehow cursed. Born in 1954 in the tiny Crimean village of Feodossya, Tsibliyev grew up the poor son of a mother who worked as a collective farmer. Even after years at Star City, there is still an air of the provinces about him, a sense that he is not altogether secure around the urbane officers who grew up in Moscow; it is easy to imagine Tsibliyev, in a bygone era, as the kind of Russian soldier called in to roust a group of protesters. His parents divorced when he was very young; when his father subsequently left, Tsibliyev never forgave him. His earliest memories were of dreaming to be a cosmonaut like Yuri Gagarin, whose 1961 flight excited Tsibliyev and so many of the other seven-year-old boys in the village school. Upon graduation in 1971, he joined the army and became a fighter pilot, spending five years flying MiGs up and down the West German border. In later years he and the American astronaut Charlie Precourt would come to the realization they had been stationed in Germany at the same time; in the event of war, they might well have flown against each other in combat.

In 1980 Tsibliyev was assigned to a squadron flying out of Odessa on the Black Sea. Though he and his comrades always expected to be sent to fly against the mujahideen in Afghanistan, they never were. In his spare time Tsibliyev did everything possible to get into the cosmonaut corps. His first application, in 1976, was rejected out of hand; too inexperienced, the rejection letter said. It would help, he knew, to become a test pilot. Five times he applied to test pilot school; five times he was rejected. After the last rejection, he was told not to apply again; by then he was considered too old for the program. He was twenty-nine. For nearly ten years there had been no new cosmonauts added to the space program, but in 1984, while in Odessa, Tsibliyev heard new applications were finally being accepted. He applied and was rejected yet again. The cosmonaut center said he was too old. But Tsibliyev kept at it and somehow got in. He was accepted as a cosmonaut trainee in 1986 and moved to Star City.

His first and only flight to Mir had come during the fall and winter of 1993–94. It hadn't gone well. Tsibliyev had been paired with one of Russia's best-known cosmonauts, Aleksandr Serebrov, a forty-eight-year-old civilian flight engineer who was then on his fourth trip into space, and who for several years held the world record for the most spacewalks. Serebrov, who grew so close to Tsibliyev he was considered a member of the family, was a garrulous, headline-loving type who listed his hobby as "fast cars." The four months the two men spent aboard Mir actually went well enough. The problems began when they left.

On the morning of January 14, 1994, Tsibliyev was ordered to fly their departing Soyuz-TM 17 around the station and take pictures that could be sent to the Americans. The "fly-around" required Tsibliyev to take manual control of the little Soyuz, using controls identical to the new TORU system. For some reason—Russian engineers would never fully explain why—his thruster control button momentarily froze. Unable to control the ship, Tsibliyev watched in amazement as it floated slowly toward the station. Inside, seeing what was happening, Commander Viktor Afanaseyev ordered his crew to head into their own Soyuz to ready for an immediate evacuation. Before Tsibliyev could regain control of his ship, the Soyuz bumped into Mir, striking it twice before bouncing away. The impact was so slight the cosmonauts inside Mir didn't feel it. Tsibliyev quickly regained control of the Soyuz. There was no permanent damage to either the station or the Soyuz, but the incident left both cosmonauts—and their ground controllers—badly shaken.

The taint of scandal spread after an incident during the pair's return to Earth. Accounts of Tsibliyev and Serebrov's landing vary, but according to officials at the TsUP, their descent capsule fell to Earth in central Asia roughly 100 kilometers from where it was expected to land. An investigation, these officials contend, revealed that the capsule had veered off course because it was too heavy. An inspection of its contents led Energia officials to levy a stiff fine against Serebrov: His sin, they charged, was overloading the capsule with personal items smuggled off Mir that he intended to sell, including postcards and other memorabilia. Serebrov hotly denied the charge. All he brought back, he argued, were a set of space suit visors, several music cassettes, including one by Elvis Presley, and two sets of space suit gloves, which he subsequently presented as gifts to Yeltsin, U.S. vice president Al Gore, Russian prime minister Viktor Chernomyrdin, and the Smithsonian.

Serebrov angrily charged that he was being persecuted by Energia's powerful chairman, Yuri Semenov, for attaining too high a public profile, a situation he traced to a goodwill trip he had taken with Mikhail Gorbachev to Tokyo two years before. The dispute was also stoked by Serebrov's vicious personal rivalry with Valery Ryumin, the onetime cosmonaut who had emerged as Semenov's "golden boy" even before being assigned to head the Russian side of the Phase One program.

Serebrov traced his problems with Ryumin to the flight of Soyuz-T8 in April 1983, a hair-raising mission in which Serebrov, Gennadi Strekalov, and their commander, Vladimir Titov, failed to dock their Soyuz to the Salyut 7 space station. An antenna used in their computerized docking somehow failed, and the TsUP, where Ryumin was then deputy director, advised Titov to attempt a manual approach, something he had never been trained to perform. According to official accounts, Titov closed the

Soyuz to within 330 meters of the station, at which point he passed out of communications with the ground, which robbed him of the crucial telemetry data the TsUP was feeding him. Without information on his speed and distance, Titov elected to abort the approach, and the three men returned to Earth without docking, having used almost all of the little ship's propellant. Today Serebrov says the Soyuz nearly collided with the station, missing a collision by barely ten meters. He charges Ryumin with "misleading" the investigating commission to downplay the TsUP's culpability. "The man is an animal," Serebrov spits. "Everyone calls him the Russian pig."

Fed up, Serebrov resigned from the program not long after his flight with Tsibliyev. Not only did the incident have implications for Tsibliyev's career, it revealed the deep-seated rift between Energia and many of the cosmonauts. Just as at NASA, there had always been grumbling and occasional finger-pointing between astronauts and ground controllers, especially when something went wrong in space. In Russia that natural tension was institutionalized. On one side were the taskmasters at the TsUP, which was controlled by Energia; on the other were the cosmonauts and Star City. While the five-year-old Russian Space Agency negotiated things like government budgets and international partnerships, few Westerners realized that the real power behind Mir was Energia, a company that essentially consisted of all elements of the old Soviet manned space program. In the early 1990s, Energia, like many former Soviet industries, had begun its difficult transformation into a private company, opening a marketing office in Washington, where it shopped Mir to skeptical Western companies as an orbital laboratory for all manner of scientific experiments. Energia has stock that is traded through private brokers—the Russian government still owns 38 percent of the shares*—and the company's executives harbor dreams that one day it will be traded on Western stock exchanges. The most powerful man in the Russian space program, Energia's autocratic chairman, Yuri Semenov, liked to call the station "my Mir," and there was some debate about who actually owned the station. "Does Energia own Mir? That's a difficult question," says Jeffrey Manber, Energia's man in Washington. "They designed it. They take care of it. The answer in a Western sense, a legal sense, is no. The answer in a practical sense is yes."

Energia's emergence as a private company only exacerbated tensions with the cosmonauts. Half the men who flew to Mir, the flight engineers,

*The Kremlin's stake in Energia is scheduled to be reduced to 25 percent in late 1998.

were Energia employees; all the cosmonauts negotiated their contracts with and received their bonuses from the company. Whenever something went awry aboard Mir, TsUP controllers like Viktor Blagov always seemed to blame it on the cosmonauts. To do otherwise, Energia's critics contended, would be to cast doubts on the technical soundness of Mir, which might scare off the European and American astronauts who were paying millions to live there. Privately the cosmonauts and their military overseers at Star City, many of them still dedicated Communists, felt that Energia put its newfound thirst for profits ahead of safety.

One man who strongly backed Star City's view was Tsibliyev's friend Serebrov, who to the dismay of Energia rebounded nicely from his resignation, accepting an appointment to the Kremlin as Boris Yeltsin's senior adviser on space activities. In his new role Serebrov emerged as a behind-the-scenes advocate for the cosmonauts, and a caustic critic of Energia. He said out loud the things the cosmonauts were too afraid to say: that cosmonauts were overworked and underpaid, serfs on Semenov's manor. Aboard Mir eighteen-hour workdays were not uncommon. "We are biological robots for Semenov," Serebrov complains, still using the word *we* when referring to cosmonauts. "He just likes to wipe his feet on us."

Serebrov fretted aloud that Energia's rampant cost-cutting would ultimately lead to a disaster in space. He pointed to the company's persistent refusal to supply Mir—as NASA had supplied the shuttle—with an on-orbit flight simulator where cosmonaut commanders could rehearse the manual docking skills they were now being asked to use. Energia's own studies indicated that a cosmonaut's hand-eye coordination deteriorated significantly after the first weeks in space, but despite Serebrov's warnings, Energia insisted that the cost of simulator software—$5,000, by one estimate—was just too expensive. Until the Mir 23 flight of Tsibliyev and Lazutkin, no one could have guessed how costly Energia's refusal would one day prove.

By early 1997, as Tsibliyev readied for his second mission, Serebrov was increasingly concerned with conditions aboard the station. The two crews who had worked on Mir during 1996 had encountered an escalating series of breakdowns: in the two Elektron oxygen generators, the Vozdukh carbon dioxide removal system, the waste-removal unit, and a host of other minor systems. The cosmonauts' main job now was helping the Americans and Europeans with their experiments, but both of the 1996 crews had been forced to sharply cut back on these activities to perform repairs; Serebrov estimated that cosmonauts were now spending as much as 75 percent of their time in orbit fixing things that broke. The TsUP, while acknowledging the problems, scoffed at that estimate, putting the number at 15 percent. Much of the problem, of course, was age; the station's major systems, designed for no more than five years of use, were a decade old, and breakdowns were inevitable. But with the addition of the two new scientific modules, Priroda and Spektr, in 1995 and 1996, Mir had also grown in complexity.

From his post in the Kremlin, Serebrov tried to persuade Energia executives to consider expanding the crews to three men from two, adding a second engineer to help with repairs. Energia's answer, of course, was no. There was simply no money to pay a third crew member. The program's main source of hard currency, in fact, came from the foreign cosmonauts who rode the Soyuz's third seat. There was no room in a Soyuz for a fourth person.

What worried Serebrov most about the upcoming Mir 23 mission was Lazutkin. A rookie engineer, he feared, was likely to be swamped by all the things that regularly went wrong aboard Mir. And though he rarely said it aloud, Serebrov didn't feel his friend Tsibliyev was at his best in complex repair situations. "Vasily is not at all mechanically inclined," Serebrov says. "He is a very diligent person, I know, but he is a fighter pilot. Honestly speaking, I was very worried about Vasily because he doesn't have an engineering kind of mind. During our flight, he kind of took it easy. I told him what to do. I knew that in a controlled situation, he would be fine. But if they were to have problems, I feared he would have a hard time managing."

Only later would Energia executives admit the extent to which Tsibliyev had relied on Serebrov during their 1993–94 mission. "The problem was, we didn't realize that during his first flight he was trying to use Mr. Serebrov as the prime person," acknowledges Valery Ryumin, the Russian Phase One chief. "He was hiding behind Mr. Serebrov's back, and this is very natural. He is flying for the first time, and his colleague was very experienced. He is simply not a leader, that's all. This is not his fault. In fact, this is our fault. We just did not recognize that."

As the Mir 23 launch date approaches, Serebrov isn't the only one fighting vague premonitions of disaster. As he lies awake at night in the quarantine rooms at the Prophylactorum, Tsibliyev too can't shake the feeling that something awful is going to happen to him in space. Two weekends earlier the commander, like many Russians an intensely superstitious man, had confided his misgivings to a close friend named Tamara Globa. Globa, however, wasn't just any friend. She was one of Moscow's best-known celebrities, a psychic and astrologer who wrote a popular column for the Russian version of *Cosmopolitan* and who frequently appeared on Russian television. Globa was a friend of several cosmonauts; Tsibliyev kept an autographed copy of her calendar in his office. He had called and practically begged her to come to Star City for a barbecue that Saturday, and despite the press of deadlines, Globa had come.

She and Tsibliyev had met during a 1992 environmental conference at the Black Sea resort of Socha. Pictures of the time show Globa, a striking woman then in her midthirties with long, straight brown hair, sitting on the grass with a smiling Tsibliyev, who looked relaxed in a purple sports shirt and white slacks. The two struck up an immediate friendship, "a childish friendship," she called it, "full of pure, innocent feelings toward each other, like two puppies rolling in the grass." She had been a friend of Serebrov's as well, having met him on a television show called *Under the Sign of Pisces,* which showcased prominent Russians born under that sign, and had put together an astrological chart for him before his 1993 flight with Tsibliyev. She had warned Serebrov how difficult that flight would be, but assured him he would survive. At the time she had kept her feelings from Tsibliyev, not wanting to frighten the rookie commander on his first flight to Mir.

But that Saturday, while Tsibliyev and a group of cosmonauts grilled *shashlyk*—mutton shish kebabs—on the snowy yellow-brick stairs of the Prophylactorum, Globa shared her own misgivings about the upcoming flight. It was a typical Star City affair. Tsibliyev, looking like an Alabama duck hunter in his camouflage suit and wide-billed camouflage hat, tended to the silver skewers of mutton while Lazutkin, smiling and joking in his big blue parka, passed around a bottle of vodka, which everyone sipped from tiny plastic cups.

"Tamara, you know I have these bad feelings about this mission," Tsibliyev whispered to Globa as he turned a spit of *shashlyk*.

"Vasily, don't worry," she said at first. "Everything will be all right."

"I know," he said. "But still. I don't know where I pick them up, I don't know where they come from. It's difficult to explain them. It's at the level of the subconscious, you know, the level of intuition."

Globa didn't want to say anything, but she too was worried. She had studied Tsibliyev's astrological charts, and they were bad. Very bad. But she didn't want to worry him, for fear that her prophecy would prove self-fulfilling. "It will be a difficult and eventful mission," she said, choosing her words carefully. She said she foresaw that the commander would have a "bad health condition" of some kind during the first month in space, but that by April he would fully recover.

"Vasily, take care of your heart and your legs," she counseled. Lazutkin had walked up to joined them.

"Many unexpected things will happen that have never happened before," she went on. The most dangerous times, Globa predicted, would be the weekend of February 22 and 23, right after Tsibliyev's birthday, and a period in June. Tsibliyev listened intently, and when it came time to leave they embraced. Globa fought back tears.

"Do not worry," she said again and again. "Everything will be all right."

Tsibliyev wasn't so sure.

Neither was Globa. "When I left him that day," she remembers, "I thought I had lost him forever."

That morning, as Tsibliyev finishes checking his appliances, Sasha Lazutkin stands inside the Prophylactorum lobby and embraces his wife, Ludmilla, for what feels like a very long time. Ludmilla, a slim, demure blonde who works as the commercial director at a Moscow clothing firm, is trying hard not to cry. Their daughter Natasha, who is twelve, is taking her father's departure well, but her younger sister, Yevgena, who is seven, is not. She is crying hard, and Ludmilla is doing everything possible to calm her while bidding her husband farewell. "Sasha, don't worry, everything will be fine," she keeps repeating. "Everything will be fine. Don't worry about us. Don't worry about us."

Lazutkin had grown up in a cramped two-room Moscow flat with his late mother, a bookkeeper, and his father, who worked in the same factory for fifty years. As a child he had always been quiet, studious, and shy, harboring secret dreams of becoming a cosmonaut. He spent his time at things he thought might make him a cosmonaut candidate, excelling as an gymnast for fifteen years, then joining a parachute club. The parachuting experiment didn't last long: First he was afraid of jumping, then he developed a fear of airplanes. It was hardly an auspicious start for an aspiring cosmonaut.

He met Ludmilla in the parachute club, when he was twenty-one and she was seventeen. He was smitten; she played hard to get. For four years Sasha pursued her—"I tortured him," she remembers with a smile—until one day Ludmilla fell while skiing and suffered a concussion. Though few of her friends knew she was in the hospital, Sasha somehow found out and appeared at her bedside bearing flowers. As she continued her recovery on International Women's Day—a national holiday in Russia, when every woman is showered with flowers and gifts—Sasha arrived at the hospital with no warning and spirited her away, throwing one of his sports coats over her flimsy hospital gown. He took her to his father's flat, where aunts and uncles and cousins were arrayed around the dinner table, and introduced her. "This," he announced, "is the woman I will marry someday."

A month later they were engaged. After the three-month waiting period then mandated under Soviet law, they were married and moved into Sasha's father's flat, the same apartment where Sasha had been raised, in a dark, graffiti-covered tenement in northeast Moscow. For years, while Sasha studied engineering and then was accepted as a technician in the cosmonaut training program, they shared the neat little flat with Sasha's father, Sasha and Ludmilla and the girls all sleeping on couches in the living room. After the elder Lazutkin died, the girls at least got their own bedroom; until the day he left to be quarantined for his first flight into space, the cosmonaut-engineer for Russia's Mir 23 mission slept on the pull-out sofa in his living room.

This mission means everything to the Lazutkins. If all goes well—and missions almost always go well on the hardy Mir station—Sasha can count on a promotion, an honor as an official Hero of Russia, more money, and best of all a new flat, paid for by the state. Ludmilla has her heart set on it. She and Sasha hug for the longest time there in the Prophy, and then he says good-bye and disappears into the crowd around the bus he will take to the airfield. Afterward, on the long drive back to Moscow with the two girls, Ludmilla finally lets loose. She cries all the way home.

Out at the airfield, there are no bands playing or exuberant formal ceremonies; these were halted a few years back when some of the cosmonauts complained that the effect was funereal. Tsibliyev, Lazutkin, and Ewald, along with their three-man backup crew, leave the buses and board the cosmonaut center's sky-blue-and-white Tupolev Tu-124 jet, which soon barrels down the runway and takes off. Their mission has begun.

Once in the air, the little jet sets a southeasterly course. Sometime after noon, it leaves Russian airspace and enters that of Russia's giant new southern neighbor, the independent nation of Kazakhstan, which sprawls across the heart of central Asia from the eastern shore of the Caspian Sea all the way to the Chinese border and the Siberian reaches. Barely two hours later, after skirting the northern edge of the vast blue Caspian, it begins its descent. Below, the lush pine forests of northern Russia have long since disappeared; in their place a frigid, treeless steppe stretches as far as the eye can see. The plane comes in low over the Aral Sea, the salty inland lake whose gradual pesticide-inflicted death is one of the world's great environmental tragedies. A small airfield comes into view. Beyond it, a city lies sullenly on the snowy plain, its only link to Western civilization a spidery thread of rail line stretching north toward the Russian border. Out on the steppe nearby sits a dingy military base, dotted with hangars and ringed by launching pads. This is Baikonur, Russia's long-secret Space City.

The snow has been cleared from the runway, and the plane lands. The cosmonauts deplane and board a bus for the short ride into the sleepy little city, whose name is now officially Baikonur, but which is still widely known by its Soviet-era name, Leninsk. To add to the confusion, neither name refers to the real Baikonur. The real Baikonur is a farming village more than a hundred kilometers to the northeast. In 1955, when the Soviets announced the construction of their new central Asian rocket facility, they indicated to the West it would be built outside Baikonur the farming village. Instead, in a Cold War–era attempt to confuse the Americans, they secretly built it here, near the Aral Sea, and erected the city of Leninsk beside it to house the thousands of technicians and military officers who came to live here. To add to the confusion, the base itself was named Baikonur.

If hell had frozen over, it might look something like Baikonur in February. In the city's central square the cosmonaut buses pass a burned-out office building, destroyed in rioting the previous winter. The state of the base and the city had been a subject of debate in international space forums even before the Americans agreed to merge their space station program with the Russians in 1993. Critics of the agreement depicted Baikonur as at near collapse, its facilities crumbling and overrun by looters. While that was far from the truth, Baikonur had endured stresses unimaginable to Westerners.

Its low point came in February 1992, when thousands of ethnic Kazakh conscripts brought onto the base to do menial labor had staged a violent revolt. The young soldiers, mostly eighteen- and nineteen-year-olds drafted into the Army's Construction Troops, were paid barely seven rubles a month, "enough to buy a stick of bubble gum," as one correspondent put it. On February 24, after even those meager checks were withheld for three months, the soldiers rioted, commandeering a stolen car and attempting to ram it into the commandant's office. In the ensuing chaos four army barracks were set on fire; afterward three bodies were found in the charred ruins, two of which were identified as those of Russian officers. Police attempted to stop the mobs from marching into Leninsk, but they were badly outnumbered. Panic swept the city as the soldiers approached and presented their list of complaints; among other things, they charged that their Russian officers had trained their dogs to attack the young conscripts for sport. Local commanders defused the situation by quickly demobilizing thousands of the soldiers and sending the rest home on leaves from which very few ever returned.

Afterward the base slowly fell into neglect. Many Russians, including some of the base's senior engineers, reluctantly returned to Russia. Those who remained behind lived in a state of constant tension with their Kazakh hosts. Today Leninsk is a divided city, with two police forces—one Kazakh, one Russian—two municipal governments, and two school systems. When the American author and space consultant James E. Oberg visited Baikonur in 1995, he found heat, electrical service, and water working only sporadically. Piles of junk, including shrouds and fuel tanks from the Russians' N-1 moon rockets, lay strewn through the heart of the base itself. "Abandoned apartments, in several groups and sometimes in entire blocks, stare windowless at the dusty sun," Oberg wrote in an engineering publication. "Gritty brown powder wafts across the streets and barren ground between the apartment buildings, whirling around groups of seated grandmothers. The dust, blown from the pesticide-laden salt flats of the dying Aral Sea a few hundred kilometers upwind, is gradually poisoning the city's inhabitants, the weakest first. In bitter unanimity, they believe that no one is going to do anything about it."

It was enough to make the older cosmonauts shake their heads. It hadn't always been this way.

* * *

As far back as the 1880s, Russian thinkers had written of the promise of rocket-powered flying machines. The most influential of the early Russian scientists was a onetime high school physics teacher named Konstantin Edvardovich Tsiolkovsky, who in a series of books between 1883 and his death in 1935 laid out all the mathematics and design theory his successors would need to build the Soviets' first multistage rockets after World War II. Spurred on by reports of rocketry advances made by the American Robert Goddard, Soviet generals quickly realized the military applications of Tsiolkovsky's work, launching Russia's first liquid-fuel rocket at an army base outside Moscow in 1933. By far the most prominent of Tsiolkovsky's disciples was a Ukrainian-born engineer named Sergei Korolev, who barely survived a late-1930s exile in Siberian prison camps to emerge as the mysterious "chief designer" of the Soviet ballistic missile program in the 1950s. Cloaked in anonymity by Nikita Khrushchev, who feared his ideas would leak to the Americans, Korolev led the team of engineers that designed and built a series of ever more powerful rockets the Soviets tested during the 1950s, a process that climaxed with the stunning flight of the first Sputnik (Satellite) in 1957.

The success of Sputnik ignited a space race with the United States that Khrushchev was determined to win. The Soviet premier put tremendous pressure on Korolev to come up with more and showier spectacles that would prove to the world just how progressive and technologically advanced his regime was, and the chief designer routinely came through, launching the first unmanned probes to the Moon and Venus and blasting a dog into space, proving that a living creature could endure spaceflight. But Khrushchev's political goals nearly pushed the infant program to the breaking point. In October 1960 an unmanned rocket designed to reach Mars exploded on the launchpad at Baikonur, killing 165 people, including Marshal Mitrofan Nedelin, an influential Russian general. Nothing, however, would derail Khrushchev's ambitions. Barely six months later Korolev launched Yuri Gagarin on his 108-minute single orbit of Earth, an achievement that stunned the world. Gagarin's flight was a model of Soviet centralized control. "From launch to landing," Oberg noted in his 1982 book, *Red Star in Orbit,* "he never touched the controls."

Following Korolev's death in 1966, the Soviet space program drifted. The very next manned flight, the launch of Soyuz 1 in 1967, ended in disaster, when the capsule's parachute lines somehow became entangled during descent, plunging cosmonaut Vladimir Komarov to his death. "Mankind never gains anything without cost," Gagarin himself wrote after Komarov's death. "There has never been a bloodless victory over nature." Gagarin's comments, underscored by his own death nine months later, pointed up a sentiment inside the Russian program that still unsettled Americans three decades later: While Russian officials said all the right things about safety, there was always the sense that they accepted the fact of death in space as inevitable, a necessary sacrifice for the motherland. When the Space Shuttle Challenger exploded in 1986, NASA grounded its shuttle fleet for three years; the Russians, by and large, reacted to their own disasters by plunging right back into space. "The Russians," notes Jim Van Laak, who serves as Frank Culbertson's No. 2 man in the Phase One program, "simply don't place as high a premium on human life as we do."

While NASA lost its momentum and direction after beating Moscow to the Moon in 1969, the Russians rebounded from several disasters—including the death of the three cosmonauts in 1971—by sending aloft seven manned space stations between 1971 and 1982. The official explanation for these Salyut stations was to prepare for an eventual flight to Mars, but the real reason had more to do with earthly considerations. Among other things, the Soviets wanted cosmonauts to monitor American missile and space launches. "It's very easy to see your missiles," says Yuri Glaskov, the No. 2 general at Star City, who flew aboard a Soviet station in the 1970s. "The way they're installed, we could look right down on them."

Still, by the 1980s the Russian program was as much about public relations as anything else. Guest cosmonauts from Bulgaria, Syria, Vietnam, and other countries eagerly trooped aboard the Salyut stations, boosting good feelings in their countries toward Russia. In 1985 the Soviets unveiled plans for their biggest station yet, Mir. (The word *mir* means "peace" but can also be translated as "commune" or "village.") When the new station's first module was launched by a Proton booster rocket on February 20, 1986, it represented an evolutionary improvement on the Salyut stations; it was the same size and of the same design, but it was the first Russian station with multiple docking ports to "park" different spacecraft.

That first 1986 component, which became the living module, or "base block," of today's Mir station, was to be the centerpiece of a far larger station that would be built over time by plugging in other modules. The first of the add-on modules, called Kvant, was launched into space a year later; an astrophysics laboratory, it carried Soviet X-ray and gamma-ray telescopes as well as British, Dutch, and German instrumentation. A problem developed when the new module's needlelike docking probe entered Mir's docking port. For some reason the module would not dock firmly; while connected to the station, it remained several centimeters out in space. An emergency spacewalk was ordered, and cosmonauts Yuri Romanenko and Aleksandr Laviekin donned space suits and crawled outside onto the hull. This was the kind of thrown-together operation at which the Russians excelled. While NASA shuttles routinely returned to Earth upon encountering mechanical troubles, it was a luxury space station operators didn't have; time and again the Russians managed repairs in orbit that left Americans experts shaking their heads in admiration. In this case, Romanenko and Laviekin took turns bravely sticking their hands down between the two spacecraft to find what was separating them. To their surprise, it turned out to be a white trash bag, left behind when a previous crew loaded garbage into a Progress supply ship docked at the same port. The two cosmonauts eventually wrenched the bag out, and the docking was completed.

Two more modules followed. Kvant 2, containing a new toilet and a shower, was sent aloft in November 1989, giving the new, enlarged station an L shape. Six months later came Kristall, which gave Mir a T configuration, an outline it would maintain for five years, until the Americans paid to outfit the final two science modules, Spektr and Priroda, which were joined to the station in 1995 and 1996, respectively.

By then the Russian program was fighting for its life. Many of the veteran engineers who had built stations during the 1960s and 1970s had retired or quit in frustration by the 1990s; cosmonauts regularly complained from orbit that the young men who replaced them didn't have the expertise to run the station as before. Only the sudden partnership with NASA in 1993, and the Americans' pledge to pay Moscow $400 million to fly seven astronauts aboard Mir between 1995 and 1998, rescued the program, and even then its survival was not ensured.

"Back when we were the USSR [space] was considered pretty important," remembers Valery Ryumin. "And then, when new people entered the government, reformers, at first we couldn't figure out their attitude toward the program. They didn't seem to be working to support the program. They didn't treat us with respect. But we fully realized it would be a disaster to let them destroy the program, the way they did destroy so many other ones. [Without the Americans] we probably would have survived, I mean, I can't think of our government being so stupid as to allow such a program to disappear. At the same time, the reformers made so many stupid mistakes over the years, it's amazing. I don't know how they manage to do things even today."

Monday, February 10
Baikonur

Tsibliyev awakens in Room 307 at the Hotel Cosmonaut around seven o'clock that morning. Outside the temperature hovers just above zero. Winter winds whistle noisily through gaps in the windows. He yawns, then stretches. Black thoughts be damned. This is it. This is the day he gets back to space.

Throwing on his clothes, the commander joins Lazutkin and Ewald downstairs for a final medical checkup, where the three men disrobe and cover their naked bodies with antiseptic alcohol. Each accepts the doctors' offer of a fast enema; there are toilet bags to be used on the Soyuz, but no real privacy. The flight to Mir is barely two days long, and most cosmonauts try to eat as little as possible rather than evacuate their bowels in the cramped little capsule. Afterward they each slip into clean underwear and walk down the hall to breakfast, which is served, as per tradition, by the backup crew. Champagne is poured and toasts are made, but Tsibliyev, Lazutkin, and Ewald barely take a sip. They don't want any additional liquid in their bladders. Tsibliyev is proud of the fact that, unlike the Americans, the Russians don't wear diapers beneath their space suits. Diapers are issued, but Tsibliyev believes that no self-respecting cosmonaut would deign to wear them.

All three men are itching to get to the launchpad. Ewald has been in training for this mission for more than two years. Lazutkin has dreamed about this day since he was a child. The night before, following their final press conference, they all sat through the traditional launch-eve showing of a movie called *White Star of the Desert,* a melodrama set during revolutionary Russia. Most of the cosmonauts have seen the movie dozens of times, but all watch it through to the end. To leave early is considered a bad omen. The story is told of a cosmonaut who once left the movie halfway through; his flight was aborted halfway through. After the movie some crews engage in *White Star* trivia contests. Tsibliyev's crew is in no mood for trivia contests. They want to get going.

In time, a priest comes into the breakfast room to bless them, which is quite a change from Soviet days, when religion never touched the cosmonaut program. Once breakfast is finished, the three men head back upstairs to Tsibliyev's room where, joined by a handful of trainers, doctors, and visiting cosmonauts, they partake in the traditional minute of silence "for the lucky road," as the Russians say. Yuri Glaskov, the round little general from Star City, says a few words, wishes them luck. Afterward Tsibliyev takes out a pen and holds the room door. The back of his door is covered floor-to-ceiling with the signatures of cosmonauts who have signed it as their last act before leaving for space. The front of the door is half covered with signatures as well, and it is here, standing in the hallway, that Tsibliyev signs his name. In their own rooms, Lazutkin and Ewald do the same. It is 11:30. It is time to leave.

Downstairs the three men stride through the lobby to the blaring strains of the unofficial cosmonaut anthem, a song called "The Grass by the House." It is a patriotic, Soviet-era song whose elegant verse doesn't translate easily into English.

The Earth can be seen through the lit window
Like the son misses his mother
We miss the Earth, there is only one
However, the stars
Get closer, but they are still cold
And like in dark times
The mother waits for her son, the Earth waits for its sons.

We do not dream about the roar of the cosmodrome
Nor about this icy blue
We dream about grass, grass near the house
Green, green grass.

In a bit of theater that's almost identical to the Americans' walk past the press at The Cape, the cosmonauts traipse out into the cold morning air past a crowd of journalists and pile into two buses, the *Star City* and the *Baikonur*; each bus carries with it a horseshoe, for luck. The slow drive across the snowy steppe to the base takes about forty minutes. There the three cosmonauts disappear into a hangar to get suited up.

At 2:20, after meeting with a group of visiting politicians and dignitaries, they emerge from their hangar in their white Russian space suits, each man carrying something that looks like a small white lunch box in his left hand. The box is a ventilator that circulates air through the suits; without the ventilator, says one Westerner who has tried on the suit, "you would sweat like hell." As a group, Tsibliyev, Lazutkin, and Ewald walk out onto a patch of asphalt. On the concrete three small boxes have been outlined in white paint. One by one, like contestants on some strange Russian game show, they step forward onto their marks.

Lazutkin, droll and unsmiling, steps onto the leftmost box, marked "BI," for flight engineer. Ewald, blocky and hunched, slouches forward onto the box marked "KI," for research cosmonaut. Tsibliyev, appearing smart and fully in command, steps in between them onto the box labeled "KK," for commander. He snaps off a crisp salute to the general standing before them. "The crew is ready to perform its task," says Tsibliyev.

The general, looking as if he had walked out of some long-forgotten May Day celebration, stands upright in a forest-green overcoat and formal military hat. The whole ceremony is over in thirty seconds. Tsibliyev, Lazutkin, and Ewald then enter the bus that takes them out to the launchpad.

After waving good-byes, the bus zips out over the asphalt in the direction of the pad. But then, when it is safely out of sight of the crowds, it stops, the doors open, and the three cosmonauts step out into the cold wind. It is time, as every serious observer of the Russian space program knows, to observe one of the most sacred traditions of the cosmonaut corps. It had been done by Yuri Gagarin in 1961 and thus by every cosmonaut since. Without additional fanfare, Tsibliyev walks around behind the bus, reaches down, fumbles around in his suit for a moment, and when finally ready, begins to grandly urinate on the right rear bus tire.

While Tsibliyev stands astride a bus tire, most of the crowd that had come to watch the launch jostles onto a second group of buses, which takes them by another route to a position barely thirty yards from the launchpad itself. There everyone piles out onto the asphalt and cheers and applauds when the cosmonauts' bus pulls up several minutes later. It is a scene that dumbfounds Americans accustomed to launches at The Cape, where onlookers are routinely kept two miles or more away from the launchpad. Here, barely two hours before launch, and within a softball's toss of the steam-wreathed rocket itself, dozens of reporters from Russia and Germany press forward toward the cosmonauts' bus, flashbulbs popping. Women in furs mill about, pointing in awe at the giant rocket. A squad of VIPs, including the usual assortment of bemedaled Russian generals and the American astronaut Michael Foale, watch in silence. There are even children scurrying about, squealing and playing, lost in the excitement of watching a rocket launch.

After a moment, Tsibliyev, Lazutkin, and Ewald emerge from the bus directly into the welcoming crowd. A chaotic scene ensues as the three cosmonauts push their way through the throng while people reach out to touch them and offer words of congratulations. Eventually they reach the barriers separating the crowd from the launch area and pass through. Then they walk across the asphalt toward the elevator, leaving the crowd behind. With a few final waves, they are inside and gone.

The crowd slowly returns to their buses, which drive them to a viewing area a kilometer away, where two grandstands have been erected to watch the liftoff. The winter sun is already setting. For two hours the crew lies on their backs inside the capsule, running equipment checks. The final countdown begins at five. At nine minutes past five, as night descends over the steppe, the giant Proton rocket suddenly erupts in a torrent of yellow flame. The spectators watch it rise for several minutes, until it is nothing but a white dot in the heavens, almost indistinguishable from the stars.

As Tsibliyev, Lazutkin, and Ewald climb upward, a pink toy rabbit leaps and dances over the commander's head. It is a gift from his daughter, and he carries it for good luck. In the crew's baggage, however, is a gift that is a far more accurate omen of things to come. It a CD of music sung by the British rock star Chris Rea, a gift Ewald has brought to give the commander on his birthday. The CD contains Tsibliyev's favorite song, called "The Road to Hell." Its chorus:

But the light of joy I know
Scared beyond belief down in the shadows
And the perverted fear of violence
Chokes a smile from every face
And common sense is ringing out the bells

This ain't no technological breakdown
Oh no, this is the road to Hell

6

Wednesday, February 12
_6:30 P.M._
Outside Mir

Soyuz-TM 25 coasts to a stop 150 meters from the sprawling Mir station, which hovers silently in space 250 miles above the Earth's surface, a giant dragonfly waiting for its newest set of occupants. Inside the Soyuz's cramped command module, Tsibliyev, flanked by Ewald and Lazutkin, activates the Kurs automated docking system. The little craft's rear thrusters fire, and the Soyuz begins easing toward Mir's Kvant docking port at a relative speed of 0.8 meters per second.

By the standards of the Russian space program during the 1990s, it has been an uneventful flight. The cosmonauts had first seen Mir earlier that day as a twinkling star in the distance, which gradually grew until they could make out the individual solar arrays, the giant space wings that drink in the Sun's rays to provide the station its power. The only problems of note had come in the first hours after the Soyuz reached orbit, when the TsUP controllers in Moscow discovered that the spacecraft's rear communications antenna had not fully extended. There was no cause for concern, they advised Tsibliyev. The forward antenna was more than sufficient to receive commands from Earth. Then, after the three cosmonauts wriggled out of their space suits, the Soyuz had inexplicably switched to its reserve engines. Specialists on the ground analyzed the problem and decided there was nothing wrong with the engines. The only explanation they could devise was that the cosmonauts' movement inside the Soyuz, especially any abrupt move from the command capsule into the living compartment, might have triggered the switchover. No one knew for sure. The Soyuz, like most Russian spacecraft, is notoriously temperamental.

By 6:49, as Jerry Linenger and his two Russian crewmates, Valery Korzun and Aleksandr Kaleri, watch from inside the station, the little Soyuz has crept within ten yards of the docking port. They are going to cut it close. In three minutes the station enters a night orbit; no one wants to dock at night if possible. Gently the Kurs system guides the Soyuz in to twenty feet. Then fifteen feet. Twelve feet.

Tsibliyev keeps his eyes glued to the command panel above him. His daughter's pink rabbit floats loosely in the gravity-free environment.

Eleven feet.

Ten feet.

Nine feet.

Eight feet.

Suddenly a warning light flickers to life above Tsibliyev's head.

It reads: "APPROACH FAILURE 05."

Immediately Tsibliyev feels the spacecraft's braking thrusters fire, automatically aborting the approach, and the Soyuz begins to drift backward, away from the station. The Kurs system, for the umpteenth time, has failed.

Tsibliyev half expected this to happen. As usual there had been problems getting the Ukrainians to deliver a radar antenna for the Soyuz, and so Energia engineers had built a new antenna to replace it. In ground tests, however, the Russian-made antenna repeatedly failed to work, apparently because its signal overlapped with the signal of another antenna used in the spacecraft's motion-control system. The engineers attempted to remedy the situation by programming new software into the Soyuz's docking computer. But after being installed in one of the Star City training simulators, the new software caused a problem. Again and again during simulated docking approaches, the Kurs system simply shut down. No one—not the cosmonauts, not the trainers, not the engineers, not the code writers—could figure out what was going wrong. Star City blamed the software and Energia. Energia blamed the simulator, arguing that the new software would work in space. Energia mathematicians were so certain the software would work in space, in fact, they guaranteed it. All during his training, however, the trainers had warned Tsibliyev to expect the Kurs system to crash.

The Soyuz is slowly drifting back . . . back . . . back, away from the station. Tsibliyev asks the ground what he should do, and at a distance of twelve yards from the docking port, the TsUP orders him to try a manual approach. Tsibliyev is ready. It is the first time he has taken the manual-docking controls since the incident three years before, and he is determined to avoid another failure. Briskly he activates the system, pushes a button igniting his rear thrusters, and inches the Soyuz back toward the docking port. Down in the TsUP the controllers are all standing. Everyone is aware of Tsibliyev's history with manual controls.

At ten feet, Tsibliyev aligns the white-cross target on his computer screen squarely on the docking port.

At eight feet everything is fine.

At four feet he feels confident.

The Soyuz nudges the station, its docking probe sinking into Mir with the slightest bump.

"We have capture," he radios to the ground. Down in the TsUP the controllers applaud.

Tsibliyev smiles. It is good to finally get the manual docking monkey off his back. The $1,000 bonus he will receive for the successful manual docking makes it even sweeter.

By the time Tsibliyev docks at Mir that evening, Linenger has been aboard the station for almost a month. After an uneventful shuttle flight he had crawled through the docking area into the Kristall module on the night of January 15 and exchanged warm hugs with his predecessor, John Blaha, the third American to live aboard Mir. Blaha spent hours showing Linenger around the station, tutoring him on the best ways to use the vacuum toilet and the two treadmills. Everyone has a different initial reaction to the station. Norm Thagard, the first American to visit Mir, thought of his neighbor's messy utility room; gear was scattered everywhere, strapped and Velcroed to the floors, the walls, the ceilings. Close your eyes, and the place smelled faintly of sweat, which was not entirely surprising. Thagard had helped his commander, Vladimir Dezhurov, hack up the shower with a machete in 1995 to make way for one of the new modules; now, even during the best of times, the crew was only allowed to bathe with moist towelettes every third day. When the coolant pipes were leaking antifreeze, Mir smelled like an automobile repair shop. Linenger thought it smelled like a musty wine cellar.

Linenger, of course, already knew the basics. Mir consists of six connected modules. The heart of the station, containing the command center, the dining area, and two curtained sleeping nooks called *kayutkas*, is the twenty-one-ton core module, usually referred to as "base block." Base block, which has the size and feel of a messy, high-tech Winnebago, is forty-three feet long and, at its widest, nearly 14 feet in diameter. It is divided thematically—though not literally—into two halves, a seven-foot-wide "working compartment" containing the station's central computers, and a wider "living compartment" dominated by the burgundy-tiled dining table. Wherever possible the Soviets had draped the station interior in soft, soothing colors. In base block the "floor" is covered with an olive-green carpet, the walls are pea green, and the "ceiling" is a creamy white. The notion of a floor and ceiling is purely arbitrary, of course. In space there is no up or down.

Despite the best efforts of the commanders, base block is usually a mess. The sidewalls are lined with the detritus of space life: laptop computers; clipboards with paper radio memos from the TsUP floating up into the air; a tool cabinet with a flipout worktable; a ham radio setup; a Sony videocamera and dozens of black lenses, covers, and extensions, all Velcroed or otherwise strapped down; and an entire wall of cassettes and compact discs, containing music as disparate as Russian folk tunes and songs by the Rolling Stones. A careless brush against the sidewalls can trigger eruptions of everything from pens and pencils to finger paintings sent up by Russian schoolchildren. Everything knocked askew has to be retrieved. Otherwise, the cosmonauts have discovered, loose items have a way of making their way into the air vents, clogging them.

Sunk into the floor at one end of base block is the commander's low-slung command center, row upon row of computer monitors, switches, and brightly lit square buttons. At the opposite end is the dinner table; its tabletop flips up to reveal a row of round niches into which food tins can be placed and warmed. A can of jellied beef tongue—a cosmonaut favorite—can be heated in five or ten minutes. Tea, coffee and fruit juices are the most common drinks, although the cosmonauts keep a store of cognac onboard for celebrations.* On the "floor" behind the dinner table is the older of Mir's two treadmills; most cosmonauts prefer using the newer model in Kristall. Facing each other behind the table, set back into the sidewalls, are the station's two closet-sized sleeping niches, the *kayut-*

*Peggy Whitson, the NASA scientist who supervised the American science program for Norm Thagard's mission to Mir in 1995, was dismayed to find that the cosmonauts actually smoked cigarettes and drank vodka aboard the station. The vodka, she discovered, was stored inside half-liter "drink bags" that were sent up on Progress supply ships under the guise of "psychological support" materials. At one point, Whitson watched an American technician jokingly show off one of the vodka bags. "I said, 'It's not a joke,' " she remembers. "If Safety knew, it would have a cow." Whitson was equally concerned that drinking vodka, a diuretic, would ruin American science data on the cosmonauts' diet.

kas. Each has a sleeping bag strapped to one wall, a mirror for shaving, and a tiny porthole to view space. All the cosmonauts personalize their *kayutkas*. Tsibliyev will pin up pictures of his wife, Larissa, and his two children, a Mir 23 patch, and a palm-sized CD player for the Russian folk tunes he loves. On base block's rear wall is the station's original, now seldom used, toilet. For sanitary and privacy reasons, no one was ever enthusiastic about using a toilet two feet from the dinner table.

Connected to base block, just beyond the toilet, is a narrow hatchway leading to the cramped little Kvant module, a dimly lit stub of a module where most crews stow their trash. Entering the hatchway, which like all Mir hatches is about three feet in diameter, requires you to shimmy through a short passageway, a sensation much like crawling through the kind of enclosed tunnel found on a child's playground. It's a claustrophobic moment, made worse by the two softball-diameter ventilation tubes that, along with dozens of black and gray cables, cram the opening. The Americans have never liked the fact that so many cables are simply slung through Mir's hatchways; they worry that in the event the station undergoes an emergency depressurization—say, if a meteorite strikes and pierces the hull—the cables would get in the way of closing off the damaged module. The Russians listened but did what they usually did when the Americans brought up safety concerns, which was ignore them. Linenger calls Kvant "the attic." It is roughly half the length of base block, nineteen feet long, and fourteen feet in diameter. Most of the time it is stuffed with white trash bags and unused equipment. On Kvant's far side is the docking port for Progress unmanned cargo ships.

On the opposite side of base block, reached by flying over and past the command center, a second hatchway leads into a round, enclosed chamber, seven feet in diameter, called the "node," which functions as the station's main intersection. There is no reason to linger in the node; it acts solely as a road to the rest of the station. All four of Mir's other modules open onto the node, as does the Soyuz escape craft, which at all times remains docked on its far side; all five of these hatchways are every bit as crammed with cables as Kvant's. To the "left" as you enter the node—assuming the base-block ceiling is "up"—is an open hatch leading into the newest of Mir's modules, Priroda, where the Americans perform some of their experiments. Entering Priroda is disorienting: It's upside down, which causes an entrant from base block to perform a complete flip to regain orientation. Compared to base block, Priroda is shiny and new. It is lined with deep storage drawers and scientific hardware, a microwave-size oven here, a laptop there. Because it is generally used by only one person at a time, it remains largely free of clutter.

A "right" turn as you enter the node leads to the thirty-nine-foot Kristall module. Linenger thinks of Kristall as an unused module; the only reason the cosmonauts enter it on most days is to use the better of the station's two treadmills. Bone loss and muscle atrophy are constant concerns for long-duration astronauts, so Russian doctors mandate that everyone run for two hours on a treadmill every day.* To use either treadmill, the cosmonauts don custom-made harnesses—each cosmonaut or astronaut is fitted to exact specifications on the ground—that hold them in place while running. The harnesses tend to be tight, and most users complain that they begin to chafe the shoulders after about thirty-five minutes of running.

At the "top" of the node is a hatchway leading into the forty-one-foot Kvant 2 module, the staging area for all spacewalks outside Mir and, more important on a daily basis, the home of the station's main toilet. A dark, cramped place, Kvant 2 is divided by internal walls into three distinct compartments: At its far end is the airlock that cosmonauts use to enter and exit the station during spacewalks, known in both the Russian and American program as EVAs, for "extravehicular activity"; when not in use, the airlock is stuffed with bundled-up space suits, gear to prepare the suits, and various tools used during EVAs. Kvant 2's middle compartment serves as a backup airlock in case there is a problem with the main hatch the cosmonauts use to enter and exit the station; over the years, the hatch has been the scene of several bits of drama that have prompted cosmonauts to use the backup airlock. Nearest to the node is an open compartment that contains the station's main toilet.

At the bottom of the node—opposite Kvant 2—is the second of the "American" modules, Spektr, which arrived during Thagard's flight in 1995. Like Priroda, Spektr is lined with shiny new drawers, laptop computers, and scientific equipment. At the far end of the module, behind a huge German-made camera sunk into the floor for Earth observations, is a small open area that Linenger has adopted as his bedroom.

The station's life-support system is simple, almost elegant—when it works. The two notoriously cranky Elektron oxygen generators, stowed

*Standard clothing for Mir's occupants is T-shirt, gym shorts, and white socks. Shoes are generally only worn to run on the treadmill. Each crew member is also given a one-piece article of clothing the Russians call a penguin suit; the penguin suit is fashioned with stirrups that fit around the wearer's feet, giving him a mild aerobic workout as he moves.

out of sight behind wall panels in the Kvant and Kvant 2 modules, produce air from nonpotable water that is distilled from the crew's own urine. Some drinking water is recovered from the air by Mir's humidity-control system; much of it, however, is brought up by the shuttle. Carbon dioxide is removed from the air by a series of scrubbers called the Vozdukh system. Electrical power is generated by the one hundred-foot-long solar arrays perched on the station's outer hull; there are eleven arrays in all, three on the base block hull, two on Kvant, two on Kvant 2; the four newest, and best, arrays are mounted outside Spektr.

To keep the station steady—to maintain its proper orientation, or "attitude," pointing toward the Sun—Mir relies on a dozen whirling gyroscopes, called gyrodynes, which are stowed in silver beer-keg-size containers in the Kvant and Kvant 2 modules. If the gyrodynes fail or need to be repaired, it is possible to maintain attitude by firing the station's main thrusters. Without the gyrodynes or the thrusters steadying the station, Mir will start to slowly spin in space, a disorienting sensation for those aboard. Losing attitude is dangerous: The solar arrays are unable to track the Sun, and thus draw new power. With no power Mir's onboard batteries begin to drain. If not immediately corrected, this problem can lead to a total loss of power; while it is possible to recover power, it is a slow and ponderous operation. For good reason the Russians call a total power outage the Coffin Scenario.

As many things as Blaha and Linenger had to discuss, there were even more topics to avoid, because Blaha's mission, thanks to a bewildering array of interpersonal clashes, structural flaws, and missed signals, had seen the near-total breakdown of the American support system. By the time Blaha began packing his things to return to Earth, his anger at NASA—and especially at Frank Culbertson—was almost blinding.

"Don't expect any help from the ground," Blaha cautioned Linenger. "You're on your own up here."

To those few who knew of Blaha's travails aboard Mir, his transformation into a scathing critic of the Phase One program was jarring. Blaha was a shuttle commander, the only one to live aboard Mir, and he loved spaceflight with every fiber of his being. He had waited nine years for his first shuttle mission in 1989, and after each of his flights, two as pilot, two as commander, he always said he felt like running back to the launch pad and climbing onto the next rocket up. To many of those who flew with Blaha, he was a beloved figure. In place of the outward rigidity of so many fighter jocks, Blaha displayed the soft edges of a child psychologist. He talked slowly, in a severe nasal twang, and used his hands a lot, touching his listener, on the arm, on the elbow. Aboard the shuttle he constantly checked with his crew to see how they were doing, what they needed, how he could help. He was meticulous, nurturing, repetitive—and notoriously long-winded. Stop and chat with Blaha in the hallway, other astronauts joked, and you risked losing the entire afternoon.

The son of an Air Force pilot, Blaha was born in 1942 in San Antonio and grew up on air bases. Graduating from Granby High School in Norfolk in 1960, he went straight to the Air Force Academy. Unlike Culbertson and so many other astronauts, he had not grown up fascinated with space; not until college did he realize that flying spacecraft was simply the next logical step after test pilot school. He made in-depth studies of the Mercury astronauts and saw that each had flown many kinds of aircraft, and with NASA in the back of his mind, Blaha did the same, flying F-4s, F-102s, F-106s, and A-37s, including 361 combat missions in Vietnam. After failing to be accepted into the first class of shuttle astronauts in 1978, he made it into the class of 1980. As the senior military officer in the class, he was its unofficial president.

For all his strengths, Blaha never emerged as a star at NASA. "Blaha was just caught in the middle of a program where other people were brighter stars, more extroverted and articulate," says Gene Kranz. "He had the potential. [Instead] he was seen as steady and dependable." The rap on Blaha was that, while friends called him thorough and detail oriented, others dubbed him high maintenance. NASA trainers, who taught astronauts how to run complex experiments in orbit, grew exhausted instructing Blaha. "John, when you train him, comes across as a dolt," says Mark Bowman, a senior engineer at Krug Life Sciences, one of JSC's training contractors. "He makes you repeat things, which is fine the first time. But the second time, when he comes back and asks the same questions, you say, 'Wait a minute, is this guy stupid? How did he get to be a shuttle commander?' I remember the other astronauts telling me, 'Just be patient with him.' But John's far from stupid. He just wants to make sure he gets it. He wants to start from scratch and go over and over and over and over." Blaha reinforced the impression of dimness by keeping his commander's ego well hidden. "I'm a real stupid person," he once told a NASA debriefing. "My IQ is not real high and I was just an Air Force pilot before I came here in 1980 and I don't know a whole lot of things. I'm not an engineer, and I'm not a scientist. I'm an operator. So I'm not a very smart person."

Some astronauts whispered that Blaha relied too heavily on subordinates on the ground, on other crew members in space—and on the women in his life. By far the most important woman in his life was his wife, Brenda, whom he had met during flight training in Phoenix in the 1960s. Brenda, who had taken her little Chihuahua-mix Dutchess with her to live with Blaha during his training at Star City, was an old-fashioned wife who meant everything to her husband: She did his laundry, cooked his meals, raised their three children, everything. The other woman in Blaha's life was his close friend Shannon Lucid, who had preceded Blaha to Mir. It was no accident Lucid had flown on both the shuttle missions Blaha commanded. On the long flights over to The Cape she often flew backseat in Blaha's T-38. They had children the same age. They talked easily. In space Lucid acted as Blaha's de facto aide de camp. "Shannon was the one who in essence took care of John during both their missions," says a NASA official who knows and admires both astronauts, and has discussed with them their friendship. "He created in Shannon a person whom he could rely on, who would do the little things for him; it was always, 'Shannon, where are my glasses?' 'Shannon, where are my checklists?' that kind of thing." When Lucid was named to be the second Mir astronaut—and Blaha chosen as her backup—some felt it angered or embarrassed Blaha. If so, he never let it show.

NASA didn't psychologically screen its Mir astronauts; even if there had been any precedent for doing so—and there wasn't, not even on the shuttle—the agency couldn't afford to flunk anyone out of the program, because it simply had no replacements. But when Al Holland, JSC's avuncular chief psychologist, traveled to Star City in 1995 to brief Lucid and Blaha on what they could expect in the way of psychological pressures aboard Mir, he did administer psychological questionnaires to both astronauts. Blaha thought the whole exercise was silly and said so; he didn't know who Holland was, and he politely scoffed at the idea that long-duration spaceflight was any different than a week-long shuttle mission.

What Holland saw on Blaha's test alarmed him: Blaha's answers indicated a high level of dependency on others, especially on Brenda. Other than outright psychosis, it was probably the single worst personality trait a person to be placed in isolation could exhibit. If Holland's analysis was correct, he suspected that the very strengths that worked for Blaha in the shuttle program—his intimate familiarity with systems, his thoroughness, his preparedness—would work against him in the Russian space program, where everything seemed impromptu and unfamiliar, and where cosmonauts were expected to think on their feet. Holland noted how, even in Star City, Lucid helped Blaha with his studies, reminded him of his assignments, relayed messages to friends. He wondered: How would Blaha fare alone in orbit with a pair of grumpy Russians?

Holland returned to Houston and thought long and hard before doing something he had never done in more than a decade at NASA: He recommended that Blaha be scrubbed from his mission. "From a medical point of view, this guy should not be flying," Holland told a meeting in Culbertson's corner office. "This is just not a good person for long-duration flight."

Of course it was a useless exercise. There was no way Culbertson could yank Blaha from the program, even if he wanted to. "All right," Holland admonished him. "But don't say I didn't warn you." Once it was clear there was no blocking Blaha's mission, Holland devised a psychological plan that might be called "the Brenda Strategy." When Blaha reached the station that September, Holland always made sure he knew exactly where Brenda Blaha was. When she traveled—and she traveled a lot while her husband was on Mir—Holland made certain he always had a phone number where she could be reached.

"If we lose comm for any period of time," Holland warned Culbertson, "we're cooked. And if anything happens to Brenda, I know we're in for a serious situation."

When the shuttle carrying a new Mir astronaut arrived at the station, it typically remained docked there several days, offloading supplies and giving Mir's newest occupant a chance to learn the ropes from his predecessor; NASA calls this the "changeover period." During the five-day changeover period in mid-January when both Blaha and Linenger remained aboard the station, Linenger received a firsthand glimpse of the pressures Blaha had endured. Blaha, in fact, got into a shouting match with his rookie commander, Valery Korzun, while the shuttle was still docked to Mir. He had spent a long day working with the shuttle crew moving containers between the shuttle and the station, and was busy changing out of his sweaty clothes in the shuttle's middeck when Korzun popped his head around a corner.

"John, what are you doing?"

"I'm changing clothes."

"You are supposed to be helping Marsha." Korzun explained that astronaut Marsha Ivins was waiting for Blaha to help her move some equipment.

"I am?"

"Yes."

"Well, Valery, I'm very sorry." And with that Blaha quickly finished changing his clothes and pushed off to give Ivins a hand. It took forever to find her in the crowded Mir-shuttle combination. When he did, Blaha told her Korzun had sent him to help. Ivins appeared confused.

"I don't need you, John," she said. "Valery and I just discussed this, and we agreed to do it tomorrow."

Maybe it was because his mission was almost over. Maybe it was exhaustion. He would never know. But for reasons Blaha could never quite explain, Ivins's matter-of-fact response absolutely enraged him. Furious, he went in search of Korzun and found him in base block.

"Valery, I. Am. Pissed!" Blaha snapped. "I have been busting my butt all day! You tell me to do something, and Marsha says I don't have to! Now you've wasted thirty-five minutes of my goddamned time!"

Blaha realized he was shouting.

"John, I'm sorry," Korzun said evenly.

"Well, Valery, that doesn't do it. I don't know why you did what you did. But I am pissed." Blaha then turned and swam back to the shuttle, where he had a private family conference scheduled with Brenda.

"John, calm down, it's no big thing," his wife counseled.

"Brenda, I can't even talk I'm so upset," Blaha said, and after five minutes he abruptly ended the comm.

Blaha simmered until dinnertime, when he joined Korzun and Kaleri in base block to eat. By that point, having tried to analyze the source of his anger, he realized what must have happened: Korzun had been using his broken English with Ivins and had misunderstood what she was saying.

"I'm really sorry, John," Korzun volunteered.

"Valery, I'm sorry too," Blaha said. "We've both been under a lot of stress."

And for the most part, that was how it worked with Blaha and his Russian crewmates. Storms appeared suddenly then passed, never to be mentioned again.

Blaha and Korzun had suffered only one other argument of note. It was a silly thing. One morning in December Blaha had been engaging in one of his favorite pastimes, talking to friends on the station's ham radio, when Korzun interrupted.

"You are not supposed to be using that frequency," the commander said.

Technically, Korzun was right. But the TsUP changed the frequencies they were allowed to use so often, friends and family members sometimes didn't get the news for days. Blaha was doing exactly what he had seen Korzun do—using an old frequency until friends back on Earth received news of the TsUP's latest change.

"Valery, I know that," Blaha replied sharply. "I'm not doing anything differently than you were doing this morning. I switched to this frequency so I could talk. And god damn it, right now you've caused me to lose this poor guy [on Earth]. He thinks I'm an idiot! You've screwed up my whole comm pass*!"

Once again Blaha was surprised by his own anger. Korzun simply gave him a blank look, turned his back, and swam off without a word. He was gone for two hours, during which Blaha realized he had overdone it.

"I'm sorry, Valery. I apologize," he said later. "We've all been working too hard."

There had been many times when Blaha wanted to eviscerate his commander. On dozens of occasions Blaha held his tongue when Korzun gave him orders in tones the American commander reserved for his infant grandson. It was the same condescending tone so many of the Russian commanders used with the Americans at Star City, and Blaha couldn't stand it. Whether it was a natural response to addressing people who hadn't mastered the language, or a question of the Americans' lack of training in Russian systems, Blaha didn't care. He tried to make Korzun understand it wasn't necessary to talk to the Americans as if they were children. At first Korzun seemed cold to the idea. But as the weeks wore on and the two men grew to respect each other, Korzun seemed to begin listening.

"When Vasily [Tsibliyev] comes, please pass it on to Vasily: Let's fix this now," Blaha remembers telling Korzun. "When Americans come up who have been in space before, maybe you need to talk to them in a little different tone than, say, you would [with] a Japanese journalist."

"He never said that," remembers Korzun. "If anything, it was John

*A "comm pass" is slang for an earth-to-ground communications session.

who [volunteered to take the subordinate role]. Every time he wanted to get on the treadmill, he asked, 'Can I do this now?' We said, 'Of course! Of course!' Yet he would always come and ask us this. Sasha and I, we felt uncomfortable. He was a veteran American astronaut, and he behaved like that. Why?" Korzun thought he knew; Blaha, despite months of Russian-language classes in the United States and Russia, was never entirely comfortable speaking Russian. "When [John and I] spoke," Korzun recalled, "I had to look him in the eye. And when I was finished, I usually said, 'John, did you understand?' And he would say, 'Yes.' And I knew he didn't. So I would speak again, slowly, maybe use some drawings, and John would go, 'Oh! Oh!' And only then would it be clear that he understood."

For Blaha dealing with the Russians was always the easy part. It was the Americans who drove him to despair. His problems began even before arriving in Star City for training in January 1995. In August 1994 he and Lucid had reported to the Defense Language Institute in Monterey, California, to begin their crash course in the Russian language. A month into their training, Blaha was getting nowhere. He could barely master the Russian alphabet: How on Earth did anyone think he would be ready to actually *speak* Russian, let alone take technical classes *taught* in Russian, in barely four months? "I remember asking one of our instructors, you know, 'What is going on? How can we do this?' " remembers Blaha. "And he said their basic course, the one they teach to diplomats and spies, is a two-year course. When they [told this] to NASA, NASA's answer was, 'These are astronauts, they're really smart. They can do it quicker.' "

Blaha's inability to master Russian made his Star City training even tougher. For eighteen months there he worked seven days a week, sixteen hours a day, without interruption. As his mission neared, the most agonizing struggle for Blaha was a strange, catch 22–like disagreement with Houston over the all-important written instructions for the experiments he would perform in space. Both his predecessors aboard Mir, Thagard and Lucid, had used as their orbital bible a five-hundred-page set of "flight procedures" that outlined in minute detail how each of their experiments was to be performed. It was a document shuttle astronauts typically trained with for a solid year before a mission. At first the Phase One office in Houston had assumed Energia would write the necessary manuals. But in late 1994, with Thagard's liftoff barely four months away, NASA science officials finally realized that Energia wasn't going to be able to furnish a usable set of procedures in time for the mission. NASA staff would have to write the bible themselves.

The job had fallen to John McBrine, a hard-partying, thirty-one-year-old Krug Sciences worker at Star City whose Herculean appetite for hard work and long hours proved indispensable to Thagard and Blaha. There was just one problem. McBrine was an exercise physiologist who knew next to nothing about space science. He and another Krug man, Matt Muller, spent weeks trying to write the procedures, piecing together instructions on faintly typed faxes sent by American scientists with what they could learn about the Russian- and American-made laboratory hardware Thagard would be working on. "We didn't know what the hell we were doing," recalls McBrine. "We really didn't. Science, training, baseline data collection—me and Matt, we didn't know what we were doing. We thought the Russians were going to do this. I was the only one stupid enough to volunteer." Eventually McBrine completed the 450-page set of procedures and had them hand-delivered to Thagard—exactly four days prior to his launch.

Blaha had watched this process in amazement and was determined not to let the same thing happen to him. Even before Thagard's mission, he and Lucid had brought the matter up with Culbertson. "Don't worry, John, we've got the matter well in hand," Culbertson had assured him. "We're going to take care of it."

But he didn't. The preparation of Lucid's procedures, in fact, proved just as chaotic as Thagard's. In February 1996, six months before his mission was set to begin, Blaha took matters into his own hands. If NASA couldn't prepare his procedures in time, he would write them himself. He began taking notes in a spiral notebook. His notes included instructions on how to perform experiments, as well as shortcuts he might take to save time. To Blaha the notebook became a kind of security blanket, a piece of portable sanity he clung to wherever he went. "They [the notes] were what saved me," he says. He was just beginning to feel comfortable when, that May, he received an e-mail message from Rick Nygren, the head of JSC's Life Sciences working group. "What he effectively said was, 'John, I understand you're trying to illegally take things on the Mir, unapproved things, you're trying to put things in your pockets and put them on Mir,' " recalls Blaha. "The inference was that I was trying to sneak some things on Mir."

Nygren, in fact, was dead set against Blaha using his notes aboard the station. As strange as it sounds, there was good reason for his concern.

The Russians were extraordinarily sensitive about the American science program, which included experiments involving everything from radioactive alloys to open flames to cancer cells. It only took one slip, Viktor Blagov's people were forever reminding the Americans, to irradiate the entire station. The TsUP thus insisted on reviewing in detail every piece of scientific hardware the Americans sent to the station, and approving every line of every instruction laying out how the equipment would be used. Blaha's notes were unofficial, and thus forbidden.

At first Blaha ignored Nygren's note. But when he began receiving similar e-mails from Nygren's subordinates, he got angry. "I wrote Frank [Culbertson] an e-mail," Blaha remembers. "I said, 'Frank, please tell your people to leave me alone. I don't want to receive any more e-mails from them. I'm an American living in Russia, training in the Russian language, and I need to prepare for my flight. I need my notes. I need them because your people did not do their job [preparing me]. I can't deal in their diddly-diddly e-mail crap.' "

The e-mails stopped for a time but then resumed. Finally, Blaha exploded. "That group of people on the ground sent Americans to Russia to do a job," recalls Blaha. "Because of the way they sent us there, with insufficient language training, we were effectively walking around with our pants down. You get used to that after a while, like an animal does. By May, after compiling all my notes, I felt like my pants were all the way up and buckled. I then watched Americans in Houston try to pull my pants back down as I got ready to go up to Mir, and I didn't understand that. They were trying in a million different ways, everyone saying, 'You can't take those things to the Mir with you. You can't. You just can't.' I never understood that. This problem became all-consuming."

From mid-May until August, on the eve of his flight, Blaha was unsure whether his notebook could accompany him to Mir. He listened to Al Holland's little lectures on preparing himself mentally for life on the station and thought: Nothing in space could be worse—nothing—than the torture NASA is putting me through with these notes. The tempest only passed when Blaha reluctantly agreed to hand over his notes for NASA's review. Jim Van Laak, the sharp-tongued former Air Force pilot who served as Culbertson's No. 2 in Houston, went through them page by page, and after consulting with the Russians, agreed to let Blaha take them to space.

The notebook incident, however, was just the beginning of Blaha's troubles. In the months leading up to his mission, it seemed that everything that could go wrong did. First his launch was delayed. On July 12 NASA announced that a gas leak had forced it to replace the solid rocket motors on the Space Shuttle Atlantis, a change that pushed Blaha's flight back six weeks, from August to mid-September. Blaha actually welcomed the delay; it gave him more time to prepare. But for Blaha, a man who relied heavily on teamwork and subordinates, the most foreboding breakdowns were not with engines but with his personal relationships. To every American who served aboard Mir, four people were key: The ops lead, who ran the Moscow-based ground team; the flight surgeon; and the two Russian crewmates. Blaha's relations with each were to prove dysfunctional or nonexistent.

The first problem was his ops lead, a thirty-seven-year-old NASA worker named Isaac "Caasi" Moore. Moore's nickname, pronounced "Casey," was *Isaac* spelled backward; a family friend had given the name to him as an infant, because his father was already called Ike. "And I've been doing everything backward ever since," Moore liked to joke. Not everyone disagreed. Moore was a glib, sarcastic Georgia Tech graduate who had spent much of his seventeen-year career in the Mission Operations Directorate as a SIM SUP, that is, a training supervisor who laid out and oversaw the complex scenarios MOD ran in its various shuttle simulators. Looking for a way to vault up the MOD ladder, Moore had eagerly volunteered for Phase One in the fall of 1995.

NASA had devised the ops lead concept after Thagard's mission to Mir in 1995. Lucid, the second American to stay aboard Mir, was thus the first to have a full-time ops lead, an energetic MOD veteran named Bill Gerstenmaier. While Gerstenmaier went to Moscow to prepare for Lucid's mission in the fall of 1995, Moore and the other new ops leads spent six months in full-time language training, in a little outbuilding at JSC. As the most senior of the group after Gerstenmaier, Moore was to take the next increment, Blaha's, and the initial plan called for Moore to serve as Gerstenmaier's assistant during Lucid's mission, which would give him crucial on-the-job training and ensure a smooth transition to Blaha's increment. It didn't work out that way. In late 1995 NASA's Life Sciences division, which was paying the ops leads, announced it didn't have enough money in its 1996 budget to support two men in Russia at once. Gerstenmaier would have to work alone, leading Lucid's ground team with no assistant; Moore would pick up tips on how to do his job where he could.

Supporting Lucid's extended increment, Gerstenmaier proved far and away the best ops lead NASA would select. The problem, at least for Moore and his successors, was that Gerstenmaier was a one-man show. He thrived by sheer force of his personality, expertly badgering the Russians, stroking Lucid, and keeping Houston informed, but never writing anything down explaining how he had done it.

Even before his first talks with Blaha about their mission, Moore knew he would have his hands full. He had worked in simulators with Blaha for years and knew well his reputation as a "high-maintenance" commander. He was not surprised to hear that Blaha was having difficulties in Russia. "I had heard a term used for the astronauts training at Star City, 'siege mentality,' and I saw it," says Moore. "They'd phone back [to Houston], ask for something they needed in eighteen hours, and get no response. You do that two or three times, you develop a mentality of 'I have to do this myself.' [As a result] John turned inward, and he didn't have the resources there to get the job done."

In May 1996, barely three months before his mission was to begin, Moore was sent to Star City for three weeks to train alongside Blaha. He badly needed to get up to speed on the science program, as well as learn his astronaut's expectations for the mission. His other assignment, as one of his supervisors put it, was to "go drink beer" with Blaha; that is, bond with him. But from the first day Moore arrived at Star City, he realized there would be no beer drinking. Moore expected to see Blaha on a daily basis. Blaha sent word that he didn't have time to see him—at all.

"I thought this was just incredible," Blaha recalls. "Here it is, in May, and I'm going to leave [for Houston] at the first of July, and they say, 'Oh, by the way, this is going to be your guy in charge. So here he is and teach him everything you can.' Here I am, and I'm busier than heck, and I didn't have any time to spend with Caasi Moore. I mean I had none. 'Well,' they said, 'let him just come to some of your [classes].' I just didn't understand it. Why didn't they give him his own [training]? Caasi had no concept of what was going on. He was joining the effort so, so late."

Overwhelmed by sixteen-hour workdays, Blaha sent Moore a message that he wanted one thing from him, and one thing only: an hour-by-hour breakdown of what Blaha would be doing his first two weeks aboard Mir. Moore groaned. Everything he needed for the job was back in Houston. At JSC there was a list of the experiments Blaha was to perform. There were also procedures for each of the experiments. But no one had yet mapped out what procedures would be done when. It was a big job, and Moore tried to do it via e-mail with his coworkers back in Houston. Two weeks into his three-week stay at Star City, he received a rough draft of what Blaha had asked for.

It was a disaster, vague, with gaping holes. Moore knew Blaha would be apoplectic. He spent his remaining week in Russia attempting to refine the draft, finishing his work at midnight of his final evening and catching a van to the airport three hours later.

Predictably, Blaha found Moore's effort lacking. "In June he sent me something," Blaha remembers. "It sort of made me aware of the huge knowledge split between the two of us. [I said] 'Now, hold it, Çaasi, what you gave me doesn't mean anything to me. When you say, from nine to eleven [o'clock], perform procedure X, you need to say, 'Go to attachment 8, paragraph this, page this and this, and perform X procedure.' He tried to tell me this wasn't a good format. And I said, 'Caasi, that's the way you're going to send it up to me in space.' I couldn't get him to accept that. Every time we met after that, it was like, I realized more and more the huge, and I mean huge, knowledge gap between the two of us."

The few times the two men met in person, this gap manifested itself in an obvious tension. At Star City, where they had managed to talk during a social occasion or two, Moore sent Blaha what the astronaut took to be an unmistakable, and unfathomable, signal that he was the man in charge.

"This is what I want from our mission," Blaha began one conversation. "And this is the way—"

Moore didn't let him finish the sentence. "John, hold on, this is the way I'm going to run your mission . . ."

Things went downhill from there. When Blaha returned to Houston in July the two met once more, in an encounter that left Blaha shaking his head. Four weeks before liftoff, Blaha took John McBrine, the Krug man, to see Moore in a conference room in Building Four. Again Moore began by outlining how he planned, as he put it, "to run the mission."

"Hold it, hold it," Blaha interjected. "When you were in Star City, I asked for a product." The schedule. "I'd like to see that now."

"Well," Moore said, producing a sheet of paper, "here's what I've got so far."

Blaha examined the list Moore gave him. It was nothing like what he wanted. It remained a vague overview of his first weeks in space. The few specific tasks it listed appeared as numerical codes.

"Caasi," Blaha said, measuring his words. "This is not what I expected." He produced a copy of an e-mail he had sent Moore, precisely outlining what he wanted. He put the two pages side by side on the table.

"This is what you gave me," the astronaut said, indicating Moore's sheet. "And this is what I gave you. They're not anything alike."

It was vintage Blaha: straightforward and punishing. Moore began to make an excuse, saying that his support staff at Lockheed Martin had let him down.

"Caasi," Blaha said, "I didn't ask Lockheed Martin to do this for me. I asked you."

Moore hemmed and hawed for a moment.

"When can I expect to see what I asked for?" Blaha asked.

Three weeks, Moore said.

Now Blaha was getting angry. "You mean to say I can't get this for another three weeks?"

Moore nodded.

Eventually Blaha's aggravation reached the point where he saw no reason to deal with Moore at all. There was simply no way for Moore to learn in three months everything Blaha had learned in eighteen months. Both Culbertson and a senior MOD official quietly asked Blaha whether he wanted to dump Moore. Blaha refused both overtures. He knew how the NASA blame game worked: Fire your ops lead at the last moment, and everything that went wrong afterward would be blamed on you. No, Blaha would stick it out with Moore. Whatever went wrong on orbit would not be his fault.

If anything, Blaha's relations with his flight surgeon, Pat McGinnis, were even worse than those with Moore. For some reason Blaha could never fathom, he and McGinnis utterly failed to bond. Other flight docs, notably Linenger's man Tom Marshburn, acted almost as personal valets for their astronauts, shadowing them through their Star City training, taking classes with them, rehearsing experiments, going over procedures, running interference with the Russians when needed. But during Blaha's training, he quickly noticed that McGinnis always seemed to be at the TsUP helping out on Lucid's flight or giving Linenger a hand training for his spacewalk in the hydrolab. "Pat avoided me," Blaha recalls. "I think he wanted to be doing exciting things with Shannon and Jerry instead of hanging out with boring old me." That March Blaha had brought up the issue with Dave Ward, the senior NASA flight surgeon at Star City. "You need to give me a flight surgeon or tell 'what's his name'—Pat—to start spending some time with me," Blaha told Ward.

But even then McGinnis never seemed to be around. Again and again Blaha sought him out. "Pat, I guess what I'm doing isn't that much fun," he told him, "but you and I gotta spend some time together." Blaha especially wanted McGinnis to attend his classes on Russian in-flight medical protocols, so that he could help him learn how to take blood samples and operate the Mir's medical equipment, such as the centrifuge. Blaha and Lucid knew so little about these tasks, they had actually scored a zero on a Russian pop quiz—a failure for which Blaha at least partly blamed McGinnis. But despite Blaha's entreaties, the young doctor managed to make only one of the classes.

In fact, it wasn't until Blaha entered quarantine at The Cape a week before his flight that McGinnis appeared and begin to grow attentive.

"Suddenly Pat McGinnis is my best buddy," Blaha quipped grimly to Brenda. "*Now* he thinks I'm important." Though Blaha was too angry to say anything to his face, McGinnis got the message quickly.

"I don't think John's too happy with me," he told Brenda.

It was the Russians, Blaha liked to say, who never let him down. The two cosmonauts he had trained with, Commander Gennadi Manakov and flight engineer Pavel Vinogradov, were good men, professionals, and Blaha had returned to Houston in July looking forward to rejoining them aboard Mir. Then, on August 9, as he sat at his suburban Houston home working on a science protocol with Matt Muller, he got a stunning phone call from one of Culbertson's men, Charlie Brown. Manakov and Vinogradov, Brown said, had just been medically disqualified for flight; there was some kind of problem with Manakov's EKG. Blaha couldn't believe it. Nor could he easily fathom the new reality Brown laid out for him: He would now be working aboard Mir for four months with two cosmonauts he barely knew, the rookie commander Valery Korzun and a flight engineer named Aleksandr Kaleri. As many cosmonauts as he had met at Star City, Blaha couldn't even place Kaleri's face. He had to ask Brown to get him a bio.

Korzun he knew, and that worried him. "He's a micromanager, Brenda," Blaha told his wife. In Blaha's mind, Korzun had two very different identities. At Star City many of the astronauts and cosmonauts taking classes in Building Three tended to take tea together shortly before noon; at those sessions Korzun had impressed Blaha as the single most Americanized of the cosmonauts. He wore blue jeans and a big smile and couldn't be nicer. He didn't mind speaking Russian slowly for Blaha's benefit, unlike Anatoli Solovyov or Vasily Tsibliyev.

But Korzun had another side, Blaha knew. That January the two men, along with Mike Foale, had performed their winter-survival training exercise together. The trio had been taken out into a remote area of the snowy forest near Star City, dumped inside a closed space capsule, and told to expect retrieval in forty-eight hours. The idea was to simulate an emergency landing in winter. On paper everything went fine. The three men built fires and lean-tos, and cut and fashioned long snow boots from rubber suits in the capsule.

But Korzun, Blaha quickly noticed, tended to treat him and Foale like children. At night they awoke every two hours to dry the sweat from their new boots, lest it freeze and cause frostbite. "Every single time we did it, even the tenth time we did it, Valery told us how to do it," Blaha recalls. "You would have thought we were five-year-olds."

Two days after learning of Manakov and Vinogradov's decertification, Blaha flew back to Star City to meet Kaleri and Korzun and attend their launch at Baikonur. At the Prophy he was pleased to find a note waiting for him from the two men, inviting him to dinner. He went, and found both cosmonauts to be charming. Korzun was as he remembered, forceful, smooth, and comfortable with Americans. Kaleri, a rising star in his second mission to Mir, turned out to be a hardworking young engineer with hangdog good looks. Still, there was no hiding his concern—or the Russians'.

"We hadn't had a single training event with John," remembers Kaleri, shaking his head. "We could tell John was worried about this. And we were worried because John was worried."

It was all too much for Blaha: the sixteen-hour days at Star City, the experience of living for the first time in a foreign culture, the notebook fight, Caasi Moore, Pat McGinnis, his two new Russian crewmates. By the time he flew to The Cape in early September for his delayed launch, Blaha acknowledges, he was near his emotional breaking point. "We all launched tired, absolutely wiped out," he says of the first four astronauts to stay aboard Mir. "What a way to go to a long-duration spaceflight."

Blaha's frustration climaxed in a dramatic confrontation with Frank Culbertson in the days immediately before launch. Blaha was working in the quarantined crew quarters when he saw Culbertson approaching.

"Frank, I don't even know why you're here," the astronaut said.

"John," Culbertson began. "I know you're unhappy with me."

"I am."

Blaha, in fact, was furious. "Frank, I like you," he said, "but this is the single most screwed-up program I have ever been associated with." And then he let loose with his litany of complaints: the total lack of training support in Star City, Culbertson's inability to make any substantive changes there, the fact that he was about to head into space for four months backed by a flight surgeon and an ops lead he hardly knew.

"Frank, Norm Thagard told you all a lot of this stuff. Y'all wouldn't listen. You kept telling us how to fix it, rather than getting it fixed. Now eighteen months later we're having the same discussion."

Blaha was just warming up. "It was forty-five minutes of a personal attack on me," remembers Culbertson. "All I could do was sit there and take it. The guy was going into space for four and a half months, and obviously he needs to vent. Part of my job is to let him vent."

On the morning of September 16, 1996, Blaha and the crew of STS-79 blasted off from launchpad 39A at The Cape. Three days later Blaha crawled into Mir, gave Shannon Lucid a quick hug, and plunged enthusiastically into his mission. His first attempts at science were disastrous. Even before Lucid and the shuttle crew left to return to Earth, he began work on his first experiment, called a binary-colloid alloy test, in which Blaha grew crystals in the Mir glovebox. The daily lives of American astronauts aboard Mir were governed by a minute-by-minute schedule, called a Form 24, that was compiled by a NASA ground team at the TsUP and approved by Russian ground controllers. Every Mir astronaut came to detest the tyranny of the Form 24, and several, including Blaha, returned to Earth with strong recommendations that astronauts be given more control over their lives. No one endured more headaches with the Form 24 than Blaha, a problem that was exacerbated by the lack of training and experience of Caasi Moore and his ground team. Moore's team, for instance, gave Blaha ninety minutes to perform the binary-colloid alloy test. To their dismay, it took Blaha five hours just to locate and assemble the necessary equipment.

"The ground, they didn't know anything about this test," Blaha recalls, his anger as fresh today as it was in orbit. "They had never done any training in it. They knew nothing. All they knew is that somewhere in a book somewhere, it says it should take an hour and a half. It says,

'Unstow seven things.' Well it took me five hours to do that. You haven't even got anything unstowed, and the whole hour and a half has gone by."

"Failure started chasing him from the very first day," Sasha Kaleri recalls with a sigh. "I felt so sorry for him. When we were saying good-bye to the shuttle crew, I remember it was painful for me to look into his face, because he was already having problems with his first experiment. John didn't even want to come say good-bye. When we said, 'Come on for maybe four or five minutes and say good-bye to your friends,' he came for exactly one minute and then went straight back to work."

Blaha encountered nearly identical problems in his first attempt to harvest wheat from the station's little greenhouse. He was supposed to photograph the experiment, but it took time to find the cameras. When he finally found them, a Nikon and a battered old Hasselblad, he discovered they didn't work properly. "The hardware was literally falling apart," he remembers. "I probably spent, oh, about ten hours of time getting these two cameras in flight-worthy shape."

By his third day in space Blaha was telling Moore and his ground team that their time lines were all but useless. To his surprise, nothing changed. Blaha got the clear impression that Moore felt easing up on the number of experiments would be viewed as some kind of failure by their superiors back in Houston. Again and again Blaha asked that the time lines be loosened to give him more time, especially to prepare the new experiments. To his amazement, Moore ignored him. "I wasn't getting the message," Moore admits today. "I just wasn't getting the message."

Within the first two weeks, in fact, Blaha earned a lasting reputation with his ground team as a whiner—a tag that infuriates him to this day. "They were completely out of sync with the Mir space station," Blaha says of his ground team. "They thought they were on a [shuttle] mission. That was the fundamental error they made. What was incredible to me was they would not take input from a person that knew differently. Their biggest error was they didn't listen to someone who had four space missions before this flight. Even when I told them about all this after day 2 or 3, their view on the ground was, 'He's just whining.' That's very unfortunate. I don't even know what you say about that. The human beings on the ground who thought I was whining ought to go sit down in a church somewhere quietly and ask themselves some very strong questions about themselves."

Unable to persuade the ground to ease its overloaded time lines, Blaha began working late into the night. Around 10:30, when he was scheduled to begin sleeping, he would start reviewing the next day's time line. By eleven he was searching for the required items. Several nights he stayed up until 4:30 A.M. assembling hardware for the next day's experiments. During this period he estimates he averaged barely three hours of sleep. By working through the night he began to keep up with each day's load of experiments. But Blaha knew it was a routine he could not maintain.

So did Korzun and Kaleri. "When John arrived at the station, he had the mentality of an astronaut who had long experience working on a shuttle," recalls Korzun. "In the beginning he wanted to work at the station the same way he had worked on the shuttle. [But] a shuttle flight is very short. The schedule is very tense. It is a short-duration run. On Mir it is a long-duration run. When we saw John working so actively, we realized he had miscalculated his energy for a long-duration flight. We spoke to John about it, but here's what John said: 'We are used to working this way; we worked this way on the shuttle.' Sasha told him, 'John, you must understand, here we have to just not work constantly, we have to find time to rest, find time to conserve our psychological energy, we have to rest in such a way we can restore our energy.' John disagreed."

"He didn't complain," recalls Kaleri. "We just saw that he had too much work to do, and he couldn't settle down properly. Days passed, and he wouldn't set up his personal things."

"John worked so hard, we said to him, 'John, you have to agree on the schedule with the ground people, taking into account the realistic time you need,' " says Korzun. "At the beginning of the flight, John didn't understand this. When he worked like this, he was very tired. His tiredness accumulated every day. I remember the day he came to us and said, 'Guys, I'm very tired.' After that we suggested to him ways to change his schedule, to have time to work and rest."

The situation came to a head on the second weekend of Blaha's mission. His flight surgeon, Pat McGinnis, had asked to take on more responsibility, and Caasi Moore gave it to him, putting him in charge of mapping out the time lines for an experiment the following week. That Sunday Moore, after reviewing the time lines himself, told Blaha that he would be sending them up shortly.

"Another *set of procedures*?" Blaha asked, his irritation clear.

"John, we haven't submitted anything yet," Moore said.

"I've got them right here in front of me!"

The Russian ground controllers, always keenly vigilant as to what experiments the Americans were performing, had laid down strict guidelines for the submission of scientific procedures. Before Blaha could be sent anything, it first had to be reviewed by the Russians. Moore, knowing he hadn't submitted anything yet, tracked down McGinnis.

"What happened with the procedures?" he asked

"We sent 'em up already," McGinnis said, explaining that he had handed the procedures directly to the Russian ham radio operator to send directly to Blaha. Moore closed his eyes, as if in pain. This was a direct and serious violation of Russian procedures, and he said so. "Pat's face just melted," Moore remembers. "We had really screwed up."

Worse, the draft procedures McGinnis had sent were incomplete. They didn't include time for Blaha to prepare the equipment. Moore realized it would be necessary for Blaha to spend another long night preparing for the experiment. Taking a deep breath, he walked down to the TsUP floor and alerted Blaha at the next comm pass.

Blaha got mad. But to the Americans' surprise, Valery Korzun was far angrier. The Russian commander had watched in dismay as Moore's ground team, rather than easing Blaha's load, seemed to pile on more and more work. That evening Korzun interrupted the American comm pass and spoke directly to the Russian shift flight director, Nikolai Nikiforov.

"I'm the commander of Mir, and I can tell you what they are doing with John Blaha over the last ten days has really been wrong," Korzun said. *"The Americans really need to get better organized."* At that moment Blaha wanted to kiss his commander. It was a gesture he would never forget.

As Korzun spoke in Russian, Moore stood on the floor of the TsUP auditorium listening. He couldn't understand much Russian, but he could tell Korzun was mad.

"This is for you," Nikiforov mouthed to Moore.

"I know," said Moore, crestfallen. "I can understand just enough."

After the pass Viktor Blagov, Nikiforov, and another flight controller took Moore aside. "I got a talking-to something fierce," Moore recalls.

"We need to offload [work from] John's schedule," Blagov told Moore. "You have him working too hard."

Embarrassed, Moore cut 25 percent of the experiments from Blaha's time lines for the upcoming week. Blaha immediately noticed and appreciated the change. But he remained amazed that it had taken the intervention of a Russian to get the Americans to realize the need for a change in the first place. "Here I was, a veteran shuttle pilot, a seasoned space flier, and I have been telling the ground the same things for ten days,"

Blaha recalls. "They never heard me. But a Russian on his first spaceflight, they heard it. That's almost incredible."

Even with the reduced workload, Blaha was approaching a state of exhaustion. The workdays aboard Mir ran fourteen hours and longer. "I can't do this anymore," he finally told Korzun. "I'm fifty-four years old, and I'm not going to make it if I continue at this pace." At night Blaha lay awake in his sleeping bag, strapped to the floor down in Spektr, and obsessed about his workload. "It just drove me into some kind of protective envelope," Blaha recalls. "I wasn't happy. I just wasn't happy. I was trying to run up a mountain, and the Russians were trying to help me, and the Americans were trying to bring me down." Many nights he called up the computerized scrapbook Brenda had made for him and looked through pictures of his children and grandchildren.

For the first time in his long career in space, Blaha was desperately unhappy. Nothing about the mission, a mission he had worked more than two years for, had gone as planned. Nothing about it was fun. He realized he was withdrawing from Korzun and Kaleri and snapping at the ground. It took a long time for him to acknowledge that something was wrong, and when he finally did he realized it was something worse than simple sadness.

It was depression. He realized he was suffering through a mild depression. The thought stunned him. Blaha had always thought of himself as a can-do guy, a fighter pilot, a positive thinker, the kind of person who helped his crewmates through whatever dark nights of the soul they encountered. The idea that *he* could be facing depression was almost too much to comprehend. Of course he told no one—not Korzun or Kaleri, not Brenda, not Al Holland, and certainly not his ground team, who he felt would use it as more evidence that he wasn't pulling his load.

Once he suspected the problem was depression, Blaha characteristically attacked it in a methodical, thought-out manner. Lying awake at night, he probed for the reasons he felt the way he did.

John, you love space, you've always enjoyed space. Why don't you love space now? Yes, working with Korzun and Kaleri had been a surprise, but they were good men, ready to listen to his suggestions. They were professionals. It was the Americans he couldn't abide. *The people on the ground have no idea what is going on. No concept. And they won't even acknowledge that this is the truth.*

When he thought it through, he realized he couldn't blame poor Caasi Moore. Moore had been thrown into the process at so late a date, no one could have gotten up to speed in time for the mission. And Pat McGinnis? Blaha could hardly blame the young flight doc for gravitating toward other, more interesting astronauts. No, the man he blamed was Frank Culbertson. There at night, alone with his thoughts, he pondered Culbertson for hours. Culbertson was a nice man, everyone agreed. But his incompetence, Blaha felt, was startling. Culbertson seemed to float above the fray, paying far more attention to George Abbey than his own astronauts. "If I was Frank Culbertson's boss," Blaha began saying, "I would put him in jail."

Korzun and Kaleri saw what Blaha was going through. "The first sign John was in a depressive state was he didn't have a desire to speak. When we saw this, we tried to get him out of this state. We spoke to him about things that had nothing to do with space. We spoke about [life on] the ground, about our childhoods; we found subjects that were dear to him. He spoke about his family. We tried to help him do his work. John always offered to help us, but since we saw the state he was in, we gave him more free time, to watch movies and [NASA videotapes of] baseball and football games. When we realized he liked the amateur radio, we worked to give him more time on that." Adds Kaleri, "We tried to calm him down by telling him a lot of other people had been through things like this."

Lying awake at night, Blaha began repeating a single thought, mantra-like. *John, this is the environment you're in. You used to love space. You sparkled in space. And now, whatever's going on, you need to accept this. Valery and Sasha are the two human beings in your life now. The ground doesn't matter. You need to accept this till the shuttle can come.*

Bit by bit, day by day, he came out of it. He started a new routine that conserved his energy and improved his spirits. Every morning after breakfast he began talking on the ham radio in base block, chatting with American amateurs in snippets of a few minutes apiece; Mir moved so quickly across the surface of the Earth it was difficult to maintain a longer signal. At night he tried to finish work at eight and watch a movie. His favorite tapes were old Super Bowls and Dallas Cowboy football games, all of which Al Holland and the NASA psychological support team had sent to the station for him.

Kaleri and Korzun realized the worst had passed one evening when Blaha lingered at the dinner table while the Russians took turns exercising on the treadmill. Up till that point Blaha had never bothered to eat meals with the two Russians, sticking instead to his shuttle-like regimen of eating when he could. "He didn't talk to us, he just worked," remembers

Kaleri. "For me the first sign he was changing to our lifestyle was on this evening. He didn't have dinner without us. At first we kept on exercising. We said, 'John, go ahead, eat.' He said, 'No, I'll wait.' And he ate with us! From that moment on, it was a totally different life for John. We discovered John was an entirely different person. He liked to talk! We started communicating with him. It was wonderful."

By the time he emerged from the worst of his funk, Caasi Moore was gone. In fact, although Blaha didn't know it, Moore had been a lame duck even before his astronaut reached the station. Moore's work during the last weeks of Lucid's mission had prompted a number of complaints to his superiors back in Houston. "There was a lot of discomfort with the ways I was doing business," Moore acknowledges. "Finally, I told my bosses, 'If I'm not doing things the way you think they ought to be done, get someone else over here." And so they did. Moore stayed for the first thirty days of Blaha's mission, at which point he was temporarily replaced by Keith Zimmerman, the twenty-nine-year-old ops lead who was scheduled to work Mike Foale's mission the following May.

Blaha was so divorced from his ground team that at first he barely noticed Moore's departure. "I didn't even care," he recalls. "It just didn't matter to me." What he did notice was the steady procession of replacement ops leads that rotated through the TsUP after Zimmerman; by the end of his mission, he had worked with seven different mission leaders, none of whom had ever attempted the job before. Blaha couldn't keep them straight and after a while stopped trying. On Halloween the ground team donned masks for a videotaped greeting for Blaha. When they took off the masks, Blaha realized he didn't know a single face—except for the ever reliable John McBrine.

The constant turnover created a tense atmosphere in which each new ops lead was forced to learn on the job. Each new man had little idea what Blaha had accomplished to date, and in several cases wrote up time lines directing him to repeat work he had already done. The defining incident of Blaha's mission involved his search through the crowded corridors of Mir for a NASA machine called a SAMS calibration device, used in calibrating the softball-size sensors placed all around the station to study the vibrations and structural stress Mir encountered in space. Blaha stayed up past midnight several nights in a vain attempt to find the device in the station's cluttered lockers.

"I don't understand why this is taking him so long," one of the Lockheed Martin engineers, Bob Hoyt, griped to McBrine. "How bad can it really be up there?"

"It could take him a week to find that thing," said McBrine.

Hoyt looked skeptical. "I could inventory that whole Priroda module in four hours."

"You're crazy."

"John's just lazy, man," said Hoyt.

Blaha repeatedly told the ground he couldn't find the calibrator.

"Oh, you're gonna have a good time with Blaha today," a Lockheed Martin man named Ed Bowers told McBrine when he arrived for work a few days later.

"Why?"

"He's mad as hell."

"Why?"

"We asked him for the calibration device again."

The request had come from the newest replacement ops lead, Jerry Linenger's man Tony Sang, who had no idea that Blaha had been looking for the device for weeks. McBrine lost his temper.

"What the fuck is going on with you people!" he said, standing. Everyone in the room—Sang, another NASA man, and five Lockheed Martin people—stared at McBrine in amazement. This was not the way NASA people behaved.

"Jesus Christ," McBrine went on, "he told you over and over he can't find it. He can't find it! Just drop the damn issue!"

By the time Linenger arrived at the station on January 15, Blaha was exhausted. He launched into lengthy diatribes about the incompetence of his ground team and of Phase One management in general. "He was hurting," Linenger recalls. "He was, in essence, depressed."

When Blaha returned to Earth on January 22, his two-and-a-half-year involvement in the Phase One program was at an end. It was something, he told every friend he met in coming months, he would never, ever try again.

7

Thursday, February 13
10:50 A.M.
Aboard Mir

"Great docking last night," Linenger is saying the morning after Tsibliyev, Lazutkin, and Ewald arrive at the station. *"A bit of excitement when we saw them backing back out. But then almost immediately they came back in manually and you could feel a pretty good jolt when they came in. So all in all it went smoothly. And it's good to see the new crew."*

"Yep," Christine Chiodo replies. *"Sounds like you've got a full house up there."* Chiodo, an assistant on Linenger's ground team, is talking through a headset microphone as she sits at the NASA console on the floor of the TsUP.

"Lots of people, lots of activity, and it seems like a different kind of place. Not so peaceful anymore."

"Oh, and I just got word from the back about the hat you wanted for Valentine's Day. Kathryn's going to get the hat. Everything's all squared away."

"I can't see anything else on my list. So, how about more news?"

"Well, Tom [Marshburn] told us that you don't want to hear anything about O. J. [Simpson], so we won't tell you about that."

"He's guilty."

"He's going to be paying more money, it looks like."

"Is he in the poor house?" Linenger asks. *"Or is he going to wriggle his way out with bankruptcy? He'll probably come up with some way to keep his money."*

"Well, it depends who you ask, I think. But I read in the paper that Fred Goldman has offered to give up any rights to the money if O. J. will just confess."

"Oh, interesting," says Linenger. *"So how's everyone there in the TsUP? Tell me about some of their adventures."*

"Well, I guess the latest adventure was last night, a sort of combined farewell bachelor party for Joe Height."

"So that's why Tony [Sang] and Tom took the late shift!"

"Well, I'll let you draw your own conclusions."

"I just did, and I'm sure I'm right."

"Yeah," Chiodo says, *"at least 99 percent."*

With the addition of its three new crew members, Mir is now a very crowded house; until Korzun and Kaleri return to Earth with Ewald on March 2, six men will live aboard the station. While Linenger beds down in Spektr and Korzun and Kaleri remain in the *kayutkas,* Tsibliyev and Lazutkin throw down sleeping bags where they can, as does Ewald. It takes a while for the new crew to adapt to weightlessness, Linenger notices. Ewald is always bumping into people in midair. But the one having the most problems is Lazutkin. He has already started vomiting. Spacesickness hits about half the people who reach Mir. Lazutkin, in fact, will endure severe headaches and periodic vomiting for much of his first two weeks aboard the station. Only Tsibliyev acts as if he were born in space. Lazutkin watches in awe as his commander effortlessly performs tight little backflips in the middle of base block, never brushing a single wall.

Linenger's first month on the station has gone smoothly enough. While an overheated satellite transmitter has wreaked havoc on the quality of Mir's communications with the ground, he has generally avoided the kind of traumatic moments Blaha endured. He sleeps strapped batlike to a wall at the rear of the Spektr module. For the most part he gets along well with Valery Korzun, although the commander is, as Blaha warned, a micromanager. Korzun, for instance, had objected to Linenger strapping his sleeping bag to the wall, because it required the American to move several pieces of unused scientific equipment. Blaha had slept on the floor, Korzun said. Why couldn't he? But Linenger insisted, saying he didn't want to stow and unstow his sleeping bag every day, and Korzun relented.

The one thing that unnerves Linenger about Korzun is the commander's obvious concern with appearances. Unlike an American shuttle commander, Korzun has little freedom in running the station. His daily schedule, like all the cosmonauts', is dictated in detail by the TsUP; every repair he attempts, every piece of equipment he thinks about moving, every bag of garbage he wants to load onto one of the Progress cargo ships, every cable he connects or disconnects—all of it must be cleared ahead of time with the TsUP. Any breach of his duties will directly impact his flight bonus. Linenger first encountered Korzun's attitude when he remarked to

Korzun that large pools of condensation were forming in the docking area at the end of the Kristall module. "Shouldn't we mention that to the ground?" he asked.

"Let's not," said Korzun. "If we do, you know, they will certainly make us work on it. And we already worked on it so much."

Linenger attributed Korzun's response not to fear of hard work—the commander worked all the time anyway—but to concern over his bonus. "It's a terrible system, just terrible," Linenger would recall months later. "Making money gets in the way of them reporting the accurate version of events on the station." It was a system that would eventually place Korzun and Linenger in direct conflict.

Like his predecessors, Linenger spends his days in the American modules, Priroda and Spektr, running NASA experiments. He has more than twenty in all, everything from sleep studies to Earth observations to studying the behavior of an open flame. The centerpiece of his science program is an experiment called Liquid Metal Diffusion, or LMD, which is designed to study how crystals grow in orbit. To run LMD, Linenger uses a laptop computer that is connected to a device that looks like a microwave oven. The idea is to insert each of five canisters containing an isotope called indium into the oven and cook it at temperatures approaching 400 degrees; radiation detectors inside the LMD will study how a radioactive tracer infuses the indium as the canister is "cooked."

The experiment goes wrong from the first day. The computer should flash Linenger a message indicating when the canister is cooked, a process that should take about five days, at which point he is to insert a second canister. But at the five-day point, Linenger discovers, there is no message. He can't be sure if the experiment has run its course. He radios down to his ops lead, Tony Sang, for advice, and Sang consults with the experiment's designers back at the University of Alabama-Huntsville. The problem seems to be in the software, but no one can be sure. Linenger spends several long days working with the software code, setting and resetting the oven clocks, but in the end he can't be sure if he has fixed the problem. He goes ahead and cooks the second sample but is never certain the experiment has been run correctly. Eventually, after wrestling with the computer and the oven for several weeks, he manages to finish three samples. It is a long and frustrating process.

At night Linenger taps out letters to his infant son, John, on a laptop. At his flight surgeon Tom Marshburn's suggestion, NASA has begun posting them on the Internet. The letters are wholesome and paternal, and they cause more than a few cynics at JSC to roll their eyes; they read like something a future politician might write. "Curiosity," Linenger wrote in his first letter, on January 23, "is what got me on this space station. Oh, sure, I went to lots of schools, did pretty well in our great United States Navy, and went through all the mechanics of the interview and application process. But the basic trait of insatiable curiosity is what got me through all that. Space is a frontier. And I'm out here exploring! For five months. What a privilege!"

On February 5 he wrote of how the station vibrated when one of his crewmates began to exercise. "Sasha is running on the treadmill, medium pace," he wrote. "I know who it is, and what he's doing not by sight and sound, but by feel. I can feel him. Frequency about one hertz. The computer and I are going up and down right now. Feels similar to being in a rowboat, near the shore, after a ski boat has gone by. Gentle, but definite swaying. The whole 13-meter 'tube' I'm in is moving. The force Sasha imparts is absorbed by the station, and it sways, resonates. If he slows down or speeds up a bit, I'll feel nothing. A peaceful float. When Shannon Lucid was on-board she had to stop running at a given pace because the station would resonate at a dangerous level."

For the most part Linenger's mood remains good; he is still in his honeymoon period. But already Marshburn has seen glimpses of the kind of egocentric behavior he had noticed at Star City. On February 7 Linenger, Korzun, and Sasha Kaleri crammed themselves into their tiny Soyuz and flew it from the Kvant docking port around to dock at the node, in order to allow Tsibliyev's crew to dock at Kvant. It was an exciting moment for Linenger, who had never flown in a Soyuz; he described it in a letter to John as "an afternoon spin in a spaceship." By the time Linenger returned to the station, NASA engineers in the TsUP were pestering Marshburn to find out whether Linenger had finished one of his experiments. Linenger had barely returned to the station, and Marshburn knew it wasn't the best time to ask. But he did it anyway. Linenger's simple reply cuts right to Marshburn's heart.

"Tom, I just did a fly-around," the astronaut said on open comm. *"I guess you didn't know that."*

It was a careless response, the kind of little sarcastic remark an overworked husband might use to put down a wife who wants him to clean the house after a long day. But to Marshburn, who had literally carried Linenger's bags for nine months at Star City, who had lugged water jugs to the gymnasium for him, who had defended him from the occasional sniping of other astronauts, it stung. "That just killed me," Marshburn remembered months later. "I felt like I'd lost his trust. I realized then that I was getting too attached, both to him and the mission."

Afterward Marshburn settled into a funk from which it took him some time to emerge. Only the gentle ribbing of his colleagues nudged him out of it. "So, Tom," a grinning Christine Chiodo piped up one day as she passed him in the hall. "Did you know we're doing a fly-around?"

Wednesday, February 19
Moscow

Congressman James Sensenbrenner is striding through the cavernous Khrunichev rocket factory, trailed by a jogging phalanx of American and Russian television crews. With all the drama of a politician who knows he is on camera, Sensenbrenner extends his right arm and points to a huge piece of misshapen steel. It looks like an unfinished submarine. This, Sensenbrenner tells the crowd, is the Service Module that forms the core of the International Space Station.

"From what I have been able to see," he says, "there has not been too much work done on this Service Module in the last year."

Sensenbrenner is a pink-faced Republican from the Milwaukee suburbs, a fifty-three-year-old millionaire attorney with a thinning white comb-over and large, rolling eyes. As chairman of the House Science Committee, he has been hammering NASA and the White House for almost four years on the Russians' inability to meet their financial commitments for building their part of the ISS. He is forever dragging the NASA Administrator Dan Goldin into hearings and haranguing him in front of the cameras. Inside NASA Sensenbrenner is considered a very powerful clown. At the TsUP, where the ground controllers call him "the Saint Bernard," the mere mention of Sensenbrenner's name causes the Russians to grimace and shake their heads.

As the cameras roll, Sensenbrenner is just getting started. "The whole reason there is a crisis today—"

To the congressman's side, the head of the Russian factory, a man who looks like an undertaker attending his own funeral, frantically tries to interrupt. Sweat is actually running in streams down the Russian's forehead and temples, drenching his stringy black hair. The cameraman from MSNBC moves into a tight focus on the perspiration.

"Excuse me, sir," says Sensenbrenner, barging onward. "The whole reason there is a crisis today is because your firm did not meet the schedule Vice President Al Gore and the Russian premier Viktor Chernomyrdin agreed to last July." He is referring to the Russian government's latest $100 million emergency financing of the troubled station.

The factory boss speaks in rapid Russian. "I am not the president of Russia," his interpreter translates. "I am the president of my company. I am not responsible for Russia."

"Nobody is accepting the responsibility for these delays," Sensenbrenner says, moving in for the kill. "But the delays are happening. They are jeopardizing this very important project."

Afterward Sensenbrenner walks out triumphantly, knowing he has achieved good television. "Obviously," he tells the reporter from MSNBC, "we have to make the final determination on the role Russia will play by the end of March; otherwise the program will slip so badly the whole program will unravel."

Thanks in part to Sensenbrenner's relentless prodding, tensions between the American and Russian space programs are running high. It's not just that the Russians have fallen behind on their funding commitments for the International Space Station. To raise the money they need to finish their modules, Yuri Koptev's Russian Space Agency is poised to unveil a new line of credit with a group of Russian banks. Sensenbrenner and others worry aloud that some of the banks may be Mafia connected, which forces NASA to dispatch a team to Moscow to analyze them. Three months earlier the veteran Russian commander Anatoli Solovyov had quit his position on the first ISS crew when an American, Bill Shepherd, was named commander over him; his resignation capped months of behind-the-scenes maneuvering, during which the Russian Duma got involved, passing a resolution that no Russian could report to a foreigner in space.

No one knows it now, but the two countries are about to have far more to worry about than money.

Sunday, February 23
Aboard Mir

Sunday is a Russian holiday, Army Day, celebrating the armed forces. On Earth it is a time for concerts and small parades. On the station everyone is working at half speed.

"Congratulations on the holiday," the morning communications officer tells Korzun. *"We want you to rest more and work less."*

Korzun smiles. This is unlikely. *"Congratulations back,"* the commander replies. *"We are going to have a military parade on the station today. The equipment will be the leader of the parade. The [air conditioner] will be first in line. All the accumulated batteries will follow it."*

After Korzun and Kaleri go over some housekeeping duties with the ground, both Linenger and Tsibliyev are offered brief private sessions talking with their families. This is already a sore subject with Linenger. He had tried on Saturday to talk with Kathryn, but his conversation was constantly being interrupted by a simultaneous chat Reinhold Ewald was having with his family on the station's second channel.

"We have a suggestion for the future," Korzun tells the ground, later that morning. *"Especially for Jerry and the foreign representatives, if the family talks are to be done, you shouldn't switch on to other [channels]. It's annoying. . . . They can get very upset about it. Please, let's be more careful about this next time. It's better not to have any talks with family at all than to have them like this."*

Lazutkin gets a quick talk with his wife during the 12:55 comm pass, while Tsibliyev tries to reach his wife again during the 2:29 pass.

"Vasya," the ground tells him. *"Larissa is not at home."*

"It's not urgent," Tsibliyev says.

"How's Reinhold?" the TsUP asks Korzun a bit later.

"He's as busy as a bee," replies the commander.

Sunday, in fact, is Ewald's first day off since arriving at the station eleven days earlier. He spends it floating through the modules filming a self-guided tour of Mir that he hopes will be broadcast on German television. He pins up drawings from German schoolchildren all over the station and makes sure his camera films every one.

"Do you want to go home?" the ground asks Korzun.

The commander sighs. *"On the one hand I want to go home,"* he says. *"On the other hand I don't want to leave the station, because at home everything is the same. And here everything is so unusual."*

At the end of the 5:40 P.M. comm pass, Korzun and the Russian crew members send down congratulations to the Patriarch of the Russian Orthodox Church. He is sixty-eight.

Late afternoon
Moscow

Three years into the Shuttle-Mir program, the smallest bureaucratic snafus remain capable of knocking key personnel out of action. Instead of readying himself for the coming week, Marshburn spends Sunday afternoon packing an overnight bag and saying quick good-byes to NASA friends in Moscow. His year-old visa is expiring. Though he had applied for a renewal the previous April, a full nine months earlier, the Russian government had never acted on his request. As Marshburn's expiration date approached, NASA was able to arrange for him to receive a limited, single-entry visa that would allow him to stay on in Moscow. There is just one hitch. To activate it he has to leave the country, then reenter. Irked, Marshburn checked flight schedules for London and other cities, but found the flights too expensive for NASA's taste. Instead he is planning to board a 6 P.M. train to Helsinki, Finland; he will take the twelve-hour trip across the border, walk around the Finnish capital for six hours, then board a return train that will bring him back to Moscow around noon on Tuesday.

That afternoon Marshburn telephones Terry Taddeo, Mike Foale's flight surgeon. Taddeo, who is at Star City training with Foale, is Marshburn's replacement should any health issues arise in his absence, and the two men spend several minutes reviewing the progress of Linenger's first five weeks in orbit.

"I don't anticipate anything will come up," says Marshburn. "They probably won't call unless a fire breaks out or something like that."

Both men laugh.

10:00 P.M.
Aboard Mir

Six men makes for a crowded dinner table, so most evenings the two crews eat in informal shifts, sucking from their tea packets and gobbling their food out of midair when they can. But this being Army Day, Korzun and Kaleri have gotten everyone to muster together around the base block table for one big meal. They have brought out some of the best new food, including cheese and sausages and red caviar, and enjoyed something approaching a nice, long banquet. Tsibliyev turns on a Russian folk song. It has been a busy weekend, and everyone is happy for a chance to relax.

Afterward they linger around the table to chat and unwind before bedding down—all except Linenger, who heads down to Spektr to begin covering his upper body with electrodes for a sleep study. Ewald and Korzun hover above their seats facing the opening to the Kvant module, across from Tsibliyev and Kaleri. Every ship that comes up brings dozens of souvenirs—postcards, posters, photos, and envelopes—and Kaleri is busy imprinting a small stack of items with a stamp authorizing that each has been aboard Mir. It is a trivial pastime, but all the cosmonauts know that those left behind on Earth treasure the items that have spent time in space.

Lazutkin flies gently past the table and eases himself through the hatchway into the tiny, darkened Kvant module. As the junior member of the crew, he has had a tough time adapting to his first spaceflight. From his first hours in orbit Lazutkin has been vomiting more or less constantly. He is nauseous for long periods every day and is beginning to wonder when, if ever, he will adapt. The others keep telling him to be patient. Things will get better with time.

Ducking into Kvant, Lazutkin opens a silvery side panel and slides out a device that looks a little like a scuba tank. The station's two Elektron units can't create enough air for all six crew members, so during changeover periods the crew generates extra oxygen with this backup system, which the Americans call the solid-fuel oxygen generator, or SFOG. To use it, Lazutkin reaches into an overhead bag and takes out a foot-long gray cylinder that looks like an overgrown saltshaker. The cylinder contains a solid chemical called lithium perchlorate, which when heated generates oxygen; a similar system has been used aboard Russian submarines for decades. Technically the cylinders are called "cassettes," but all the cosmonauts call them candles. The SFOG eats three candles a day, and it is the crew's custom to load the last one before bedtime. "Time to burn a candle" someone will say when it is time to reload a cassette. Lazutkin doesn't know it now, but it is a fateful choice of words.

Loading a candle into the SFOG is simple, a process Lazutkin likens to jamming a mullet ball into an old-fashioned musket. Taking out the cassette, he hovers above the SFOG, unlatches the top, and shoves in the gray cylinder, which he feels set nicely into place. Relatching the top, he turns a red star-shaped dial on the container's outer wall to activate it. After several moments the scent of fresh oxygen fills his nostrils, and he realizes the device is working properly. He hovers above it a second, just to make sure, waiting for the rush of warm air he expects. He spreads his hands over the container, and within moments he feels the warmth.

"I was ready to fly back," he remembers. "It was the normal procedure. No one ever worries whether it is working."

Suddenly he hears an unfamiliar sound, a quiet hissing. It sounds like an old babushka shushing a newborn.

SSSHHHHH . . .

Lazutkin's face hovers barely eighteen inches above the top of the container. He peers at it intently. And then, suddenly, sparks fly out of the top. Before he can say or do anything, a small column of orange-pink flame shoots from the container. He flinches.

For a long moment he stares at the flame as if it is a dream. This is not supposed to be happening. Lazutkin is strangely calm. It is as if he were in a laboratory and one of his experiments has suddenly gone awry. The flame, he thinks, looks like "a small, tiny, baby volcano." He estimates its initial length at between nine and twelve inches. It has the angry roar of a blowtorch. "I was just watching," he remembers, "thinking: This is unusual. It shouldn't be doing this. Why is it doing this? My first idea was, [that] I did something wrong."

Too stunned at first to say anything aloud, Lazutkin fights the powerful impulse to grab the fiery container. He cannot grasp what is happening. He thinks to shout out a warning but is possessed by the strange idea that he doesn't want to frighten anyone. Instead, very quietly, almost to himself, he says aloud, "Guys, we have a fire."

No one hears him.

* * *

Hovering at the base-block table about ten feet from where Lazutkin is marveling at the "baby volcano" he has somehow created, Reinhold Ewald is the first to react. "I saw flame spitting out of the device, literally into Sasha's hand," he remembers.

"Pozhar," Ewald says, mouthing the Russian word for fire.

At first Tsibliyev, who is in the air just across the table from Ewald, his back to Kvant, doesn't think anyone has heard the German's words. "I see Ewald's face, I read his lips, he says the word so softly, I didn't think anyone hears him," Tsibliyev remembers.

Turning, Tsibliyev sees the fire erupting in front of Lazutkin and repeats the word, this time loudly: *"Pozhar!"*

"I said, *'Pozhar,'* but I didn't think anyone believed me," Tsibliyev recalls.

Valery Korzun does. Korzun, hovering above and to Ewald's right, cannot at first see into Kvant. Lowering his head to peer inside, he instantly sees flashes of bright orange and white flame erupting all around Lazutkin. In a split second, he pushes off from a side wall and flies across the table, cutting through a gap between Tsibliyev and Kaleri. In moments he is past the toilet entrance and into Kvant.

"Pozhar! Pozhar!" Korzun hollers as he passes. Smoke, grayish and white, is already enveloping Lazutkin.

"Korzun flew in like this giant hawk," Lazutkin remembers with a smile. It was so like the commander, the strapping, macho Cossack coming to his smaller friend's rescue. As Korzun settles at his side, Lazutkin thinks to reach out and switch off the red-hot canister, but it has no effect. The oxygen from the canister is obviously fueling the fire, creating the blowtorch effect. The flame is now shooting up into the open air in the center of the module, flashes of sharp red and pink, at a 45 degree angle in front of him. He cannot be certain, but it seems to be nearly two feet long and growing.

Lazutkin jerks a wet towel from a holder on the wall and throws it onto the flame, which instantly engulfs it. Flaming bits of towel swirl up and around the module. Lazutkin ducks back, fearing his hair will catch on fire. Korzun, hovering at his side, immediately realizes the flame is too big to be smothered.

"Get the fire extinguishers!" he says.

* * *

Fire in a zero-gravity environment is not something human beings know much about. Both Linenger and Shannon Lucid, in fact, ran experiments in which they observed an open flame in a self-contained glovebox. It is gravity that causes a flame on Earth to flicker upward; in zero gravity, fire expands in all directions at the same speed, creating a flame that looks like a burning ball. The fire that erupted in front of Sasha Lazutkin looked nothing like a ball, however. Oxygen roaring out of the SFOG sent it shooting outward much as it would on Earth.

Everyone in base block is startled by Korzun's sudden call for fire extinguishers. Hovering at the dinner table, his back to the fire, Sasha Kaleri turns to see it and immediately realizes what is happening. To him the fire appears a reddish shining in the air; he sees sparks cascading through the module around Lazutkin. Much like Lazutkin, he resists a powerful impulse to leap into the module and attempt to smother the fire. "Two people are already a crowd," he remembers thinking. When Korzun calls out for a fire extinguisher, Kaleri has a small problem: The postcards and envelopes in his hands. As fast as he can, he jams several into niches beneath the table and others into a nearby sack.

Sitting beside Kaleri, Tsibliyev, who served on the fire brigade at his Crimean grade school, doesn't need to be told to grab fire extinguishers. There are two attached to the walls in base block, and the moment Korzun soars into Kvant, Tsibliyev flies over and grabs one. The other he reaches just as Kaleri tears it from its holder. Kaleri takes one of the fire extinguishers and passes it through the hatch to Lazutkin, who quickly passes it to Korzun.

"Sasha, quickly, leave the module!" Korzun barks. Lazutkin dips his head and propels himself through the airway into base block.

As Korzun turns back toward the fire, he sees glowing bits of molten metal and other flaming particles floating out toward him. The fire is growing larger every second, its outer edges flicking toward the far wall of the module. No one has to tell Korzun what will happen if the fire somehow burns through wall panels and pierces the hull: They will all die in minutes as the station's atmosphere whistles through the hole.

Smoke begins to sting his eyes. The fire extinguisher in his hands has two settings, one for foam, the other for water. Korzun switches on the foam, and as the smoke grows thicker and darker around him, he points the extinguisher at the flame.

Nothing.

Nothing is coming out of the extinguisher. "At first I thought neither foam or water was coming out," Korzun remembers. "I thought it was just gas. I couldn't tell what was happening, because it was so dark."

Unsure whether the fire extinguisher is working, he drops it. It floats off into the gathering murk. The smoke is growing thicker. He realizes that he needs an oxygen mask. Turning, he ducks and propels himself out of the module.

"Everyone to the oxygen masks!" Korzun shouts.

All five of the cosmonauts tumble toward the far end of base block in a chaotic tangle of arms and legs. Russian curse words—"Shit! Damn!"—accompany the flying scrum. Lazutkin, streaking past the others, is the first to reach a mask. He doesn't put it on, thinking he won't need it.

Korzun's order to don oxygen masks takes Kaleri by surprise. The flight engineer has assumed the fire is already under control. He lunges toward the far end of base block, followed by Korzun, who reaches his mask in two or three seconds; later the commander will not remember retrieving or donning the mask.

"Where's Jerry?" Korzun asks. Someone says he is in Spektr. "Bring him in here!" Korzun says, springing back toward Kvant. "We all need to be together. Okay. Now, everyone travel in pairs!" In laying out firefighting practices, the Russian trainers at Star City have emphasized how crucial it is to travel in pairs. On Earth, someone who faints or is overcome by smoke will keel over, presumably hitting the ground and prompting those nearby to rush to the rescue. In microgravity, an unconscious person will simply float in space, motionless; unless someone is hovering alongside, you may never know that individual is in trouble.

"Sasha!" Korzun shouts to Lazutkin. "Prepare the ship!"

Korzun's order is for Lazutkin to prepare one of the two Soyuz escape craft for evacuation. Lazutkin immediately swims off toward the node, where the Soyuz that he, Tsibliyev, and Linenger would use to evacuate the station is docked. There is just one problem: The Soyuz reserved for Korzun, Kaleri, and Ewald is located at the end of Kvant, on the far side of the steadily growing blowtorch in the middle of the module. Simply put, there is no way to get to the Soyuz without putting out the fire. As Korzun recrosses the dinner table with a second fire extinguisher, he sees thick black smoke beginning to pour out of Kvant into base block.

* * *

It is at about this time, as the five cosmonauts in base block are scrambling for their oxygen masks, that the station's fire alarm—a loud, piercing buzzer—finally goes off. According to Kaleri, the nearest sensor to the fire is located near the node; the alarm does not go off until the first wisps of smoke cross the length of base block and approach the node. The alarm triggers an automatic shutdown of the station's thundering ventilation system; this is intended to prevent the system from blowing smoke into the other modules. In the event, it is only partially successful. Smoke is soon pouring into base block.

The alarm jars Linenger down in Spektr, where he has already strapped himself to the wall in anticipation of sleep. He is midway through another letter to his son, John, when the alarm goes off. In a flash he untangles his legs from the bungee cords securing him to the wall, flies down the length of Spektr and into the node, where he runs headlong into Tsibliyev and Ewald, who confirm that there is, in fact, a fire in Kvant.

"Is it serious?" Linenger asks in Russian.

"Seryozny!" someone answers. "It's serious! It's serious!"

Crawling through the node, Ewald slices away from Linenger into Kristall, where there is a container of oxygen masks he is familiar with. The Russian oxygen mask works on the same principle as the SFOG, using a chemical reaction to create a flow of oxygen across the wearer's mouth. Ewald pulls the ring atop a circular container and lifts out the topmost mask, then straps the mask across his face. It covers his mouth, nose, and eyes, protecting him from smoke inhalation. Attached to the bottom of the mask is an oxygen bottle. Flipping a switch on the container releases a breath or two of oxygen. To activate the full flow of oxygen, Ewald takes several quick breaths; the humidity from his breath is supposed to activate the oxygen flow. But as Ewald pants into the mask, he realizes nothing is happening. There is no air flow. The mask, like the "candle" spouting fire back in Kvant, should feel warm if the proper reaction has occurred. Ewald's mask stays cold.

Without thinking, he grabs for a second mask. "At a time like this, you don't argue with the device," Ewald recalled months later. The second mask works. In seconds he feels a warm flow of oxygen across his mouth and nose. He turns and flies back into base block, where he is immediately met by an ominous sight. Thick black smoke is quickly filling the module. It has already shrouded the table where he was sitting moments before.

Through the gathering murk he can just make out Korzun fighting the fire in Kvant. Of the fire itself, all he can see through the smoke is a yellow glow.

Linenger too experiences problems with his oxygen mask. It fits onto his head but won't fill up with oxygen. Smoke is already entering the node as Linenger fiddles with his mask, trying to make it work. He holds his breath for several long moments, then grabs for a second mask, flinging the other aside. Tsibliyev, who has easily donned his own mask, watches as he takes several quick breaths and, to his relief, finds the second mask works as planned.

Leaving the node, Tsibliyev takes Linenger into Priroda to fetch the fire extinguishers there. Linenger grabs for one but is startled to find it is secured to the wall.

"It won't come off," he says to Tsibliyev, who has found the second extinguisher will not come loose either. Both men give the extinguishers a quick tug. Nothing.

Months later, NASA officials analyzing the fire will be deeply disturbed by this incident. The problem of immovable fire extinguishers in Priroda will even be raised in a congressional hearing by the NASA inspector general as evidence that Mir is unsafe. In fact, according to Korzun, the problem was a simple but dangerous oversight. When Priroda was blasted into space and delivered to dock with Mir in 1996, its fire extinguishers were secured by transport straps. For some reason, none of the crews who worked aboard Mir in the intervening nineteen months ever released the straps.

This oversight effectively disables the two extinguishers Tsibliyev and Linenger have their hands on.

"Jerry wanted to talk, to ask me about it, he was saying, 'What? How?' " remembers Tsibliyev. "I said, 'We don't have time to discuss it. Drop it. Let's go to Kvant 2 and get 'em there.' "

Shooting quickly back through the node into Kvant 2, Tsibliyev grabs one of the two fire extinguishers there and hands it to Linenger.

"Give it to Korzun," he says.

The second fire extinguisher Tsibliyev leaves on the wall. Training rules dictate always leaving one behind, in case a fire should break out in the module. Linenger takes the extinguisher from Tsibliyev and ricochets back into base block, where he is met by the sight of thick black smoke pouring into the module from Kvant. Handing the extinguisher to Kaleri, he cannot see his hand in front of his face. Tsibliyev follows right behind and immediately hears Korzun shouting for more fire extinguishers. Tsibliyev and Linenger turn around and fly quickly to Kristall, where they recover one more extinguisher, which they hand to Kaleri in base block.

Lazutkin, meanwhile, hearing Korzun's order to prepare for emergency evacuation, heads through the node and into the Soyuz in which he, Tsibliyev, and Linenger would return to Earth. "We were like Pavlov's dogs," he remembers. "We have been trained to fulfill the [commander's] orders. If you have a command, don't think. Do it." Lazutkin hunches over and begins to detach the dozen or so cables draped across the entrance, including the six-inch-thick white ventilator tube.

"What's happening?" Tsibliyev asks Lazutkin after a moment.

Lazutkin glances back into base block. "It's completely dark," he says. "You can't breathe." Together the two men shut the capsule's door to prevent smoke from seeping in.

While his comrades swarm through other parts of the station, Korzun reenters Kvant to fight the fire, alone. He hovers directly by the near wall with base block, his feet sticking through the hatch into the crawlway between the two modules. Kvant is now totally dark, smothered in smoke. To Korzun, who cannot see his own hands, the flame itself appears only as a bright white glow, perhaps two feet long, beneath him in the murk. He takes the second fire extinguisher, turns on the foam, and shoots it in a stream at a point where he believes the fire is hottest. It's difficult to tell, given the visibility, but after thirty seconds or more he doesn't believe the foam is having any effect. The flame is shooting out so fiercely, it seems to be blowing the foam away. Korzun sees glowing particles of foam swirling around him in the dark. "It didn't seem to be working at all," he remembers. "The fire was too strong."

Korzun turns a knob on the fire extinguisher and switches his stream to water, spraying it all around the glowing white flame. He is struck by how eerie the situation is. With the ventilators shut off, the station has gone quiet. The only sound, other than the occasional muffled comments of Kaleri or one of the others behind him in base block, is the sound of the fire. It hisses at him —"like fried eggs in a frying pan," Korzun says later.

At first Korzun doesn't think the water is working either. He directs the stream toward the center of the hissing white glow but can't be certain it is hitting the flame. And then, after a minute or so, the fire extinguisher gives out. There is no more water. Korzun turns and slips back through the hatch into base block.

"I need more fire extinguishers!" he shouts.

Kaleri hands him the second unit from base block, and Korzun quickly ducks back into Kvant to face the fire.

None of the astronauts admit to any real fear as they fought the fire. "You trained with the Soyuz to believe you can escape in your Soyuz in all circumstances," says Ewald. "Return to Earth is assured in your mind. Even in the worst circumstances, in face masks, you think like that. You can come down. What I thought, after some seconds, after some action, I thought this would be the end of my two-week science mission. I wouldn't get any results from my science. [When I returned to Earth] I would get a big hug, a big clap on the shoulder, but the results would have been zero for my flight. This was the end of my mission." Ewald pauses in his recollection, pondering the loss of scientific work he had trained nearly two years to complete. "Sounds professional, right?" he asks.

But in the midst of Ewald's reverie, as he hovered amid the smoke in base block, a numbing realization struck. "Our Soyuz was on the far side of the fire," he recalls. "It was quite clear that we would have to go through the fire to our Soyuz." As Ewald hovers there in base block, now thoroughly filled with thick black smoke, he briefly considers making a dash for the second Soyuz, to begin powering it up for evacuation. But no sooner did the thought occur than Ewald banished it from his mind.

"I'm in the military hierarchy on board; I do as I'm told," Ewald remembers. "I'm not there to make up my mind and do things out of a heroic feeling. You do what the commander tells you to do. Even if I could have been some help, you do what the commander tells you, so I stayed [in base block], not because I was a coward, but because it is best to do things in an orderly way."

When Korzun returns to the fire with his third fire extinguisher, the glow beneath him seems somehow smaller. He begins blasting water directly at it. At about this time he hears Linenger behind him.

"How are you?" the American shouts through his mask. "Are you okay?"

Linenger is the crew's only doctor. It is his unofficial duty to check on Korzun's health, but the commander is unable to focus on what he is saying. He is too busy directing the stream of water.

"Yes, Jerry, I am fine!" he shouts back over his shoulder. "Stay there in base block!" After a few moments, Korzun adds some additional comments, ordering Linenger to keep a close eye on the crew. Smoke inhalation is a real danger.

It is at about this point, Korzun remembers, that the size of the white shining beneath him appears to be shrinking. He keeps the water on it, but inch by inch the flame appears to be dying out. Korzun doesn't remember making anything like an expression of victory or relief. His pulse is still racing, and he is breathing heavily inside his mask.

"Jerry!" Korzun hollers.

Linenger floats up to a position directly behind Korzun. "I need you to prepare help in case anyone is injured," the commander says. "Check all the American FA [first aid] boxes, see what they have in them and what can be used. Check all the Russian FA boxes, as well. See what's in those." Linenger immediately springs across the length of base block to the node, then turns and shimmies into Kvant 2, where the medical supplies are stored.

A moment later the shining beneath the commander seems to disappear. Korzun takes a deep breath but keeps the stream of water pointed downward.

"It's over," Korzun tells Kaleri. "I think it's over."

8

11:50 P.M.
Aboard Mir

The smoke is slowly dissipating in base block as all six crew members, still wearing their oxygen masks, gather for the next comm pass. The damage appears to be limited to the SFOG itself, which is destroyed, and the nearby wall panels, which are badly scorched. The hull is intact. Still, Korzun is unsure what to do. He needs guidance from the ground on whether it is safe to remain on board; no one is sure what gases have been released into the station's atmosphere. Both Soyuz escape craft stand ready for immediate evacuation. They are flying across North America now, moving into range of the radar station NASA operates at Wallops Island, Virginia. The quality of transmission over Wallops is never strong, but it is all they have. When they come into range, Korzun begins to speak into his headset microphone, describing what has happened.

There is no reply.

Korzun repeats himself, in Russian, then in English. On the other end of the line there is only static.

"TsUP, guys, we cannot hear you," he says, speaking in clear, deliberate sentences. *"We have had a fire aboard! We managed to extinguish the fire. The oxygen canister began burning. We extinguished it after using the third fire extinguisher. The crew is wearing oxygen masks. The pressure is O_2 equal to one-five-five. PCO_2 is equal to five-point-five. There are nine masks left. After we take off the used masks, if we feel worse, we will put a new set of masks on and go to the [Soyuz] capsule. We will be situated in the capsules. The smokiness of the atmosphere is below average. But we don't know the level of toxic gases."*

Finally there is an acknowledgment from the ground. *"I've got your information,"* says an unidentified voice.

"Any questions?" Korzun asks.

The comm breaks up again.

"Speak, Valera!" the ground says.

"We have ten minutes left before the masks we are wearing begin to finish," Korzun continues. The commander, in fact, has already removed his mask and taken several tentative breaths of air. *"Now we are taking the masks off and checking how we feel. If we feel worse, we'll put on a new set of masks. We have nine left. And we will go to the capsule. Then we will be waiting in the capsules, waiting for when the atmosphere will be cleaner, little by little. . . . The next communication will be at 4:16."*

Another voice gets on the line. *"We approve your plan,"* he says. *"Everything is right."*

"Okay, I understand," Korzun says. *"We'll be controlling the situation. The only problem is we are fighting CO_2. The capsules are ready for us to move into them. Meanwhile, I'm reporting, I have taken the mask off. Till now I feel normal."*

"Are all your filters turned on?"

"The filters, yes, the filters are switched on," replies Korzun.

"Did anything else burn?"

"No, it's normal now," says Korzun. *"Sasha Lazutkin is on duty [in Kvant]. When I left it fifteen minutes ago, it wasn't burning. But the flame was so big that the metal at the end of [the SFOG] melted. All of it. And a little bit of the interior [of the module] around it. The panels were not touched, and all the rest [is fine]. The only thing that it was, was the [SFOG]. We'll get the reserve canisters and in case of loss of pressure, we'll use reserve oxygen canisters in the base block."*

"We've got it," the ground says. *"Everything is all right."*

And with that, the pass is suddenly over. Korzun is nonplussed. What are they to do now? No one on the ground mentioned evacuation. For the moment the commander is unsure whether to head into the Soyuz capsules or not. The next pass, over the ground station at Petropavlovsk, on the Pacific peninsula of Kamchatka, is set for 4:16, about four hours hence.

"I guess we wait," Korzun tells Tsibliyev.

No one sleeps. Gradually the smoke clears, helped by the station's five different atmosphere-cleansing systems. At first everyone lingers in base block, discussing how best to proceed. Then the oxygen masks begin to run out.

"I'd like to go to a second round," says Linenger, his voice still muffled by the bulky mask.

"We can't," Korzun replies. "If we use them, we won't have any left."

By Korzun's count, they have used nine oxygen masks. Nine remain. Donning another six masks would leave only three. Russian guidelines mandate that the crew can remain on board only if there is at least one mask per crew member. If they go to another round, at least part of the crew will be forced to evacuate.

They decide to conserve the air in the masks as long as possible. Korzun orders everyone to be quiet and still in an effort to save oxygen. The smoke is thinnest in Kvant 2, so everyone but Korzun gathers there. Linenger sets up a first aid station, laying out tracheal tubes and a portable ventilator, and gives each crew member a thorough examination. By 2:00 all the oxygen masks have run out, and Linenger dispenses white 3M surgical masks for everyone to wear. The smoke has left a layer of grime throughout much of the station, and the crew spends most of the next two hours wiping and cleaning every surface. Around four, when everyone gathers in base block for the upcoming comm pass, Linenger dispenses soap packets he has prepared. Everyone washes up, then changes clothes and hands their T-shirts and shorts to Linenger, who stuffs them in a bag.

4:16 A.M.
Aboard Mir

"We can hear you well."

"Our situation is as follows," Korzun begins. He is floating in base block with everyone else. *"Everything has been normalized. The smoke has disappeared. It still smells of burning. The crew is wearing masks that prevent harmful gases from penetrating. . . . Medical examination of the crew has been conducted: pressure, pulse, lungs. The crew's condition is normal. . . . The oxygen pressure is one five five. In the future we will use canisters. We will use the [second SFOG] in base block, which is in reserve. We will observe the security measures while turning it on. But maybe you have some recommendations in terms of using it. Now, Vasya will speak about the condition of anti-fire devices on board, and we will answer all the questions you are interested in."*

"Okay, Valera," the TsUP replies. *"Do you think that the crew feels satisfactory?"*

"Yes, their condition is good. There have been no injuries. Everybody feels good. We don't [need to] waste any time with that. The doctor has conducted a full examination. Everything is under control."

"Also, we would like to receive from you the exact time and location of the fire."

"22:35 is the beginning of the fire," Korzun explains. *"The canister got on fire approximately one minute after the installation. Sasha Lazutkin controlled the activation. But the fire was so big and active that even the use of the fire extinguisher did not have practically any effect in the initial stage. It's good it was there. We used three fire extinguishers during the fire. And two were still left prepared for the future."*

"Vasily Vasilyvich, go ahead," the TsUP says.

"We used five fire extinguishers," reports Tsibliyev, *"out of which three were used completely, and two were prepared. Now five of them are still left in the complex. Nine oxygen masks were used."*

"Okay," the TsUP replies. *"How do you estimate the possibility of another fire now?"*

"Right now it's [fine]," says Korzun. *"But the reason why we asked for recommendations on canisters is because we don't understand the reaction. The fact is, there was some uncontrolled reaction during burning. The body of the canister was burned. . . . And even the metal was melting on the circular closing device. The temperature was that high. Now of course, while turning it on we will use fire extinguishers that are ready. And if there is any sign of a fire we will use extinguishers in the foam mode."*

"Guys, we haven't looked at this question from this point of view," the TsUP replies. *"Up until now, you don't use canisters until a special order is given."*

"Right," says Korzun. *"Here's what Sasha thinks. If the new canisters were stored on Earth for a long time, maybe it's better to use old ones that were stored in conditions of weightlessness."*

"What is the serial number? We have old canisters that were stored on board a long time."

Sasha Kaleri breaks in. *"No, we don't understand the reaction. Maybe it's some kind of redistribution of the density of the charge, or something like that. We have to look into the history of the storage."*

"Okay, we've got that. Did we understand you correctly, that the old one, the one that was stored a long time, got on fire?"

"Well," says Kaleri, *"they were from a container that's in [Kvant]. Behind the panels."* He reads off some serial numbers.

Korzun cuts in. *"Guys, we also have a question on the chemistry of this sub-*

stance. It didn't burn to the end, because water was put on it. Does it mean that there are no toxins there? And what happens when you put water on it?"

One of the senior Russian doctors, Igor Goncharov, gets on comm. *"Guys, we will give you the precise information on toxins later. And here's another thing. Please put on masks by all means."*

"We're using masks," says Korzun.

"Now, dairy products are recommended. Take more milk and curds."

"Yes, yes."

"You can take vitaron. Two capsules."

"Okay."

"You can also [take] carbolen. If you have headache symptoms and so on."

The doctors congratulate Korzun and the entire crew for a job well done and urge them to sleep. Everyone tries. No one sleeps soundly.

Monday, February 24
10:00 A.M.
TsUP

Twelve hours after the fire, Tony Sang steps off the NASA van in front of the TsUP with no hint of what transpired aboard Mir. He walks through security and heads to the NASA suite, where he spends forty-five minutes with his team going over questions and answers to submit to Linenger at the morning's first comm pass. At no time does Vladimir Solovyov or any of the Russians interrupt to tell him what happened.

Not until a matter of minutes before the comm pass, in fact, when Sang walks down to the floor, does he realize something is wrong. Everywhere, thronging the aisles, hunched over consoles, whispering urgently among themselves, he sees new faces. In Houston Sang worked in Life Sciences, and he immediately sees he knows many of the new men. They are the Russian life-support specialists. In an instant he realizes something is wrong.

Vladimir Solovyov stops him as he prepares to don his headphones. "There's a problem," the Russian says. "We need the time."

Sang starts to ask what happened, but the pass is already beginning.

He stands in silence and listens as Korzun's voice crackles over the speaker. Sang's Russian is rudimentary—he could find his way through a Moscow grocery store—but he can't understand what Korzun is saying. Then he hears a word he knows: *pozhar.* Fire.

He glances over at Solovyov and arches his eyebrows, as if to question what he heard. Solovyov nods. There was a fire. Sang picks up a phone at his console and speaks to one of NASA's translators, who is transcribing the pass up in the NASA suite.

"Has there been a fire?"

"Yes."

"What's the crew saying?"

When the pass ends, Sang manages to get a few details of what happened. He immediately returns to the NASA office, phones Culbertson at home in Houston, waking him from a deep sleep, and briefs him on what little he knew.

"Where is the crew?" Culbertson asks. "And are they okay?"

"Everyone's still on board," says Sang. "They're on gas masks."

"When did this happen?"

"Last night about 10:30."

"What?"

"Yeah, last night."

"And when did they tell you?"

"This morning after I got in."

Culbertson is floored. An American astronaut goes through the worst space fire in history, and NASA isn't told for more than twelve hours?

"Why did they wait til now to tell us?"

"I don't know, Frank."

News of the fire hits the Phase One office with the stunning force of a midnight tsunami. Nothing like this has ever happened before, and Culbertson, Jim Van Laak, and the rest of NASA are forced to admit they are wholly unprepared to deal with its implications. The Phase One office, after all, had been conceived as a minimalist science-support operation, not a mini–Mission Control. A central tenet of NASA's contract with the Russians is that NASA would guarantee the safety of cosmonauts aboard the shuttle, and the TsUP would guarantee the safety of astronauts aboard Mir. It simply wasn't feasible, Culbertson was always telling people, for NASA to "look over the Russians' shoulders" and second-guess their safety standards. To doubters Culbertson asked a simple question: Would NASA let Russians stand around Mission Control second-guessing shuttle flights? Of course not. But this arrangement, combined with the lack of any real safety scares during the first three Mir missions, had rendered Phase One complacent. For the first time since American astronauts began journeying to the station two years before, the fire woke Culbertson and Van Laak to the fact that *someone could actually get hurt up there.*

While NASA employed hundreds of engineers to pore over every detail of a routine shuttle launch, it had almost no structure in place to analyze Mir. There was a Russian-American working group responsible for Mir safety issues, but until the fire its NASA cochairman, a laconic fifty-six-year-old Apollo veteran named Gary Johnson, did not even consider the astronauts' safety *aboard* Mir part of his job description. "Whenever our astronauts were aboard the Mir, the Russians were responsible," Johnson recalls. "Our only safety analysis was for docking missions. Due to the ground rules established at the beginning, I myself did not closely follow the Mir operations after the shuttle left. [The Russians] didn't share a lot of information about the operation of their station or the status to it."

NASA, in fact, as Culbertson was forced to admit in the wake of the fire, knew next to nothing about Mir's inner workings. No one had attempted anything like a basic safety assessment of the station before the White House first announced the Shuttle-Mir missions in 1993. NASA's subsequent efforts to catalog Mir's internal systems had been piecemeal and slapdash. The first such effort came in late 1993 when a freelance technical writer named David Portree, having just finished authoring a NASA report on the dangers of orbital debris, volunteered to write a Mir guidebook for use in Houston. Given a NASA contract so small he couldn't afford a trip to Moscow, Portree discovered that NASA had almost no usable information in its files on Mir. Instead he culled his research from Russian newspaper articles, books, and from talks with outside experts on the Russian program. The only input Portree received from the Russians themselves was a single pair of one-hour interviews with cosmonaut Sergei Krikalev, who was training in Houston for his shuttle flight.

Not surprisingly, the report Portree produced in November 1994 was long on general descriptions of Mir and its various systems and agonizingly short on the kind of detailed technical information engineers and astronauts needed to assess its safety. "Sources for [this] document," Portree noted in his introduction, "were extremely limited." But despite its limitations, Portree's book was embraced by NASA engineers, who dubbed it "the bible"; during the early years of the program, it was NASA's principal source of information on the station. Portree himself cringed at the nickname. "We knew almost nothing," he remembers. "I hate to say it, but it's true."

NASA administrators realized this, and in late 1994, barely six months before Norm Thagard's liftoff, assembled a team of young engineers to initiate a more systematic study of the station. It was called MEAT, for Mir Environmental Assessment Team, and its members spent many frustrating months attempting to pry technical information out of the Russians. The first problem MEAT encountered was the fact that little in the Russian program was written down. There were virtually no technical manuals to consult, no books. The only place the Russians seemed to store information was in their heads. Typically there was a single seventy-year-old engineer who served as the resident expert on each Mir subsystem. These elderly engineers, schooled in the Soviet manner, jealously guarded their technical secrets, rarely sharing them even with their own subordinates. To them the idea of divulging technical information to a bunch of fresh-scrubbed young Americans was anathema. "Some of them thought we were spies," remembers Keith Zimmerman, who worked on MEAT before going on to be Mike Foale's ops lead. "The older Russians especially didn't like us. They didn't want to tell us anything."

When the Americans pressed for information, the Russian engineers' reply was invariably "Let me check with my superior." Sometimes this resulted in a trickle of data during second, third, and fourth meetings; more often it didn't. Sometimes a Russian would simply refuse to divulge any information at all, insisting it was one of Energia's "proprietary" secrets. Every now and then MEAT got lucky; a friendly engineer, usually from the TsUP's younger generation, would slip them a report or a schematic drawing with the whispered admonition, "You didn't get this from me." Tony Sang, another member of MEAT, encountered Russians who lectured him that his requests for schematics weren't covered under NASA's 1994 contract with the Russian Space Agency—and they weren't. The contract, in fact, only required the Russians to divulge information about failures aboard the station; as a result, when something did break down, the MEAT engineers swarmed it like barracudas drawn to raw meat. At other times they tried pumping trainers at Star City for information, only to find, to their dismay, that Star City and the TsUP talked so little that the trainers were sometimes teaching cosmonauts how to use Mir systems that had been dismantled years before.

NASA's ignorance of Mir had become apparent when the station encountered minor breakdowns during Lucid and Blaha's missions. The two Elektrons were forever conking out, and from time to time a greenish coolant would seep from some of the station's internal pipes. Because the Phase One office had no technical analysts or safety experts of its own, Jim Van Laak had turned for help to the NASA division that did, the Mission Operations Directorate, MOD, whose eighteen flight directors oversaw all aspects of shuttle missions. And therein lay a problem. For three years MOD had repeatedly dodged all efforts to get its people more involved in the Shuttle-Mir missions. Its flight directors had consented to oversee the shuttle missions that docked with Mir, but when it came to helping out on anything involving an astronaut *aboard* the station, they resisted. To Culbertson and Van Laak's irritation, many of MOD's flight controllers simply didn't agree with the central thesis of the Shuttle-Mir program, that there was much to be learned from the Russians. Instead they focused their efforts on the International Space Station.

"There's always been a disconnect there," acknowledges Phil Engelauf, the senior shuttle flight director who served as one of MOD's point men with Phase One. "To be honest, I don't think watching the Russians launch another Progress, or do another EVA, is really going to make me all that smarter, especially for the resources required. I'm not going to learn anything materially new about spaceflight operations. I think we're pretty much on the flat part of that learning curve."

In fact, though MOD's warrenlike offices in Building Four sat just across the grassy lawn from Phase One's fifth-floor suite in Building One, the two groups were miles apart in philosophy. Part of the problem was NASA's structure. The agency is set up as what business experts call a matrix organization; that is, its divisions act as stand-alone businesses, keeping their own budgets. Among other things, the MOD bosses didn't want to spend their precious cash on activities like Phase One that they considered frivolous. "Van Laak wants free help" is how Engelauf puts it. "Every time something happens aboard Mir, Phase One comes to us and says, 'Why can't you prevent that from happening?' Phase One really wants MOD to get into the Russians' shorts. Our first reaction is, this is the Russians' spacecraft, we are guests. We have a guy aboard, and we'll support him, but there's very little we can do. We help them sometimes. If it's trivial, if it's easy, [our attitude is] 'Yeah, well, Jim, we'll come to your meeting.' But for anything else, we go talk to the bosses upstairs."

MOD's innate skepticism of Phase One was hardened by the kind of pointless bureaucratic infighting that sometimes mars NASA's effectiveness. During the 1990s MOD has undergone a series of wrenching budget cuts, and men like Engelauf looked askance at the resources lavished on Phase One. Engelauf considered Culbertson "a little empire builder." Worse, he was another empire-building astronaut. "Culbertson knows what he wants, and tries to seduce you into wanting it, too," says Engelauf. "Van Laak thinks only he knows the answers, and he just can't get the rest of you to understand it." Van Laak returns the sentiment. "You know the old saying, 'Your shit don't stink?' " he asks. "Well at NASA we say, 'Your excrement don't off-gas.' That's the way the flight controllers feel." To Van Laak, Engelauf was "a dinosaur standing in the way of progress."

It was against this poisoned backdrop that Van Laak had been seeking MOD's analytical help for more than a year. Six months before Linenger's flight, he had gone to Jack Knight, then head of MOD's systems division, and urged Knight to get his people more involved monitoring the Mir program. In Van Laak's mind there was a compelling reason to do so: Many of the Russian life-support systems aboard Mir, including the lithium perchlorate "candle" involved in the fire, were to be used aboard the International Space Station, the ISS. "I was trying to wake up MOD and other folks," says Van Laak. "I kept telling them, 'Come and see this hardware in action. Would you rather be reading some dry technical description or following this day to day?' The answer I got from Jack Knight was 'Love to do it, got no budget, leave me alone.' "

"Do you have any people following these Russian subsystems for ISS?" Van Laak had asked Knight.

"No."

"Are there are plans to move out in that direction?"

"No, that's a Russian concern."

Van Laak, to his lasting regret, had let the matter drop. But now he badly needed MOD's expertise. A few days after the fire, when he got his hands on downlinked video of damage inside the Kvant module, Van Laak showed it to Randy Stone, the head of MOD. Stone was alarmed. "We have to find out what's going on on the Russian side of this situation," Stone said. Sensing a slight change in the atmosphere, Van Laak got back to Jack Knight. "Jack, I know you don't have any people who are experts on this," Van Laak said, "but I need everything you've got."

Knight sent over a half-dozen MOD engineers who sat down in Van Laak's office, watched the fire video, and unanimously agreed something had to be done. The problem was, none of them knew the first thing about Russian systems. "I was taken aback by their lack of knowledge," Van Laak recalls. "For the first time everyone realized there was a gaping hole in MOD as to expertise on Russian systems. Having found that there were no clothes in the closet, there was initially a lot of rationalizing that 'oh, well, we don't need clothes anyway.' But of course we did."

What was needed, Van Laak realized, was a full-time NASA systems expert to monitor Mir in Moscow. When Knight resisted the idea of detailing someone from MOD, Van Laak said, "Jack, I'll buy it. Tell me how much it costs." Van Laak ultimately paid MOD about $1.5 million—roughly half of Phase One's cash reserves—to hire three young engineers to rotate through the TsUP to monitor Mir systems. The first engineer wouldn't arrive in Moscow until May, a situation that wouldn't help the gradually overwhelmed NASA team at the TsUP in the troubling months ahead.

The blinders NASA donned toward Mir safety prevented it from heeding several loud warnings about the risk of fire aboard the station. While the Americans had little experience with fire in space—there had been one "microfire" aboard a shuttle in the early 1980s, the result of a short-circuited cable—Russian space stations had suffered several serious fires over the years, few details of which were ever disclosed publicly, much less to NASA. A fire aboard Salyut 6 in 1978 churned out enough thick black smoke that the station was nearly evacuated. Several European astronauts who had flown aboard Mir had returned to Earth surprised by the Russians' relative lack of concern toward fires. Little of the station's electrical wiring, for instance, was protected by fireproof housing.

"They don't see the need for such precautions," the French astronaut Jean-Loup Chretien once told a NASA briefing. "After all, they've had several fires aboard their space stations and found . . . they're easy to put out, they're no big deal."

On October 15, 1994, an oxygen generator identical to the one that burst into flame in front of Lazutkin had caught fire. The fire was apparently smaller than the one Lazutkin ignited; a cosmonaut named Valery Polyakov managed to smother it with an extra uniform. As usual the TsUP blamed the cosmonauts, saying that the crew had forgotten to turn on the generator's exhaust fan, which had caused the unit to overheat and catch fire. At the time the TsUP made no public announcement of the incident, which was only disclosed by a Russian technical publication well after the fact. The Russians' silence was understandable. The 1994 fire occurred just five months before Norm Thagard's inaugural flight to Mir, and the Russians presumably wanted nothing disclosed that would spook the Americans.

James Oberg, the NASA orbital dynamics specialist who moonlighted as an expert consultant on the Russian space program, learned of the 1994 fire through personal contacts in Russia and asked several NASA safety officials, including Gary Johnson, whether they were concerned with Russian firefighting capabilities. Johnson, though sympathetic to Oberg's fears, told him there was nothing NASA could do; under the Phase One guidelines, safety aboard Mir was purely the Russians' concern. In a written response, the NASA public affairs office told Oberg: "After reviewing this information as well as information provided by the Russians about their on-board fire suppression and warning systems, NASA is satisfied with the safety and reliability of Russian hardware."

Oberg wasn't. Despite NASA's official confidence in the Russians, several agency officials confided to him that they weren't at all sure they were being told everything the Russians knew about the 1994 fire or other fires aboard previous stations. "The big deal is not the fires in space, but the smoke screens right here on Earth," Oberg wrote in a March 1995 *Space News* column on the eve of Thagard's flight. "Now, once again, with American lives quite possibly dependent on full disclosure of all safety issues, the time for incomplete information is long gone."

Johnson and other NASA safety officials agreed, but there was little they could do: Their hands were tied.

Aboard Mir

By Monday morning, twelve hours after the fire, Linenger is growing anxious to talk to someone on the ground. All six crew members are still wearing the 3M masks, and as the only doctor aboard, he has dozens of questions about what gases and particulates the fire may have released into the station's atmosphere. Benzene and hydrogen cyanide are real concerns. Was there asbestos in the oxygen canister or the surrounding panels? He has no idea. In fact, he needs to know everything he can about the SFOG system in order to gauge what its demolition means for Mir's air supply.

At midmorning Linenger begins asking Korzun for permission to talk to Tony Sang on the next comm pass. "Jerry, just wait, we'll get to you," Korzun tells him.

Linenger emphasizes the importance of getting immediate information on toxic materials from the ground. "You understand, Valery, we really need to do this now," he says.

One comm pass comes and goes and Linenger waits, getting antsy. He knows next to nothing about the masks through which they are breathing. Can they wear out? Could they actually be soaking up contaminants instead of keeping them out? He just doesn't know.

"I really need to talk on this one, Valery," he tells Korzun as the next pass approaches.

"Okay, Jerry, okay," Korzun says. "Next one. We'll get to you."

An hour later the next comm pass begins. Linenger hovers in base block as Korzun talks with the ground. He notices the commander isn't even discussing the fire or its aftereffects. With two minutes to go in the pass, in fact, Korzun is engaged in an animated discussion of the logistics of a press conference the TsUP is attempting to set up for Ewald with German television and print reporters. All at once Linenger realizes what is happening. Korzun and the TsUP don't want him to get on open comm and discuss the fire with reporters around to hear it. "It was very obvious the Russians didn't want me to talk about the fire," Linenger recalls. "It was as if someone had whispered up to Korzun that this was a touchy situation, and they didn't want the press to get involved."

Seeing the end of the comm pass approaching, Linenger finally loses patience.

"Hey, Valery, there's two minutes left. I need to talk *now,*" he says, the urgency in his voice clear.

"No, Jerry, just wait, we have to do this," Korzun says.

"No, Valery, now."

At this point Linenger activates his microphone, breaks into the comm channel and speaks directly to the ground. *"We need to talk about this now,"* he tells the ground.

"Jerry, just wait, we will get to you soon," the ground replies.

Korzun's face registers an expression of shock and amazement. It's a look Linenger has never seen before. Realizing he has crossed a line, Linenger signs off. The moment the comm pass is over, Korzun rips off his headset and angrily turns to the American. "Don't you *ever* do that again! What's the matter with you? You can't say that to the ground!"

Linenger tears off his headset and shouts back. "We *need* to talk about this now! I mean, what's going on? Why are we not talking about this? This is important! You should know this. You're the commander!"

"No! No! *I* am the commander. You cannot do this! Don't ever do that again!"

"It's ridiculous to be talking about some stupid press conference! We've got health problems up here! We're up here endangering ourselves, and you're talking about a press conference! I have to know more, I need to know this, I need to know that. We have to know what's in the air. We don't know anything!"

The intensity of Korzun's anger surprises Linenger. "We were really screaming at each other, real nose-to-nose stuff there for a minute or two," he remembers. Then Linenger breaks from the argument and floats down to Spektr. A moment later Korzun follows him into the module, and the argument continues, with the Russian commander repeatedly berating Linenger: "Don't you *ever* do that again!"

When Sasha Kaleri enters the module, Korzun leaves. "We all need to calm down," Kaleri tells Linenger. "I'll talk to him. He understands. He's just under a lot of stress."

It is at this point, just hours after the fire, that Linenger begins to suspect the Russians are attempting to cover up details of the fire, a feeling that solidifies later in the week when he finally sees the official Russian press release announcing it. The TsUP terms the roaring fire in Kvant a "microfire" that lasted ninety seconds, a figure that Viktor Blagov and other Russian officials will stick to for months. The ninety-second estimate will eventually emerge as a major point of contention between Linenger and almost everyone else involved. According to Linenger, all six crew members gathered after the fire and discussed its duration. All, he insists, agreed that the fire lasted fourteen minutes. It is a figure that Linenger will repeat in press interviews he gives in coming months.

The difference, of course, is more than semantic. A ninety-second fire is a minor event the Russians believe they can safely play down; a fourteen- or fifteen-minute fire is a far more dangerous incident that could conceivably lead to an American reassessment of the Phase One program. NASA critics will later seize on the disparity as evidence that the Russians covered up safety hazards aboard Mir in an effort to keep alive NASA's $400 million contract for the Mir missions. In fact, there is some evidence that is the case. That Monday, the day the TsUP announced the fire, was a uniquely sensitive moment between the two countries. That same afternoon, just five days after Congressman Sensenbrenner's televised appearance at the Khrunichev factory, Yuri Koptev of the Russian Space Agency stepped before a bank of microphones in Moscow and announced that the Russians would, in fact, delay delivery of the crucial Service Module of the International Space Station. In the wake of Koptev's announcement there was widespread speculation that NASA, under pressure from Sensenbrenner and other congressional critics, might back out of or somehow alter the Russian partnership.

To this day, NASA is at a loss to explain the difference in the fire-duration estimates. Frank Culbertson has repeatedly claimed that the difference was an honest error, that an interpreter had mistakenly translated "fifteen minutes" as "1.5 minutes." In fact, Valery Korzun—a commander who was preoccupied with appearances—now acknowledges that he was the source of the ninety-second estimate. He also admits that this estimate was wrong. "It was very difficult to calculate the time, so we said ninety seconds," Korzun says today. "I really don't know how long it was. It was not more than three minutes. Three or four minutes extinguishing the fire. One or two minutes clearing the ventilation systems. I think the maximum time would be about four minutes."

While no one on board now confirms the ninety-second estimate, no one confirms Linenger's fourteen minutes either. "I didn't do an [exact] timing," says Kaleri, "but if I remember now, remember the places I've been, it was between three and five or six minutes."

"I don't know how long it was, it could have been ten minutes," says Lazutkin. "If you heard ninety seconds, that's not true. It's not two or three minutes either. I really don't know. Everyone has his own different recollection."

"I think about nine minutes," says Tsibliyev. "It's difficult to say. Nobody was watching the clocks."

3:17 P.M.
Aboard Mir

"Just wanted to let you know that the toxicology community is assessing what could possibly be in the air," Sang is saying.

Linenger finally gets to talk to the ground that afternoon. To his surprise, NASA knows little more about the status of Mir's atmosphere than the crew. The TsUP, Sang says, is reasonably sure there was no asbestos in the SFOG, but Blagov is trying to get a detailed list of its components to make sure. "The [SFOG] is a Russian Department of Defense item and its components are classified," Sang e-mails Linenger later that afternoon, "but Blagov believes he can get it."*

To NASA's dismay, there is no air-analysis equipment on board; to test the atmosphere, the eleven air samples Linenger has taken must be returned to Earth. The first chance to return the samples to Earth won't come until the following Sunday, when Korzun, Kaleri, and Ewald are to pilot their Soyuz to a soft landing in Kazakhstan. Even that process won't furnish a definitive reading on the air quality, because a number of harmful gases that might be present could deteriorate and disappear in the days it will take for the samples to reach Houston. In the meantime Linenger has recommended that all six crew members continue to wear the 3M surgical masks, even while sleeping. He is worried they won't be enough.

"The air looks good," Linenger says. *"I've taken all the samples I can do, I've done everything I can think of. But if there is any big danger that we should know about, I'd like to know ASAP."*

"Understand," says Sang.

"Crew is in good health. Everyone is doing fine, we're out of the woods that way, but I just don't want any long-term effects. I want to stay cautious. So we're wearing masks. The air seems good quality right now."

"Right."

"Could be a little bit of smoke inhalation also. The lungs all sound good. There have been no changes . . . Everyone is working hard to get things back in order."

* The Mir astronauts send e-mail via ham radio with the help of software called the packet system. They refer to sending *packets* back and forth, and sometimes use the word as a verb: "I'll packet that to you tomorrow."

Tuesday, February 25
Aboard Mir

By Tuesday, while Linenger is able to return to his experiments, he remains concerned about the station's atmosphere. *"Tox data's a little slow coming from this side, but at this point our main concern is CO_2 and hydrogen cyanide,"* Marshburn, having returned from Helsinki, tells him. *"The systems up there are probably capable of catalyzing the CO_2. I doubt that much hydrogen cyanide was released."*

"Okay, that's kind of how we feel up here," Linenger says. *"The masks at night are pretty tough to wear because of the buildup of CO_2 inside. . . . This first couple of nights sleep was pretty difficult for everybody. I think we need to look into these masks. I think they're a great thing to have; on the other hand, we have to be careful [because] there is a CO_2 buildup, and I experienced it myself. You wake up in the middle of the night suffocating wearing the mask. And other people had the same problems, so our sleep was pretty bad for the first couple of nights."*

Wednesday, February 26
Aboard Mir

By Wednesday Linenger is getting a strange feeling. It's been three days since the fire, and things are returning to normal aboard the station. What is odd is that no one in the NASA hierarchy has contacted him to talk about what happened. Once his initial anger passed, he understands and accepts the fact that the Russians need to play down the fire's severity; it is an embarrassment to them, and it is an integral part of the Russian institutional character to lie or obscure embarrassing incidents. But NASA? Linenger wants to believe the agency is looking out for his best interests, his health. Yet he is keenly aware of the political importance NASA places on the Phase One program, and is starting to fear that the agency might be conspiring in the Russian coverup.

"Who's involved in the medical command?" Linenger asks Marshburn that morning.

"Info went to Frank for him to distribute on the U.S. side."

"Are you talking to your medical folks? Because I think this thing is critical. I think it's very important that someone at a very high level understand that we want to use this [toxicology] information first to assess the health impact, second to assess Mir systems. I want someone to fight hard for the health of the crew and it should be at the [topmost] level, and I want Frank to get the word from me that I think this is critical that someone right now proactively determines what needs to be done."

Linenger mentions that he is sending down a full report on the fire. *"I think it's important that we get some lessons learned, so I don't think it's good if we hush up too much. I saw a packet file that said there was a small, non-threatening fire on Mir, and everyone is just fine. It was definitely life-threatening."*

"Yes," Marshburn replies, *"we got the transcription of the first pass over Wallops and we understand that this was a big deal."*

"[This should go] to someone at a very high level, not just you and Tony. I just want to make sure that people are taking this seriously."

The rest of the week passes quickly. There is no way to gauge the toxicity of the air until the samples can be returned to Earth, and NASA representatives aren't due in Moscow to join the Russian fire investigation until the following Monday. In the meantime, communication with the ground continues to be plagued by deafening bursts of static. There are frequent snarls, delays, and pileups in e-mail traffic to and from the station, especially Linenger's e-mail, which prompts several complaints from the American to his ground team

"By next week things should settle out," Linenger tells Marshburn on Saturday evening, March 1.

"Well, until the next thing happens."

"Don't say that!" Linenger says. *"I'm starting to get superstitious."*

Sunday, March 2
Aboard Mir

One of the most frustrating things for everyone aboard Mir is the deteriorating quality of voice communications with the ground. That Sunday, even as Kaleri, Korzun, and Ewald undock their Soyuz and return to Earth without incident, comm drops in and out all day long. In midafternoon Tsibliyev snaps at the ground about the "ratty" comm, demanding that the TsUP fix this "comm circus" or stop communicating altogether. The station normally bounces its radio signals off a pair of Russian satellites, but transmissions to one satellite, called Altair, were shut down in January when a transmitter aboard Mir overheated and was shut down.

For all the Russians' problems, the rattiest comm comes over the two American relay stations, at Wallops Island, Virginia, and Dryden, California. NASA had volunteered the use of Wallops and Dryden early in the Phase One program, hoping to free up more personal time for the astronauts to talk with their families. It hadn't worked out that way. In time the Russians began using passes over the two American facilities for operational communications, especially during the two- to three-week period every two months when the station's orbit sent it increasingly over North America. The problem was, neither station had been rigged for anything other than occasional use. At Dryden, near Edwards Air Force Base in the California desert, two six-foot-long antennae had simply been clipped onto the side of a satellite dish. "It was a really fast-and-dirty setup, really kind of thrown together," recalls Mark Severance, the thirty-five-year-old NASA engineer who volunteered to work on improving the system in the fall of 1996.

A team of engineers from NASA's Goddard Space Flight Center in Maryland had been examining the problem for months when Severance arrived on the scene. But the young Texan had unique experience to bring to bear: He had been conducting ham radio sessions with Russian spacecraft since the mid-1970s, when as the thirteen-year-old son of a NASA contractor he caught the space bug following the Apollo-Soyuz missions. As a teenage hobbyist Severance rigged an antenna on the chimney of his family's Fort Worth home to listen in on communications between Soviet spacecraft and the tracking ships that zigzagged the Caribbean from their base in Havana harbor. Severance began taking Russian-language classes in high school—the only school in Fort Worth that offered them—so he could better understand what the Russians were saying, and by the time he enrolled at Southern Methodist University in Dallas had chatted with scores of friendly cosmonauts, all of whom seemed pleasantly amazed to find a cheerful Russian speaker to talk to as they crossed North America.

That fall, while working as a NASA engineer in Houston, Severance heard of the difficulties with Dryden and Wallops and volunteered to look into the situation. Analyzing the signal between Mir and the two ground stations, he quickly recognized two problems that should have been obvious to any amateur ham operator. For one thing, the NASA receivers at Wallops and Dryden weren't adjusting the signal to account for a simple auditory concept called the Doppler Effect. "This is something that ham guys correct for on every pass, and yet we weren't doing it, and I didn't understand why," remembers Severance.

The second problem was even more basic. The NASA receivers were rigged to catch a signal broadcast at the very narrow level of five megahertz, the width of a standard ham radio signal. But Severance knew from years of hobby experience that the Russians broadcast a much broader signal, on the order of thirty megahertz. This difference in "bandwidth" meant the two NASA stations were only receiving a fraction of the Russian signal. Severance measured the signal and found he was right. The Russians were broadcasting at thirty megahertz. This was something any fourteen-year-old hobbyist could have seen, but for some reason NASA hadn't.

Pleased but confused, Severance presented the bandwidth problem to a group of a dozen Goddard engineers at a meeting in Houston in January. They didn't believe him. The Goddard men, mostly NASA veterans in their forties and fifties, showed Severance a piece of paper called a technical specification that said the Russian signal was at five megahertz. "Look at the requirement documents," one of the Goddard men told him. "It says five megahertz."

"Forget the docs," Severance said. "Look at the performance. You're clipping the audio."

But the Goddard men didn't believe him. If a technical spec said the Russians were broadcasting at five megahertz, they were broadcasting at five megahertz. No one had thought to question the piece of paper, which was clearly in error, and no one had thought to independently measure the width of the Russian signal.

Nothing happened. Severance waited while the Goddard engineers returned east to pore over the data. For a month they kicked around a variety of technical remedies that Severance just knew wouldn't fix the problem unless the basic issue of bandwidth was addressed. From Houston Severance bombarded the Goddard team with memos that he felt were being ignored. He complained to friends and to at least one former astronaut. Still, nothing happened. Up in space Linenger and his Russian partners could barely hear anything through the static over Dryden and Wallops.

Finally, on Valentine's Day, Severance went directly to Culbertson and Van Laak in an effort to cut through the red tape. Culbertson, chagrined, immediately freed up $300,000 to begin rigging a new set of antennae that would clear up the signal. The Goddard team belatedly climbed on board, and on the Monday morning after the fire, just hours after Korzun's inability to talk to the TsUP during a Wallops pass, Severance began work on the new antennae. The new setup won't be ready until June, but by then it will prove very useful.

Monday, March 3
Moscow

NASA's Phase One safety director, Gary Johnson, arrives in Moscow in the morning and by the end of the day is closeted with his Russian counterparts, eager to join the fire investigation. He is surprised when the Russian safety cochairman, Boris Sotnikov, smiles and informs him that while the Russians are thankful for the offer, NASA's help won't be necessary. "The investigation's already over," Sotnikov says cheerfully. "Here's a copy of the initial report."

Johnson's surprise quickly turns to irritation. According to Sotnikov, the Russian investigation concluded the previous Wednesday, just three days after the fire. Johnson can't imagine NASA closing out an investigation of this importance in three months, much less three days. His irritation grows as he studies the Energia report. Flipping through it, Johnson notices that the investigating commission is made up almost solely of senior Energia officials with responsibility for safety aboard the station—hardly a group, Johnson suspects, likely to find fault with its own efforts. He registers a mild protest but, other than that, feels powerless to alter the process.

The Russian report blames the fire on the oxygen canister, speculating that moisture could have clogged the device's air filter, causing it to overheat and burst into flame. "The commission," it concludes, to no surprise of Johnson, "holds that the cassette failure is an isolated event." It recommends a more thorough investigation after the burned-up unit was returned to Earth.

Johnson wants to know more, much more. If moisture caused the fire, where did it come from? And more important, what is to stop this from happening again? The Russians seem to think the candle was manufactured improperly. But when Johnson asks to visit the Moscow-area factory where it was made, the Russians refuse. It is a military plant, Sotnikov advises, off-limits to outsiders. Johnson wants to know more about how the machine creates the heat that activates the lithium perchlorate. Apparently, it works on the same principle as a pistol; when the red starlike activator is turned, a finger of metal is rammed into a pan containing some kind of chemical, which ignites to create the heat. To Johnson's dismay, Sotnikov won't tell him what the chemical is, saying it is a "proprietary secret" of Energia's. Someone slipped Johnson a formula that suggests the chemical contains magnesium. When Johnson asks if that is true, one of Sotnikov's engineers angrily demands to know where Johnson obtained the formula. He won't say.

Johnson will stay on in Moscow for a week and even sit in on Korzun and Kaleri's debriefings, but he will return home with little real insight into what happened aboard the station.

On Monday, as Johnson arrives in Moscow, Viktor Blagov is in Houston to brief Culbertson on the fire; that afternoon, in the large Phase One conference room, the silver-haired Russian lays out the TsUP's plans for dealing with the generating unit. For now, of course, they have told the crew to avoid using the oxygen candles at all. Only one of the station's two Elektron units is still working, but the TsUP plans to crank it up to full power twenty-two hours a day to furnish oxygen for the three remaining crew members. In the unlikely event that the crew is forced to resume using the candles—and Culbertson makes clear NASA doesn't want that to happen—Blagov has directed them to use only cassettes manufactured after 1995; he suspects that the cassette that burned was an older unit. There are 240 remaining cassettes on board, and Tsibliyev and Lazutkin will spend two days later that week cataloging their dates. New fire extinguishers and gas masks will be sent up on the next Progress, scheduled to lift off from Baikonur at the end of the month.

Culbertson's follow-up questions are rudimentary, a telling indication of NASA's lack of basic knowledge of Mir safety systems.

"Do you train specifically for postfire activities, for survival?" Culbertson asks.

"Yes," Blagov says, and gives details.

"The gas masks have an oxygen generator. Is it like these candles?" For all Culbertson knows, the gas masks could burst into flame.

"It's a cold O_2 generator," one of Blagov's aides says. "When the crewman breathes, CO_2 and humidity activate it."

For two hours the NASA men pepper Blagov with questions, and he does his best to answer them as clearly as possible. The bottom line remains, however, that for all the talk and all the reports, no one has a clue what caused the fire. And while Culbertson's men focus on that question, no one in the room bothers to discuss the odd docking test that even then the cosmonauts aboard Mir are preparing to attempt the next morning.

Aboard Mir

Monday morning, Linenger breaks from his experiments to chat on comm with Marshburn, who mentions in passing that he and Sang have just received the astronaut's report on the fire.

"Wow, that's a long, long time," remarks Linenger, who had sent the e-mail report the previous Wednesday.

"Yeah, no kidding."

"You know, let me talk a bit," says Linenger, *"[about] the feedback I've been getting. I would say it's very important to get accurate feedback [during] any kind of crisis of any sort. Because basically I was getting a strong feeling up here that I was on my own and getting no—no support from you at all. [It's] the big management support, that is what I was feeling isolated from. And I got a few letters yesterday—[a little] feedback—and it really helped."*

The delay in receiving the fire report, Linenger goes on, is especially infuriating. He actually discussed the fire via ham radio with Dave Leestma, the head of Flight Crew Operations, over the weekend and assumed Leestma had read his report; now he realizes he hadn't.

"That is totally unacceptable," says Linenger. *"Understand, that thing came down on telemetry days ago, so it's been sitting in the system down there just rotting*

away for a few days. And our management should get that information fast. . . . Again, that causes mixups. I'm thinking they're not responding to what I wrote. What's going on?"

Marshburn says he'll try to fix the problem. Linenger signs off asking for some pretzels to be sent up on the next Progress. *"You know, those big fresh ones from the mall would be great,"* he says. *"And also some Pringles with onions or peppery stuff."*

Monday morning, Tsibliyev hauls the TORU docking computer out of its storage in Priroda and seats it into a notch in the base-block floor. Comm was so poor the day before, the docking experts were forced to wait until today to fully brief him on Tuesday's test. Linenger sees the new contraption in base block and asks what it's for. Tsibliyev describes the general outlines of what they will be doing, but fills in few details. Later Linenger will wish he had asked more questions.

9:30 A.M., March 4
Aboard Mir

Tsibliyev rises early and slips into his formal royal-blue jumpsuit for the TORU test. For all the complex physics and mathematics involved, the plan for the morning's test is simplicity itself. The ground is to maneuver Progress-M 33, a sturdy little eight-ton bumblebee with two long solar-array "wings," to a position about seven or eight kilometers above the station; Mir will be flying "upside down" in relation to the Earth below, so the Progress will actually seem to the cosmonauts to be approaching from below. The TsUP will then release the Progress to fly on an arcing track that if uninterrupted would send the spacecraft whizzing by one side of Mir. As it approaches, a camera mounted on the Progress will beam a picture of the station onto the Sony monitor set before Tsibliyev. One part of the Kurs radar has been left on, to furnish the commander with range and speed data on the oncoming ship. Using the black TORU joysticks, he is to take control of the Progress and maneuver it toward

Mir, braking its speed and slowly bringing it within one hundred meters of the docking port at the end of the Kvant module, directly behind where he sits in the middle of base block. Once comm is reestablished, the ground will take over and help him guide the ship to final docking. "The ground sends it flying at you," explains Linenger. "You're supposed to catch it. You're not supposed to let it hit you."

Despite his ignominious "bump" with Mir using manual controls three years earlier, Tsibliyev will later say he was not nervous about the test. He has performed this maneuver in the training simulator many times. By midmorning he and Lazutkin have positioned themselves in base block with a long instructional memo the TsUP has sent explaining the test in detail. Linenger is dispatched down to Kristall to watch for the Progress as it approaches. The ground's part of the plan goes as expected. At a distance of about seven kilometers, it releases control of Progress-M 33, which begins bearing down on Mir at a speed of seven meters a second. At its current rate of speed, Lazutkin calculates, it should reach the station in roughly fifteen minutes.

The first sign something is amiss comes when the telemetry data running down one side of Tsibliyev's screen indicate the Progress is five kilometers away and approaching. At roughly this point, the camera aboard Progress should activate, giving Tsibliyev an image of Mir. But at the moment the image should appear, it doesn't. Tsibliyev's screen remains black. At first he isn't too concerned. The camera is bound to come on. At one point there is some movement on the screen. A horizontal white line appears for a few moments, then winks out. The screen remains black.

As the minutes tick by, Tsibliyev begins to grow uncomfortable. Lazutkin flips through the instructional memo, searching for any clue to explain why the screen remains black; he finds nothing. By about 10:15, when the Progress is approximately four minutes away, Tsibliyev tells Lazutkin and Linenger to stay glued to the windows. The big cargo ship is out there somewhere, heading straight for them, and he has to know where it is; at this point the only way he can do that is for one of the other two men to physically spot it out in space.

"It was the most uncomfortable [time]," Tsibliyev remembers months later. "I felt as if I'm sitting in a car, but I don't see anything from the car, and I know there is this huge truck out there bearing down on me. You don't know if it's going to hit you or miss you. It's like a torpedo, and you're in a sub."

"Where is it?" Tsibliyev begins asking the others. "Do you see it yet?"

"No," says Lazutkin, peering out the big base block window behind the commander.

"No, nothing," Linenger says over the intercom. There are three windows in Kristall, and he is floating between all three, scanning space for any sign of the Progress.

They wait.

"Do you see it?" Tsibliyev shouts a few moments later.

"No, I don't see it," Linenger breathes over the intercom. Lazutkin, floating at the base block window, shakes his head. "Nothing," he says.

Several moments pass.

"Do you see it?" Tsibliyev asks again.

"I don't see anything," Linenger replies, hurriedly shuttling between his portals.

Static fills the Sony monitor. Linenger can tell from the tone of Tsibliyev's voice the Russian is growing anxious. More time goes by. Somewhere out there a fully loaded spaceship is bearing down on them.

"Where is it?" Tsibliyev demands. He turns to Lazutkin. "What should I do?"

Lazutkin has no advice.

They wait.

At the two-minute point, Tsibliyev begins to sweat.

"Do you see it?" he shouts again.

"No, I don't see it," Linenger says. Lazutkin concurs. "Nothing," he says.

"Find it!" Tsibliyev orders. "Find it!"

After several more moments, during which he floats back and forth between Kristall's three portals peering into the inky blackness of space, Linenger hears Lazutkin's voice over the intercom. It is filled with tension. "Jerry, get back in base block quick," he says.

Lazutkin has spotted the Progress. Until this point Mir's massive solar arrays had blocked his line of sight. But now he sees the ship approaching fast, slightly below the station. From his vantage point in base block, it appears to be heading for an imminent collision.

"I see it!" Lazutkin says.

"Where is it?" Tsibliyev asks.

"It's close!" Lazutkin replies.

This is as technical as Lazutkin's response gets. "I see it in full size," he remembers months later. "All the solar arrays, the antennae, everything. That's when I told Jerry to go to Soyuz."

Linenger propels himself down the length of Kristall as quickly as he can. Reaching the node, he sees Tsibliyev sitting at the console, jerking at the TORU joysticks. The Sony monitor still shows nothing but snow.

"What's it doing?" Tsibliyev shouts.

Lazutkin turns to Linenger. "Get in the spacecraft," he says quickly. "Get ready to evacuate."

As Linenger turns, he sees Tsibliyev furiously manipulating the TORU joysticks. He realizes the Russian is attempting to fly Progress blind. Swiftly, Linenger folds himself into the Soyuz capsule and immediately begins pulling out the various cables and ventilation tubes that connect the craft to Mir. Floating up into the node, he sees Tsibliyev still in base block, sitting before the monitor. He appears to be on the verge of panic.

"What's it doing?" the Russian shouts.

Lazutkin's reply is unclear. Crouched at the mouth of the Soyuz, grabbing and disconnecting cables as fast as he can, Linenger glances over his shoulder to see Tsibliyev jump back from the console and check the portal himself. Then he springs back to the monitor and pulls at the black joysticks once more.

"What's it doing?" Tsibliyev shouts at Lazutkin.

9

10:10 A.M.
NASA Operations Room, TsUP

Tony Sang and his team crowd around the monitors in the NASA suite to watch Tsibliyev redock the Progress. Sang isn't especially worried about the maneuver. From conversations with Viktor Blagov he understands—incorrectly—that the Russians have handled these kinds of long-distance manual dockings on several previous occasions with no problem.

In front of him, Sang's monitor shows video being shot by the camera aboard Progress-M 33. Gone are the overlaying targeting sights that they normally would see; the sights have been unavailable since communication with the Altair satellite was cut off. Still, the moment the image from Progress flickers onto his monitor, Sang realizes something is amiss.

"This doesn't look right," he says.

The screen should show Mir floating in space as the Progress approaches. Instead the monitor in NASA's office shows Earth in the distance. There is no sign of the station.

"This doesn't look like any Progress docking I've seen before," Sang muses aloud.

As the minutes tick by, Sang and his group keep waiting for Mir to come into view.

"Where is it?" someone says.

After about ten minutes, with no sighting of Mir, the picture winks out. Several moments after that, as Mir once again comes into range of a Russian ground station, Tsibliyev's voice comes over the comm. Sang and Tom Marshburn watch intently as one of their interpreters, a temporary replacement they don't know well, busily scribbles down what the commander is saying. As usual they have no sense of the words or even the tone of the cosmonaut's message.

"The commander is really excited," the interpreter says.

"What do you mean?" asks Marshburn.

"I don't know, but he's really, really excited," the interpreter replies. "Something's going on."

New to the job, and to the technical terms Tsibliyev is using, the replacement interpreter is unable to decipher precisely what has happened. Sang realizes Tsibliyev is angrily complaining about some kind of malfunction on his screen; apparently it isn't working. Curious, Sang and Marshburn hustle down to the floor to find out what is going on.

10:19 A.M.
Aboard Mir

It misses.

Barely fifteen seconds before impact, as Linenger scrunches himself into the Soyuz to prepare for emergency evacuation, Tsibliyev's screen suddenly activates, and he realizes the Progress will not hit the station. The screen, broadcasting from the camera aboard the Progress, shows Mir, uncomfortably large and close. But from this vantage point Tsibliyev sees the Progress will pass underneath the station, narrowly avoiding a collision. Crouching by the base block window, Lazutkin watches the ship sail by harmlessly. He guesses the distance at two hundred meters or less.

Emerging into the node, Linenger sees Tsibliyev dramatically sag in relief. All the pent-up energy in the commander's shoulders seems to drain from his body as he leans heavily on the TORU controls. For the longest time no one says anything.

10:34 A.M.
Aboard Mir

Tony Sang's interpreter is right. Tsibliyev is angry.

"I will repeat," the commander says at the beginning of the pass. *"We*

watched it visually. . . . There was no picture for a long time. At 10:19 it appeared. . . . It started moving away, under us. We were close to it. We were like 200, 220 meters close to it, judging by its size. . . . We managed to apply the brakes. The speed was around two meters [per second], and then it started moving away very fast. And that's the last thing that we saw. Now there is no picture again. . . . We couldn't observe anything for a long time. There was no picture . . ."

Vladimir Solovyov himself gets on the comm. *"Did you have control of it?"*

"I started braking and switching off the angle mechanisms. It passed by at a very high speed. It wasn't possible to see where to go. I touched the handles intuitively. We didn't collide with it. . . . There was no picture. And there is no picture now. And it's hard to say how to control it. Only when we started braking, a picture appeared."

For the moment, the TsUP is primarily concerned with locating the errant spacecraft. *"[It's] somewhere underneath,"* the comm officer says.

"But I don't have anything," Tsibliyev says. *"Nothing can be seen. Just the mist."*

"Read from the screen," the comm officer suggests.

"I can't see anything anyway," Tsibliyev snaps. *"It's not us. We saw through the window that it started moving to the side of [base block]."*

The rest of the comm pass is spent attempting to find the Progress. After signing off, Tsibliyev turns to Linenger and Lazutkin and launches into a lengthy tirade directed at the TsUP's incompetence: "Jerry, what was I supposed to do? What could I do? The screen shows nothing! Nothing! What could I do?" It takes a while for the commander to settle down, and when he does he heaves a long sigh.

"Guys," he says, "I never want to do that again."

11:00 A.M.
TsUP

Forty minutes after Progress-M 33 nearly rammed the station, Sang and Marshburn still have no idea what happened. When the comm pass ended, the two sought out Solovyov on the floor and asked. The old

Russian calmly explained that yes, there has been a problem. No, they don't know what it was. The Progress sailed by Mir harmlessly.

"How far away was it?" Sang asked.

"It was pretty far away," Solovyov said. There was no danger.

Sang, his interest piqued, pressed for a moment.

"It would be great if you could get back to me on that," he told the Russian.

Solovyov never did, and Sang never followed up. Sang's team, in fact, had no idea how close the station had come to a catastrophic collision. "The Russian ground control team gives us no indication that anything big had happened," remembers Marshburn. "They tell us they just decided not to complete the docking. They cite some software problems. It was no big deal. . . . We reported that back as the sum total [of what had happened]. We had no idea."

In fact, Marshburn says, no one on the NASA team learned of the danger Linenger had been in until several weeks later, near the end of Linenger's mission in May. "I found out when Jerry was talking to a reporter, when he tells [the reporter] about it," says Marshburn. "I was amazed." Sang says he didn't learn there had been a near miss until long after Linenger had returned to Earth. "Obviously, yes, we should've known more about it," admits Sang. "I could've asked around."

Sang's bosses in Houston knew little more than he did. For all the talk after the fire about NASA becoming more involved and knowledgeable about Russian systems, the Phase One team neither anticipated the docking test's risks nor, incredibly, discovered them afterward.

"We knew very little [ahead of time]," acknowledges Frank Culbertson. "I knew they were going to redock the Progress. I didn't know they were going to do it manually. They didn't really talk about doing a test. The Russians, obviously, felt it was not an operational concern of ours, and why did we need to know details?" Later, when told the test had been a long-range manual acquisition, Culbertson expressed surprise. "We asked a lot of guys to look at it [afterwards]," says Culbertson. "For some reason no one was all that alarmed. No one brought any concerns to me. In retrospect, we could've asked more questions."

The one man who was in a position to enlighten Culbertson's team, Linenger, inexplicably didn't. He and Marshburn had a long conversation during a 7:52 comm pass that Tuesday evening; nowhere in the NASA log is there any indication that Linenger even mentioned the near collision that morning. Nor is there any indication he brought it up in subsequent e-mail traffic. "I assure you," says Jim Van Laak, "if Jerry had communicated anything like, 'My God, Vasily was desperately inputting in the blind, it was fifty meters away!' we would have put everything to a blinding halt while we figured out what was going on. But we had zero input from Jerry in that regard. I don't understand Jerry at all."

Linenger reluctantly acknowledges his failure to inform the ground team what had happened. "There was a miscommunication there," he says. "[I] assumed that they must know. How could they not know about it? Bad assumption, I guess."

No one knew it at the time, of course, but NASA's failure to remain abreast of what the Russians were doing was to ultimately have severe consequences for everyone concerned.

What went wrong that morning of March 4? To this day, no one is certain.

The TsUP, however, developed several theories. "There was a fault in the camera" aboard the Progress, says Vladimir Solovyov. "It was [either] out of order, or the crew forgot to switch it on. I think the camera worked properly, but the crew forgot to turn it on." Does the crew agree with this hypothesis? Solovyov smiles. "What do you think?" he replies. "Who would agree? [And] we can't prove it objectively." Is it possible, as Tsibliyev suggests at one point, that atmospheric factors such as cloud formations might have somehow interfered with the camera's transmission? Solovyov laughs. "No, that's out of the question. [But] it's impossible to duplicate conditions 100 percent on the ground. We can have different interferences that can't be imitated exactly."

Afterward, when technicians at the TsUP reestablished contact with the Progress, they checked to see whether the camera had somehow been disabled. "It was working," remembers Blagov. "All systems were working perfectly." Blagov, like Solovyov, says he believes Tsibliyev simply forgot to turn on the camera. "It's hardly believable that the camera was working before, it was working after, but it was not working for the docking," he says. "It [would be] a miracle. Unless you believe in UFOs and fairy tales." Another possibility, Blagov says, is that Tsibliyev failed to press the camera button—a small white button the size of a matchbox—with enough force. "The button should be held down," he says. "If it's only touched, it wouldn't work."

That may be. But the most plausible explanation for the camera's failure was one neither Solovyov nor Blagov would admit publicly, because it would have implicated decisions made on the ground. According to several NASA officials who later analyzed the incident, the Russians privately suspected that the Kurs radar signal that supplied Tsibliyev with telemetry data may have somehow interfered with the camera signal. This, of course, would have been the TsUP's responsibility to foresee, not Tsibliyev's. The obvious solution, as the Russians studied the problem that spring, was to turn off the Kurs signal altogether, robbing Tsibliyev of all speed and range data. It was a fix that would ultimately lead to disaster.

Whatever the precise cause of the incident, which came to be known in NASA circles as the Near Miss, the Russians relentlessly played down its significance to Culbertson, Van Laak, and the few other Americans who bothered to inquire. Part of this, of course, was simply habit. Like bureaucrats everywhere, Solovyov and Blagov tended to downplay everything that went wrong, and became irritated when anyone at NASA or in the press confronted them with tough questions. This was standard behavior during the Soviet years, when bad news was routinely covered up, and both Solovyov and Blagov are products of the Soviet system. But another factor in the Russians' obfuscation was a purely post-Perestroika development: money. No one in the Russian hierarchy wanted to do or say anything that might scare off NASA and its millions, which were, after all, the only thing propping up the Russian space program.

Surprisingly, though, the most significant factor in the Russians' downplaying of the Near Miss was their refusal to believe the crew had ever been at risk. The Progress, they calculate, was set on a course toward Mir that sent it flying past the station at a distance of two hundred meters or more. "All crews can [say] what they want to say, instead of what's happened in reality," says Solovyov. "Several times [Tsibliyev and Lazutkin] were trying to tell us fairy tales about how the big truck was going to kill them, how it was going to crush them. Many times we tried to explain things to them, how they were in no real danger at all. We have the concrete information [on the ground]. We knew it missed them at some distance, half a kilometer. Nothing horrible has happened, despite these terrible stories."

Solovyov attempts a weary smile. "I know the case with Tsibliyev," he says. "He is sure Mission Control wants them to come back as dead heroes. It's very far from the truth."

Wednesday, March 5
Aboard Mir

The Near Miss marks a clear turning point for Tsibliyev. In the days that follow he grows increasingly irritated with ground controllers, frequently snapping at them. Off comm, alone with Lazutkin and Linenger, he begins excoriating the ground, complaining that Solovyov, Blagov, and their staff are a dangerous bunch of know-nothings. "Vasily's attitude, and he said this more than once, was 'Those stupid idiots! They almost killed us!' " Linenger recalls. "He was losing confidence in the ground. You have two situations in a row when you could have died. You realize the ground can't help you. They can't do anything to help. You've got to take care of yourselves." Tsibliyev would deny it all later, but Russian and American ground logs, as well as the recollection of many of those involved, indicate that, emotionally at least, early March brings the first questions about the Russian commander's stability.

While Tsibliyev spends the next few days discussing what went wrong with Solovyov and other ground controllers, the fact is he doesn't have time to dwell on the failed test's implications. The very next morning the Elektron unit in Kvant 2 has to be shut down when an air bubble is found blocking the flow of water in the unit's electrolysis canal. Ordinarily a shutdown of the cranky Elektron is no cause for alarm, but this time is different. The station's only other Elektron, in Kvant, has been out of commission for weeks and needs replacement parts that must be sent from Earth. Without the second Elektron up and running, the crew has only one remaining way to create oxygen: a second SFOG in base block, the backup oxygen generator, a duplicate of the one that caught fire.

Lazutkin immediately goes to work in an attempt to fix the malfunctioning Elektron. He has already been working long hours on several other balky systems, including the station's urine-reclamation unit. As he works, Linenger is struck by the differences between the rookie engineer and his predecessor, Sasha Kaleri. Where Kaleri calmly and deliberately went about his repairs, making sure to finish and stow his work behind panels at the end of each day, Lazutkin seems to ricochet between tasks all day long. Panels stand open at all hours, wires and hoses jutting out like the innards of a dying surgical patient. "Lazutkin was behind the eight ball from the start," Linenger remembers. "He was learning as he went. He basically didn't know what he was doing. I didn't feel he was getting to a point where he could feel confident and plan his own day out. He was just swamped the whole time."

In an attempt to keep up with their repair and maintenance work, Tsibliyev and Lazutkin, in a routine that will persist for months, begin staying up late at night, sometimes until four and five in the morning. Linenger wakes many mornings to find the Russians sound asleep in their curtained *kayutkas*, stirring only for comm passes, not dragging themselves to work until ten or eleven o'clock. Tsibliyev begins taking afternoon naps. Linenger feels the commander is undisciplined.

Friday, March 7
Aboard Mir

Linenger's frustration with the station's communications problems is approaching the exploding point. He is supposed to get two comm passes a day with Sang and Marshburn, but rarely does. When they do talk, their words are often wiped out by bursts of static. Sometimes Linenger can't hear the ground; sometimes the ground can't hear him. The situation is exacerbated by problems with his daily schedules, the Form 24s. Time and again Linenger discovers his Form 24s contain scientific procedures that are incomplete, misleading, or simply wrong; an experiment that he is given an hour to complete may take five hours. The Russians marvel at NASA's insistence on scheduling its astronauts' activities so minutely, but NASA insists. It is the NASA way. The agency has done this on the shuttle for sixteen years and hasn't yet learned that long-duration spaceflight may require a more flexible approach. Nor has it learned anything from John Blaha's mission. Linenger, in fact, has many of the same problems Blaha had. Whenever there is a problem with the Form 24, he is forced to wait for a comm pass to work out the problem with Sang. To Linenger, who prides himself on efficiency and organization, this is massively frustrating.

He complains to Sang, who acknowledges the screwups and tries repeatedly to fix them. The problem, Sang tries to explain, lies in the Russian process for approving the forms. Each Form 24 is based on a time line, which Sang and several Russian planners draw out on a five-foot-long piece of white paper. Each day's time line is constantly updated until five days before its use, at which time it is approved by Viktor Blagov and radioed up to the station. The Americans have asked for and been repeatedly denied final sign-off approval of the all-important forms. Sang updates the Form 24s by talking with two Russian women, whose names he knows only as Nadia and Vladia. The changes he makes with Nadia uniformly make it onto the Form 24; the changes he makes with Vladia don't. He appeals for help to Blagov, who insists he will talk to Vladia. But the problems persist.

By Friday morning Linenger is at wit's end. Omissions and incorrect directions are wasting hours of his time; writing down questions to ask Sang and Marshburn wastes even more time. And if there is anything that bugs Linenger, it is having his time wasted. After mulling over the situation for days, he makes a momentous decision, which with no warning he springs on Sang and Marshburn at the Friday morning pass.

"I think we should stop doing voice comm," Linenger says. *"I think we ought to stick with radiograms back and forth. I'm sending that information down to you [in an e-mail]. So in the future why don't we just not schedule voice comm. I think we'll be more effective doing it the other way."*

Marshburn and Sang, sitting beside each other at the NASA console on the floor, exchange amazed glances. Sang is so stunned, it takes him a moment to reply.

"So am I to understand," he says, *"you don't want to use voice comm any more?"*

"No voice comm. I think we ought to use the time to send a radiogram back and forth. And the data just gets scrambled too much, and for science I'm not sure it's the smartest thing to do. And they've got lots of operational things they need to use the comm for. So I really think we ought to stop voice comm, and I've got a note to you [coming] about that. I think it will work a lot better."

"Okay," Sang says, unclear exactly what he should say; he wants to object, but it's clear Linenger has made up his mind. Instead he says, *"We'll try that one out."*

Afterward Sang takes off his headset and buries his face in his hands. To him Linenger's decision is a stinging personal rebuke. "My jaw about hit the floor," he remembers. "I thought Jerry was saying I was a failure. I felt like a total failure." Sang, much like Marshburn a month before, begins to sink into a mild funk that lasts for days. He is convinced he has ruined the entire mission. How can they possibly finish all the science, all the experiments, without voice comm? "I blamed myself for the longest time," he says.

They call Culbertson in Houston to break the news that Linenger is going incommunicado.

"Why?" Culbertson asks.

"I really don't know," says Sang. "Yeah, comm has been bad, but we've lived with that since the beginning. This is nothing new."

"Well," says Culbertson, "just let it go for a while. My guess is he'll come out of it in a few days."

Marshburn tries to downplay the whole matter, hoping to make Sang feel better.

"Are you worried?" Culbertson asks the flight doc.

"No, not at all. Frank, comm sucks. It really is bad. Altair is down, it's really, really ratty. Now, if the end of April comes around and he's getting ready for the EVA and he's still not talking, then I'll be worried."

Culbertson isn't too concerned. He has seen astronauts withdraw after a few weeks in space, most notably Blaha. Even in his Navy days Culbertson had witnessed how sailors' behavior changed after a month or two at sea. The newness of time at sea, or in space, wears off. The honeymoon ends. Work becomes a grind. He figures Linenger's decision is the result of a momentary burst of frustration. No doubt in a day or two, maybe a week, he will be back to normal. "It appeared to be an attempt to get attention to do things differently," Culbertson recalls, "and he was upset that he wasn't getting the sweeping changes that he was looking for."

Sang and Marshburn spend the rest of the day mulling what to do. They decide to stay upbeat; in a follow-up e-mail to Linenger they strike an ebullient tone. "Hey buddy—they installed the main engines in Atlantis today!" they begin the note, updating Linenger on the next shuttle mission. "[As for] COMM, we have not received your downlink on this yet but here are our thoughts. Voice comm has been terrible from our end, too. We won't expect to hear from you and will ask the ground control team to packet up during that time. We are always in the control room during a comm pass, and sometimes when the capcom sees us there he feels obligated to call you to comm even though we've explained that we don't need to communicate unless you need to. We understand your desire to not be interrupted, and continue to explain this to each of the shift capcoms."

"DON'T NEED TO STAND BY AT ALL," Linenger radiograms down two days later. "WORK ON STREAMLINING THE PACKET PROCESS. . . . [TODAY IS] FIRST TIME SINCE THE 5TH I GOT ANY PACKET. AGAIN—PRIORITIZE. SINCE WE HAVE IN THEORY TWO COMM PASSES A DAY, THIS SHOULD BE EASY. NO VOICE BETWEEN US AT ALL. MAKES US WORK HARDER TO GET IT RIGHT THE FIRST TIME."

Though it doesn't immmediately occur to anyone at NASA, the Russian psychologists at Star City believe Linenger's decision is a direct result of the stress he endured during and after the fire.

Saturday, March 8
Aboard Mir

The failure of the Elektron system means the crew will be forced to rely on the backup oxygen canisters, which are identical to the ones that caught fire two weeks before. It is a prospect that makes everyone in space and on the ground uneasy, but in the end they have no choice. It is the canisters or nothing.

Now that the ground is preventing them from using the older canisters, Tsibliyev isn't sure they have enough left to last them more than a few days. "Jerry, have you heard how many candles we have?" he asks Linenger, who is surprised at the question. Linenger smiles as he recalls: "It was a crackup. I mean, I couldn't believe this. They thought the Americans had better information than they were given. Vasily, none of us, we didn't have any idea what was really going on. We were always trying to piece together what is actually happening to us. It was ridiculous. We couldn't trust what the ground told us."

Saturday morning, the crew crowds around the backup SFOG in base block to install the first canister since the fire. As Tsibliyev and Linenger stand by with blankets and fire extinguishers, Lazutkin gets ready to install a new candle. All three men make sure they are positioned on the far side of the candle, near where the Soyuz is docked at the node; they want to make sure they will be able to evacuate in the event of another fire.

A little before ten, Sang and Marshburn join a growing crowd of Russian technicians on the floor of the TsUP. The mood is tense.

"Are you ready?" the shift flight director asks.

"We're not kids," snaps Tsibliyev. *"We've burned these things before."*

"Burn it."

"Getting ready," Tsibliyev says. Lazutkin bends over and opens the filter. On the TsUP floor the tension is thick.

Lazutkin inserts the canister.

Nothing happens.

Lazutkin carefully extracts the canister and inserts it once more.

Nothing.

The astronauts exchange glances.

"Do the following," the shift flight director says. *"Do not start again. Get the filter out and check it."*

Lazutkin obliges. The filter seems fine. He puts it back and tries the canister once more.

Again, nothing.

Lazutkin tries two more times, and still the canister will not work. Tsibliyev says they will keep trying. *"Other guys told us it sometimes starts after twenty tries."*

On the sixth attempt the canister suddenly engages.

"Warm air is flowing," Tsibliyev announces. *"Jerry is standing by with a blanket."* Marshburn notices several Russians snicker and wonders whether the remark is intended as a dig at Linenger.

Anxious seconds tick by. Sang and Marshburn hold their breaths. It is at this point, everyone knows, that the earlier canister had burst into flame.

This time nothing happens. There is no baby volcano.

"Congratulations," the shift flight director says. All across the floor, smiles break out.

Later that day, after Tsibliyev has a private family conference with his wife, Larissa, Linenger surprises Sang by getting on voice comm, reminding him that he still hasn't seen any data on what if any toxins the fire released into the station's air supply. *"If we have another fire, I don't have anything from the first one yet, so you know that kind of stuff should be at the top of [your] priority list,"* says Linenger.

Sang has just received the data. The air, he says, appears to be fine. He volunteers to send up the results in a radiogram.

"We got your letter today [about voice comm]," Sang continues. *"I can understand where you're coming from."*

"I think it's the only way to go," says Linenger, reaffirming his decision to refuse all further voice communication. *"I think it's a rational choice. If we want to try and fix it, we should shoot for the next flight. . . . We shouldn't waste any time anymore. You know, John Blaha had the same thing, and I know the reality. It's just not going to work."*

As if to second Linenger's decision, the comm goes out briefly.

"We have to start improving the alternate means as best we can," Linenger says when he comes back.

"Oh, I copy that," says Sang.

"Yeah, I'm sure you'd do the same thing. During the day I wrote down notes and by the time we get to our comm session I basically can only tell about a tenth of what I've written down. Basically I'm wasting about 45 minutes per session getting prepared for it and then we get nothing out of it. And I think you don't hear half of what I'm saying. So I'm not sure you're getting good data down. But I just think we ought to cut our losses here and just try to do something and do it right. So, that's where I'm coming from."

"Okay, I understand. Just a heads up, the management both on Russian and NASA sides are probably going to have us still schedule two passes per day and give you the option of whether to talk or not."

"That's unacceptable. I will not participate in a failure. . . . We have to use any time we have to do radiograms so [we can] have comm about science back and forth. . . . I don't want to participate in failures."

"Understand."

Later that day, Tsibliyev and Lazutkin get their first glimpse of the spectacular Hale-Bopp comet from the windows in base block. It is not an omen of good things to come.

The NASA team's assumption that Linenger will quickly resume voice comm proves misguided. The following week, in fact, Linenger refuses to take part in his weekly PMC (private medical conference) with Marshburn.* Up till this point, the astronaut's self-imposed isolation has been

*Under their contract with NASA, the Russians agreed not to listen in on these private medical conferences or on the astronauts' private family conferences (PFCs). However, no one on the American side believed these calls were private; at one point, Marshburn passed a Russian conference room where he could hear an American PFC blaring loudly on the stereo. He began calling the PMCS "MCs."

at worst a temporary nuisance. But refusing the PMC is a far more serious matter. NASA needs detailed medical data on its astronauts, especially those in the brave new world—to Americans—of long-duration spaceflight. And especially those on an eleven-year-old space station whose atmosphere may or may not entirely clear, and which is prone to intermittent leaks of all manner of toxic fluids.

"Tom, you can't let him do that," Marshburn's boss at JSC, Roger Billica, tells the young flight doc when Marshburn calls for advice. A call goes out to Dave Leestma and Bob Cabana at the Astronaut Office, and their reply is categorical as well: "No way."

Up at the Phase One offices, Jim Van Laak is beginning to get angry at the difficult loner they've put in space. "We just have to draw a line; he's not up there on holiday, free to do as he chooses," Van Laak grouses. "There are rules here, and there are rules we have to enforce."

"He's completing isolating himself," Sang tells Culbertson, who agrees. The situation is growing more serious than Culbertson had thought. He begins asking around about Linenger, seeking out friends or anyone else at JSC who thinks they know the man. He finds almost no one. "And I really tried," Culbertson remembers. One exception is Scott Parazynski, an astronaut who trained with Linenger. Parazynski isn't much help. "That's just the way Jerry is," he tells Culbertson.

Culbertson, like the ground team in Moscow, decides to treat Linenger gently for the moment and hope he will eventually decide to leave his cone of silence. It falls to another Mir astronaut, Mike Foale, to take more direct action. Foale sends an e-mail message to Linenger urging him to get back on voice comm. Foale thinks he understands Linenger and his view of the best way to work within the Russian system. The two men had engaged in long debates at Star City over the purpose of the Shuttle-Mir program. Foale argued that it was a way for Washington to draw Moscow into a Western-style partnership while pumping money into an industry to keep its technology and its scientists from drifting off, perhaps to renegade countries like Iraq. Linenger didn't buy it. To him, the Russian contract was a contract pure and simple. The Russians were being paid $400 million for their services, and by God, they ought to provide them. Foale thought it was hardly a mind-set that would endear Linenger to his crewmates.

"I knew he had this worldview that was getting in the way of his relationship with the Russians, and with everyone on the ground," Foale recalls. "Plus, he wants to be an efficient *worker*. He wants to *produce*. I told him [in the e-mail], 'I honestly believe that not talking to the ground is a serious mistake. I know you're doing it for the very best of reasons, but you must understand, it's having a very negative effect you didn't intend.' " (According to Foale, Linenger didn't respond to his message. As Linenger remembers it, he did.)

It falls to Marshburn to tell Linenger that he will have to continue the PMCs, which usually take place on Wednesdays. The news does nothing to improve the astronaut's mood. On Friday March 14 comm fails when Linenger is on the verge of reading some of the letters to his son on a National Public Radio broadcast. He has been looking forward to the occasion for days. "Unbelievable," he notes in a terse e-mail to Sang. "Can't execute a ten-minute comm pass."

Sunday night, March 16
Moscow

In a fashionable two-room flat in the east Moscow neighborhood of Izmailovo a party is going on. There is vodka and music and hors d'oeuvres and plenty of good cheer. It is the hostess's fortieth birthday, but as friends and family swirl through the little flat, the birthday girl sits on a couch and nervously eyes the phone. She is expecting an important call. When it comes, she snatches up the receiver expectantly.

"Can you hear me okay?"

Tsibliyev's voice sounds as if he speaking from inside a distant tin can. Technicians at the TsUP have been working for days to make sure the patch from Mir goes through with as little static as possible. Lazutkin is listening in.

"I can hear you fine, it's okay," Tamara Globa says. *"How are you, my little ones? I'm always thinking of you, missing you. I was worried."*

"What you told us about, Tamara, everything has happened like you said it would," Tsibliyev says after some small talk. *"The worst month, February, is already over . . . and now everything is fine."*

"Vasya, how do you and Sasha feel now?"

"Good."

"Have you seen the comet?"

"The comet can be seen very well. It's very beautiful. Very big."

"It can be seen only once every two thousand years."

"I know," says Tsibliyev. *"We already agreed that next time, in two thousand years, we are going to look at it again."*

Globa chuckles. *"Next time you will tell me about it, by all means."*

"What's happening there on Earth?"

Globa mentions a recent devaluation of the Russian ruble, then reminds Tsibliyev of one of her predictions. *"Remember,"* she says, *"an unusual time period is supposed to start for you soon."*

Tsibliyev laughs. *"It has already started."*

"Has it? . . . I mean besides the problems that you have now. I think they are going to go away soon and something that will be quite interesting for you will start. Despite all these difficulties and danger it should be a wonderful flight. . . . Vasya, I also wanted to tell you, there will be one more difficult event. It will be at the end of March but it will not be like the one you had in February."

"Thank God."

"Sasha, take care of your health. And you, Vasya, pay attention to your legs. Understand? Yes?"

"Do Svidanya, Tamara."

"Do Svidanya."

Wednesday, March 19
8:33 P.M.
Aboard Mir

"Since you haven't mastered the art of using a toilet yet yourself," Linenger is saying, *"maybe I can give you some tips."*

After several attempts, Linenger has finally gotten the chance to read some of the letters for National Public Radio.

"Urinating for us men is a piece of cake. Turn the lever at the end of the suction holes, give it 30 minutes or so to create a bit of vacuum, grab the funnel, and fire away. Wipe it clean at the end, so it's ready for the next guy. Son, you can do it.

"Now, the business end is a bit trickier. Positioning is the key. If you're not

exactly lined up—if you're not perfectly centered over the two-inch hole, and if you don't pull yourself firmly into the seat and make a good seal, well it ain't pretty. And, as unbelievable as it might sound, maybe even a bit worse than your diapers. The more you put into the propulsion end the better. When you're done, you peer in. Make sure you have good separation. Use the tissue sparingly, because a five-gallon tank fills up quickly, and there are only so many tanks on board, and you definitely don't want to wake up some morning panicking and find that all the tanks are full.

"Well, that's about all the fatherly advice I can give you for one night, John. Now that I think about it, diapers aren't so bad. I'd stick with them until you hit two at least. Love ya, John. Good night. Sleep tight. Dad."

Linenger reads four of his letters, and Marshburn remembers it as one of the most emotional moments of his mission. Floating down in Spektr, Linenger has just finished reading and is still on open comm when a loud pop can be heard.

"Tony, we just lost all power," Linenger says to Sang. *"It's pitch black in here. I think a relay just shot down."*

The sudden power outage in Spektr climaxes what has been a disorienting day—literally—for the crew. That morning, one of the station's primary angular-rate sensors blew out. The automatic switchover to a backup sensor took three minutes, but by that time the station had already lost its "attitude," that is, it began slowly spinning in space. Linenger watched in amazement as the station's walls began rotating around him. Using a set of commands radioed by the TsUP, Tsibliyev managed to stabilize the station by firing its main thrusters, but even so he was unable to keep all of Mir's solar arrays pointed at the Sun. The station switched to battery power, and the TsUP switched off the gyrodynes to conserve energy. An hour before Linenger suddenly lost all power in Spektr, even as ground controllers figured out how to replace the broken sensor, a low-power alarm had sounded in Spektr.

It is the kind of problem the TsUP quietly wrangles with on a regular basis, but it is startling to the young members of Sang's ground team. Power flickers off and on in Spektr all that night. The next day, when Tsibliyev goes to replace the malfunctioning sensor, he discovers that his access to it is blocked by a newly installed gyrodyne. In the end, it takes several long days and nights to fix the problem, which Tsibliyev and Lazutkin manage by stringing a set of cables to a properly functioning sensor in the Kristall module. Months later, Sang vividly remembers the incident.

"I actually got to talk to [Jerry] that night," he recalls with a weary smile. "It was a treat."

Sunday, March 23
TsUP

Culbertson arrives in Moscow for meetings this afternoon, and the first thing he does is hustle down to the floor of the TsUP to talk to Linenger. Sang and Marshburn had e-mailed the astronaut ahead of time, alerting him to the fact that Culbertson would want to get on comm as soon as he arrived. Linenger replied with an e-mail for Culbertson laying out many of his complaints, including the ongoing problems with the Form 24s. Among other things, he is irritated that the Russians interrupt his science to draw blood for medical work. "If they want to do [an experiment] on me—they should ask Tony if he can fit it in," Linenger writes. "We are partners, not underlings. Tony is the Ops Lead—fully capable of that role—and should be treated as such. Frank: you put together a very good Ops group. They need authority to do their job properly. They give me a plan. I execute. We both go to bed happy, knowing we are doing good work together as a team. Our science is important—it should be treated as such. (Without that, why are we up here? Just working to try to keep the station going; see how long we can hold out?)"

Marshburn escorts Culbertson to the NASA console and gets him set up with a headset. After a bit, Linenger's voice comes through loud and clear, filling the little auditorium.

"Frank, I've got to hit you with something right off the bat," Linenger begins.

"Go ahead," Culbertson says.

"All right. I've got a problem with the Form 24 every damn day and I can't tolerate it anymore. So starting Wednesday I'll tell you all the things I'd like to do. Starting Wednesday I want the Form 24 to say 'Linenger work per radiogram' and I want Tony to send me something directly telling me what the plan is. I've got Tony Sang—who knows what the hell he's doing—I've got a capable group, but by the time [the Form 24] goes through the system I get a plan that is a mystery book every time. I wasted three hours yesterday running data takes that didn't need to be

done. Tony gives me sample numbers, the things never arrive. There's no quality control of anything, and I can't tolerate it anymore."

As he sits at Culbertson's side at the NASA console, Marshburn's eyes widen. Linenger's voice is high-pitched and shrill; he sounds as if he is on the verge of some kind of breakdown. Culbertson, on the other hand, is hit by a sense of déjà vu. The anger, the demands, the sense of needing control over his situation—this is exactly the kind of outburst he had heard from Linenger during his training at Star City. It isn't easy to listen to, but Culbertson realizes it may be therapeutic to let his overstressed astronaut vent awhile.

"Okay," Culbertson replies, keeping his tone soothing. *"So you're getting something different from what you expect on the Form 24?"*

"The Form 24 is a worthless piece of paper and what we need is Tony writing something that says, 'Jerry, tomorrow we want you to QUELD. I want you to run sample A and B and do it blah blah, and that's what I need. The Form 24 ties Tony up, it takes him all day to get it through. . . . I don't know how long to get it through."*

"Okay."

"The worst, it is absolutely worthless, it is so simple to do this right, and we're just, you know—I've got a great team down there, and I feel isolated as hell up here from them."

Culbertson sees a small opening. *"Well,"* he says. *"They're anxious to start talking to you again, too."*

Linenger will have none of it. *"Well I don't care about voice comm. We're talking clearly here but normally we'd be interrupted here right now and they'd say, 'Okay, that's enough,' or we'd have terrible comm and we wouldn't hear each other."*

Furthermore, Linenger goes on, the Russians don't need to know every detail of what he is doing. *"For my procedures, they absolutely don't need to know how many times I need to push ENTER on a given QUELD run. I mean, it's worthless to them. We're professionals up here. When I'm running QUELD and it's using up power, I tell the commander I'm using up 290 watts, is that okay? They don't need to know that detail. It's wasting time."*

Linenger is just gathering steam. *"The second thing is telemetry. I've got files that have been up here for five days. . . . If Tony doesn't get it for five days, I'm sitting up here frustrated. And when I do telemetry down, for example, when I sent the fire report down, it sat there. I put it in the middle of [German] files so that I'd*

*An experiment in which Linenger bakes alloys in his onboard furnace.

know when it got to the ground. The next day Reinhold [Ewald] told me they got all the files—his group [did]. They already gave him feedback on it. [My] message got to you five days later. . . . It was sitting on someone's desk or circulating among the Russians for five days before our group gets it, and that is just unacceptable."

"Yeah," Culbertson says, *"that's one of the things I have to work on."*

"I'm aware of that," Linenger says. *"And as to the other thing, Tony just said they tried twice this week [to receive my e-mails] and had no joy getting it down, and did I have it cued up? I get that feedback, you know, three days, four days later. And I know when someone goes into the telemetry system because it shows up very plainly on my screen. And nobody touched that telemetry system . . . [So] if they don't get it, for God's sakes they ought to call the commander and say, 'We've got a problem with telemetry—check the cable!' Or if they've got another system problem they ought to take care of the problem."*

Linenger isn't finished. *"Let me tell you what I'm blasting all this for. 'Cause when I came on board, I told myself I'm going to talk up and make things change, or I'm not going to do a damn thing. And it's getting very close to that point. And Tony Sang probably feels the same way, 'cause they work hard down there to do a good job and I get lousy information up here."*

"Okay," Culbertson says. *"We'll see what we can do to fix that, Jerry. I understand the problem clearly, and we'll work it out."*

"Okay, thanks, and I'm sorry—I'm just blasting away," Linenger said. *"[Now] on the fire stuff. I saw one little inaccuracy. A little article [I read] said 90 seconds for the fire, and [it] was more like at least ten or twelves minutes minimum. If you're the people that are looking at the safety of that, [and] said [it] burned for 90 seconds, that's just dead wrong. So someone ought to correct that."*

"That was based on your report, Jerry," Culbertson said.

Here Culbertson is clearly mistaken: He is mixing up Linenger's report of a fifteen-minute fire with the Russians' report of a ninety-second fire. The misunderstanding, however, underscores Culbertson's relative lack of attention to the matter.

"Ninety seconds? It burned?"

"Minute and half is what you had in the report," Culbertson says. *"I know the whole situation lasted a lot longer."*

"Well, I gotta correct myself there, I made a mistake. It was definitely ten, twelve minutes it burned. We had flames for ten to twelve minutes."

"Okay."

"And if I did that, I apologize. . . . You talk to anybody here, and it took us 90 seconds just to get the gas masks on . . ."

Here the TsUP interrupts. It is time to begin wrapping up.

"Okay, Jerry, thanks for the info."

"Okay, sorry Frank, but starting Wednesday I need Tuesday night a plan I can execute, because the one thing that does—" He starts over. *"Marsha [Dunn, an Associated Press reporter] was asking what bothers me at night, and what bothers me at night is when I do a lousy job during the day. When I don't have a clear plan. It's so garbled up by the time it gets through the system, it's worthless."*

"Okay, well, I'm sorry it feels that way to you. Actually you're doing a good job from our perspective, and I understand the things you're dealing with."

"Okay, we're executing. I don't disagree. We're doing the science. But I'll tell you, it's painful, and it's wasting a lot of my time."

"We don't want it to be any harder than it has to be," Culbertson says, signing off.

Culbertson puts down the headset and turns to Marshburn, who is clearly stunned by what he has just heard. For the first time since the fire, he is frightened.

"Frank," the flight doc says, "I have *never* heard anything like that before."

Culbertson has. He reflects on the similarities between Linenger and Blaha. Blaha was a commander, accustomed to giving orders and having them followed. Linenger is a naïf and a perfectionist, and like most perfectionists, a complete pain in the ass. "I hoped Jerry felt better after that," Culbertson remembered months later, "because I sure felt like shit."

Culbertson rises from his seat, and with Marshburn at his side, strides straight into Viktor Blagov's office. By the time Culbertson leaves the TsUP that night, NASA has permission to uplink a new set of English-language Form 24s on which Sang can have final approval. With any luck, Culbertson hopes, it will settle his astronaut down.

While Culbertson's visit to Moscow clears one problem plaguing Linenger's mission—the Form 24 snafu—it plants the seed of another, more insidious issue. In furtherance of the "cover-up" he already suspects, Linenger now sees Culbertson's mistaken assertion that the ninety-second estimate came from the astronaut's own report as evidence that someone—either at the TsUP or at NASA—has doctored his report on the fire. He suspects it is the Russians, scrubbing away any taint of scandal or incompetence from their beloved space program.* The day after talking with Culbertson, he e-mails Sang: "Need to know if your copy of [fire] report said 90 sec. burn. If so, bring your copy with you and ask for immed. voice comm. Have them get me. Also, byte count. (Asked for 2 weeks ago—never rec[eived]; Why not?)" Linenger wants to see the byte count, that is, the length of the report in the computer, so he can see whether the report has been tampered with.

The next day, March 25, when Sang hasn't responded, Linenger e-mails again: "ASAP: Tony: grab both my reports on the fire. Read thoroughly. Then voice up to me directly (have someone grab me) whether it said 90 sec. of burn. Yes or no; and the byte count on both."

Sang checks the reports and in coming days repeatedly assures Linenger there is no mention of ninety seconds in them, and no sign of tampering. Marshburn sees what Linenger is doing and grows worried. He doesn't think there is any conspiracy. What he suspects—what he fears—is that Linenger is exhibiting signs of paranoia.

Sang too grows concerned, especially after making a quick trip to Germany for a debriefing with the German space agency on Reinhold Ewald's mission. "So," he asks Ewald on the drive back to the airport in Munich, "how long *was* the fire?"

"Ninety seconds," Ewald says.

"Really."

Of all the people examining the ebb and flow of Jerry Linenger's moods, none was more fascinated than Al Holland, the avuncular NASA psychologist who padded around JSC in khakis and gym shoes. This, after all, was exactly what had drawn Holland to NASA thirteen years before, the chance to analyze astronauts during spaceflight. And it was exactly what NASA for ten of those years had refused to let him do.

NASA, it has been said, is not an organization of scientists, but of pilots

*Viktor Blagov confirms that the Russians routinely read all e-mail traffic between the Mir astronauts and their ground teams. Linenger's messages were downloaded to a computer in a part of the TsUP that the American ground team was forbidden to enter. Anywhere from thirty minutes to three days after the downlink, a little Russian man whom Sang and Marshburn dubbed "Beaker"—for a *Sesame Street* character with a high, squeaky voice—would bring them a computer disk containing the messages. "Sometimes he would show up, sometimes he wouldn't," remembers Marshburn. "But we were never allowed to go find him."

and engineers, two groups for whom comfort and expertise are found in the precise analytical tools of mathematics and physics. Human behavior cannot be measured by a slide rule, which is just one reason the pilots and engineers who run NASA have long distrusted anything that smacks of psychology. As far as most astronauts are concerned, the only person more loathsome than a flight surgeon is a psychologist, for the simple reason that the last thing an astronaut wants is to be disqualified for flight by some couchbound know-nothing guessing that he might not perform well in space. For Holland the defining moment of his NASA career had come early on, when the former Skylab astronaut Joe Kerwin turned to him during a meeting after Holland made a suggestion for dealing with a shuttle crew.

"Son," Kerwin said, "you gotta understand, the crews won't be happy until the last psychologist has been strangled on the entrails of the last flight surgeon."

It had always been this way, although during the 1960s a series of psychologists within the agency had persevered against this bias and compiled reams of data about astronaut performance, moods, and efficiency during the Mercury, Gemini, and Apollo programs. Their findings, however, were always closely guarded. When JSC in 1984 hired its first full-time staff psychologist, a flight surgeon named Patricia A. Santy, Santy was amazed to find that all of the historical data had somehow vanished. "I discovered that all the work of my predecessors had disappeared into a black hole," Santy writes in *Choosing the Right Stuff: The Psychological Selection of Astronauts and Cosmonauts,* an academic text she wrote after leaving NASA. "No records existed at NASA about their work. There was no record of psychological selection on the medical charts of astronauts. There were no archives that housed all the data collected; there was, in truth, a complete absence of behavioral sciences in *any* recognizable form in every part of the Space Program from selection and training to flight. When I questioned why this was so, I was accused of wanting to 'destroy' the agency and described as 'dangerous.' "

By the mid-1980s, in fact, when Holland joined NASA, the only role for psychologists in the manned space program was in the biannual selection of astronauts, when an outside consultant or two was called in to make sure none of the applicants were psychotic. Beyond an outright case of psychosis, however, the psychologists had very little input into the selection process. Their focus was purely on what NASA calls the "select-out" process—meaning who was mentally unfit to be an astronaut—and never the "select-in," meaning who was best fit to fly. (After years of lobbying, Holland was allowed to begin "select-in" analysis in about 1994.)

Holland was an industrial psychologist working in Houston when he received a contract to design a screening process to select console operators in Mission Control. The project was ultimately canceled, but in the meantime Holland befriended Gene Kranz. Kranz was interested in psychology but not necessarily for astronauts; he asked Holland to look at ways NASA could retain its best people. Holland came back with a procedure called validated personnel selection, a series of psychological and skills tests that every new NASA employee—and astronaut, Holland suggested—would take. This was hardly a radical idea; it was, in fact, exactly the kind of thing IBM and General Electric had been doing for years. The NASA hierarchy wouldn't hear of it. "The attitude was just 'no, no, no, no, no—no tests,' " Holland recalls. "To do that, people would have to assess each other. Whoa! No way they would ever assess each other."

After joining NASA full-time to work with Santy, Holland promoted his testing idea for years, to no avail. He came up with a stream of similar ideas—"team training" for shuttle crews, a more rigorous astronaut-selection process—but got nowhere. "For years it was basically impossible to do anything," he remembers. "I couldn't even get appointments to see most people. No one would see me." The only time he was allowed to get involved with astronauts was during the selection process, and then he rarely got the sense his ideas were listened to. He tried to get involved in the Space Station Freedom planning, crunching Soviet data and American Antarctic studies in reports on "habitation issues," or how astronauts could best deal with the isolation of long-duration space missions. But for the most part the only work NASA would fund was basic research, which Holland, Santy, and others wrote up in obscure articles with titles like "Multicultural Factors in the Space Environment: Results of an International Shuttle Crew Debrief." "We didn't have hardly anything to do during the shuttle era," Holland says. "We were like Maytag repairmen. Seven to fifteen days on the shuttle. I mean, why do you need us? Anybody can get along for two weeks."

Everything began to change with the 1992 announcement that a single NASA astronaut would visit Mir. For Norm Thagard's initial flight in 1995, Holland and a panel of Russian psychologists drew up a weekly questionnaire Thagard and his cosmonaut crewmates would fill out to help the doctors study crew interaction. It was standard stuff: How are you sleeping? What is your mood? How are you adapting to space? Holland worked on the proposal for months. NASA killed it. "No one appreciated how important psychology was at that point," Holland says. "We were babes in the woods."

After the Shuttle-Mir program was expanded in 1993, Holland finally managed to convince his bosses that a full-blown program of psychological support was needed. Ironically, he based his argument not on the usefulness of his work but on the fact that the Russians were doing it; if NASA didn't look out for the psychological health of its Mir astronauts, Holland argued, the Russians would, and no one wanted that. The program Holland devised for Thagard's first mission, and honed in later increments, was on its face commonsensical: uplinks of American news broadcasts, personal family conferences, and e-mail, all tools to make the astronaut feel less isolated. By the time John Blaha went to Mir, Holland had stocked the station with dozens of American movies, heavily oriented toward science fiction and stories of people persevering against long odds: *Scott of the Antarctic, Alien, Die Hard, Independence Day,* a half-dozen *Star Trek* movies, *A Time to Kill.* One of NASA's newer innovations could be found on Linenger's IBM Thinkpad. It was a software program called the Crew On-orbit Support System, or COSS, which included science procedures, computer games like Myst, training simulators, and family snapshots. COSS was intended to be the astronaut's best friend; when an astronaut wakes up on Father's Day, for example, a computer message led him to a video of his family wishing him a happy Father's Day.

Meanwhile on the ground, Holland arranged "cultural training" for all the Mir astronauts and their families, everything from coaching them on how to order in Russian restaurants to briefing fourteen-year-old Danny Thagard on Russian dating mores. "This changed my life," says Holland. "I wanted to do this for years, [but] the agency wouldn't entertain it. Suddenly, now we had things to do." For Holland the only disappointment came when the decision was made against performing any psychological screening of the Mir applicants. The reason was simple: Thagard, Lucid, and the others were the only volunteers NASA had, and it couldn't afford to lose even one.

Analyzing astronaut behavior aboard Mir was more problematic. Linenger and his predecessors had at least two questionnaires they filled out on crew interaction and behavior: a weekly report they filed anonymously for a San Francisco researcher named Nick Kanas, and a second report that went straight to Holland's group. The problem, of course, was this approach relied on the astronaut's own input, which was hardly objective. John Blaha had managed to hide his depression from Holland and everyone else on the ground until his postmission debriefings.

The Russian Institute for Biomedical Problems, called IBMP, employed dozens of doctors whose sole responsibility was monitoring the mental health of cosmonauts. The tools IBMP used ranged from the mundane to the bizarre. Its psychologists paid close attention to body language: If two cosmonauts didn't look at each other during a videoconference with the ground, it might mean they weren't getting along. If a cosmonaut crossed his arms, it might be a signal that he was withdrawing or growing defensive. The IBMP doctors also used a homemade voice-stress analyzer to study speech patterns; the Russians always insisted they never used this device on the Americans aboard Mir, but Holland thought they did.

Both programs had good reason to follow crew behavior closely. Though they furnished few details, and never discussed the incidents with outsiders, the Russians told Holland that cosmonaut psychological problems had prompted the TsUP to prematurely end three missions in the last twenty years.* While NASA had launched its share of prickly astronauts over the years—Wally Schirra's obstreperousness during Apollo 7 is often cited—it had endured only one known psychological incident of any real severity, the "mutiny" of the last three astronauts to occupy the short-lived Skylab space station in the winter of 1973–74.

That third Skylab crew—Commander Gerald M. Carr, pilot William R. Pogue, and science pilot Edward G. Gibson—had gone on a one-day work strike at the end of its sixth week in space after voicing many of the same complaints that had driven Blaha to despair: overwork, primarily, and the ground's failure to accurately time-line its daily activities. The problems arose when the ground insisted on modeling the third crew's time lines after those of the second Skylab crew, a famously hardworking bunch headed by the astronaut Alan Bean. But the third crew had been a bitchy group to begin with: Carr complained that Skylab's soap was like shampoo; Pogue carped that the towels felt like steel wool; all three hated the toilet, which was put halfway up a wall. "I don't know how that was designed," Pogue complained at one point, "but I'm sure it wasn't by anyone who took a crap and noticed his posture." At one point all three

*According to NASA sources who have spoken at length with Russian psychologists, these missions were Soyuz 21 in 1976, shortened because of unspecified "interpersonal issues"; Soyuz T-14 in 1985, "mood and performance issues"; and Soyuz TM-2 in 1987, "interpersonal issues [and] cardiac irregularity."

men had to be reprimanded—the first time a NASA astronaut was ever publicly reprimanded—when a nauseous Pogue threw up and Carr, instead of packaging and sealing the vomitus for analysis as required, elected to flush it out the trash airlock. "We won't mention the barf," Carr said, unaware that his comments were being tape-recorded by the ground.

That NASA faced exactly the same problems with Blaha and Linenger spoke to the agency's lack of institutional memory; twenty years after Skylab, while anecdotes from the mission remained fresh in some participants' mind, NASA had few details of the Skylab crew's behavior in its archives. In preparation for the Mir missions, a young flight surgeon named Mike Barratt was forced to track down the Skylab crews' medical records by obtaining them from one of the retired astronauts. Nor were the Russians any help. In twenty-five years they had published only a smattering of academic articles about the behavior of cosmonauts. "We basically had to start from scratch," says Holland.

While no one ever ruled out the possibility of an astronaut experiencing a psychotic episode in orbit, it was the kind of behavior exhibited by the third Skylab crew—defensiveness, mild paranoia, and general orneriness—that Holland tended to focus his efforts on preventing. Every astronaut was different, of course. But in a series of briefings he gave to each Mir astronaut, Holland laid out a laundry list of symptoms that struck many long-duration crews: boredom, irritability, mild depression, reduced initiative and motivation, anxiety, fatigue, reduced concentration, sleep disturbances. The critical period came four to six weeks into a mission, when the newness of space wore off. At that point, Holland told each astronaut, borrowing a phrase he had once heard an Arctic explorer use, it was time to "put on your blinders and head north," that is, dip down into whatever inner reserves of willpower every person has and put them to use. Be aware of the mental aspects of long-term spaceflight, Holland preached, and be prepared to fight them. Sunday nights held the greatest potential for depression; even astronauts were capable of dreading a coming workweek. "Celebrate at any excuse," Holland urged.

The most common psychological problem seen in the shuttle program was what Holland sometimes called the "you-they" scenario, where an antagonistic relationship developed between the astronauts and the ground, usually because the crew felt overworked or ignored. To prevent this, Holland tried to emphasize full and open communication between everyone involved; the more everyone talked, the fewer misunderstandings that arose. In fact, the crew was only part of an equation that Holland divided into three distinct components: the crew member, the ground team, and the organization behind them. "Most of the problems originate with the organization," Holland says. "Do [the crew and the ground] feel supported by management? Are they talking enough with management? That's where 80 percent of the problems originate."

It was the very problem that plagued Blaha and both his predecessors, and now was afflicting Linenger, who felt Culbertson and the higher-ups in Houston were ignoring his concerns about the fire. As the spring wore on, Holland spent long hours talking to Tom Marshburn and poring over Linenger's e-mails in an attempt to decipher the astronaut's behavior. Eventually he decided it was best to simply leave Linenger alone. "I felt we could still operate as it was," Holland recalls, "that this was not the beginning of a long period of [mental] degradation where his position would deteriorate. I thought it was a reflection of Jerry's personal style. I felt it wouldn't be the best thing for Jerry [to continue], and it wouldn't be the best thing for the mission. But it was workable."

To make sure, Holland urged that Linenger be given more and better-quality family conferences with his wife, Kathryn, and more opportunities to chat with friends on the ham radio, which Linenger seldom used. Short of yanking him off the station—which wasn't possible anyway—it was all they could do. Within days, however, the question of Linenger's behavior would be all but forgotten, as the station plunged into a set of new and far thornier problems than it had encountered in months.

10

Wednesday, April 2
Aboard Mir

The first hint something is wrong with the crew comes on Wednesday night, even as Tsibliyev and Lazutkin begin grappling with a sudden malfunction in the station's coolant systems. Specialists in the TsUP have detected a pressure drop in one of the pipes carrying coolant to the Vozdukh carbon dioxide removal system, suggesting a leak somewhere in the system. The affected coolant "loop," as it's called, is one of two redundant lines behind the wall panels in base block.

It is the second such leak the crew has found in the past week. On March 27 Lazutkin had gone into Kvant 2 to pick up some buckets and cleaning supplies. Lifting one of the buckets, he noticed the side of the container was covered with a film of liquid.

"Vasya, what's this?" he asked Tsibliyev. Both men ran their fingers through the slimy liquid and sniffed. It smelled like coolant used in the Elektron and Vozdukh systems, a suspicion they confirmed with a litmus-paper test. Lazutkin removed several wall panels and inspected the piping behind them. It was dark inside, but using a flashlight he found a large drop of coolant clinging ominously to one of the pipes. Bulging slightly in the zero gravity, it looked like an egg sack in some fetid celestial spiderweb.

Tsibliyev alerted the ground, which checked the pressure in the piping. If a serious leak existed, there should be a drop in pressure. There was no pressure drop. However, the threshold for a leak indication is fifty to sixty milliliters of coolant, leading Viktor Blagov to believe that the leak, if any, was tiny. The liquid on the pipes, Blagov told Tony Sang, may just have been some spillage from the last occasion the coolant system was refilled; they couldn't be sure. Blagov had sent up orders for the crew to shut down the Kvant 2 cooling loop, mop up the liquid, dry the pipes, and check the area frequently for reoccurrence.

The next day, however, Lazutkin found a crack in one of the Kvant 2 pipes, which the crew filmed and downlinked to the TsUP. Coolant was clearly seeping out of the pipes, and faster than the ground had thought possible. With the cooling loop shut down, the module began to heat up, and that weekend, while Russian engineers put together plans to fix the leak, temperatures began rising in the module. On Monday Tsibliyev reported it was "getting pretty hot in Kvant 2." Another one hundred milliliters of coolant was found on the pipes, suggesting a leak rate of eight or nine milliliters per hour. To cool down Kvant 2, the TsUP radioed up a series of computer commands to reorient the station so that the module would be out of direct sunlight. That, however, put base block directly into the Sun. The crew was warned base block would begin heating up quickly.

It has. By this afternoon the temperatures inside base block have spiked up toward 90 degrees. Searching for the suspected second leak, this one in base block, Tsibliyev and Lazutkin spend much of the morning and early afternoon on their hands and knees, bathed in sweat, probing around behind wall panels. It is grueling work, made worse by the heat. Around three they think they find the leaks, two areas of corrosion on some interior piping that seem to be seeping fluid. Lazutkin takes video of the corroded joints and daubs on some putty-colored sealant.

"It's clear this is the source of the leak," Tsibliyev radios down during a midafternoon comm pass. *"We are trying to keep the delta pressure constant, just as you instructed earlier."*

By evening base block has become a sauna, and an odious one at that. It smells like the inside of an auto shop, which is not surprising; the coolant leaking into the station's atmosphere is a chemical compound called ethylene glycol, better known as antifreeze, the same antifreeze used in millions of car radiators around the world. NASA toxicologists are already familiar with ethylene glycol, having analyzed its properties following a similar but smaller leak during Shannon Lucid's mission a year earlier. Tom Marshburn had worked a portion of Lucid's mission and was also familiar with the greenish liquid. He discovered that not much is known about ethylene glycol's effects on humans; most studies have been performed on animals. But ethylene glycol can clearly be fatal if drunk in large enough quantities. People who have committed suicide by drinking antifreeze die of kidney failure. Children who have drunk smaller quantities of the sweet liquid by mistake have suffered kidney damage. The kidneys tend to remove about a third of the chemical; the rest is metabolized by the liver into formaldehyde and formic acid, which quickly eats away at the kidneys and causes brain damage. It is an agonizing way to die.

Marshburn isn't too worried that any of the three crew members will ingest ethylene glycol directly. Wearing the 3M surgical masks—the same ones used after the fire—should greatly reduce the chances of accidental ingestion. Nor are fumes that much of a worry. The crew could probably breathe fumes for months without suffering any effects. The main worry is the station's drinking water. That week the Russian doctors rule that the crew cannot drink any of the condensate water on board until tests can be run to make sure glycol levels in the water are safe; luckily, there remains enough other potable water for the mission's duration. Because there is no way to test the water aboard the station, samples must be taken to Earth, which can't be done until the next shuttle arrives in mid-May.

A few minutes before seven Wednesday evening, Linenger surprises the ground by appearing for a comm pass and talking to Marshburn, briefing him on some mild chemical reactions Tsibliyev and Lazutkin are experiencing after spending all day up to their elbows in ethylene glycol. Their symptoms include rhinitis, an inflammation of nasal tissues, as well as several "skin burns" where strong portions of the chemical had touched the Russians' skin. He has performed preliminary exams on both men, Linenger says. Pulse is fine. Nasal discharge is generally clear. Lungs are clear. But, Linenger cautions, he strongly advises that both men use 3M masks and rubber gloves when working on the chemical-streaked pipes. From what Linenger says, it sounds as if neither man has been doing so.

"Jerry sounds great!" one of the interpreters scribbles in the NASA log. "Sounds upbeat, happy to chat."

But the Russian ground controllers aren't happy to learn that Tsibliyev and Lazutkin apparently haven't donned their protective materials. *"Guys, be more careful when you work with thermal loops,"* the shift flight director Viktor Chadrin tells Tsibliyev on a pass later that evening. *"Follow all the protective measures. Don't take risks."*

"We do that," Tsibliyev snaps back.

"Jerry told us of some problems [during] the previous comm pass."

"Like what?"

"You had nasal irritation and redness on your arms."

Tsibliyev turns to Linenger and makes a comment that everyone at the TsUP can hear. *"Jerry, don't mention that any more,"* the commander can be overheard saying, *"or we won't mention it to you any more."*

Standing next to Chadrin's console, Marshburn cringes. There is no mistaking the anger in Tsibliyev's voice.

Thursday, April 3
Aboard Mir

The situation inside the station is growing worse by the hour. By Thursday morning, temperatures in base block have risen to more than 90 degrees. The humidity is rising as well, and antifreeze fumes are growing thick. After daubing sealant on the leaking pipes in base block, pressure levels indicate there is at least one more leak in the station, maybe more. Tsibliyev and Lazutkin don their protective gear and plunge down behind the panels in Kvant to search for them.

Later in the morning Russian doctors get on comm and tell Tsibliyev that the crew must refrain from all exercise until conditions return to normal. This is a jarring order for all three men, especially for Linenger, who craves his daily runs like a smoker craves nicotine. For the two Russians a halt to all exercise has less to do with habit than safety. To survive the high G-forces of their eventual descent in the Soyuz, the Russian doctors have repeatedly emphasized how important regular exercise is to prevent potential heart problems.

While Tsibliyev and Lazutkin spend the day bathed in sweat, clawing through the station's innards in search of coolant leaks, Linenger stays out of sight down in Priroda, working on his experiments. Tsibliyev is growing increasingly irritated by Linenger's insistence on sticking to his own routines. While the two Russians work like slaves, Linenger rarely if ever offers a hand. What really angers Tsibliyev, though, is when Linenger eats meals and uses the treadmill, as per his daily schedule, even as the Russians work to repair the leaks. On Thursday and Friday the commander makes several sarcastic comments about Linenger on open comm. One of the first comes Thursday night, when the commander, describing the repair work to the ground, pledges to continue the work the next morning.

"Our only request [is], don't schedule this work when Jerry is doing physical exercise on the treadmill," he tells the TsUP at day's end. *"Whatever happens, he'll be in time on [the] treadmill for exercise."*

Up in the NASA suite, the interpreter Mike Malyshev turns to Marshburn as he finishes jotting down his translation of the commander's comments. "Look at this," he says. "Vasily is mad at Jerry. He's obviously pretty upset."

Marshburn is immediately concerned. It does look like Tsibliyev is angry. He hustles downstairs to ask one of the Russian biomed technicians—a man he knows only as Sasha—about the pass. Together the two men go over the transcript of Tsibliyev's comments.

"What do you make of this?" Marshburn asks. "Do you think Vasily is angry?'

Sasha downplays the commander's comments. "No, Vasily is upset, but probably not at Jerry," he says. "We think he believes Jerry is a good guy." He points to a Russian phrase Tsibliyev used to describe Linenger: *Un maladietz.*

"*Un maladietz,*" the Russian explains, "means 'he's a good boy.' "

Whether or not there are strains emerging between the two men, Sang and Marshburn realize Tsibliyev and Lazutkin are slowly being overwhelmed by all the repair work. They ask Blagov whether he would like Linenger to take a break from his experiments and help out. "We talked to the [TsUP] management on the fact that you have not been involved in much [if any] of the Mir [repair] activity," Marshburn e-mails Linenger. "Mr. Blagov said he is agreeable for you to participate in some of the repair work and Tony does too. If you are agreeable we would like you to participate in some of the repair. . . . Please give us your thoughts."

Though he says nothing to the ground, Linenger is incensed when he reads the message. Completing his experiments means everything to him; it isn't his love of science, but the fact that NASA will judge the success of his increment by how many experiments he completes. What angers him is the feeling that he is being asked to pay for the mess Tsibliyev and Lazutkin have made of their duties.

"I basically looked at [Sasha and Vasily] and thought: You guys have lost control of your lives and your schedule and your ability to work effectively," he recalls. "The ground made things so much worse. It never understood how overwhelmed they were getting. The ground is just jerking them back and forth, you know, 'Do this one minute, do this the next minute.' And if you look at who was really working the most up there, I was—*I* was. Yeah, they were staying up till four in the morning. But they were sleeping till eleven. They'd sleep all day. They weren't doing the work they should.

"[And so] when they told me to stop my science and help on repairs, I [thought]: Bullshit! Helping them isn't going to help anything! You can only get one hand behind a panel! It does no good! It was all diplomatic bullshit. The Americans were trying to look like good guys, helping out the poor Russians by asking me to help. Why should we stop this science, for which we're paying $400 million, and help out? It was worthless. The leaks aren't going to stop. The U.S. science was the only thing that worked."

Friday, April 4
Aboard Mir

Things get worse on Friday. Temperatures in base block are spiking up toward 100 degrees as Tsibliyev and Lazutkin, outfitted in 3M masks, goggles, and rubber gloves, crawl back behind the panels in Kvant searching for leaks. But then, right after lunch, they appear to catch a break.

"I have found a location that I tend to ascribe the leak to," Tsibliyev radios down at 1:19. *"It is in Kvant, right near the docking assembly."*

High up on one of Kvant's sidewalls, far back in a maze of shadowy coolant pipes leading to the Vozdukh carbon-dioxide removal system, the commander finds a set of interior panels coated with ethylene glycol. He cannot see the suspected leak site without cutting out large sections of piping, and he can only reach the area by sticking his entire arm into an otherwise inaccessible niche behind the panels.

"It looks like the leak is right under a fan," Tsibliyev tells the ground. *"The access is extremely poor, you understand. We can take out the fan and wipe off that spot."*

"That's not a fan," the ground tells him a bit later. *"That's an air cooler."*

And therein lies the problem. To patch the leak, Tsibliyev is forced to remove the fan and isolate the coolant pipes, forcing the ground to switch off the Vozdukh carbon dioxide removal system. By late afternoon, as the hot, sweaty work continues, carbon dioxide levels are beginning to rise throughout the station. Tsibliyev suggests switching over to the supplemental carbon dioxide removers, a set of lithium hydroxide canisters that can be used much like the "candles" in the backup oxygen generator. The trouble is, they only have five of these canisters left. More are due to arrive with an incoming Progress on Tuesday, but until then they only have enough canisters to use one every thirty-six hours; normally they would use three every two days. Until the Progress arrives, they will have to live with higher-than-normal CO_2 levels.

Tsibliyev is growing angrier by the hour and complains constantly about the ground. "Vasily wanted to be a martyr, he wanted someone to feel for him," Linenger recalls. "He had a fun-loving attitude; he was looking for the time things would get fixed and he could relax a little. He said point-blank to me, 'What am I doing up here? This is nothing like the last time [I was in space].' He always blamed the Russian economy, 'We don't have enough spare parts, we can't do anything.' He would go on for thirty, forty minutes, basically raving and cursing the ground, and I would basically close my ears, because I don't want to go down that spiral with him. It was like Captain Queeg. He would rant for thirty minutes, just screaming, and he didn't even realize I wasn't listening. He would just go on and on at how lousy the ground was. Everything made him mad. If the ground said, 'Brush your teeth,' he got mad. If they said, 'Work on the Elektron,' he got mad. Vasily was basically losing it."

Saturday, April 5
Aboard Mir

Saturday is another long workday. After Lazutkin redirects some air ducts, channeling cooler air from other modules into base block, the temperatures there fall into the upper 80s. But with the Vozdukh system still out, the carbon dioxide levels have risen to worrisome levels. That morning, after everyone gets a good night's sleep, Tsibliyev and Lazutkin go back to work attempting to patch the leaking pipes in Kvant; Linenger disappears into Spektr to begin a long furnace run. In the afternoon Tsibliyev informs the ground it will be necessary to cut out some steel girders to properly patch the latest batch of leaks. Blagov says hacksaws should arrive on Tuesday's Progress.

As the two Russians work, the commander makes a number of caustic comments about Linenger's refusal to help out. "It was a problem," recalls Lazutkin. "We all were a crew, like a family. Each member of this crew, this family, we have our responsibilities. At the same time, there are crew responsibilities that are not written down. You have to live with somebody and react to that person's needs, see when he needs help, not just when he asks for it, but when he needs it, you have to be attentive. You have to pay attention to each other, and not just to your own needs. Jerry paid attention to his own needs. We were just learning how to work together, the Russians and the Americans. Three people of one nationality, that's fine, we could handle the situation. But here there was a guy of a different nationality. We had to be patient. If Jerry were Russian, we wouldn't let him do things like that." Lazutkin sighs. "Jerry didn't behave like a Russian."

After the mission, Linenger would insist he regularly asked Lazutkin whether he could help out on repairs. "Sasha had his turf, and this was it," Linenger says. "I didn't feel welcome helping out. I didn't get the sense he knew what he was doing. He didn't like exposing his ignorance, you know? I'd go to help and he'd say, 'No, I'd rather do it myself.' "

The rising carbon dioxide levels are the main reason the ground has forbidden all three crew members from exercising; if any of them runs for one hour on the treadmill, he will exhale enough carbon dioxide to push the levels even higher. That afternoon Linenger, ignoring the ground's directive, goes to exercise on the base block treadmill. Tsibliyev objects. The NASA logs for that day note: "Jerry was trying to exercise earlier today, but Vasily [said he] had to 'almost push him off the treadmill' to make him quit."

"Jerry told us that he wouldn't feel good if he missed his exercise time, and that he was going to exercise anyway," remembers Lazutkin. "So what kind of feelings do you think we had? We realized exercise could do damage to his health, and ours. And he would be increasing the CO_2 [levels], which could mean the entire mission would be [ended] prematurely. We would have to go home if they got any higher."

Linenger reluctantly agrees to curtail his exercise period, but later that day brings the matter up in a comm pass with Marshburn.

"What you all are doing is exactly what you need to," Marshburn says. *"No less than two liters of water per day, and that would be the treatment on the ground as well. There's really no other thing [to do]. Just to let you know, even though the levels of ethylene glycol are kind of high by [NASA] standards, the standards are set to prevent performance decrement from headaches and so that means there's no permanent tissue damage going on."*

"I noticed you didn't get my last [e-mail] down," says Linenger. *"I guess you know most of [that] stuff already. I wrote in there basically where the leak was, above a ventilator, and it sprayed out about 800 milliliters gathered up, so there's quite a bit that was actually spewed into the air."*

"We didn't get that. I think the temperature is the biggest factor. It really makes that stuff [volatile]. Otherwise ethylene glycol won't get into the air too much. So when you get it cooled down that'll help a bunch."

"I think the ventilator thing got it into the air, there's no doubt in my mind. The leak was right above a turning ventilator, and stuff is all over the place. So I would revise what you think we got."

"Okay," Marshburn rogers. *"Anything else for us?"*

"CO_2 and exercise. Any words you got on that I'd be interested in. You'd hate to be working all this time then go into a bedrest thing for five, six, seven, eight, ten days."

"Yeah, they're concerned about the heat load as well."

"I'm not concerned about that," says Linenger. *"Is that the main concern or is it the CO_2?"*

"Both that and CO_2 generation. Vasily has said they're pretty low on CO_2 canisters. Now when Progress gets there, that changes everything, of course."

"Doesn't change everything. I understand the effectiveness of these [canisters] isn't that great because of the humidity [up here]."

"That's true," says Marshburn. *"And the heat doesn't make it very effective, either."*

"When do we start to be able to do anything again, I guess is my question. Even walking. I'd sure like to walk or use the expanders where I don't get my respiratory rate up and don't blow a lot of CO_2."

"Okay, I'll ask."

Linenger emphasizes how badly he wants to exercise. He needs to stay in shape. *"I just hate to see everything going down the tubes after all the work I've been doing,"* he says. *"The other [question] I have is a general question. Is there any concern or consideration of what Mike [Foale] is going to be doing here? I mean, I understand I'm here and we're having problems, but is anyone looking down the road for Mike?"*

"Yeah, no question about it," says Marshburn. *"It's all very high, top-level discussions, and Frank has been talking to us several times a day."*

"I think that's important. There's enough little things going on, that you start wondering, you know. It's a difference with me up here. I'm not concerned with myself, I guess, but the next guy coming up ought to seriously think what we're doing here, and make sure we understand a lot of these different things are happening."

"Yeah, fully understand, and I can guarantee Frank is asking and is being asked the same questions."

When the comm pass ends, Linenger silently curses his inability to tell the ground what he honestly feels. "Frankly, I thought it was time we all got the hell out of there," Linenger remembers. "But I couldn't say anything on comm because the Russians are listening in. And I got the distinct impression that Frank Culbertson didn't want to hear about it."

Marshburn puts down his headphones after the conversation with Linenger and rubs his eyes. He had tried to sound confident. It was important, he thought, not to let Linenger know how worried he had become. The breakdowns, the ethylene glycol in the air, Linenger's paranoiac symptoms over the fire report, and now the obvious tensions with Tsibliyev—it's all more than the young flight doc ever expected to witness. He had worked portions of Blaha's and Lucid's missions, and never felt that his NASA superiors placed that much importance on the flight surgeon's role in long-duration spaceflight. Not until the weeks after the fire, when Culbertson began peppering him with phone calls, did Marshburn feel his job was valued. "I never realized what the role of a flight surgeon at NASA was till then," remembers Marshburn. "Honestly, I didn't think anybody gave a shit what I did till Frank started calling me. I guess they cared, I just didn't know it."

But there is a downside to his newfound importance. It is at about this point that Marshburn has a nightmare he will never forget. In the dream Linenger has sent down a confidential memo. For some reason it's on a CD-ROM. When Marshburn slips the disc into his computer, there is no video, only sound. The sound of screams. Screams and screams and screams and screams. And then comes Linenger's voice, plaintive and unmistakable. He is wailing, over and over, "Get me out of here!"

* * *

In his office down from the TsUP floor, where a stereo speaker allowed him to listen to each comm pass, Vladimir Solovyov noticed the obvious signs of strain between Tsibliyev and Linenger. "I have been flight director for ten years, [and] the doctors say I have a good feel for the psychological side of things," Solovyov recalls. "Listening to the conversations, I felt there is some problem between the Russian crew and Linenger. I felt it in the way they were speaking to the ground, [and in] how they organized their work. Drop by drop, the situation was getting worse and worse. Several times [Linenger] didn't want to clean the station, several times he didn't want to vacuum. As I say, the situation just got worse and worse. We had been trying to repair the situation, for example, we wanted to give Linenger more time to talk to his family in Houston. And we told Tsibliyev, 'Please don't offend the American.' "

"On the station we've had twenty-seven crews, and we never had an episode like this," Viktor Blagov says of the friction between Linenger and Tsibliyev. "To be honest, I never want another person like Linenger aboard. He causes a huge amount of problems."

At first Blagov and Solovyov attributed Linenger's problems to inexperience. "I think he was just very frightened by the incidents aboard the station," says Blagov. "He pushed the notion that the station was very dangerous." But the more they listened, the more they came to believe the problems lay in Linenger's character. "Here we think of him as not being a gentleman," says Blagov. "He is impulsive and inclined to be panicky. He didn't like to speak to the Russian crew. He could be silent for days." At one point, says Blagov, "we asked Vasily, 'Do you want Frank Culbertson to talk to him?' But in a day, Vasily said, 'Stop worrying about it, everything's fine.' I think Vasily decided to just let it go."

As Blagov and Solovyov grappled with the situation, a more systematic—and impartial—analysis of the crew's difficulties was under way at Star City. The Russian psychologist who had warned against flying Linenger, "Steve" Bogdashevsky, was intensely concerned about behavioral changes he had seen in all three men since the fire. Over the years Bogdashevsky had developed his own system for classifying the personalities of cosmonauts using a mix of Jungian analysis and a lesser-known system called "socionics," which he said had been developed "by a nice lady in Lithuania." In his system Bogdashevsky classified Mir's occupants in three areas, which led him to identify sixteen basic personality types. Linenger was Type No. 4, a "logical, intuitive introvert." Bogdashevsky's colleagues had nicknamed this type "Robespierre." Tsibliyev was Type No. 15, a "logical, intuitive extrovert," what Bogdashevsky called the "Don Juan" type. The two personalities were nearly exact opposites, and mixing them together on the station had led to problems in the past.

In Bogdashevsky's analysis Lazutkin was the key to crew harmony. Lazutkin was Type No. 3—"Hugo"—an ethical, intuitive extrovert. "Sasha should have been the intermediary between Vasily and Jerry," says Bogdashevsky. "Everything would have been fine. But he couldn't manage to do that, to be the bridge to Jerry. I had conversations with Sasha before the mission, and I told him, 'You have this mission, this task.' But judging by everything, this huge egotism of Jerry's amazed Sasha to such an extent that he decided, as I understand it, not to fulfill this mission, but even to hurt Jerry, to ignore him."

The problems between Linenger and Tsibliyev, however, were only part of a larger problem Bogdashevsky felt the crew was facing: exhaustion. Ever since the first Elektron breakdowns a month earlier, Tsibliyev and Lazutkin's daily routines had been wrecked. Many nights the two Russians stayed up past midnight tinkering with the Elektron or other malfunctioning equipment. Bogdashevsky was a keen believer in the theory that a lack of sleep was a key indicator of stress, and he saw that both Russians were getting less and less sleep.

"We were first alarmed by the fire," he says. "Usually it takes a year or more to fully relieve the stress after something like this. It's scary, and you could see it. The fear was in them. It really changes a person's behavior. They became more cautious. They didn't feel as relaxed. We started picking up nuances we didn't pick up before. They became more demanding to the ground. For instance, [if] Vasily has a question. The ground says, 'Wait a minute.' And they became irritated. After the failed docking, it became clear to us that the psychological state of the cosmonauts was becoming worse. I wrote a paper with an unfavorable psychological prognosis, and I called for everybody's attention to change their attitude toward the crew. But the attitude remained the same. The attitude can be characterized as a sweat-sucking system. [The TsUP] just makes them work harder and harder."

Bogdashevsky's first warnings to the TsUP came in a report written on March 23. His preliminary diagnosis for both Tsibliyev and Lazutkin was exhaustion. Further, he felt Tsibliyev was suffering from something he called "ostheno-neurotic syndrome," a related condition. "When a person is osthetic," Bogdashevsky explains, "he gets tired faster and gets irritated. It depends on the person, the mind. One person can get depressed. Another person, his blood pressure changes. It depends." In Tsibliyev's case, it has led to increasing irritation, both at the ground and at Linenger.

Long hours of repair work during the week of March 20, Bogdashevsky warned the TsUP, has seriously disrupted both cosmonauts' sleep schedule and biorhythms. "Our medical group, which consists of psychologists and specialists in work and rest, wrote nine reports, and all the reports warned that this was dangerous," says Bogdashevsky. Foremost among Bogdashevsky's recommendations was more sleep and less work for Tsibliyev and Lazutkin. This didn't happen; if anything, the two Russians' workload actually increased as Mir suffered more breakdowns in coming weeks. Worried that Tsibliyev was growing overwhelmed, the doctors urged that the TsUP help the commander "prioritize" his work. "This was especially important for Vasily, because he is a man who likes order," says Bogdashevsky. "He cannot live without it. Everything must go on its own little shelf."

As for Linenger, Bogdashevsky advised the temporary suspension of the American science program. Instead of performing experiments, he suggested, Linenger should be put to work unloading the Progress cargo ship, which was due to arrive that following Tuesday. "Jerry was having what we called an 'alarm reaction,' which means he was mad at NASA," says Bogdashevsky. "He had a lot of emotions. He said a lot of things. We saw his alarm level was extremely high. That's why we asked him to help unload the cargo ship. We did this to try and get him involved in the team's work, so he could see the differences that the crew was dealing with, so he would take them personally. Plus, the experience of mankind shows that the best medicine is work. When a person works, he doesn't have time for unpleasant thoughts. It distracts his mind. We felt it was good for everyone to be distracted together."

While seeing the wisdom of getting Linenger more involved in repairs, Viktor Blagov ignored Bogdashevsky's recommendation to ease Tsibliyev and Lazutkin's workload. The fact was, the TsUP regularly ignored the Star City doctors; the Institute for Biomedical Problems had medical jurisdiction over Mir flights, and Blagov tended to dismiss Bogdashevsky and his people as knee-jerk apologists for lazy cosmonauts. Blagov, in fact, was tired of Tsibliyev's whining and often said so. His philosophy was to work the cosmonauts hard, and when they complained, work them harder.

"We do not work in ideal conditions in space," Blagov recalled months later. "It's tough up there, I know. But I'm against treating cosmonauts as babies who need a lot of care and feeding. They are adults who receive a lot of training and a lot of money. That's why they have to work. Their flight costs the country a lot of money. Every hour they do not work in orbit costs thousands of dollars. If I were in orbit, I would work quietly without complaint. The cosmonauts, when they complain that they are tired and overworked, you cannot always believe them. How can you be tired in weightlessness? They should work and stop complaining. As for psychological pressures, I don't think there are any. There are always complaints that we plan too much work. The cosmonauts complain that we plan more work than they can fulfill. I think we do not give them *enough* work!"

Sunday, April 6
Moscow

That weekend Sang and Marshburn pick up the first disturbing hints that conditions aboard the station have deteriorated to the point that the Russian psychologists may consider canceling Tsibliyev and Linenger's spacewalk, the first-ever joint EVA between a Russian cosmonaut and an American astronaut. In Sang's mind this would be disastrous; the EVA is the highlight of Linenger's increment. He decides to call Kathryn Linenger, who remains in Star City. It is a delicate call, and he tries to select his words with care.

"Kathryn, I don't really know Jerry," he begins.

"Yes you do," she says.

"No. I don't. Look, he cuts off comm and snaps at us. Now, I need to know, do you know if Jerry and Vasily are getting along? Because frankly, his EVA is in jeopardy."

Sang explains what he and Marshburn suspect is happening with the psychologists, and Kathryn volunteers to bring up the matter with Linenger in an upcoming personal family conference. For some reason her answer, when it comes several days later, doesn't assuage Sang's concerns at all.

"He says there's no problem," Kathryn says. "Jerry says they're getting along fine."

Amid all the troubles, there are two bits of good news that Sunday. Down at Baikonur, Progress-M 34 blasts off just after eight Moscow time, carrying parts to fix the Elektrons and a supply of new lithium hydroxide (LiOH) canisters in case the Vozdukh carbon dioxide remains down. If all goes as planned, the Progress should reach the station Tuesday night. Linenger is even more thrilled by an e-mail from Sang. "Just talked to [Dr.] Goncharov, et al.," Sang writes. "They're OK with you going ahead and doing walking and light running on the treadmill, and working with expanders in Kristall for an hour tomorrow. We'll hash out a schedule [how much time per day] and have the shift flight director tell Vasily about this."

Monday, April 7
TsUP

By Monday morning Sang and Marshburn realize they have to do something quickly or the EVA will be canceled. Sang finally takes his concerns to Blagov. He and the old Russian have developed a curiously close relationship. At one point Blagov even offered Sang a job, one that paid nearly six hundred dollars a month. Sang politely declined.

"Viktor, I need to know who's at fault here," Sang says. "Is it Jerry, or is it Tsibliyev, or is it both of them?"

"Well," Blagov replies, smiling. "these aren't the best kinds of guys to have up there to do the EVA."

"What kind of astronauts do you prefer?"

To Sang's surprise, Blagov names John Blaha. Blaha, for all his shortcomings, was a disciplined shuttle commander who kept up regular communication with the ground and spoke better Russian than Linenger.

"I really believe Jerry should do the EVA," Sang says. He is hard-pressed to come up with a reason. But he knows how much the spacewalk means to Linenger, and for all his "man in the can's" snappishness, he feels duty-bound to make sure it goes forward.

"The EVA *is* at some risk," Blagov says. He volunteers that he has just telephoned someone at Star City to find out how Tsibliyev and Linenger's hydrolab training went. This is bad news. Sang knows what Blagov doesn't: The training was minimal, and what little time the two would-be spacewalkers had together in the tank didn't go well.

By Monday afternoon Sang and Marshburn have gotten together a draft letter they intend to send to Linenger. It contains a strong, direct warning that Linenger find a way to repair his relationship with Tsibliyev, or his EVA will be canceled. They take it to Blagov, who reads it carefully.

"This is a good letter," Blagov finally says, smiling. "And you will not send this up."

Sang snorts a laugh. As tired as he is, Marshburn forces a smile. It is just too strong, says Blagov. Temper it, he suggests.

That afternoon Sang sends up a watered-down version of the letter. It orders Linenger to immediately halt all but essential science activities, but stops short of alerting him to the ground's concern about the EVA. "When you start adding up all of the operations for pre-EVA stuff, the EVA itself, post-EVA stuff, the other routine science stuff and pre-pack activities, you don't have any more time for science," Sang e-mails Linenger that day. "On Thurs. the cosmonauts have a full day of unpacking, so it is important that you participate. . . . You will be involved in unpacking the Progress on Wed. and Thurs. and most likely involved in the repair of the Mir systems. After all you are a flight engineer."

As if the relationship between Tsibliyev and Linenger isn't troublesome enough, Monday morning brings a new problem: blood tests. Astronauts never like giving blood; it reminds them of the age-old canard that they are guinea pigs rather than space fliers. But giving blood aboard Mir raises far more worrisome questions for Linenger. Under decrees laid down by the Russian doctors, all crew members are to have blood drawn every three months, in a procedure called the "MK-12." The blood is wiped on strips of paper and inserted into an antiquated onboard machine called a Reflotron, which analyzes the blood in sixteen separate tests, producing a data stream that is then downloaded to doctors in the TsUP for study. The American flight surgeons, including Marshburn, recoil at the Russians' continued use of the Reflotron. To them it represents a level of medical expertise not seen in Western countries since the 1950s.

The machine's age aside, the procedure for using the Reflotron has one major downside as far as Linenger is concerned. To get all the blood the machine needs for its sixteen tests, it is necessary to prick as many as five, six, or even seven fingers. That's a lot of stabbing to do, and Linenger is worried about soreness in his fingers, especially with the EVA in three weeks. Marshburn agrees and has been wrangling with the Russian doctors about the test for months. They adamantly refused to exclude Linenger from the test. Marshburn argued that Linenger at least be allowed to take the blood from a vein rather than his fingertips, but the Russian doctors would still not budge, not even after Marshburn secured a letter from the Reflotron's maker, the Daimler-Benz company in Germany, stating that blood from the veins is every bit as useful as blood from the fingers. On top of everything else, both Linenger and Marshburn have serious doubts about the Reflotron's accuracy, especially after its long storage at temperatures approaching 100 degrees.

Still, the Russian doctors insist on going ahead with the MK-12 as planned. Linenger's fingers will be pricked, and pricked and pricked and pricked, until the Reflotron is sated. Linenger tells Marshburn that he has no intention of going ahead with the tests, and Marshburn backs him. The only test they agree to even consider is one for hemoglobin, which Linenger acknowledges may have some medical value.

Monday morning Tsibliyev and Lazutkin are surprised by Linenger's refusal to do the entire test.

"You have to do it, by all means," Tsibliyev says.

But Linenger refuses. Finally, Tsibliyev throws up his hands. "Okay, then you have to solve the problem with the ground yourself."

"How's MK-12 going, any questions?" a Russian doctor asks Tsibliyev at the 9:13 pass.

"No questions so far," the commander replies. *"The only thing is, Jerry does not want to perform the test completely."*

"It's worthless what you're doing," Linenger tells Lazutkin. "I'll do the hemoglobin, that one makes sense. None of the others make sense." Linenger recalls: "You just had to laugh. You're in the middle of all this stuff happening, and they're asking you to poke your finger to check your cholesterol. It's really not the right thing to be doing, but the ground, they're just clueless. It's like a war. The enemy's shooting at you all the time, and you're supposed to get on the radio and ask what we're supposed to do, and the generals down there tell you to shine your shoes."

Tsibliyev and Lazutkin go ahead and perform all the tests. The next day the commander is sorry. One of his fingertips becomes infected.

That afternoon Tsibliyev is surprised and irritated when Linenger asks permission to use the Kristall treadmill. While Linenger has received the go-ahead to resume exercise, no one has told the two Russians they can exercise as well. *"Jerry came over today, again [and said] 'Let me do my exercise. . . . I need to exercise—period,' "* Tsibliyev tells the ground that afternoon. *"[I said], 'Okay, go walk on the treadmill a little, just don't run on it.' And all this because his doctor allowed him, and we know nothing about it. We are forbidden to exercise, and he is allowed to exercise. I just don't understand this. This is a very interesting situation. A message he got says he may exercise. So, he may and we may not. He lives exactly per his daily schedule and we have forgotten when we bedded down on time last . . . let alone [had] meals on time."*

That night Tsibliyev and Lazutkin work past midnight cutting and repairing the pipes in Kvant. The next morning, with the Progress due to arrive late in the day, the commander again awakens in a foul mood. It doesn't take the ground long to realize the focus of his anger: It's Linenger again. The American slept blissfully through the night down in Spektr while the Russians continued working.

"We worked with a saw til three o'clock but we succeeded," Tsibliyev tells the ground at the 8:16 pass. *"It will take lots of time and effort to clean this area. It is full of slime."*

"If you need some help, we can get your third member involved," says the shift flight director, indicating the ground's willingness to get Linenger involved.

"Let's see, he is always ready to work, especially on a treadmill," Tsibliyev replies.

The commander is even harsher during a comm pass two hours later. He and Lazutkin are talking between themselves, apparently unaware that the TsUP can hear their every word.

"It's good that there is at least one white man on the station—Jerry," Tsibliyev says. *"He jogged, he took a bath and now he is having lunch. He conducted an experiment. And we are like fools working since last night."*

Lazutkin concurs. *"Why, since [early this] morning, I've been hungry."*

"I wolf down a piece of meat because I felt my eyes were coming out," says Tsibliyev.

Suddenly the TsUP interrupts. *"We have been hearing you from the very beginning about the white people and about the fools, and about a piece of meat being wolfed down,"* the comm officer says.

"Oh, and we didn't think you heard us," says Tsibliyev, who clearly doesn't care who heard him. *"We were just talking between each other."*

As the Progress nears for its scheduled early-evening docking, the commander continues taking open-comm gibes at Linenger throughout the afternoon. At a 2:21 P.M. pass, Tsibliyev tells the ground that he and Lazutkin won't be exercising today, but that Linenger already has. "Vasily says he is extremely happy for him," notes the NASA log.

"Jerry ran a little bit today [on the treadmill]," Tsibliyev tells the ground a little later, noting that Linenger's exercise session has pushed the station's carbon dioxide level just above recommended levels. *"If we [exercise] a little bit, that would be dangerous. That's why I say there is at least one person here that exercises."*

The dinner hour brings still more comments from Tsibliyev, who instead of eating with Linenger is preparing video recorders for the Progress docking.

"That's great," says the TsUP, congratulating Tsibliyev for his hard work. *"Good for you."*

"Good for Jerry," rasps the commander. *"War is war, but dinner comes according to the schedule."*

"Should we say, 'Bon appetit' to him?" asks the ground.

"He has already finished. Why say it to him?" Everyone, including the ground, has a laugh at Linenger's expense.

11

Linenger has remained so absorbed in his experiments that he hasn't the slightest hint of the sarcastic comments Tsibliyev has been making about him to the ground. The reason he doesn't, it appears, is that he insists on staying down in Priroda and Spektr running his experiments when Tsibliyev and Lazutkin speak to the ground. Tsibliyev is free with his sarcasm behind Linenger's back but says nothing to his face. Later, after returning to Earth, Linenger will repeatedly express dismay at the perception that he and Tsibliyev weren't getting along.

"When [I] came back, we compared stories, and man, the ground had a totally different concept of the reality of what was going on up there," Linenger told a June 6 NASA debriefing session. "We got along fine the whole time. . . . I mean I fight more with my wife. When I heard that, I [thought]: Where in the world did you get an idea that we were butting heads? . . . I had no clue that anyone thought that."

Tuesday, April 8
Aboard Mir

Tensions are escalating as the fully loaded Progress nears Mir that evening. For days specialists at the TsUP have been radioing up instructions for manually docking the ship if the Kurs system again fails. Another manual docking, of course, is the last thing Tsibliyev wants to attempt after the near miss a month earlier. Still, the TORU system is locked in place as he and Lazutkin peer out the base block windows while the Progress approaches. At 6:25 P.M. the Progress fires its first thrusters, slowing its approach. An hour later a second set of thrusters ignites, slowing it still more. By 7:40 the Progress is eleven kilometers out, gliding toward the station like a catamaran on a glassy sea. Everything appears ready. Ten minutes later, as the Progress passes the three-kilometer mark, more thrusters fire. Four minutes after that there is a fifth and final firing. Soon the Progress maneuvers into a final approach and, with Tsibliyev's hands hovering above the black joysticks, docks without incident. This time the Kurs unit works.

The crew awakens early the next morning to begin unloading the ship. The shift flight director informs Tsibliyev in an early pass that Linenger has been instructed to help out; the commander indicates he'll believe it when he sees it. Then, a few minutes before noon, Tsibliyev's tone suddenly changes. *"Jerry expressed his desire to work,"* the commander tells the ground. *"It is not a problem now. We shall use him."*

By Wednesday morning, Linenger has finally gotten the clear impression from Sang's letter that everyone—Russians and Americans alike—wants him to pitch in on repairs and on unloading the Progress. In fact, Sang's decision to shelve most of the NASA science program gives him no other choice. Linenger decides he is ready to finally help out Tsibliyev and Lazutkin, but only to a point.

"Got your update note. Understand," he e-mails Sang on Wednesday. "But still think we need to keep going strong [on science]. . . . Stopping everything I don't like. Up to now [today] only one head into Progress anyway—so I'd be of little help. Instead, I've been very productive with our science program. Bottom line: Better to give me stuff [to do]. If I can do it, I will. If I'm useful elsewhere, I'll try to squeeze it in or postpone it. We are a pretty good team up here."

Sang and Marshburn try to drive the point home in a comm pass that afternoon: No science, they tell Linenger, at least no major experiments, until the crisis passes. Linenger reacts petulantly. Somehow, while the two Russians spend their days crawling through the station's coolant-drenched innards in search of leaks, he has convinced himself that he has in fact been helping the Russians all along.

"I think everyone on board takes some pride in that we are getting something constructive [accomplished] rather than just repairing things and staying alive," he tells Marshburn and Sang that afternoon. *"If you are worried about any psychological bonding, or things like that, we have been working together as a team. If I need help, I ask for help. If they need help, they tell me they need some help with something and in general I help out. During the Progress [docking], I was turning wrenches to help fasten the thing on, spotting out of the window and all that kind of stuff. So, as far as crew interaction, I think we as a team up here do pretty well, and the commander handles it pretty well. . . ."*

". . . Okay," says Sang.

"Okay."

"Hopefully, you can understand that. We are not going to cancel all science."

"Okay," Linenger says. *"But, uh, I am able. I am working pretty efficiently. I get up early in the morning some times and start things up, so I can get a little more done than scheduled. We are moving along good, and like I say, I think people respect what I am doing and I respect what they are doing, and when we need help we ask each other. So we have good relationships. We are working good together."*

An hour after Linenger signs off, the following note appears in the NASA log. "Vasily says they have to keep a constant eye on Jerry to see what he is doing at all times. [They] told him to let them know if anything catches fire again." Even from hundreds of miles away in space, the commander's sarcasm is withering.

Tsibliyev spends most of Wednesday afternoon rummaging around in the Progress, which is packed so tightly it will take days to unload. He hauls out the new Orlan space suits he and Linenger are to use for their EVA later in the month. But by early evening he still can't find the saws and tubing he and Lazutkin need to begin repairs on the Elektron and Vozdukh systems. At one point, Solovyov gets on comm and tells the commander to look underneath the crates of apples and oranges. Eventually, Tsibliyev finds the tools and promises to start work the next morning.

"The worst times are behind," Solovyov tells the crew later that day, *"so start getting back on track with the daily work schedule and exercise. This is my order. The [repairs] are important, but the EVA is of paramount criticality, too."*

"Okay," replies Tsibliyev. *"We'll not do exercise today, with your permission."*

"That's fine, but tomorrow, we'd like you guys to start a new life together."

"Okay, sleep is refreshing, but way too short."

"At least eight hours a day, please," says Solovyov. *"No more three-hour nights, please. Bed down at 11 P.M."*

"Okay, understand all."

"Do not be too tough on Jerry," says Solovyov. According to both Russian and American flight logs, this is the Russian flight director's only open-comm admonition to the commander to ease up on Linenger. Solovyov hopes the crew will begin getting along better once the leaks are fixed, and conditions begin returning to normal. For the moment at least, Tsibliyev gets the message.

"Oh, no, everything is fine up here," the commander replies. *"Jerry says*

things up here are much better than he thought they would be. And actually he's been very helpful up here as an M.D., looking after us closely, examining [us] when required, recommending the right things to do, from his medical perspective."

"Okay, priority number one, crew rest and work schedule," Solovyov concludes. *"TCS and Elektron maintenance ops come second. No science. This is my official strong request."*

As Mir's problems multiplied in April, Culbertson and Van Laak began hearing the first murmurs of discontent from their superiors in Washington. NASA critics, including the maverick NASA engineer-author James Oberg, started weighing forth with public comments criticizing Mir's safety and age. The decision whether to go ahead and send Mike Foale to Mir—and no one at this point was seriously advocating a cancellation—would ultimately be NASA Administrator Dan Goldin's call, but the most influential voice in the agency's regular preflight safety reviews would be an independent review commission chaired by the former astronaut General Thomas Stafford. Van Laak and a number of other NASA officials spent much of the month briefing Stafford's people.

Other concerns came from inside the agency. Fred Gregory, the former astronaut who headed NASA's safety organization at headquarters, worried aloud about sending Foale up without both Elektron systems working. Following a March 19 teleconference, Gregory sent Culbertson a letter indicating he would have serious reservations about continuing the Phase One program without two fully functioning Elektrons aboard the station. NASA was scheduled to ferry a new Elektron unit up on the next shuttle, but Gregory's language—that there be two "functioning" Elektrons before Foale could stay—painted Culbertson and Van Laak into a corner. The only way that could be achieved was if Tsibliyev and Lazutkin could somehow install the new Elektron during the changeover period—an unlikely event, given the complexity of integrating a new unit into Mir's aging systems.

"If taken literally, we couldn't possibly do this," remembers Van Laak. "So we asked Fred, and he said, 'Yes, that's what I meant.' . . . Fred is stubborn as a mule. We had a terrible two-hour telecon with him, where he insisted if both Elektrons weren't working, we couldn't go. . . . Finally, after that, we said, 'This is stupid, let's go convince Fred that all we need to do is leave an operational Elektron when the shuttle leaves.' And so, in a follow-up teleconference, Culbertson and Van Laak convinced Gregory that it wasn't possible to have two "functioning" Elektrons by the time the shuttle left Foale aboard the station. What they convinced Gregory to approve was having two "functional" Elektron units, the newer one of which could be installed later. "We ended up with this wordsmithing thing," Van Laak remembers. "It was just an embarrassing moment. Fred gets himself boxed into a corner, and he was embarrassed. To help him get out of his corner, we had to kick down one of the walls."

The real problem, as far as Culbertson and Van Laak were concerned, was the corrosion on Mir's internal pipes. This suggested a structural defect that might get worse in coming months; the corrosion, NASA would later discover, was being caused by condensation wearing through spots where Mir's aluminum piping touched stainless steel grounding straps. Culbertson asked Charlie Precourt, the commander of Foale's upcoming mission, to perform a thorough checkout of the station when his shuttle docked in May. "I remember thinking, 'Is anyone going to ask me about this?' " Linenger recalls. "[I wanted to say], 'Hey guys, you've got a problem! You can't pretend you don't have a problem!' "

When hints of the Foale debate reached Mir, Linenger noticed a change in his crewmates' behavior. Both Tsibliyev and Lazutkin began volunteering less information about the station's problems. "Without the Americans they cannot keep the Mir going—we all knew that," Linenger recalls. "Vasily is still pro Russian space program, so it was always sort of a tense discussion whenever we talked about the station's health. We did talk about it, but as things got worse, Vasily started getting to the point of, 'Well, should we tell Jerry about this?' I remember going up to Sasha one day [and saying], 'Hey, Sasha, what are you working on?' [He says] 'Ah, nothing.' And I'd look in and there would be this great glob of ethylene glycol." Linenger snapped a few pictures of the corroded areas but stopped short of a wholesale photographic study. From the looks he got from Tsibliyev and Lazutkin whenever he brought out his camera, he felt like a spy.

NASA's quiet debate infuriated the Russians. When a Russian reporter cornered Blagov and asked him about the rumors that Foale might not fly, Blagov lashed out. "We would ask the Americans: 'What kind of experts are you to think about deserting this unique space station?' " he asked. "We're telling them, 'Look, you guys were here when everything went fine, and now you want to leave us when we face the smallest difficulty. It's simply indecent.' " Blagov belittled NASA's concerns about the fire, repeating the ninety-second estimate. "There was no danger for the crew at all," he fairly spat. "We think the Americans overreacted to that largely because of their own tragic experience—fire on the Apollo spacecraft during ground trials and the Challenger disaster."

For the first time American reporters and legislators began inquiring into Mir's difficulties. The Wisconsin congressman James Sensenbrenner of the House Science Committee passed an amendment requiring NASA Administrator Goldin to personally vouch for Mir's safety before sending up another astronaut; the measure never came to a vote in the Senate. "We compromised safety before," Sensenbrenner carped to *Time* magazine, "and we got the Challenger disaster." Culbertson tried to sound upbeat. "I think they're in a very good and stable situation right now and certainly able to continue the mission," he told a *Washington Post* reporter that week. "It's very true the Russians are going through a difficult time right now. [But] just as we expect them to stick with us on the International Space Station . . . I think it's important that we as partners stick with them during a difficult time with Mir."

It was just what the Russians wanted to hear. "If they didn't want to send him, I would have understood," Blagov recalls. "But the Russians would never do such a thing. If we are partners, then we should not forsake each other. That's what we tried to convince them."

Friday, April 11
Aboard Mir

That morning, after myriad tests and stops and starts, Tsibliyev and Lazutkin manage to get the patched-up Vozdukh system running. Saturday is Cosmonaut Day in Russia, in honor of Yuri Gagarin's first fight, and the system is fixed just in time for official congratulations and greetings from Valery Ryumin and Yuri Semenov. But no sooner have the two signed off than things begin to fall apart again. First the urine-processing system inexplicably shuts down. And then, during tests of the Elektron system, the cooling loops repeatedly fail to hold pressure. Ethylene glycol is still seeping from somewhere inside the system.

"It is definitely leaking," Tsibliyev reports in the early evening. *"Also, we may need up to two weeks to get organized prior to the EVA."*

"You are kidding," says the crew communicator.

"Oh, it's very messy. Really. . . . Two days to organize this place is not enough. And no, I'm not kidding."

"Stop all work," the shift flight director orders a bit later. *"Just do cleanup, get organized and maybe while doing that you'll accidentally bump into the leak. If you are lucky."*

"Okay," Tsibliyev says. *"But I don't know how we're ever going to be able to get back to normal up here."*

Among the food items Linenger has requested from the ground are pretzels, and to his delight, he spies a bag of Rold Gold pretzels nestled in the Progress as the crew continues unpacking that weekend. Linenger grabs the bag and is just about to open it when Tsibliyev stops him.

"Hey, don't eat those!" the commander says. "Those are for the commercial."

In the midst of one of the more difficult periods Mir has endured in its eleven-year history, the TsUP has scheduled Tsibliyev to perform not one but two television commercials, one for Rold Gold, the other for an Israeli milk company. Linenger can tell Tsibliyev is embarrassed to be filming the commercials, which is one reason, he suspects, the commander insists on doing them late at night. The first to be filmed is the milk commercial, which requires Tsibliyev to gobble globs of milk floating in the air; there is no speaking part.

"I couldn't believe it," Linenger remembers. "In the middle of all this, they want him to do a milk commercial. All of a sudden, at 11:30 at night, we need to do this milk commercial. 'Stop working on the oxygen generator, we have to do a milk commercial!' I could see he was sort of ashamed that the Russian space program had degenerated into doing milk commercials, so I said 'Good night' so he could do it. Then later I came back out and saw Sasha filming him. He looks at me and turns beet red. It was embarrassing for both of us."

Tsibliyev's embarrassment turns to irritation the next morning when the TsUP, acting on directions from the commercial's director, asks him to reshoot one scene. The commander, it turns out, hadn't been smiling.

* * *

Tsibliyev and Linenger do manage to bond a bit on the one subject on which the American is an expert: the crew's health. If American astronauts have little use for NASA flight surgeons, Russian cosmonauts live in mortal fear of their own doctors, whose judgment of their fitness for flight determines the trajectory of their careers. Tsibliyev doesn't trust the Russian doctors who are telling him that ethylene glycol is harmless—a mistrust that leaps exponentially after an incident one morning this week. All three crew members are badly shaken when Lazutkin wakes up with an alarming change in his face: His eyes are the size of golf balls. The condition, apparently caused when Lazutkin wiped his eyes with a glycol-streaked finger, lasts only a few days and causes no permanent damage. But it does nothing to build trust between the crew and the Russian medical establishment. Nor do the results of the MK-12 blood tests. That week Russian doctors send up data from the tests with a note saying that the crew has a clean bill of health.

"Thank God our blood tests are fine," Tsibliyev tells Linenger. He hands him the results, a series of numbers that mean nothing to the Russian.

Linenger studies the numbers for several long moments.

"Vasily, how can they say that?" he says.

The numbers, measuring cholesterol and a host of other blood work, don't make sense. If half the figures are accurate, Linenger realizes, the crew should be dead. After reading the numbers himself, Marshburn concurs.

"Every result but one or two," Marshburn tells Sang, "are incompatible with life."

Alerted to the obvious discrepancies, the Russian doctors quickly apologize, theorizing that the ancient Reflotron must have been affected by the station's high temperatures. But the incident has a galvanizing effect on both Tsibliyev and Linenger. Neither is inclined to any longer trust the doctors, or anyone else in the TsUP.

"Basically they were gonna tell these guys they were healthy, no matter what," Linenger recalls. "These two guys felt like they were sacrificial lambs, that the space station was the only thing that mattered, that their lives didn't matter at all. The Russians were going to keep them up on that station no matter what. No matter what."

Monday, April 14
Aboard Mir

Monday morning, with coolant fumes still thick inside the station, Lazutkin and Tsibliyev begin tearing apart the Kvant module in search of what they fervently hope are the last ethylene glycol leaks. The work is delayed for hours by hundreds of excess food containers the crew has piled into the module to load onto the newly docked Progress. Much of the accumulated trash ends up down in Priroda, which causes Linenger to have fits. It all smells, and he is forced to move things around to get his experiments done. While Tsibliyev begins hauling other bags of trash into the Progress, Lazutkin starts removing wall panels and rooting around behind them for signs of the leak.

"Crew reports the inner depths of Kvant look like a mangrove swamp," the NASA log notes just before noon. *"Looks like a rat-friendly environment, with lots of dirty water. Will definitely take a lot of effort to dry up and clear."*

For three days the two Russians wipe up the pools of water that have condensed behind the walls of Kvant. The work goes slowly, in part because Lazutkin and Tsibliyev need to be careful where they put their hands. *"Electrical conduits are dangerously close to water puddles,"* the commander reports at one point. By Wednesday Linenger has joined in, a fact that Tsibliyev, to the NASA ground team's surprise, makes sure the TsUP knows about. *"Jerry helps us a lot,"* the commander volunteers, according to the NASA log. *"Psychological compatibility is present. No grounds for any psychological concerns. Crew interaction is at the best, and it just does not get any better than this."*

This, of course, is a jarring departure from Tsibliyev's stream of earlier comments about Linenger. Down in the TsUP, Sang and Marshburn exchange knowing glances. Tsibliyev has clearly gotten the message: Clean up your act, or the EVA—for which the commander will receive a $1,000 cash bonus—will be canceled. Late in the day Linenger asks Marshburn about the spacewalk's status.

"The committee is going to decide Wednesday or Thursday," Marshburn says, *"but right now it looks like a go."*

Despite all the months of preparing and training to do the EVA, Linenger has mixed feelings when he hears the Russians have approved the spacewalk. "On the one hand, you absolutely want to do it," he remembers thinking. "But at the same time I'm looking at Vasily thinking, 'I hope he can keep his head on straight.' You don't want to push him over the edge."

Thursday, April 17
Houston

The guards at the front gate of the Johnson Space Center wave the car by without checking its credentials, which is a huge relief to Byron Harris, because he doesn't have any. Harris is a veteran investigative reporter with WFAA, the ABC television affiliate in Dallas, and he has come to Houston with a camera crew to put a climactic ending on one of the most ambitious stories he has attempted in twenty-three years at the station. It involves what he believes to be corruption at the highest levels of the Russian space program.

Harris's strange journey had begun two years earlier when, during a dinner with James Oberg, the NASA engineer and author, Oberg had mentioned hearing about a group of expensive new private homes being built for cosmonauts and senior officials at Star City. Among those few Westerners who knew of the houses, Oberg said, no one could figure out where the Russians could come up with the money to build them—unless funds were somehow being siphoned off from the space program itself, perhaps even from the $400 million NASA contract. Oberg had mentioned the homes in a piece he wrote for a little-read engineering magazine, but no one in the mainstream press had taken notice. When Harris began studying whether to broadcast a lengthy piece on the Russian program in early 1997, Oberg reminded him of the homes, and Harris resolved to check out the rumor.

Taking several days off from a trip to Stockholm, Harris had arrived in Moscow in early March to a chilly reception from the ABC News bureau, which was worried that he might do something to anger the Russian government, on whose good graces the bureau placed a high value. Harris forged ahead nevertheless, picking up a driver and interpreter and heading straight for Star City with no credentials or official clearances. Arriving at the front gate, he slipped the guards ten dollars and to his surprise was allowed to drive directly onto the base. For more than an hour, the driver nervously navigated Star City's narrow roads in search of the homes. Harris spotted some dachas that seemed to be far more modest than Oberg had described. Taking a risk, he began stopping pedestrians and asking them if they knew anything. His break came when one woman said she knew of the houses and motioned in their direction. "Your friend," she said to the Harris's interpreter. "He is a spy, no?"

The area the woman pointed out led them to a high wall, where Harris's driver finally lost his nerve and demanded they leave the base before they were arrested. Harris reluctantly consented, but on the way out, his interpreter thought to ask the front-gate guards, "Do you know where the cosmonaut cottages are?" Ever helpful, the guards directed them to a nearby road that ran through a forest outside the base. Harris's car sped down it and into a clearing in the woods. "All of a sudden, 'Bam, there they are, these huge houses,' " Harris recalls. "It looked just like an American subdivision." More than a dozen new homes, each with an estimated value of $350,000 to $500,000, lay out beyond the forest. Several were under construction.

Harris returned to Russia a month later to try and nail down who owned the houses. Masquerading as an American businessman shopping for a dacha in the area, he questioned workers at the construction site, talked up Moscow real estate agents, and pored over ownership records at a local land office. Two of the largest homes, it turned out, belonged to Star City's top two officials, Generals Klimuk and Glaskov; both men, Harris knew, made about $12,000 a year, hardly enough to build a half-million-dollar home. Water and sewage services appeared to come directly from Star City. Officially, no one at the base would discuss the homes. Unofficially, at least one person did, confirming Harris's suspicions that the whole development was being built with government funds. When Harris returned on April 10 to do a final "standup" in front of the homes, he got another eyeful of evidence: army trucks moving supplies around the construction sites. As his cameras rolled, a long black Volga limousine appeared from a backroad to Star City and pulled into one of the homes.

There was only one dangling thread that bothered Harris. "People at the housing complex were extremely fearful of the Russian Mafia," he remembers. "The lady at the local land office mentioned it as well. Everyone was scared of the Mafia out there." Oberg had encountered similar warnings after writing about the homes two years earlier. After his article was published, he says, he ran into Vladimir Solovyov in a cafeteria at the Johnson Space Center. According to Oberg, Solovyov was angry about Oberg's article.

"You've heard about the Russian Mafia, haven't you?" the Russian asked.

Oberg nodded.

"It is dangerous to write about these subjects," Solovyov said. He repeated himself for emphasis. "It is dangerous to write about these subjects."

What, if any, role elements of the Russian Mafia played in erecting the houses Harris couldn't say. But by mid-April he had everything he needed except one thing: a chance to confront Russian officials about the houses. And so on April 17, tipped that Yuri Glaskov, the round little general who was the No. 2 official at Star City, was in Houston, Harris hurries down to JSC to confront him. Stealing glances at a picture of Glaskov that Oberg had found, Harris sets up his camera outside Building One and waits. When Glaskov emerges, Harris sticks a microphone in his face and asks about his new house.

"There are no such houses at Star City," Glaskov answers via his interpreter. When Harris produces a picture, Glaskov brushes it off. "I don't know anything about it; I don't live there." When Harris presses, Glaskov bridles. "This has nothing to do with the space program, and I don't want to talk about it." When Harris tells him he has proof that he owns one of the houses, Glaskov cuts off the interview. "I've been working all my life and my wife is a pilot, and we've made enough money to get a house to live in. I don't think this question is appropriate."

When Harris's piece runs, it will be picked up by ABC's *Nightline*, which will invite NASA Administrator Goldin on to defend the Russian homes. He can't. Several days later the House Science Committee will ask the NASA inspector general's office to begin a full investigation.

Friday, April 18
Aboard Mir

Operating the toilet remains a headache. Because Lazutkin is continuing repairs on various parts of the urine-reclamation system, the crew is obliged to flip a different combination of switches to activate the toilet almost every day. Linenger is having trouble keeping straight what switches to flip.

"Jerry does not know which toggle switches to use, or probably he does not care," Tsibliyev reports during an afternoon pass. *"We ask him if red LED [lights] come on and he says no, then we all go [in] there together and see that they [are on]. So probably we'll have to accompany him to the head every time he uses it."*

Monday, April 21
TsUP

The EVA is eight days away, and Tony Sang is worried. It's not just the strains between Tsibliyev and Linenger that bother him. It's Linenger's training—or lack thereof. The American has never been on a spacewalk before. At Star City, where EVAs are rehearsed in a massive underwater training facility called the hydrolab, Linenger did almost all his training "runs" with Mike Foale; he and Tsibliyev managed exactly two sessions together. "The first run was a disaster," Linenger later told NASA debriefers. Communications went out, the two men were unable to hear each other talk, and the session was prematurely ended. Worse, the wrong equipment was put in the water for them to work with. A second run was judged satisfactory, although it ended with clear tension between Linenger and his Russian trainers. During that session, Linenger noticed that Tsibliyev was performing one of their tasks out of sequence. He mentioned it to the commander, but Tsibliyev ignored him. Afterward, when Linenger brought up the incident with their trainers, one of the Russians sternly rebuked him for questioning Tsibliyev's action.

"You do what the commander orders you to do," the Russian told Linenger. Sang, who witnessed the exchange, did not like the feel of the meeting, or of Linenger's relationship with Tsibliyev or the trainers. At the time both he and Matt Muller, the Krug Life Sciences trainer who served as Linenger's de facto valet at Star City, felt Linenger needed more and better EVA training. "It was an issue," recalls Muller. "We lobbied for more [training]. . . . The word came back, 'It'll be enough.' "

Now, with the spacewalk just a week away, Sang brings up the matter of Linenger's training again, this time with the head of NASA's EVA working group, Richard Fullerton.

"Check with the guys in the Russian EVA group," Fullerton advises. "We'll do what the Russians say."

It was the classic NASA response all during the five-year life of the Shuttle-Mir program. " 'It's the Russians' responsibility'—you just heard that over and over," remembers Muller. The problem, Muller and others observe, was that for any NASA official who disagreed with the Russians on a safety issue, confronting anyone in the Russian program was purely a "lose-lose" proposition. The Russians almost certainly wouldn't give in to the requested change; they amended their safety rules about as often as their constitution. Thus, any American who brought up a safety matter and couldn't get it "fixed" would find himself in the unenviable position of having "dropped the ball." This was a strong disincentive for anyone at NASA to make waves; complain about a safety issue, and you're likely to be ignored, until the moment something goes wrong, at which point everyone will blame you for failing to follow through.

"No one would take those kinds of risks at NASA," says Muller. "Do you think anybody in NASA stepped forward to review Russian firefighting procedures after the fire? No way."*

But Fullerton's casual response wasn't a case of a NASA official shirking responsibility. Just as Phase One hadn't paid any real attention to safety aboard the Mir until the fire, so too it hadn't asked the Russians for permission to thoroughly review safety aspects of the EVA. Fullerton wasn't getting more involved in Linenger's preparations because no one had asked him. It was, in fact, the first spacewalk involving a NASA astronaut for which NASA did no formal safety review. Only afterward would the Phase One office realize the mistake and ask the Russians to be involved in safety reviews for subsequent joint spacewalks.

And so Sang remains troubled. He knows Linenger hasn't had enough training. But no one else seems worried. "There were a lot of structures outside [on the Mir's outer hull] that we didn't know anything about, which scared me," he remembers. "[Using] the NASA way, we would map the outside [of Mir] to a tee and rehearse the hell out of it. We never did that. We weren't doing things the NASA way. That was the point."

*NASA formed a small committee to investigate the fire, headed by an MOD engineer named Mark Ferring. "I spent a week on this, maybe two," remembers Ferring. "As I recall, we really couldn't make heads or tails of what happened."

A week before the EVA, the Russian and American psychologists give their final consent to the spacewalk in a joint teleconference. "I thought it through carefully, and I basically concluded that they could do the EVA, and it would be okay," remembers Al Holland. "I talked with [Tom Marshburn], and he was confident there was no problem. The Russian psychologists, they agreed. There's a little bit of friction, sure. We sort of all came to the conclusion that maybe Jerry wasn't a team player like others had been. But we thought things would be okay."

In fact, as the week before the spacewalk wears on, the friction between Tsibliyev and Linenger clearly affects their preparations. The EVA itself isn't especially complicated. The two men are to venture out to the end of the Kristall module and set up a suitcase-size piece of equipment called the Optical Properties Monitor (OPM), which contains dozens of samples of materials like silica-fused aluminum that the Americans hope to use outside the International Space Station; by storing them in the OPM for a period of months, NASA scientists will get a sense of how the materials handle space. Linenger and Tsibliyev will also install a small radiation sensor and retrieve an experiment that measured the number of foreign particles that struck Mir over the last few months.

As they prepare for the EVA, the problem isn't that the two men argue over what they will do once they get outside the station; it's that they don't talk much at all. "We wanted to," Linenger told his debriefers later. "[But] it was one of those things you just kind of keep putting off, because some other system's failing. It was like, you need oxygen, so we can't talk about our [EVA] plan here. We better go fix the oxygen generator." Instead Linenger spends his evenings alone poring over the EVA data and time lines prepared by his Star City trainers.

"Normally, the crew spends hours going over the EVA, how they'll use tools, where they will go, what they will do; they frequently inspect and go over suits together," says Jim Van Laak. "In Jerry's case they never really sat down and did any of that. They didn't do the four- or five-hour blow-by-blow. It was, 'Any questions? Here's what we're going to do.' Because Jerry wasn't talking to us, we didn't know he wasn't getting the kind of detailed review he wanted."

The centerpiece of a standard EVA preparation aboard Mir is the suit checkout. Ordinarily an astronaut or cosmonaut prepares his own space suit. Linenger didn't, but not for lack of trying. Several nights in a row Linenger beds down in the far end of Spektr while Tsibliyev and Lazutkin stay up past midnight preparing the suits. What little he can see of their preparations, however, worries him. At one point, the two Russians appear to be cannibalizing parts from one space suit to install in another.

"Where does this go?" Linenger hears a voice say.

"I don't know," comes the reply.

Linenger gets the distinct impression that Tsibliyev doesn't know much about assembling the suits and is relying on Lazutkin. In fact, the American begins to suspect that the reason the commander doesn't want him involved is to hide his own lack of knowledge. He is concerned but says nothing to anyone in space or on the ground.

None of this is apparent to NASA. A few days before the spacewalk Richard Fullerton arrives at the TsUP, but his role is purely to observe. He recalls little discussion of either the technical or psychological aspects of the spacewalk. "It was a fairly leisurely type deal," says Fullerton. "Everything went pretty well. I [only] heard stories afterward. I was like, 'That would have been nice to know.' " Not until he attended Linenger's subsequent debriefing on the spacewalk did he learn, Fullerton recalls, that "Jerry and Tsibliyev didn't talk before the EVA. The guys have to work together, and they didn't work together. If you don't talk ahead of time, and don't choreograph, it's not going to be pretty."

7:53 A.M.
April 29
Aboard Mir

Linenger and Tsibliyev crouch in the airlock at the end of Kvant 2 and face the battered yellow-and-white outer hatch. Though the two men haven't talked much about the EVA, the hatch is one thing Tsibliyev has mentioned again and again. "Jerry, just be careful about the hatch," the commander keeps saying. From the nervous tone of Tsibliyev's voice, Linenger has gotten the sense the hatch is so unreliable it could open if bumped, which is absurd. But now, looking closely at it for the first time, Linenger sees why the Russian is worried: The hatch is held on by a set of clamps called C-clamps. To Linenger it looks unstable, completely jury-rigged. His first thought is that if a cosmonaut wanted to commit suicide, he could blow this hatch in a second, killing everyone aboard.

The problem, Linenger knows, is the hinge that connects the hatch to the hull. The hinge was damaged during an infamous spacewalk in July 1990. Two cosmonauts, Anatoli Solovyov and Aleksandr Balandin, had been forced to perform an emergency EVA to repair a group of loose thermal blankets on their Soyuz return capsule; unless the blankets were refastened to the hull, it was feared the two men might burn up on reentry. The problem was, Solovyov and Balandin had not been specially trained for the spacewalk. They prepared instead with videotapes sent up in a Progress and by watching televised practice sessions beamed up from the Star City swimming pool.

It wasn't enough. On July 17, 1990, the two cosmonauts had crouched in the Kvant 2 airlock, exactly where Linenger and Tsibliyev now stood, preparing to leave the station. Before exiting the hatch, they had taken a pressure reading in the airlock. Either their handheld pressure gauge malfunctioned, or they misread it, because when they bent to open the hatch, there was still some air remaining in the airlock. The hatch immediately slammed outward on its hinges with terrific force.

The two cosmonauts then proceeded with the EVA, which proved dicier than anyone had expected. Fixing the thermal blankets took far longer than anticipated, and the spacewalk degenerated into a repair marathon that stretched past six hours. The space suits Solovyov and Balandin wore had only been rated for six and a half hours of use; when the two cosmonauts reached that point, the ground urgently ordered them to return to the airlock. Leaving their tools and ladders at the work site, Solovyov and Balandin were forced to scramble back across the length of Kvant 2 in total darkness, an exceedingly dangerous transit.

It was only when they reached the airlock and crawled inside that Solovyov realized the hinge had been damaged. The hatch wouldn't close behind them. By this point the cosmonauts had been in a vacuum for nearly seven hours, and it was imperative that they find a way back inside the station. Clambering back outside the airlock, they tried the seldom-used backup airlock farther down Kvant 2, which to their relief opened and closed behind them. The EVA lasted seven hours and sixteen minutes.

The outer hatch, however, remained open to space. Solovyov and Balandin tried to fix it during a second spacewalk a week later, but it still wouldn't close tightly. Then they discovered that a piece of the hinge cover had broken and lodged between the hatch and its frame. Removing the broken piece, they were finally able to close and repressurize the hatch. Several months later a new team of cosmonauts returned and found the hatch impossible to permanently repair. Instead they attached a set of clamps to secure it in place.

It is this set of clamps that Linenger and Tsibliyev are staring at uneasily seven years later. To his relief, the commander opens the hatch without incident and crawls outside onto an adjoining ladder just after nine o'clock. Linenger begins to follow. Outside the Sun is rising. The Russians have planned the EVA at a sunrise so as to get the longest period of light. But because of that, Linenger's first view of space is straight into the blazing Sun. "The first view I got was just blinding rays coming at me," Linenger told his postflight debriefing session. "Even with my gold visor down, it was just blinding. [I] was basically unable to see for the first three or four minutes going out the hatch."

The situation only gets worse once his eyes clear. Exiting the airlock, Linenger climbs out onto a horizontal ladder that stretches out along the side of the module into the darkness. Glancing about, trying in vain to get his bearings, he is suddenly hit by an overwhelming sense that he is falling, as if from a cliff. Clamping his tethers onto the handrail, he fights back a wave of panic and tightens his grip on the ladder. But he still can't shake the feeling that he is plummeting through space at eighteen thousand miles an hour. His mind races.

You're okay. You're okay. You're not going to fall. The bottom is way far away.

And now a second, even more intense feeling washes over him: He's not just plunging off a cliff. The entire cliff is crumbling away. "It wasn't just me falling, but everything was falling, which gave [me] even a more unsettling feeling," Linenger told his debriefers. "So, it was like you had to overcome forty years or whatever of life experiences that [you] don't let go when everything falls. It was a very strong, almost overwhelming sensation that you just had to control. And I was able to control it, and I was glad I was able to control it. But I could see where it could have put me over the edge."

The disorientation is paralyzing. There is no up, no down, no side. There is only three-dimensional space. It is an entirely different sensation from spacewalking on the shuttle, where the astronauts are surrounded on three sides by a cargo bay. And it feels nothing—*nothing*—like the Star City pool. Linenger is an ant on the side of a falling apple, hurtling through space at eighteen thousand miles an hour, acutely aware what will happen if his Russian-made tethers break. As he clings to the thin railing, he tries not to think about the handrail on Kvant that came apart during a cosmonaut's spacewalk in the early days of Mir. Loose bolts, the Russians said.

Loose bolts.

No one on the American side has ever been able to get a straight answer from the Russians on what caused the handrail to come loose; the best guess is metal fatigue, which is an especially frightening idea, suggesting that other parts of the hull may be less than stable. Whatever the cause, the Russians afterward began using what they call a "dual tether" protocol, meaning spacewalking cosmonauts are always supposed to have two tethers clipped to parts of the hull while outside the station. The problem is, Mir wasn't designed for a dual tether protocol, so in many places there simply aren't two good spots to which tethers can be clamped. Sometimes a spacewalker has nothing but a single tether and his hands holding him to the station.

Loose bolts.

Linenger tries not to think what would happen if his tethers or a handrail should somehow come loose: Unless Tsibliyev can somehow pull him back, he will float off into space to die. This, the spacewalker's ultimate nightmare, has never actually occurred. In the annals of manned space travel it has almost happened only once, to a Russian crew in December 1977. A cosmonaut named Georgi Grechko was finishing an EVA outside the Salyut 6 space station when, as he returned to the airlock, his partner Yuri Romanenko asked to take a quick look outside. According to most accounts, Romanenko pushed hard against the hatch and lost his balance. Before Grechko could react, Romanenko floated out into space, his hands and legs frantically cartwheeling in a fruitless effort to grab some part of the station. Grechko, glancing down, realized that Romanenko's safety tether had somehow come loose and was trailing behind him like the tail of a kite. Thinking quickly, Grechko grabbed the tether and hauled Romanenko back into the station, apparently saving him a lonely and ignominious death. Grechko now denies that this incident ever took place, saying it is a myth rooted in a "bad joke" Romanenko told upon returning to Earth.

This kind of accident isn't all that likely aboard the shuttle. Were an American astronaut to somehow float off into the space, the shuttle could simply turn around and pick the astronaut up. Hypothetically, the Russians could do the same thing, rushing into the Soyuz to pick up a "lost" cosmonaut or even maneuvering the Mir itself by its thrusters. But powering up and moving the Soyuz takes time, and the Mir's maneuverability is cumbersome at best. American EVA experts have decided that a NASA astronaut lost from Mir would be unlikely to live. "Basically they would be history," says Jim Van Laak. "We consider it an accepted risk."

And if the worst should happen? If Linenger should somehow lose his grip and fly out into space unable to be retrieved? He may or may not drift away from the station. There are no winds in space. Unless he pushes off from the hull, he could simply float there in space, perhaps just inches out of Tsibliyev's grasp. Fifty or even fifteen feet away from Mir's outer hull, with the station in full sight, he would have no way to return. He could see Tsibliyev, talk to the commander, and be helpless to come back.

Death would come slowly. The Russian space suit Linenger is wearing contains about eight or nine hours of usable oxygen; how long it would last would depend on how much oxygen he had already used. His battery would probably go out first, shutting down the Orlan suit's carbon dioxide removal system. If he wanted to end it himself, he could simply allow the carbon dioxide to overcome him, and he would gently fall asleep forever. A switch inside the suit, however, would allow him to easily purge it of carbon dioxide, flushing it with additional oxygen.

The best bet is that he would eventually die of hypoxia, the slow draining of oxygen from the bloodstream. There, drifting alone in space, his mind no doubt filled with images of his wife and child, he would get headaches first—small ones initially, then worse. He would grow confused. His mind would get fuzzy. He might hallucinate. The strange thing about hypoxia is that at some point, well past the point of no return, a feeling of euphoria would wash over him. It would, most doctors agree, be an absolutely wonderful way to die.

"Jerry, just wait," Tsibliyev is saying, as the two men cling to the ladder. "I'll go first."

The Russian stops for a second to admire the view. This is Tsibliyev's sixth EVA—he did five with Serebrov in 1993—and the experience of walking in space never fails to move him. "There is suddenly this huge planet below you," he remembers. "Inside the station, you cannot see it, only parts of it. When you get out, you really see it, the whole thing, it's so unusual, so dramatic, so emotional, you have to be a little scared." With a final glance downward, Tsibliyev surges forward in search of the spot to place a small radiation dosimeter, the first of several tasks on their list that day.

For the longest time Linenger remains frozen. Nothing is familiar. Nothing looks as it did in the swimming pool at Star City. And everything is falling. Slowly he inches along the handrail, clamping and unclamping his tethers every few feet. With Tsibliyev almost out of sight ahead of him, he continues like this for several minutes, until the handrail suddenly stops. Raising his head to look around, Linenger sees he is surrounded by all manner of structures the Russians had never told him about. Solar arrays tower over him like statuary. Clipped everywhere, to the handrails, to arrays, everywhere, is a thicket of little sensors and experiments.

"Vasily, which way can I go?" Linenger asks. He points off to one side. "Can I go this way?"

"No," the commander replies, waving his hand. "Solar panel. Watch out."

"Can I go this way?" he asks, pointing to what appears to be a path through the panels.

"No. Solar sensor."

Linenger's anxiety rises as he examines the cluster of giant winglike solar panels he has entered. The edges are sharp—*razor sharp* is the term he will later use in his debriefings. He is certain that if he bumps into one of the arrays, an edge will cut and puncture his space suit, instantly killing him. The outside of Kvant 2, in fact, is by far the most crowded exterior surface of the entire station. Because its outer hull is closest to the airlock, Kvant 2 is covered with all manner of Russian and American experiments. Richard Fullerton calls it "a pincushion."

NASA's stance on Linenger's EVA, in fact, is strangely at odds with directives given the only other astronauts to walk outside Mir. A year earlier, during Shannon Lucid's mission, two shuttle astronauts, Rich Clifford and Linda Godwin, climbed out of the Shuttle Atlantis to attach experiments onto the docking module at the end of Kristall. Both the Russians and NASA had forbidden Clifford and Godwin from venturing off the docking module into the field of experiments and solar arrays farther up the hull of Kristall. "They said it wasn't safe," recalls Clifford, who remembers agreeing wholeheartedly once he got outside Mir and glanced up the sides of Kristall. "There are appendages all over Kristall," he says. "Some of them were visibly sharp. Snag points. Sharp edges. Not a clear translation path."

But a year later, no one has raised questions about sending Linenger, a first-time spacewalker, out onto the station's crowded outer hull. No one has mapped the arrays and experiments for him. No one has shown him the safest, or for that matter any, transit routes across the hull. He is on his own, and he is frightened.

Tsibliyev hustles on ahead, leaving Linenger to fend for himself. Slowly the American inches forward, clipping his tethers to whatever handrails he can find and taking care to avoid the solar arrays. Finally, midway up Kvant 2, Linenger reaches the end of the Strela arm. The arm is a forty-six-foot-long pole that, with the use of a hand crank at its base, can be telescoped out to its full length. To get over to the docking area at the end of Kristall, where they are to install the OPM, their plans call for Linenger to physically mount the end of the arm, as he would a horse; Tsibliyev, using the crank, is to extend the arm and swing Linenger out and across open space to the docking area. The idea is roughly the same as fly casting for trout. The boom is a fishing pole in the commander's hands; Linenger is the hook. Once Linenger is swung safely across to the docking area, he is to retether his end of the pole to the station's outer hull. Tsibliyev will then crawl his way along the arm to join him.

The slow-motion ballet begins as Linenger starts untethering the end of the Strela from the outer hull of Kvant 2. Meanwhile Tsibliyev has made his way along the length of the module to the outer hull of base block, where the base of the Strela arm is anchored. As Tsibliyev readies the arm, Linenger clips the unwieldy OPM unit to a hook at the end of it. Then he gingerly shimmies himself onto the boom beside it, hugging the slender steel rod with his knees and forearms.

Slowly, Tsibliyev swings the boom free, sending Linenger arcing out into open space. For Linenger, leaving the solid footing of the station's outer hull behind, the impression of free fall is almost unbearable. Fighting a brief surge of panic, he is seized by the idea that the boom is about to break, sending him spiraling off into the vastness of space. "I'm just out there dangling," Linenger later told his debriefers. "[It was] very uncomfortable out there. [But] again, you just overcome it. You say, 'Okay, if it breaks, it breaks.' "

It gets worse when Tsibliyev begins extending the boom. To lengthen the arm, the commander has to forcibly yank on a set of handles, as if pulling a wooden stake out of the ground. Each yank, if successful, frees one more segment of the arm, thus lengthening the boom. For Linenger, hanging out at the end of the arm in open space, the yanks are nightmarish. Each time Tsibliyev pulls, the American feels a sudden jerk, and he involuntarily tightens his grip.

Then things get even worse. As the boom extends out toward its full length, Linenger notices it is beginning to sway, as if in a breeze. As the commander extends the arm still farther, Linenger feels the whole boom vibrate under him, then begin to slowly swing back and forth. He wants to scream. After several long moments of this, the boom is finally extended to its full length, and Tsibliyev begins attempting to maneuver Linenger across open space to the docking area at the end of Kristall.

This is where the real anxiety begins for Linenger. The boom is so long, and the solar arrays so large, that Tsibliyev cannot physically see Linenger for much of the time he is clinging to the end of the arm. The commander swings the arm by instinct in the direction of Kristall, while Linenger attempts to give him directions. But, Linenger learns almost immediately, conventional directions don't mean much in space.

"To the right!" Linenger says at one point. "From you, to the right!" But Tsibliyev is standing at a 45-degree angle to Linenger. His right is somewhere beneath the American's knee. Tsibliyev begins craning his neck to spot Linenger, who tries in vain to give more directions.

"I need to go out two feet more!"

"No, no, I need to go out farther to miss this solar panel!"

It is no use. Tsibliyev cannot follow his directions. Gradually, the Russian begins to swing the boom over toward Kristall, but its swaying and vibrating are giving Linenger fits. By this point, the boom is so long it begins to swing on a wider and wider arc. Linenger is certain an S-curve has developed in the pole, limiting the commander's control over it. He is convinced the whole boom is about to snap.

He continues helplessly swinging back and forth, as Tsibliyev moves the pole across the face of the station. Then, at one point, Linenger turns his head and realizes he is about to crash into a sharp-edged solar array.

"Vasily, stop!" he says.

The commander stops moving the pole, but his momentum brings Linenger, by his estimate, within six inches of the array.

He exhales.

It is at roughly this point, with his knees squeezed around a vibrating steel pole dangling out in open space, swaying crazily across a field of knife-edged machinery, that the slapdash nature of the entire Russian EVA process strikes Linenger with the force of a two-by-four.

"It's risk upon risk is what you start feeling," Linenger later tells his debriefers. "When you go out the hatch and see C-clamps, and then you get on the end of the arm that's [bending], you don't have a lot of confidence that that thing's not going to break either. . . . You've got a lot of risk on your mind, and you really have to compartmentalize it all the way and do the job. And I was surprised I was able to do that. I was able to do that, and I'm not sure I was trained to do that. But I would suspect some people would not be able to do that."

It dawns on Linenger how little he really knew in advance about this spacewalk. "There's nothing orchestrated at all about the EVA," Linenger says. "It was winging it, basically, the whole time. It's nothing like the shuttle, where you say, 'Okay, there's going to be a handhold here, and then you go from there, and you go to point B.' "

Eventually, despite all the fits and starts, Linenger lands on the end of Kristall, just beside the docking port used by the shuttle. For the first time since leaving the hatch, he is able to anchor his feet under a rail, grab another rail with his hands, and feel steady. The handholds are solid here. He secures the Strela arm and waits for Tsibliyev to shimmy across it, which the Russian accomplishes with no trouble. They begin connecting the OPM to the outer hull at 10:14. They have been outside for just over an hour.

From his vantage point inside the station, Lazutkin tries his best to videotape the EVA, but the windows are small and don't give him the chance to film much. Out at the end of Kristall, painted orange to stand out against the gray-streaked station, Tsibliyev and Linenger look like thick white tadpoles, slowly spinning this way and that, crawling all around the OPM, a fat egg floating in space. Each man appears stiff and lifeless, arms and legs and trunks rotating as one, like plastic action figures in the hands of some giant cosmic child.

And then, just as Linenger thinks he is beginning to master the endless sensation of falling, night falls like a guillotine. Outside Mir there is nothing subtle about the movement from day to night. One second the area around the two spacewalkers is lit as if by spotlight. The next moment the light winks out, and they are engulfed by the darkest night Linenger has ever experienced. "Blackness," he writes in a letter to his son, describing the moment. "Not merely dark, but absolute black. You see nothing. Nothing. You grip the handhold ever more tightly. You convince yourself that it is okay to be falling, alone, nowhere, in the blackness. You loosen your grip. Your eyes adjust, and you can make out forms. Another human being silhouetted against the heavens."

They flick on their visor lights. Russian EVA protocols call for them to stop working during night passes, unless necessary. Linenger, halting work on the OPM hookup, stands absolutely still. There in the dark he once again begins to feel disoriented. Slowly, inch by inch, the sensation of falling is somehow changing—to what, he isn't immediately sure. Then, ever so slowly, he begins to feel as if he is falling forward. As the minutes tick by, he feels as if he is slowly being stood on his head. There in the pitch black of space, still feeling as if he is falling off a cliff, he begins to fight the almost uncontrollable urge to stand upright. His head tells him he is being ridiculous. There is no "upright" in space. But his emotions tell him otherwise. Bit by bit, space is slowly standing him upside down.

Finally, after a half hour spent tumbling forward, the Sun returns and Linenger now faces the awkward feeling that he has been turned upside down. He forces himself to continue. By and by, they finish installing the OPM and climb back across the Strela to the outer hull of Kvant 2. There Linenger detaches the debris-catching experiment, called the Particle Impact Experiment (PIE), and sticks it under one arm. For Linenger one final rush of anxiety comes as they finish. He is standing in the middle of Kvant 2's maze of hulking solar arrays, sensors, and boxy experiments with no clear path back to the airlock. In fact, he realizes, he has no idea where the airlock is. He sees what he thinks may be a promising path but immediately finds his way blocked by a large box.

"Vasily, how can I get over this experiment over here?"

"It's going to be tough," says Tsibliyev. "I'm going to start taking stuff back into the airlock. See you inside."

Linenger is dumbfounded. He has no idea which way the airlock is, and Tsibliyev clearly has no intention of showing him the way. He watches as the commander pushes off from where he had been tethered, and begins turning his head, apparently in search of the airlock. It dawns on Linenger that the Russian doesn't know where he is going either. "I could tell he had no clue," Linenger remembers.

Surrounded by solar arrays and a maze of experiments, Linenger searches for something—anything—to orient himself. After several moments he spots a window. He makes his way carefully toward it, peers inside, and sees the familiar confines of Kvant 2. Getting his orientation, he realizes Tsibliyev is going the wrong way.

"Vasily, airlock's that way," he says. "See you inside."

Linenger gingerly crawls toward the airlock and within minutes joins

Tsibliyev inside. They have been outside the station for five hours. Carefully closing the hatch behind them, Linenger heaves a giant sigh of relief. Later, when they have wriggled out of their space suits, Lazutkin cooks them a meal. The next day Linenger triumphantly e-mails Sang: "EVA pretty much flawless."

There is one strange footnote to the first joint Russian-American spacewalk. Months later, when Tsibliyev returned to Earth, he had a long conversation with Vladimir Solovyov in which he described what happened during the EVA. According to Solovyov, Tsibliyev said that at one point he and Linenger got into a heated argument—over what, he didn't say.

"He said they almost had a fistfight," Solovyov recalls. In fact, Solovyov says, Tsibliyev got so angry at Linenger that he lost his temper and swung his gloved fist at him, striking him on the helmet.

Presented with Solovyov's allegation, both Tsibliyev and Linenger deny that anything like this happened. Neither the NASA or Russian comm logs contain any suggestion of quarreling between the two men. The station, however, was out of communication with the TsUP for long periods. Whether there was an argument is something that may never be known.

By the time Linenger returns to the safety of base block, there are only three weeks left in his mission, most of which he will spend finishing up experiments, packing, inventorying and cleaning the station. He remains testy. The day after the EVA, Marshburn asks him to find some butterfly bandages to give to Tsibliyev and Lazutkin. "I had to move garbage to get at the med stuff," Linenger says in an e-mail. "As usual, the screw on this panel doesn't align properly, and it is a chore getting it back on. 35 minutes later, I am not sure what I accomplished. If they need more butterflies, they will ask me. . . . This is micro management at its worst."

Cleaning and straightening the station is an arduous process that reminds Linenger just how filthy Mir has become. "You would be appalled at the lack of simple preventive maintenance," he e-mails Culbertson on May 6. "As an example, I filled two jumbo ziplocks with debri [sic], nuts, bolts, you name it—from the screen of one ventilator (at least one inch thick with lint alone). The vacuum cleaner is so weak that I basically had to pull handfuls out and breathe in the dust."

It is during the cleaning process that the pent-up tensions between the three men finally burst into the open. Surprisingly, Tsibliyev isn't the catalyst. While straightening Spektr and Priroda, Linenger is trying to stow all the clutter he can and move trash out to other modules. One of the largest pieces of debris, an unused electronics board, he shoves into Kristall. Not long after, Lazutkin brings it back. Irked, Linenger returns it to Kristall. Once more Lazutkin brings it back.

"This stuff is not staying here," Linenger finally snaps when he sees Lazutkin returning the boxy board.

"Yes it is," Lazutkin says. "You have your two modules, but I have the rest of the station."

When Linenger goes to take the board and move it to Kristall, Lazutkin physically stops him, and for several moments the men actually engage in a spirited tug-of-war over the equipment.

"Okay," Linenger finally says, and the two men agree to take the matter to Tsibliyev for adjudication; the commander says he will think about it. For two days Lazutkin and Linenger don't speak, flying wordlessly by each other in the station's tight byways. Eventually, according to Linenger, it is Lazutkin who breaks the ice and apologizes.

By the time they finish cleaning, the station is free of much of the detritus that tends to clog the hatchways. Linenger inspects it all with a touch of ambivalence. He wants to be able to show Foale and the incoming shuttle crew just how dilapidated Mir is, and he realizes his own efforts may undercut his arguments. "It's not the real Mir they see," Linenger says of shuttle crews. "The Russians, they clean up, put new filters on. It's like a homeless shelter at Christmastime. If you think that's the way it looks the rest of the year, you're a fool."

After a seamless liftoff just before dawn on May 15, the Space Shuttle Atlantis carrying Linenger's replacement, Mike Foale, coasts up to the Kristall docking port and docks with the station two days later.

"Michael, welcome to the doorstep of your new home," Linenger radios Foale as Atlantis enters its final approach.

"Well, Jerry, if you're looking at me, hi, I'm waving through the window," Foale replies.

Once the docking is completed, Commander Charlie Precourt, Foale, and the five other shuttle astronauts all pile into Kristall and exchange bearhugs with Mir's three weary crewmen. Someone hands Linenger a bag of pretzels, and he rips it open and immediately begins eating. Tsibliyev is much happier to see the new Elektron unit, repair kits, and LiOH canisters the shuttle has brought in its cargo bay, not to mention dozens of boxes of new food, including apples, oranges, chocolates, ice cream, and foie gras, courtesy of the European Space Agency. Precourt, tasked with quietly rendering an assessment of both the two Russians and the station, watches Tsibliyev and Lazutkin closely. "They were tired of the workload; it was obvious they were both just exhausted," he remembers. "When the shuttle came, they kicked back and relaxed. Relaxation just melted into their faces."

Where Linenger is concerned, Precourt is in a delicate position. Houston has chosen to ignore the astronaut's pointed hints to delay sending Foale, and Precourt is worried that Linenger will be upset. To his relief, Linenger shows no sign of being offended. In fact he takes the commander on a detailed tour of the station, during which Precourt photographs spots of corrosion and repair on the cooling loops. At night Linenger talks for hours on end, regaling Precourt and the STS-84 crew with tales of the fire, the Near Miss, and all the equipment failures.

"Jerry just wanted to talk," recalls Precourt. "He talked and talked and talked and talked. The events just came spewing out of him. Finally, he had someone to speak English with." Linenger tells his tales with the drama and awe of a man who has just crawled out of burning wreckage on the side of a highway. To himself, Precourt thinks Linenger sounds exactly like what he is, a rookie. The unspoken subtext of many of his remarks is that Mir is now too dangerous for Americans. "You know, I've been around fighter jets and spacecraft for twenty-five years," Precourt recalls, "and these things just happen. You have a fire in your engine, you just don't quit. You fix it and go on. You just go on."

On May 18 Linenger gives a warm hug to Tsibliyev and a quick handshake to Lazutkin—relations between the two men remain strained—and floats into the shuttle, leaving Mir for the last time. "I stand relieved of my duties on the Mir," he enthusiastically radios Houston from the shuttle. "It's good to be back on U.S. soil." After finishing his inspection, Precourt leaves Mir with little doubt that Tsibliyev and Lazutkin could safely navigate the remainder of their mission, which to no one's surprise has been extended until August. "I got a really good feeling that these guys were going to be fine," Precourt recalls. "I had no second thoughts about leaving Mike at all."

As he settles into his new surroundings, Foale too is unconcerned. "I'm not worrying about this," he told a *Washington Post* reporter several days before liftoff. "The difficulties you may have heard about on the Mir, these are typical in a program that's very mature. The station has been up there a long time and there's a lot of work to do to keep repairing, maintaining it. . . . To be honest I actually look forward sometimes to these off-nominal situations, as we call them, when things don't quite go per plan. It makes life more interesting."

Foale's mission, of course, was destined to be far more interesting than he ever thought possible.

PART TWO

1992 to 1996

12

At the dawn of the 1990s, NASA was an agency adrift. The glory days of Apollo were but a distant memory, the new frontier of the shuttle era was long conquered. For three years following the Challenger disaster in 1986, NASA had struggled just to survive, to get back into space. By 1990, while the shuttles were finally flying again, there was a sense that they—and NASA—had no clear mission, no real reason for being. Where once NASA drew headlines for its achievements in space, the agency now drew attention for all the wrong reasons: investigations into cost overruns on its long-delayed Space Station Freedom, congressional committees that routinely pilloried NASA administrators for bloated budgets, allegations of fraud, and spacecraft that always seemed to be breaking down. Every few months yet another blue-ribbon committee trotted out yet another set of recommendations for establishing the agency's priorities. From California to Florida to Washington to Texas, morale at NASA centers was near an all-time low.

In 1989 the new Bush administration attempted to refocus NASA by establishing an advisory committee called the National Space Council. Inside NASA news of the committee's formation was greeted by snickers; the snickers turned to guffaws when Vice President Dan Quayle was named the new council's chairman. To run the council's small staff, Quayle hired Mark Albrecht, a thirty-eight-year-old former CIA analyst who for six years had served as California senator Pete Wilson's national security specialist on the Senate Armed Services Committee. Albrecht was a typical Hill rat, a squat, bearded infighter with a Ph.D. from the Rand Corporation who tooled around in a battered old Mercedes sedan. He knew next to nothing about NASA.

But if Albrecht was a neophyte, he was a determined neophyte. That spring of 1989 he scouted his new environs by soliciting ideas from the leading names in space circles—Carl Sagan; Norm Augustine; the heads of Boeing and Lockheed—and quickly deduced that what NASA needed was a single overarching goal for the future, something like John F. Kennedy's challenge to put a man on the moon. After long talks with Bush's chief of staff, John Sununu, and Richard Darman, Bush's budget director, Albrecht devised a wildly ambitious goal for Bush to announce in a speech on the steps of the Smithsonian on July 20, 1989, the twentieth anniversary of man's first steps on the Moon: NASA, Bush declared that day, would establish a manned base on the Moon and send an astronaut to Mars within thirty years. To Albrecht's amazement, NASA administrator Richard Truly, the levelheaded if uninspiring astronaut brought in to head the agency after Challenger, initially rejected the idea, even though it promised NASA $30 billion over the next three decades. The fact was, no one in the space program took Bush's speech seriously.

Bush's plan, in fact, was pure blue sky; what it needed was details, and the White House turned to NASA to provide them. Albrecht sat down and explained what the White House wanted to Aaron Cohen, the kindly old engineer who was then director of the Johnson Space Center.

"Most of all we want alternatives, plenty of alternatives," Albrecht told Cohen.

"What do you mean, alternatives?" Cohen asked. From the blank look on the JSC director's face, Albrecht could tell he wasn't getting through.

"Alternatives," Albrecht repeated. "I mean, there has to be more than one way to do this. Give us a Cadillac option, then give us the El Cheapo alternative, with the incumbent risks. Talk about all the different technologies that could be learned."

Albrecht got a bad feeling when Cohen's weekly status reports began crossing his desk that August. "We didn't like the reaction we got from NASA," he remembers. "It had an 'uh oh' quality to it." The NASA reports seemed to be full of lofty verbiage but few technical outlines or alternatives for what a lunar base and a Mars mission would actually look like. Albrecht kept emphasizing that the president wanted to see lots of alternatives, but, he recalls, "we kept getting nothing but grunts back. We kept getting these blank stares. We were beginning to get the strong feeling that we wouldn't be getting any alternatives."

Albrecht's concern deepened when NASA was unable to meet its ninety-day deadline. Truly was traveling, he was told, and the plan couldn't be sent to the White House without Truly's approval. Days went by, then weeks. On the eve of the Space Council's November meeting, Albrecht still hadn't seen NASA's plan. When it finally arrived, Albrecht's heart sank. There was, as he feared, exactly one alternative: Complete

NASA's beloved Freedom space station, then build a lunar base, then send a man to Mars. In his only nod to Albrecht's demand for alternatives, Cohen had supplied three time lines, listing costs for a Mars mission in fifteen, twenty-five, and thirty years. "We looked at each other and said, 'God damn it, we cannot do this!' " Albrecht recalls. "There are no alternatives. There's just one plan, split three ways."

The incident embodied all the qualities critics so disliked about the people who ran NASA: proud, headstrong, independent, disdainful of outsiders, and in love with their own big-ticket space projects. "These guys," Albrecht complained to Quayle, "don't feel like they work for the president of the United States." And they didn't. For eight years the Reagan administration had more or less ignored NASA. In the resulting vacuum Truly and his predecessors had taken guidance from the members of Congress who controlled their budgets, especially a senior aide on the House Appropriations Committee named Dick Malow, a man nicknamed "King Malow."

For two frustrating years, during which Bush's grand plan was slowly forgotten, Albrecht wrestled with Truly in a vain attempt to return the agency to White House control. Then, in November 1991, a man named J. R. Thompson resigned as the agency's No. 2. The job was a presidential appointment, and it fell to Albrecht to propose a successor. Albrecht again called all the usual suspects—Sagan, Augustine, top brass at Lockheed, Boeing, Martin Marietta—to fill out a list of potential replacements. What he heard surprised him: No one, he was told, would take the job as long as Truly remained NASA Administrator. "You're not going to get anyone good to come and work for Dick Truly," one aerospace leader told Albrecht, "because Dick Truly is not up to the job."

Even Albrecht, a critic of Truly's, was surprised at how widespread the anti-Truly feelings had become. When he briefed the vice president on his findings, Quayle told him to ask still more questions: Was there support for Truly's ouster? After talking with George Abbey, who was serving as the Space Council's in-house wise man, Albrecht returned with the answer they had hoped for: There was. One day that December he gathered three former NASA administrators—Jim Beggs, Thomas Paine, and Jim Fletcher, who called in from a hospital bed—in the office of Quayle's chief of staff, Bill Kristol. As Quayle listened, each of the three repeated Albrecht's message: Truly had to go.

It was a significant moment in NASA history. Imagine Henry Kissinger, Cyrus Vance, and James Baker coming to the White House for a secret meeting to sack the current secretary of state. Albrecht wondered aloud whether any of the three chiefs might volunteer to carry the message to Truly himself. Each declined. After conferring with the president, who considered Truly a personal friend, Quayle reluctantly summoned Truly to his office. There the vice president offered Truly the chance to make a gracious exit. He could have any open ambassadorship in the world—Quayle suggested one of the former Soviet republics, Kazakhstan or Belarus, where astronauts were still considered heroes—in exchange for his resignation. Truly seemed to take the news well and said he would think about it. Afterward he met with the administration's personnel chief, Connie Horner, and promised to mull his choice over the coming weekend.

But the following Monday, Truly sent a message no one in the White House expected. He wouldn't resign. "Then he went into utter radio silence for a week, maybe two weeks," remembers a Quayle adviser. "It was this very awkward period." Albrecht had been certain Truly would fall on his sword when asked by his commander in chief. The next thing Albrecht knew, he received a phone call from the president's chief of staff, Samuel Skinner. Truly had barged onto Skinner's appointment schedule and was apparently coming in to plead his case. Quayle and Albrecht simply couldn't believe Truly's audacity. "This was messy, messy, messy—it was ridiculous," remembers a former Quayle aide. "I am still mystified why he decided to grab a grenade and go into the Oval Office to fall on it."

When Truly came to his office, Skinner repeated the message that he needed to resign. But even then Truly refused to go. "I want to hear the president say it," Truly told Skinner. "I want to hear it from the president's lips." Amazed, Skinner turned to Albrecht to resolve the increasingly unseemly spectacle. No one wanted to ask the president to fire Truly. Bush detested firing anyone personally; when it came time to fire his own chief of staff, Sununu, he'd had his son George Jr. do it. "Everyone was telling [Albrecht], 'You can't make the president fire somebody! He hates this!' " remembers a former White House aide.

But there was nothing Albrecht could do. And so, at five o'clock on a Monday afternoon in mid-February, Truly disappeared into the Oval Office. A half hour later he was gone, having agreed to tender his resignation. No sooner had he left than Albrecht was summoned to Skinner's office. "What the hell is it with Dick Truly!" Skinner demanded.

"What do you mean?" Albrecht asked.

"He came in here and begged and whined and said, 'These people are telling you the wrong things, you're listening to the wrong people.' This is outrageous! The president had to ask for his resignation!"

The next morning Bush was striding through a White House corridor on the way to a meeting when he spotted Albrecht, who many in the White House were blaming for the whole Truly mess. The president gave Albrecht a look he had never seen before and hoped never to see again. "Your job," the president intoned as he passed, "is to get me the best NASA Administrator in history, and do it before Truly's resignation is effective."

Albrecht had no doubt about his own fate if he failed. Truly was to resign on April 1; that gave him barely forty-five days to find a replacement.

"You must be having a wonderful day," Connie Horner remarked when she saw Albrecht later that morning.

"Connie," Albrecht ventured, "who was the fastest person confirmed in this administration?"

"Jim Baker," Horner replied. "He went through like a bullet train."

"And how long?"

"Forty-eight days."

Albrecht had forty-five. The time was February 1992: NASA had no direction, no clear future, and now it had no leader. It was, more than a few observers felt, at rock bottom. But as Albrecht began his search for Truly's successor, one thing was certain. NASA's new leader, whoever he or she was, would be chosen by the White House and remain under its firm control.

Mark Albrecht had no idea whom he could armtwist into taking the NASA Administrator's job. Who would want it? It wasn't just that Bush's approval ratings were down. It was the fact that, if Bush lost to Bill Clinton in the November election, anyone he chose faced a term at NASA's helm of barely eight months, followed by unemployment. "Whoever picked the timing ought to be shot," one congressional source sniped to the *Washington Post*.

Still, in a matter of days, Albrecht managed to put together a short list for Quayle to peruse. Everyone on it was a known quantity—executives at Hughes Aerospace, Boeing and Lockheed—except one.

Quayle was going down the list when he came to the new name.

"Who is Daniel S. Goldin?" he asked.

Albrecht held his breath. Goldin was an obscure middle manager at the TRW aerospace conglomerate in Southern California, whom he had met a couple of years before. Goldin had dropped by the White House to pitch an idea to produce a smaller, cheaper version of a massive NASA project called the Earth Observation System. The EOS, as it was called, was a hugely ambitious scheme to launch six fifteen-ton observatories into orbit by century's end; it was so big, critics called it "Battlestar Galactica." At an estimated cost of somewhere over $30 billion, the EOS would cost roughly as much as the entire Apollo program. Richard Darman was calling it "the $34 billion thermometer." Albrecht was even more critical, calling the EOS "the absolute poster child of the civil space program run amok." Goldin told Albrecht he had conceived a cheaper version of EOS, using a series of smaller satellites, but when he had pitched the idea to NASA officials, they had threatened to cut off contracts to TRW if he pursued it. "I just thought you'd like to know that," Goldin had said.

Though he was a total unknown to official Washington that spring, Daniel Saul Goldin was destined to go down as one of the most energetic and effective Administrators in NASA history. Born into an Orthodox Jewish family in the South Bronx in 1940, Goldin had been a skinny, bookish kid with bad eyes who loved Flash Gordon and Buck Rogers serials. Upon graduation from City College in 1962, he joined NASA's Lewis Research Center in Cleveland, only to leave five years later for TRW, where he remained for twenty-five years. At TRW he managed a grabbag of missile- and consumer-technology programs until the late 1970s, when he disappeared into the company's "black" programs—its classified missile and satellite business—a job about which he cannot speak to this day.

Along the way Goldin often crossed paths with NASA executives and developed strong opinions about what was wrong with the agency. A colorful character who wore cowboy boots and shot from the lip, he was an unabashed romantic on the subject of spaceflight, insisting that it was man's destiny to conquer space. But NASA, Goldin declared in the sound-bite quotes reporters would come to love, was too set in its ways, "too stale, male, and pale"; it needed to do things "faster, cheaper, better." Once in office, Goldin would prove a whirling dervish, a "manic workaholic"—in the words of the *Washington Post*—who jetted all around the country, meeting with NASA employees, members of Congress, and seemingly anyone who had ideas on shaking up the agency. Critics called him "that lunatic" or "Captain Chaos." But Congress was impressed. "I detect a backbone in this nominee," then-Senator Al Gore observed at Goldin's confirmation hearings.

From the outset, Goldin had been Albrecht's top candidate for the job. After getting the go-ahead from Quayle, Goldin was brought to Washington to make the ritual rounds of congressmen. "The general reaction [to Goldin]," says one backer, "was, 'Jesus, who the hell was that guy? He's great! Where did you find him?' " Finally, after Goldin fought off a last-minute bout of second thoughts, Albrecht was poised to submit his nomination to Congress. There was just one last problem. One night Goldin mentioned to Albrecht that, by the way, did it matter that he was a registered Democrat?

Albrecht nearly choked. "Dan, you are to tell no one this," he said. "Do you understand? No one."

Albrecht hung up and phoned Quayle.

"I've got fabulous news," he told the vice president. "Dan Goldin is a registered Democrat."

"You are kidding me."

"No, I'm not."

And then Dan Quayle chuckled and mentioned the obvious. In that case, Goldin ought to sail through his confirmation hearings in the Democrat-controlled Senate. And he did. At 10 P.M. on the night of March 31, 1992, Daniel S. Goldin was named the ninth Administrator of the National Aeronautics & Space Administration.

After three years of bickering with Dick Truly, the White House finally had its own man at NASA's helm. Goldin, however, was slow to exert his new authority. Truly, born and bred at NASA, had been a favorite of the rank and file, and Goldin was warily regarded as an outsider, a change agent in an agency that hated change. To tutor him in his first months on the job, Albrecht assigned the taciturn George Abbey; the two men, polar opposites in personality but united in their determination to shake up NASA, were soon inseparable. Almost every night that spring Goldin, Albrecht, and Abbey huddled in the Space Council's White House offices to discuss strategy. "No one understood more than Dan Goldin how much he needed to be arm-in-arm with the White House, if only for his conservation," recalls Liz Prestridge, a senior Space Council aide. "It was one against twenty thousand, and without the proper political cover, the twenty thousand would have had his head."

Then, on June 1, Goldin's small circle abruptly changed, when Albrecht resigned to join private industry and a combative forty-year-old congressional staffer named Brian Dailey took his job. Dailey was a tough-talking nuclear weapons expert with a doctorate from the University of Southern California who immediately sympathized with the hazing Goldin was enduring at NASA. On his first day on the job, June 1, he and Goldin were driven out to NASA's Goddard Space Flight Center in suburban Maryland, where President Bush was scheduled to give a speech. Two NASA scientists escorted them to a conference room, then left them there, failing to tell them when the president arrived. While the NASA brass hobnobbed with Bush, Goldin and Dailey waited in the conference room, until eventually they caught on and scrambled out to the auditorium where the president was preparing to speak.

For all Goldin's travails, Dailey faced a far more serious problem his first week on the job that June. Russian president Boris Yeltsin was due in Washington in two weeks for a summit meeting, and something approaching panic was sweeping senior White House staffers as the realization spread that nothing much of significance was on the agenda. There would be the usual announcements of nuclear weapons cuts, and Bush would be able to announce an increase in aid to Russia. But these were hardly earthshaking developments. Though it was to be Yeltsin's first formal summit meeting with Bush, the *Washington Post* was already bemoaning what it termed the city's "ho-hum" attitude toward the Russian leader's visit, noting that the summit carried none of the "electricity" of Mikhail Gorbachev's visits to Washington in previous years.

Bush badly needed something electric. Most polls showed him trailing Bill Clinton, while the independent campaign waged by Texas billionaire H. Ross Perot monopolized the press corps' attention. The Bush reelection effort, dubbed by columnists Roland Evans and Robert Novak that summer as "the worst-conceived presidential campaign in memory," was near its low point. Yeltsin was to arrive on Monday evening, June 15, and the previous week Bush had received his worst reviews of the campaign during a miserable three-day trip to Latin America. Bush had to be rushed from an open-air speech in Panama City when demonstrators ignited a wild melee, forcing Panamanian police to fire volleys of tear gas that dispersed the crowds and brought tears to the president's eyes. It was an ignominious beginning to a trip that only got worse after Bush arrived in Rio de Janeiro for the United Nations Earth Summit, where demonstrators and third-world delegates had a field day lambasting the U.S. leader's environmental policies. When Bush returned to Washington on Sunday, June 14—the day before Yeltsin's arrival—the *Washington Post* termed it "one of the more troubling weeks of his presidency."

Inside the White House that week, there was enormous pressure to somehow enliven the Yeltsin summit and provide Bush the boost he needed in the polls. A call went out to every level of the administration to find something—anything—Bush could do with Yeltsin that might provide a rosy photo opportunity or a new policy coup. Space was a natural venue for superpower cooperation, and up in the Space Council offices in Room 426 of the old Executive Office Building, Dailey spent his first days on the job wracking his brain for something he might propose for the summit's agenda to help the president. He asked George Abbey to study the possibilities, and after several days Abbey returned to him with a plan.

At a summit the year before, Mikhail Gorbachev and Bush had tentatively agreed to a ceremonial swap of cosmonauts and astronauts: A NASA astronaut would fly aboard Mir, while a cosmonaut would fly aboard the shuttle. But with the coup attempt against Gorbachev that September and the disintegration of the Soviet Union three months later, the idea had fallen by the wayside. A week before Yeltsin's arrival, Abbey and Dailey were walking through the lobby of the old Executive Office Building when Abbey said, "Brian, I think we have an idea of something we could do with the Russians." Dailey stopped and listened as Abbey outlined his plan.

"You'll have to get this done, but here's the idea," Abbey began. Why not revive the astronaut-cosmonaut swap? It would be a dramatic demonstration of the new spirit of superpower cooperation Bush was extolling. It was simple and direct, the Russians would no doubt go for it, and best of all, it would be a tangible thing for Bush and Yeltsin to jointly announce.

"Is it doable?" Dailey asked. "You think we could really get Houston to agree to it?"

Abbey said he would take care of the astronauts. "Let me work the White House side," Dailey mused. "I don't know if we could put this on the summit agenda, though. I mean, it's late."

It was. Most of the agenda items for the summit had been in place for weeks, if not months. Still, Dailey phoned Goldin later that day and pitched the idea. Goldin loved it. There was just one problem. On paper the Space Council staff wasn't supposed to initiate new policy; that was the job of council members themselves. To preserve appearances, the two men agreed that Goldin would send Dailey a letter requesting his input on the idea. That way a paper trail would exist that suggested the idea had emanated from NASA, when in fact it had been a Space Council idea.

Dailey next ran the idea by Quayle, and the vice president loved it too. Dailey began to get excited. Here, it seemed, was the perfect plan, a way to help the president politically—and not coincidentally, Dailey himself. But no sooner did Dailey sense momentum mounting for the idea than he was confronted by the seemingly insurmountable task of placing the item on the summit's agenda. The State Department was the official gatekeeper for new action items, and in a phone call later that week to Reginald Bartholomew, an undersecretary of state, Dailey was told it was simply too late to do anything this big.

"The State Department got very mad at me," Dailey remembers. "I was new, and it was last minute. You have to understand, this was unheard of, to get something in a summit discussion in four days. Bartholomew fought me all the way. [He said] 'We're going to wait, this is too fast, too soon. . . . Maybe we can explore this later.' "

Dailey took a hard line. "God damn it, we've got to do it now," he remembers telling Bartholomew. "We've got to force it."

But Bartholomew would not be forced. At the weekend, with Yeltsin scheduled to arrive in Washington Monday evening, the idea seemed stillborn. His only remaining option, Dailey felt, was to take the idea directly to Bush himself. But the president was still in Latin America and wouldn't return until late on Sunday. Then, attending an embassy dinner on Friday evening, Dailey spotted Kathy Sawyer, the *Washington Post*'s science reporter, and got an idea. He had only recently met Sawyer and wasn't sure she could be trusted, but he was out of options. Deciding to take a risk, he took Sawyer aside and gave her the whole story "on background," meaning he could only be identified in the newspaper as a senior administration source.

Sawyer's story ran on page A4 of the *Post* Sunday morning. "U.S.-Russian Space Mission Considered; Docking of Shuttle at Mir Station Would Symbolize Cold War's End" read the headline over what Dailey, emitting a sigh of relief, saw was a generally upbeat story. "Proponents argue that such a joint endeavor would be a way for the Bush Administration to demonstrate vision and leadership at a critical point in [an] election year," Sawyer wrote. "However, the [administration] sources cautioned that various factions within the administration must agree on the best set of options before an accord is approved."

Monday morning Quayle called Dailey into his office. "Great press," the vice president said. Bush had read the story and immediately embraced the idea. The next thing Dailey knew, Reg Bartholomew from State was on the phone, reluctantly acknowledging that the matter could be placed on the summit agenda. It was a momentous decision. No one knew it at the time, of course, but in four short days, with the sole purpose of boosting the president's reelection chances and with almost no input from NASA, Dailey and Abbey had set the space agency on a historic new course.

If they were to proceed with the swap, a host of matters needed to be resolved. Both the Russian and American astronauts would need to be trained in the use of each other's instruments and technologies, which would take months if not years. Just working out the logistics would take months of negotiations. And for that matter, how would the shuttle actually dock with Mir? For all its flights over the previous eleven years, the Shuttle had never actually docked with anything.

But before anything could be negotiated, Dailey and Goldin had to figure out who to negotiate with. Administrators and astronauts in the two countries' manned space programs had been fairly close for a period in the 1970s, during and immediately after the Apollo-Soyuz mission in 1975. But the Soviet Union's invasion of Afghanistan in 1979 had severed almost all those ties; during the 1980s the two space programs collaborated on only a smattering of medical and scientific committees. Further complicating Goldin and Dailey's task, the "Russian space program," as Westerners understood the term, was no longer wholly under the control of the Kremlin. The Soviet design bureau that built Mir and supervised its daily operations was in the process of being privatized into an aerospace company they were now calling NPO Energia. For several years Energia executives, spearheaded by the young American who ran the company's new Washington office, Jeffrey Manber, had lobbied Congress to allow Energia rockets to carry U.S. satellites into space. Yuri Semenov had actually testified before a pair of congressional committees, urging NASA to look seriously at purchasing his company's launch services. So far Semenov's efforts, which were opposed by American rocket makers, had led to only one contract of note. A NASA veteran named Sam Keller had negotiated a $10 million contract with Energia to study the feasibility of using the Russians' Soyuz descent capsule as an escape craft for use on Space Station Freedom.

Dailey, however, found the idea of negotiating directly with Energia untenable. "Dan, this is totally unacceptable," he told Goldin. "We cannot be in a position of the American government working with a Russian aerospace company. It's got be government-to-government."

Goldin agreed. The obvious choice of negotiating partner was the infant Russian Space Agency, run by a former Soviet bureaucrat named Yuri Koptev. But Koptev's agency had only been formed that February, two months after the collapse of the Soviet Union, and neither Goldin nor Dailey had any real sense of what it actually was. Was it to be a full-fledged space agency like NASA? Or was it an advisory group like the National Space Council? When Dailey was introduced to Koptev in a pre-summit meeting, he tried to draw out the Russian on his plans.

"What've you got in the way of staff?" Dailey asked.

"Myself," Koptev replied, "and three staff members."

This was not promising. If NASA was to have any chance at avoiding a head-to-head negotiation with Energia, it had to find a way to bolster Koptev's little agency. Dailey began that effort on the first morning of the summit, Tuesday, June 16, when Yeltsin, Bush, and a crowd of aides gathered for breakfast and meetings at Blair House. Dailey scanned the room and didn't like what he saw: Koptev was nowhere to be seen. He crossed to where Quayle was sitting. "This is very bad," he whispered in the vice president's ear. "We need to get Koptev in here."

Quayle rose and approached Yeltsin. Later that morning Koptev appeared at Blair House. Afterward he and Goldin met at the NASA Administrator's Watergate apartment and found they saw eye-to-eye on a number of matters; Goldin left the meeting with the feeling that Koptev was a man he could work with. Once Bush and Yeltsin formally announced plans for reviving the astronaut-cosmonaut swap, the NASA chief agreed to lead an American delegation to Moscow that July to work out the details.

July 1992
Moscow

A week before Goldin and Dailey left for Russia on July 9, an American advance team arrived in Moscow. The team was headed by one of Goldin-Abbey's first hires, the former astronaut Bryan O'Connor, who until Abbey's call inviting him to return to NASA two months before had been a Marine Corps commander at the naval airbase at Patuxent River, Maryland. O'Connor had just rejoined NASA on June 1 as the No. 2 man in the Office of Space Flight and was floored to be asked to head NASA's team in Moscow. "I don't have any international experience," O'Connor cracked, "other than the time I flew with a Mexican guy on the shuttle." But O'Connor was a favorite of Abbey's and could be counted on to take orders with little backtalk.

With a month to prepare for his trip, O'Connor canvassed everyone he could find at NASA with Russian experience, including Arnold Aldrich and other old hands who had worked on Apollo-Soyuz. Aldrich told him not to fret, Moscow was clean and safe, there were police everywhere, and the only security problem he might encounter was the KGB bugging his hotel room. O'Connor wasn't too worried. The astronaut swap was a done deal as far as he was concerned; all he had to do in Moscow was smooth out details, such as working out the technical challenges of docking a shuttle to Mir.

O'Connor had barely checked into his Moscow hotel, the Penta, when he realized that nothing was as he had been told. The secure, gleaming city he expected to see had been replaced by a grubby, dirt-streaked metropolis thronged with beggars and street gangs. Lurid stories circulated of murders in Moscow's finest hotels, talk of the Russian Mafia's influence was rampant, and the Penta staff advised O'Connor to have his people travel in pairs and stay off the streets at night. Even that advice didn't help. O'Connor was following one of NASA's senior science officials, John Uri, through a Moscow park one afternoon when a pack of street people suddenly emerged from a group of bushes, surrounded Uri, and began rifling through his pockets. O'Connor ran toward his companion, shouting and swinging a bag containing a ceramic Russian doll. Though Uri's attackers quickly scattered, both men were left shaken by the incident.

But it was a tour of Energia's main factory outside Moscow that really stunned O'Connor. O'Connor wandered into some of the giant hangars and found them all but empty, their sole occupants forlorn clusters of three or four bored-looking engineers. His guide spoke of the thousands of people at work in the Russian space program.

"So where is everyone?" O'Connor wondered aloud.

His first meetings with Energia officials were no more promising. O'Connor immediately developed a strong dislike for his Russian counterpart, Valery Ryumin. When the two were introduced, Ryumin eyed O'Connor with what the astronaut took to be thinly veiled condescension.

"How many days have you been in space?" Ryumin asked archly.

"Sixteen."

Ryumin, O'Connor knew, had been in space 362 days. From that point on, O'Connor recalls, "Ryumin treated me like dirt. To them the number of days spent in space is a measure of manhood." Today Ryumin smiles when O'Connor's name is mentioned. "When we met the Americans for the first time," he says, "it didn't take long to find out who was professional and who was just a randomly chosen person."

The enmity between the lead negotiators, O'Connor and Ryumin, would not dissipate in coming months. "I never trusted him, and he never appreciated me," says O'Connor. He would bad-mouth people and complain about everything—and drink vodka. He was just a jerk."

Even worse, from O'Connor's point of view, was his counterpart at the Russian Space Agency, a pasty-faced bureaucrat named Boris Ostroumov, whom the older Americans remembered from Apollo-Soyuz. According to O'Connor, Ostroumov sometimes fell asleep during negotiating sessions. "Ostroumov was a zero," O'Connor recalls. "You could smell vodka [on his breath], you could see it in his eyes. This was in the workday."

Still, nothing prepared O'Connor for the tenor of his first meetings with Energia executives. Ryumin and his aides, O'Connor discovered, were far from sold on the idea of an astronaut swap. Energia, in fact, had adamantly opposed the swap idea when Gorbachev first floated it to Bush in 1991. "Energia's opposition was strongly backed by the Soviet foreign ministry at the time," recalls Bob Clarke, a senior counselor at the American Embassy in Moscow who was later hired to head international relations at NASA. "Their feeling was 'What do we get out of it?' "

Scattered around conference tables at Energia headquarters, the Russians threw up dozens of objections to the proposal. Ryumin was blunt and caustic. Russian cosmonauts would not fly on the shuttle, he lectured O'Connor, because it was patently unsafe. For one thing, it had no ejection system; such systems had saved the lives of more than one cosmonaut over the years. The fact that the shuttle had an auto-destruct was astounding to the Russians. The system, which enabled NASA ground controllers to blow up the shuttle in the unlikely event it wandered off course and threatened to crash into a populated area, was anathema to the Russians, who angrily said no such system had ever been incorporated into Russian spacecraft.

And that was only the beginning of Energia's objections. The whole idea of swapping a single, six-day flight on a "dangerous" shuttle for three months on the ultrasafe Mir struck the Russians as comically unfair. It would take one hundred shuttle trips, Ryumin said, to equal one stay aboard Mir. And why would they want to do such a thing? Safety considerations aside, taking an American aboard the Mir for free would elbow out other foreigners—the French, say, or the Bulgarians—who would pay hard cash for the opportunity to spend time in space.

But perhaps Ryumin's biggest objection, one he threw at O'Connor in their very first meeting, arose out of a seventeen-year-old incident that the Americans had long forgotten. During the Apollo-Soyuz docking in 1975, astronaut Tom Stafford had piloted his Apollo capsule to what the Americans felt was a perfect docking with the Soviet Soyuz. The Russians, however, insisted the docking was far from smooth. In their collective memory, Stafford's Apollo had hit the Soyuz so hard it had nearly ruptured its hull. What, Ryumin demanded of O'Connor, would prevent the shuttle—a far larger vehicle—from ramming the Mir just as hard?

O'Connor, already overwhelmed by the realities of simply working in Moscow, was dumbstruck by the Russian objections. Heavily jet-lagged, he was getting no sleep at night. When Goldin and Brian Dailey arrived in Moscow several days later, he wasted no time telling them he had serious doubts about NASA's ability to follow through on the agreements Bush and Yeltsin had signed the month before. In a limousine ride out to the Energia offices, O'Connor ticked through his concerns one after another.

"I just don't have any idea how a place that looks like it's falling apart, weeds growing up in the buildings, people not being paid, tremendous morale problems, no safety organization we can see, you know, this is a hell of an environment," O'Connor told Goldin. "Safety is the primary thing. I have real concerns about how we can ensure a good, safe operating environment for our people."

Not to mention the safety of American astronauts aboard Mir. "I've got to tell you," O'Connor continued, "I'm very concerned about the safety implications of what I've seen. To be launching our people on such an old system as Soyuz, operating on a space station the pictures of which look like a pretty rough operation, it doesn't seem even to a casual observer that their safety program measures up to ours." He had real doubts, O'Connor said, that the astronaut corps would back any kind of partnership with a program as dilapidated as the Russians'. There had to be some way, O'Connor went on, to delay any agreement with the Russians to enable NASA to take a fuller look at safety concerns.

Goldin said nothing. But Dailey, the only other person riding in the limousine with the two NASA men, was growing furious as he listened to O'Connor. Everyone he had spoken to, including no less a figure than George Abbey, felt Mir was safe. And if it was, by God, what was the problem? Dailey had worked long and hard on this project, and had far too much riding on it, to let some weak-willed flyboy derail the president's wishes. Dailey and O'Connor had never met, but there in the back of the limousine Dailey lost his temper and ripped into the former astronaut.

"Look, pal," he said, looking straight at O'Connor. "This is not your damn choice, this is the White House. The White House wants this done, and it will be done. You either get in line or get out of the way. If this is not a question of safety, and everyone's told me it isn't, then you forget about all this other stuff. This will be done. You got me?"*

Dailey's ferocity startled O'Connor, who fell silent as the limousine approached the Energia complex. Only as the three men stepped out of the car did a stark realization hit O'Connor, just as it would strike scores of NASA executives in coming months. The plan to put American astronauts aboard Mir had little to do with progress in space, and even less to do with advancements in science; all that was just public relations blather to be peddled to an American public that wasn't paying attention anyway. This was about politics. This was about what the White House wanted done to buoy Bush in the polls. And there was one more cold fact that

*Both Dailey and O'Connor confirm details of this conversation.

O'Connor learned in that limo ride: Anyone who had doubts about using the American space program to advance a political agenda would not be around for long.

The room, an outer office at Energia headquarters lined with pine paneling, fell silent as Yuri Semenov rose to address the NASA delegation. The chief designer was a big man, six-foot-two and 210 pounds, with a fine head of bright white hair and a commanding presence; behind him, Koptev sat meekly, saying nothing. Sitting behind Goldin, O'Connor had high expectations for this first meeting. He hoped Semenov would be able to cut through the tangled web of objections his deputies were throwing up.

But from Semenov's first words, it was clear the chief designer had not bought into the new spirit of international cooperation. For over an hour he extolled the virtues of the station he insisted on calling "my Mir," and spoke of how he needed money to pay Energia's twenty thousand employees. If NASA wanted to work aboard Mir, Semenov lectured the Americans, it would have to pay. And while he didn't mention any dollar figures, he left the NASA men with the impression that he expected the bill to be in the millions. "This was almost comical, even at the time," O'Connor recalls. "It was so obvious that my whole notebook full of plans and expectations was so out of whack with what we were hearing from Semenov. He didn't want to trade, to cooperate. He wanted cash money. He just went on and on, 'We are a company now, we want you to buy our services. All this country-to-country stuff is bogus. Talk to us directly like you would Boeing or Lockheed.' It was very haughty, looking down their noses at us."

As Semenov's lecture droned on, O'Connor, Goldin, and the others exchanged amazed glances. Everything Koptev had told them would be possible wasn't. Only then did Goldin begin to grasp the structure of the Russian space program: Koptev's fledgling Russian Space Agency was a barely functioning figleaf for Semenov's Energia, now a private company, which firmly controlled the station. Afterward everyone walked downstairs for the grand luncheon Semenov had assembled. It was to be a uniquely Russian spectacle, a full midday banquet packed with two or three hours of toasts and vodka and more toasts and more vodka. Waiters fluttered between tables piled high with kielbasa and cheeses and vodka and champagne. At least one member of the NASA delegation, Irene Fursow, the lead translator, realized that the sumptuous spread had a Potemkin's village quality; she overheard the Russian waiters, apparently running low on supplies, ordering food and drink from sidewalk kiosks out in the streets.

"We can't do this," Dailey hissed to Goldin, as the two men took their seats at the head table. "We've got to have a serious talk with these guys, and we have to have it right now."

Goldin agreed. As the waiters began serving appetizers the NASA administrator rose and, along with Dailey, told Semenov they needed to talk. Together the three men left the room. The old-timers in the NASA delegation, led by Sam Keller, who as the agency's point man for Russian relations advocated direct negotiations with Energia, were mortified by Goldin's breach of decorum. "Goldin is a very rude person, in the sense that he does not pay much attention to the niceties," recalls Keller. "Ten minutes into this lunch he stands up and says, 'Well, we've wasted enough time here, let's go to work.' He stood there and insisted that Semenov and his senior people go and start their negotiations."

Behind closed doors, Goldin sternly lectured the chief designer on the realities he faced in the New World Order. "It was pretty intense, but not angry," Goldin remembers. "I told him, 'We are going to do this with you or without you. Our people and your people have agreed that we should do something in space, and you, Mr. Semenov, are not entitled to a veto.' I wanted Semenov to understood there were bigger guns involved than Yuri Semenov and Dan Goldin and Yuri Koptev. . . . He had an unrealistic expectation. He thought, 'Here comes the U.S., let the good times roll.' It wasn't for [him] to tell us what to do."

That night, at a reception at the American Embassy, Goldin decided to drive home his point with a bit of macho gamesmanship. At one point he approached Semenov and pushed his face up to within inches of the old Russian's nose. "Mr. Semenov," he said, speaking slowly, "we had a very interesting meeting this afternoon. Have you thought about it?"

Goldin stared directly into the chief designer's eyes. Semenov met his stare for a long moment, then looked away.

"I got you," Goldin said with a smile.

From that point on, Goldin largely refused to deal directly with Semenov. If the chief designer needed something, Goldin had him channel his request through Koptev. It was in this way that Goldin almost singlehandedly "created" the Russian Space Agency, which until that point Semenov had all but refused to acknowledge. A level below Goldin, Bryan

O'Connor, assigned the task of negotiating a Memorandum of Understanding laying out plans for the astronaut exchange, tried the same trick with the cold, blustery Valery Ryumin and got nowhere. Whenever Ryumin confronted O'Connor with a negotiating point, the astronaut turned to his Russian Space Agency counterpart, Boris Ostroumov, for a decision. It was useless. Ostroumov never ventured an opinion, simply deferring to Ryumin. He was, in O'Connor's words, "a complete wuss."*

Despite this, by the end of the week-long trip a basic plan of action had emerged. NASA would send a single astronaut to Mir in the spring of 1995; the astronaut would fly to the station aboard a Russian Soyuz and be picked up by an American shuttle, which would dock with the station using an American-made replica of an androgynous docking "collar" one of the NASA men, Arnold Aldrich, had discovered moldering in an Energia warehouse. (The collar had been designed for use by Russia's Buran shuttle and had been mothballed along with the Buran.) In return the Russians would send two cosmonauts to fly aboard the shuttle.

Goldin was satisfied. At a trip-ending banquet, he gave an eloquent toast to the new spirit of cooperation between the two programs.

*Goldin's decision to freeze out Semenov created an immediate rift in the NASA delegation. Sam Keller felt Goldin's rudeness toward the chief designer was dooming any hopes for lasting cooperation; Keller thought it was ridiculous to negotiate with a Russian Space Agency that as yet had no real power. Tensions between Goldin and Keller came to a head when it came time to draft the Memorandum of Understanding. Goldin directed one of Dailey's Space Council staffers, Gerald Mussara, to draw up the memorandum; Mussara began doing so while sitting in the back of Goldin's limousine as the other Americans toured various Russian facilities. Taking suggestions from George Abbey and others, Mussara finished the paper the next day. But no sooner had he finished than Mussara was told that Keller's people were working on a memorandum of their own. "It got to the point where Sam wouldn't even talk to us," asserts one Space Council staffer on the trip. "[His people] wrote talking points for Goldin they wouldn't share with us. It got to this ridiculous, childish level. It was so horrible, it was embarrassing."

"At some point I concluded that this was ridiculous," remembers Mussara. "We could not have two draft agreements. I went to [Keller's people] and said, 'Look, there's no reason why this cannot be an open, collaborative process, let's just see what you're doing and put together one document. We're not doing our jobs here. This is ridiculous.' "

Keller's aides reluctantly turned over a copy of their proposed memorandum, but even that bit of compromise failed to head off one last comic standoff. Mussara, having finished a final draft that incorporated some of Keller's language, was lingering outside the second-floor business center in the group's hotel, the Radisson, when he spotted Goldin and Dailey striding toward him. When he reached down to pick up a copy of the agreement from a side table to show to Goldin, Keller appeared and grabbed the same piece of paper. For several seconds, as their bosses approached, Keller and Mussara engaged in a spirited tug-of-war, both men trying to rip the paper out of the other man's hand. Not six months later Sam Keller, who had been appointed the agency's Russian relations chief just that summer, retired from NASA after thirty-two years.

"What do you think of our new boss?" the NASA translator, Irene Fursow, asked Yuri Koptev as the two eyed Goldin.

Koptev smiled. "Irene," he said, "we can work with the devil."

The two shared a chuckle. "Seriously," Koptev went on. "You know what we have lived through. We can live through this."

July 1992
Houston

"What assignment from NASA would it take to make you stay at this point?" Norm Thagard was asking Dave Hilmers. The two veteran astronauts were hanging around the office they shared on the third floor of Building Four one afternoon, discussing their career plans. Both men had fully recovered from their hectic, fifty-five-experiment flight that January.

"Nothing," Hilmers said. "Nothing they could give me would make me stay at this point." Hilmers had already been accepted at the Baylor College of Medicine in Houston. He was going to be a doctor.

Hilmers looked at Thagard.

"What would it take to get you to stay?"

Thagard thought about the question for a long moment. He was a wiry former Marine, a dour little man with thinning hair and watery eyes. A member of the original shuttle class of 1978, Thagard had been an astronaut now for fourteen years. At forty-nine he was at the point where he couldn't expect to keep being chosen for flights over younger men. Everyone knew the Astronaut Office was no place to grow old. In truth, with four flights under his belt, there were no new mountains for Thagard to climb at NASA, unless he wanted to go into administration—and he didn't. He had begun talking with his alma mater, Florida State University, about a position as an engineering professor. There was, however, one last assignment that might intrigue him.

"I guess I would stay if I could fly in the Russian program," Thagard said.

Another astronaut, Mark Lee, stuck his head in the door. "You know

Dan asked for volunteers last November." Dan Brandenstein was head of the Astronaut Office.

Thagard turned in his seat. "You're kidding."

No, Lee said. In the wake of the Gorbachev-Bush summit the previous summer, Brandenstein had mentioned it at one of the regular Monday morning tag-ups; at the time, Thagard had been in training for his mission and had missed the meeting. Maybe a dozen people had volunteered, Lee said. Story Musgrave had. Shannon Lucid, too.

Thagard wanted to kick himself. Other astronauts might think a turn aboard the aging Mir station, not to mention a year or more training at Star City, was agony. Not Thagard. The chance to be the first American to go through the Russian training program? The chance to be the first American aboard a Russian station? The chance to set the American record for days in space? Thagard thought it all sounded like a wonderful adventure.

One last adventure in a life full of unlikely exploits. Norman Earl Thagard had risen from a poor southern boyhood to achieve prominence in more separate careers than most American households. He was an electrical engineer. Also, a fighter pilot. And a medical doctor. And an astronaut. None of which could have been predicted when he was born in the summer of 1943 in the sleepy northern Florida town of Marianna. His father was a bus driver, and his young mother ran off soon after the birth of Thagard's younger sister. He saw his mother once after that, at his high school graduation, and not again for another thirty-five years. He was raised instead by his paternal grandmother, Mary Sue Cooper, in a little frame house in Jacksonville. His father became a long-haul trucker, a ghost who passed through Florida every few months.

Somewhere, maybe it was from his entrepreneurial uncle Cary, a seventh-grade dropout who made a career leasing trucks, Thagard picked up a relentless tenacity. From his first years in elementary school he was the classic small-town overachiever, fascinated by amateur electronics, fighter planes, and, even before Sputnik, space. He read Asimov and Bradbury and Analog and in seventh grade began writing science fiction stories about space travelers and weird aliens that he sold to the other kids for a nickel apiece. In ninth grade he picked up a library book describing how to build a "foxhole radio" and built it himself, earning a ham radio license soon after. By the time he was fourteen he had his entire career mapped out: He would be an electrical engineer and a fighter pilot and a doctor.

By the time he graduated from Jacksonville's Paxon High in 1961, he had added another goal: to be an astronaut.

It was all ridiculously ambitious for a skinny kid whose family was so poor he went through high school working thirty-five-hour weeks bagging groceries after school. His father had been in a serious accident his sophomore year, but the money from the insurance settlement paid his first year's tuition at Florida State. His freshman year he joined the Naval Reserve and Air Force ROTC to make some money. After completing his master's degree, having worked his way through school as a disk jockey, he entered pilot training and gained his wings in 1968. In Vietnam he flew F-4s out of Chu Lai, returning to the United States after a year in combat, an ambitious twenty-seven-year-old determined to finally make some money. After working for an electrical utility in San Antonio for a time, he decided to go to medical school, and was in his final year at the University of Texas Southwestern Medical School in Dallas in 1977 when his wife, Kirby, mentioned that she heard an interesting radio advertisement: NASA was hiring new astronauts.

"I've got to send off for an application," Thagard said, excited.

"I already did."

Chosen as a mission specialist in 1978, Thagard flew four shuttle missions between 1983 and 1992. Gene Kranz called him "a bread and butter guy," meaning he was the kind of astronaut who stayed out of the headlines but made missions run smoothly. One astronaut called Thagard a "torpedo:" Wind him up, point him at a target, and forget about it; Norm Thagard would get the job done. As efficient as he could be, however, Thagard had a reputation as a grump. It wasn't entirely true. Like so many men of his generation, Thagard had a sense of humor he rarely chose to display outside his home; there, sitting before the television with Kirby and their three boys, he could recite every line of most Mel Brooks movies. Around the Astronaut Office Thagard could be candid about a mission's shortcomings, which also earned him an undeserved reputation as a bit of a whiner.

A couple of days after his chat with Hilmers and Lee, Thagard arrived early for work, as was his custom. At 6:30 he was ducking into the third-floor break room to get his morning coffee when Dan Brandenstein hailed him. "Have you got a minute?" Brandenstein asked.

"Sure."

"I just need to know. Would you be interested in flying the Russian mission?"

Thagard's response was immediate and unequivocal.

"Absolutely." He smiled broadly and mentioned the conversation with Hilmers. He thought he had missed the boat.

No, Brandenstein said. If you want it, it's yours.

"I want it," said Thagard.

"You'll be the prime then," Brandenstein said. There would be no backup, he went on. Given the press of shuttle flights, there was simply no way they could spare a second astronaut.

Thagard immediately plunged into preparations for the mission, joining a working group that was looking at what experiments they would be doing in space. At some point, Brandenstein told him, he would be assigned to Russian-language courses at a school in Monterey, California, but Thagard was impatient to begin. He bought a Russian textbook with his own money and joined an informal group at JSC that had already started twice-a-week Russian classes. Thagard quickly took to the new language, though others were having real trouble, especially Shannon Lucid, who had begun the classes even though she had no role yet in the program. In six months of coursework Thagard wasn't sure Lucid had learned a single word of Russian. "You know, Shannon," he began kidding the Oklahoma-raised Lucid, "you speak Russian the way it's commonly spoken on the streets of Tulsa."

Thagard had been expecting a formal announcement of his new assignment by Labor Day. When it didn't come, he was told to expect it in October. When Halloween came and went with no announcement, he began to grow concerned. He had been pushing to start his Monterey language classes, but that too seemed always just over the horizon; at the moment, Brandenstein told him, there was simply no money to pay for it. At that point Thagard began asking blunt questions: When was the announcement coming? Was there a problem with his selection? Officially Brandenstein assured him there was nothing wrong. But, as so often happened in the Astronaut Office, Thagard got the real answer via rumor and off-the-record conversations.

Yes, there was a problem. George Abbey wanted someone else to fly the mission. Thagard tried to remain calm. But this was exactly the kind of political game he and other veteran astronauts had feared would reappear when word reached Houston that spring that Abbey, given up for dead four years before, had miraculously resurrected himself as the power behind Dan Goldin's throne. It wasn't that he had ever clashed with

Abbey. Thagard was too smart for that. Like everyone else who had survived Abbey's rule in the 1970s and 1980s, Thagard had learned to keep his mouth shut. He had never publicly complained about Abbey, had never crossed him, had never been caught drunk in public or fooling around with another astronaut's wife. In his early years, in fact, Thagard had gotten along well with Abbey. Even before his first flight in December 1982, Abbey had selected him for a second flight, at the time an unheard-of compliment. But as the years went by Thagard had never become one of "George's Boys." He had never been a Bubba. He had done well in space, though, and that still counted for something. He still flew.

Now, in the fall of 1992, Thagard learned for the first time that he was not Abbey's choice for the historic Mir mission. Abbey, it was said, wanted Bill Shepherd, the shuttle commander who had been a member of Frank Culbertson and Blaine Hammond's Maggot class. Thagard couldn't believe it. What did Abbey have to do with this? Why was he even allowed to voice an opinion? Thagard suspected he knew the answer. The Russian program was Abbey's baby, and he wanted the NASA astronaut to be a star. Shepherd was the NASA equivalent of Robert Redford, while Thagard was but a hardworking character actor, a Brian Dennehy, a George Kennedy. Character actors, Thagard knew, rarely got chosen for a starring role.

February 1993
Washington

The chain of events that would ultimately lead to the merging of the Russian and American space station programs began late on a Thursday afternoon that February, when Dan Goldin, who had kept his job after Bill Clinton captured the presidency that November, received a sudden summons to the White House. There one of the new Clinton administration's budget advisers informed the NASA Administrator that the agency was in line for a stunning 20 percent cut in its upcoming budget. As matters now stood, Goldin was told, there was simply no alternative: They would have to kill the space station. "The blood drained out of my face," Goldin remembers.

Nine years after it had first been proposed by Ronald Reagan, Space Station Freedom was both NASA's shining hope for the future and its single biggest headache, a mammoth bureaucracy-within-a-bureaucracy that had managed to consume more than $8 billion in taxpayer money without ever launching a single component into space. It was the poster boy for every politician who decried six-hundred-dollar toilet seats, Exhibit A for every NASA critic who attacked the agency's obsession with bloated, overrun-plagued megaprojects. Every year or two it seemed to survive another congressional effort to kill it; later that year it would win its narrowest legislative victory by a single vote.

As far back as 1958, NASA administrators had proposed lifting a manned space station into Earth orbit, but John F. Kennedy decided to go to the Moon instead when the space program proved unable to agree on exactly what a space station was. That was always the problem: Between the politicians and the engineers, no one could ever agree on precisely what a space station should look like. By 1969, at the height of the Apollo program, NASA engineers were planning on a one-hundred-man station to be in orbit by 1980. They ended up instead with Skylab, an experimental ministation that hosted three astronaut crews over a period of six months in 1974 and 1975. With the end of the Moon missions in the 1970s, the Nixon administration decided that, before the United States could build a permanent station, NASA first needed a reusable vehicle to supply it. And so the space shuttle was born.

In the early 1980s NASA Administrator James Beggs began lobbying the Reagan White House for a station as "the next logical step" for NASA. Reagan, a space nut himself, greenlighted the idea over the objections of aides including his budget director, David Stockman, who like many in Congress simply couldn't see where the money would come from. NASA promised to build the station for no more than $8 billion—a number it seemed to pluck from thin air—and when by the late 1980s it unveiled new projections of costs as high as $31 billion, Congress howled. Still, egged on by politicians from NASA states like California, Texas, Alabama, and Maryland, the program lived on, and by the early 1990s was employing more than twenty thousand engineers and technicians in thirty-seven states.

Somehow the program grew even as the station shrank. In 1986 NASA had laid out plans for an eight-man station working with nineteen scientific instruments, a "garage" for repairing satellites, four separate laboratories, and a hangar for building spaceships to fly to Mars. But after six years of constant budget fights with Congress, the eight research functions NASA envisioned had been whittled down to two. The main lab and habitation modules, which would now house four people and not eight, had shrunk from forty-three feet in length to twenty-six. Space programs from Canada, Europe, and Japan had joined the project, each agreeing to contribute a module or a piece of equipment. "You have the [bureaucratic] superstructure of a battleship sitting on the structure of a P.T. Boat," a congressional aide sniped to the *Washington Post* in 1993.

The Clinton administration's new budget chief, Leon Panetta, had been pushing to kill the station even before Clinton took office. Still, when the White House moved to do it that wintry Thursday in early 1993, Goldin begged for time to study the situation himself. "I said, 'I'll tell you what,' " he recalls. " 'Give me a few days, I understand the problem. Let me prepare a working budget.' " Goldin returned to his office that evening, stricken. Without a space station, NASA had no future. Without a station, they could forget about Mars. He and Abbey summoned a half-dozen NASA staffers from around the country to a private home in Alexandria, Virginia, where they spent a furious weekend rethinking their options to shrink the existing station plans. By Sunday night they had three new ideas, which they dubbed Plans A, B, and C.

Monday morning, having stayed up for seventy-two hours straight, Goldin returned to his office and asked an aide to bring him a box of her son's Lego building blocks. He couldn't risk taking the new plans to a printer; they might leak and cause a panic. Instead Goldin sat down in his office and used the brightly colored Legos to build primitive mock-ups of Plan A and Plan B. For Plan C he grabbed a single cardboard toilet-paper holder. The next day he returned to the White House, used the mock-ups in a briefing for the president's senior staff, and, to his surprised relief, walked away with their go-ahead to fully develop the three new, cheaper options as part of an emergency redesign effort. The White House deadline was June 7. He had ninety days.

Even as Goldin scrambled to save the space station, the Clinton administration's new national security adviser, the professorial Anthony Lake, was grappling with a pressing problem of his own. Two and a half years earlier, in November 1990, a newspaper in India had disclosed a pact in which the Soviet Union had quietly agreed to sell cryogenic rocket engines to the Indian government. The Bush administration swiftly objected, noting that the engines would enable India to mount nuclear warheads on intercontinental missiles, which would no doubt prompt its western neighbor, Pakistan, to try the same; both the Russians and the Indians denied the rocket engines would be used for weapons purposes. For two years the State Department pressed first Gorbachev and then Yeltsin to stop the sale, but amid the turmoil of the Soviet Union's dissolution Russian leaders never followed through on airy promises to do so. In April 1992 the United States finally lost patience, announcing sanctions against Glavkosmos, the Russian government agency spearheading the sale. It was the first nasty dispute between the United States and Yeltsin's fragile new Russian government.

Now the whole ugly mess had fallen into Tony Lake's lap as an early test of the Clinton administration's resolve to stop the spread of nuclear weapons. As Lake pondered his options, it was clear the big-stick approach hadn't worked. Maybe, he thought, they should try a carrot instead. Since the behavior they would seek to influence was in the aerospace sector, it made sense that the Americans offer incentives in the same area. To that end, Lake formed a special interagency task force called the Space Cooperation Working Group, which included representatives from the U.S. Trade Representative's Office, the Departments of Transportation, Defense, and Commerce, and NASA. Subgroups went to work brainstorming ways the United States could increase Russia's share of the commercial satellite market, as well as projects the two space programs could collaborate upon.

As it turned out, the Russians grabbed for the carrot even before it was offered. In Moscow, Yuri Koptev read the announcement of NASA's redesign effort and immediately saw an opportunity to raise some badly needed cash. What if, he asked Yuri Semenov, they could get NASA to pay Energia to build its station? Semenov liked the idea, and during a visit to Washington in March Koptev arranged a meeting with Goldin in the NASA Administrator's office. "Dan, what if we were to join you in an international space station?" Koptev said as he sat in one of Goldin's plush wing chairs.

"I'll never forget when Yuri dropped the bomb that day," Goldin remembers. "I was flabbergasted." Koptev's offer presented a solution to a conundrum that had been nagging Goldin almost since the moment he took office a year before. "I was worried we had no experience dealing with [a] station whatsoever—none," he remembers. "We had no major experience in logistics. We knew nothing about the hazards. For Apollo we had Mercury and Gemini [to learn from]. For a space station, which is more complicated, we had nothing." The Russians, Goldin realized, had what the Americans didn't: experience. Just as important, they had the expertise to build a set of cheaper space station modules. Koptev followed up the meeting with a March 16 letter. "Billions of dollars" could be saved, he wrote, if the Russians and Americans could merge their space station efforts.

Everything began to come together in the first week of April, when Clinton and Yeltsin held their first summit meeting in Vancouver. Sitting in the Spanish-style residence of the president of the University of British Columbia, Clinton proposed an attractive deal: If Yeltsin stopped the transfer of engine technology to India, the Clinton administration would not seek to block the sale of the engines themselves. Further, Energia would get a shot at the American satellite market. More important for NASA, although this part of Clinton's proposal was not disclosed publicly, Russia would also be given the chance to "participate" in the American space station; details of this participation were not spelled out, though Energia executives assumed it meant they would sell the United States two or three of their modules. Yeltsin accepted the offer, although everyone involved, with an eye toward a referendum on Yeltsin's leadership later that month, strongly denied the obvious linkage between the two actions. A commission headed by U.S. vice president Al Gore and Russian prime minister Viktor Chernomyrdin was formed to oversee the new cooperative efforts.

That spring thousands of NASA and aerospace-industry workers followed every twist and turn of the space station redesign effort. Goldin structured it as a sort of beauty contest, with three separate design teams cloistered at the agency's Arlington, Virginia, offices—just across the Potomac from Washington—and charged with coming up with the best and cheapest new design. Even though word of possible Russian involvement in the new, smaller station leaked to the *New York Times*, no one outside the White House and the upper reaches of NASA seemed to believe Energia would be allowed to play a role of any significance. The idea was just too radical. Behind the scenes, however, Goldin and Abbey were beginning to believe that both time and money could be saved by bringing the Russians in as equal partners.

The Russians weren't making things easy. A small delegation from

Moscow visited Arlington in late April to throw out ideas on how Energia might join the American station effort. In what appeared to be their main proposal, the Russians urged NASA to pay Energia $7 billion to build the new station. In a subsidiary proposal the Russians offered to let NASA buy one of its science modules for $35 million. Both ideas were dismissed as nonstarters. Things went no better on the political front. By early June Yeltsin still had not followed through on his pledge to alter the Indian missile deal. When Chernomyrdin scheduled a trip to Washington later that month to meet with Gore, the White House decided to get tough. In a letter sent to Yeltsin about June 20, Clinton and Gore threatened to institute new sanctions against a variety of Russian agencies, a move that could cause the loss of more than $1 billion in prospective American contracts. Unless the Russians rescinded the objectionable parts of the Indian missile deal, the sanctions would take effect on July 15.

An American delegation led by Lynn Davis, an undersecretary of state, and Leon Fuerth, Gore's national security adviser, arrived in Moscow on June 30 with an offer they hoped the Russians could not refuse. The delegation presented Yeltsin's top aides with two options: Either endure the stinging sanctions, or accept an equal participation in a new International Space Station. Not only could the Russian government expect to earn several hundred million dollars by helping launch the new station's elements into space aboard Proton rockets, Fuerth for the first time outlined an idea, advanced by Goldin and Abbey, for a series of "rehearsal" shuttle missions to Mir in which NASA astronauts would live aboard the station for extended periods. All told, Fuerth suggested, the new package could be worth as much as $950 million to Russia, as opposed to the Indian deal, which promised anywhere from $210 million to $350 million.

Still, the Russians wavered. Hardliners in the Supreme Soviet were set on proceeding with the Indian missile agreement, and Yeltsin's political position remained unsteady enough that he was forced to accommodate their needs. The following week Warren Christopher, Clinton's baleful secretary of state, pressed the issue once again at a Group of Seven economic summit in Tokyo. This time Yeltsin seemed to agree. But the matter remained unresolved until a Russian delegation led by Koptev arrived in Washington on July 15. The following morning, with sanctions poised to begin, the Russians finally caved. India would get its engines, but not the ability to produce more of them. While vague on specifics, Russia promised to be better about this sort of thing in the future. In return it received the right to make eight commercial launches of American-made satellites through 2000. And NASA, in a second agreement Goldin and Koptev quietly signed on the morning of July 17, agreed to "define and determine the feasibility of a cooperative spaceflight program." No additional details were disclosed.

It was the beginning of a new era for both space programs. Other than a few scattered newspaper articles, no one outside NASA even noticed.

July 22, 1993
1:49 P.M.
Dzhezkazgan, Kazakhstan

The muddy-brown Russian military helicopter broke through the clouds just as a wall of ominous gray thunderheads appeared from the northwest. Mike Barratt, a thirty-four-year-old NASA flight surgeon, peered down onto the sunbaked Kazakh steppe. In his headset he could hear the high oscillating tone of the Soyuz capsule's pulse code beacon and the clear, confident voices of the cosmonauts as they spoke with the lead helicopter. There on the ground, two miles ahead, Barratt could just make out the capsule, a tiny thimble on the vast featureless steppe, visible only because of its billowing white parachute. Three helos and a pair of all-terrain vehicles were already on the ground, their occupants racing to erect a scaffold above the fallen capsule.

Here in a remote area of central Asia, three hours southeast of Moscow, Barratt was one of three Americans on the first NASA team to witness the Russian space program recover one of its Soyuz capsules. All but unnoticed by the outside world, Barratt was in the vanguard of an American army of doctors, engineers, and astronauts that would begin to invade Russia in the coming months. Already Norm Thagard and other astronauts had toured Star City, to get a sense of what training there would be like. Two cosmonauts, Sergei Krikalev and Vladimir Titov, had been living in apartments in Houston since November 1992, preparing for their shuttle missions. Bureaucrats and scientists in Washington and Moscow were hard at work planning the experiments Thagard would attempt aboard Mir.

By the time Barratt's helicopter bumped to a landing on a small salt flat, the rain had begun, great gusting walls of water that drenched everyone in the recovery operation. Barratt, hugging his camera in a vain attempt to keep it dry, scrambled out and ran toward the landing site 150 meters away. The scene had a surreal quality. With lightning flashing in the southeast and thunder rumbling all around, doctors and soldiers and reporters were running to and fro; to his amazement, Barratt watched as Kazakh villagers began to arrive in rickety trucks, motorcycles, and cars, aging grandmothers in tattered shawls, mothers carrying infants, even men on horseback. There, in the middle of the chaos, the descent module stood upright, a rare occurrence; four out of five capsules, the Russian doctors had said, came to rest on their sides.

One of the Russian doctors, a man named Vorobiov, had climbed onto the scaffolding, a green lightweight metal tripod, and was forcibly lifting the last of the three cosmonauts from the Soyuz and helping him onto an escape slide. The cosmonaut was then hustled into a waiting six-wheeled all-terrain vehicle, which Barratt noticed was missing one of its tires. The vehicle took the three cosmonauts a hundred yards to a pair of waiting medical tents. There, medical assistants helped the commander, Gennadi Manakov, out of the vehicle and watched as he plopped into a faux-leather armchair someone had set up in the mud. Manakov was pale and soaking wet, but as reporters flocked to his side, he laughed easily, answered their questions, and signed some documents. A white-coated Russian doctor hovered at his side, fanning the cosmonaut's face with one hand and taking his pulse with the other.

The rain began to slow. Manakov rose from his seat, the interviews complete, and staggered toward one of the medical tents, nearly falling at one point. Barratt followed the commander inside and Manakov offered his hand, but the Russian doctors quickly separated the two, pushing Barratt away. Reporters continued to traipse into the tent, tracking mud all about, as the doctors took the cosmonauts' pulse. Everything about the scene surprised Barratt: the villagers, the reporters, everyone in close contact with the unsteady cosmonauts. Immune system deficiencies are common after long-duration spaceflight, but the Russian doctors seemed to think nothing of allowing runny-nosed babies within sneezing distance of the cosmonauts. To Barratt the contrast with NASA's rigorous postflight procedures was jarring.

An hour after landing, the three cosmonauts, all now walking unaided, limped out to board a waiting helicopter, which spirited them to an airfield in the nearby city of Dzhezkazgan. Lingering at the site, Barratt watched as Russian specialists began to dismantle parts of the descent capsule in preparation for its return to Moscow. Two of the Soyuz's thrusters were carefully disassembled; the unused rocket propellant, about fifty canisters each the size of a shotgun shell, was pushed into a plastic bag and lobbed into a hastily dug hole for detonation. A Russian specialist shook out a forty-foot-long trail of black powder leading to the hole and attempted to light it with a match. In the drizzle it wouldn't light, so the specialist ran back toward the hole and shook out a small pile of powder directly beside it. He tossed a match and ran. A moment later the rocket cartridges ignited, erupting into an angry five-foot column of fire that warmed Barratt's face at a distance of forty feet.

Afterward the Americans walked to a waiting helicopter and were ferried back to the airfield, where they boarded a Tupolev jet for the three-hour trip back to Star City. The cosmonauts were already on board, and by the time the aircraft took flight the vodka was flowing. There were toasts and laughter and much relief, especially for the doctors, whose jobs all depended on the cosmonauts' safe return. At one point Dr. Vorobiov approached Barratt and demanded to know why he hadn't asked permission to enter the Russian medical tent. Barratt said he hadn't known it was necessary. He noticed Vorobiov was not entirely sober. Softening, the doctor volunteered that he never slept the night before a capsule recovery. Barratt said he understood; he rarely slept the night before a launch. The American sensed a moment of solidarity with one of his new comrades.

Suddenly, Vorobiov took a step back and in broken, raspy English said to Barratt, "I am Russian doctor. You are American doctor. Now we must fight!"

Barratt stared. Vorobiov, a squat, powerfully built man in his fifties, stuck out a large hand, and Barratt realized he intended to arm-wrestle him, in a standing stance, right there in the aisle of the airplane. As a number of Russians scrambled to the pair's side to act as attendants, Barratt took the doctor's hand, and Vorobiov furiously threw himself into the contest. Barratt managed to hold his own for about fifteen seconds, but Vorobiov's arms were as thick as Barratt's legs. With one final push the Russian doctor yanked Barratt's arm down to his side, nearly wrenching it from its socket. In a flash Vorobiov was on him, delivering a suffocating bear hug that lifted Barratt off the floor.

"You," the Russian happily announced, "are strong physician!"

July 24, 1993
Arlington, Virginia

The Russians arrived in the last days of July, two dozen bureaucrats and engineers shuffling into the Crystal City Marriott just across the Potomac from downtown Washington. They appeared nervous and wary, even more so when they learned a group of American naval intelligence officers was meeting in the hotel at the same time.

Now that politicians in both countries had quietly mandated that Russia and the United States were to merge their space station programs, someone had to figure out how. On the American side that job fell to a prototypical NASA good ole boy named David Mobley, a lean, fifty-four-year-old engineer from southern Alabama who had worked at the agency since 1961. Mobley was a plain talker, temperamental, and a tad idiosyncratic; assigned to one of the station redesign teams that spring, he had spent most of one afternoon debriefing a woman who swore she had been abducted by aliens, but whose resulting ideas on space station design Mobley found "fascinating."

The initial meeting between the two teams, held in a Marriott conference room, was as formal and stiff as anything convened at the United Nations. A follow-up gathering of technical teams in a nearby office building was no better. The Russians, led by Viktor Legostaev, an Energia vice president, made two proposals. The first was a suggestion that the Americans build a station around the existing Mir complex; the second was for NASA to pay Energia to build a new station for the Americans to use. Mobley politely shot down both ideas. At that point things broke down completely. "We couldn't agree on what time of the morning to meet, what building, who would attend, nothing," Mobley remembers. "Turns out they were angry because we wouldn't pick them up in the morning. They were mad that they had to walk. This went on for days."

Late one afternoon Mobley was angrily pacing the lobby of the building, waiting for some colleagues to join him for dinner, when a group of the Russians passed by on their way back to the Marriott for the evening. "Good meeting," one of them said. "See you tomorrow."

Mobley lost it. "Bullshit," he said in his thick Alabama drawl. "It was not a good meeting. If it ain't no better than this, why don't you guys just go home."

The Russians seemed nonplussed. "We are going back to the hotel," one of them said.

"I don't mean the hotel," Mobley shot back. "I mean Moscow." As the stunned Russians hustled out the front door, Mobley froze, thinking, *What have I done?*

But the next morning Mobley noticed a distinct change in the meeting's tone. It was, in fact, a cause-and-effect that dozens of NASA astronauts, doctors, and engineers would see in coming years; the Russians seemed to take American requests seriously only after an American erupted in an emotional outburst. In this case Legostaev agreed to an agenda and, to Mobley's surprise, stuck to it in coming days. Both sides laid out their "flight elements," that is, an inventory of every vehicle and module they could put into space; to Mobley, the Russian inventory served as a shopping list. He liked what he saw, but his questioning was handicapped by the fact that the Russians funneled all their whispered comments through a pair of English speakers—KGB men, Mobley figured—who were clearly not engineers. Mobley needed to talk directly to the engineers, but he had no idea who they were. Mobley instructed his deputy Chuck Daniels to bring to the next meeting pads of paper and Magic Markers and place them in front of every Russian. Mobley planned to ask a basic engineering question. Whoever picked up a Magic Marker and drew his answer on paper, Mobley bet, was an engineer. To his surprise, the ruse worked. One Russian drew his answer, and this was how Mobley was able to identify the man who became his principal counterpart, Leonid Gorshkov.

Mobley, hoping to streamline the process, maneuvered Gorshkov and another Russian, Pavel Vorobiev, into a small side room and there met with the two for the next week. What Mobley wanted, he had decided, was the Russian command module, the heart of the station; if a Russian command module could be inserted into the new, smaller station design the Clinton administration had tentatively approved in June, there was serious money to be saved. Growing excited, Mobley walked over to an adjacent building and presented the new configuration to the NASA station redesign team, which included scientists and engineers from Europe, Canada, and Japan. Everything went smoothly until after the meeting, when one of the Japanese representatives, Takao Kato, took Mobley aside.

Under the configuration Mobley envisioned, the new Russian module would rest beside the Japanese module. The Japanese government, Kato said, would never allow its module to sit next to the Russians'. He explained that since the end of World II, Japan and Russia had been embroiled in a squabble over which country owned four small islands off the north coast of Japan.

Afterward Mobley and Chuck Daniels repaired to a nearby Hamburger Hamlet to figure out what to do. The new configuration was dead; the Japanese would never go for it. As they sipped their beers, Daniels took a napkin and began drawing a totally new configuration. He came up with a station whose modules lay in a T-configuration; the Russian and Japanese modules were set well apart. Mobley took the new idea to the three Russians and for two weeks of closed-door meetings tried to convince them to go along. In the end the Russian objections to the T-configuration weren't technical but economic: The new layout included only one Russian module, and Energia wanted to sell NASA a second module, costing an estimated $300 million. Mobley wouldn't do it. "I knew I had them cornered when Vorobiev started quoting the U.S. Constitution and the Bill of Rights," Mobley remembers. "He ran out of technical arguments, so he started in about fairness and treating people equally."

On a Friday in mid-August the Russians gave in and signed a document outlining a new International Space Station in the T-configuration. The Russians would supply the command modules and other equipment. In return NASA agreed to pay Russia $400 million to let five—later seven—astronauts live aboard the Mir space station in a test program that would give the two space programs time to learn how to work together. The following Monday Mobley led an American delegation to Moscow, where Yuri Koptev and Yuri Semenov swiftly approved the plan. Two weeks later Al Gore, Viktor Chernomyrdin, and Goldin stood before a Washington press conference and formally unveiled it. The Russians and Americans, they announced, were going into space together.

Talks on just how the new International Space Station—the ISS—and the Mir missions would be operated continued through the fall in Washington and Moscow.* By far the most memorable of these meetings occurred on

*Negotiations on the final contract between NASA and the Russian Space Agency stretched into the next spring. Energia officials marveled at many of the standard clauses NASA inserted into the contract. One, for instance, forbid the signer from doing any busi-

the first weekend in October, when Goldin and a NASA delegation arrived in Moscow just as a violent coup attempt against the Yeltsin government broke out. A final agreement between the two countries November 1 was greeted with bipartisan approval in Washington.

Houston was another story. While politicians and engineers sketched out plans for the new ISS, the astronauts and flight controllers at JSC were startled to find themselves suddenly thrust into business with the Russians, their longtime competitors. Gene Kranz, then head of MOD, and Dave Leestma, head of Flight Crew Operations, shot off letters to Goldin in September demanding to have some input into the design of the new station, which, after all, they would one day be asked to run; Kranz got into a loud argument with one of Goldin's top aides, General Jeremiah Pearson, who lectured him that "there will be no more memos."

The development of the International Space Station, a story that falls outside the scope of this narrative, was to consume wide swaths of the NASA bureaucracy for the next six years. What almost no one seemed to focus on that fall was the more immediate consequence of Russian-American space cooperation, the missions of the American astronauts to Mir.

"This whole Mir thing just kind of crept in, almost unknown to us all," recalls Kranz, who retired as head of MOD in March 1994. "We saw these things coming, but we never thought it would become a plan in itself. To be blunt, the Shuttle-Mir missions were never really something that I took too seriously. I thought Shuttle-Mir would die, because of a lack of a mission and a crisp, clear reason for doing it."

Even the man named to oversee the newly formed Phase One office had misgivings. "At the beginning, I thought, 'What are we doing this dumb-shit thing for?' " recalls Tommy Holloway, then deputy chief of the space shuttle office. "Later I came to realize, if we're going to do ISS, then Phase One offers a tremendous opportunity. Better to work on this stuff now than reading about it in books, I suppose."

ness with the Soviet Union, Cuba, or Vietnam. "I said, 'What does this mean? How can I avoid doing business with myself?' " recalls Alexander Derechin, Energia's head of international marketing. "They said, 'Oh, forget about that, it has nothing to do with you.' " Another clause obliged the signer to follow all federal regulations concerning sexual harassment in the workplace, which left the Russians scratching their heads. But the facet that most perplexed Energia executives involved a clause in which each country pledged not to sue the other in the event their spacecraft somehow collided. "I said, 'What is this thing, 'cross-waiver liability'?" remembers Derechin. "They said this means that if shuttle hits Mir, they pay us nothing. And if Mir hits shuttle, we pay them nothing. I said, 'Please explain, how exactly would Mir hit shuttle?' This did not seem fair to me."

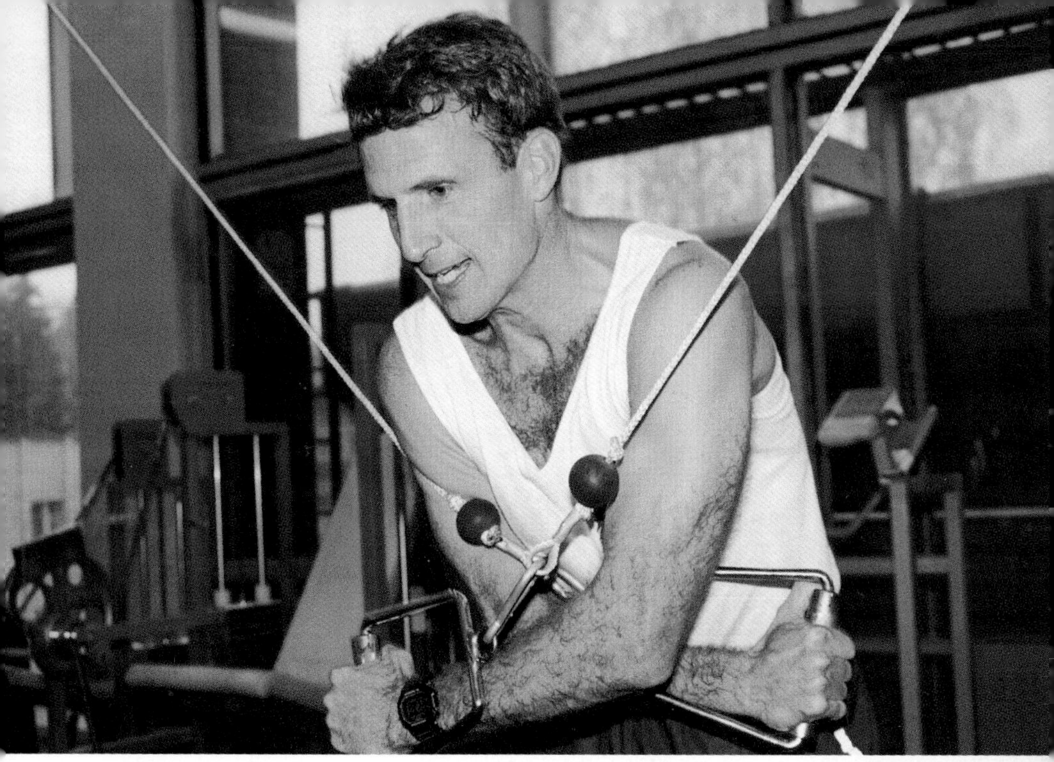

Jerry Linenger: The fourth American to live aboard Mir, the forty-two-year-old Navy doctor perplexed NASA by cutting off voice communication with his ground team, infuriated Russian officials by criticizing Mir's safety, and exasperated his cosmonaut crewmates by refusing to help out with repairs.

George Abbey: As director of the Johnson Space Center, Abbey has been the all-powerful lord of the astronaut corps for almost twenty years, and a figure who inspires fear and loathing among the astronauts.

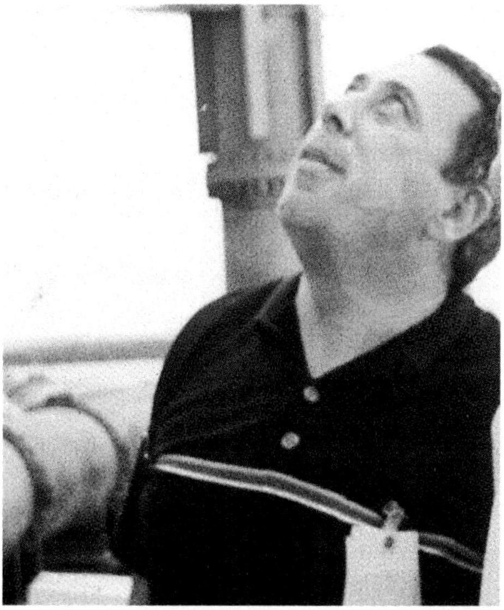

January 12, 1997: Linenger (*above*) departed for launchpad 39B, where at 4:27 A.M. the Space Shuttle Atlantis lifted off (*below*), ferrying him to the Russian space station. His mission went forward over the objections of Russian psychologists, who felt he was ill-suited for a long stay in space.

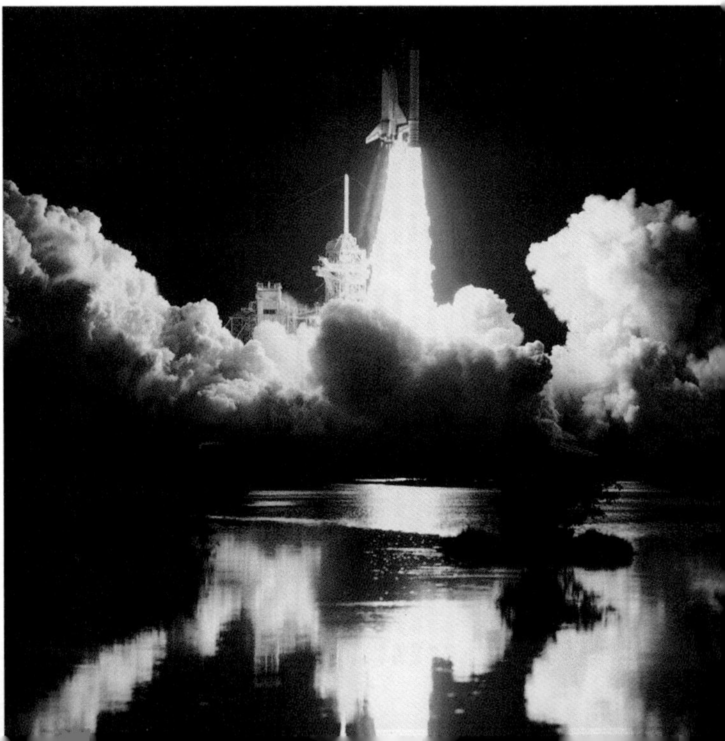

Frank Culbertson: The astronaut who headed the Shuttle-Mir program was blindsided by the Russian station's sudden troubles.

Dave Wolf: Suspended in astronaut purgatory following his role in an FBI sting operation, the hard-partying Indianapolis native was NASA's only choice for the final Mir mission.

Blaine Hammond: The butt of jokes inside NASA, the Astronaut Office's safety chief was a lonely critic of the Russian program.

Vasily Tsibliyev: The Russian commander, hoping to rebound from his embarrassing first mission to Mir in 1993, was thoughtful and hard-working, yet somehow cursed.

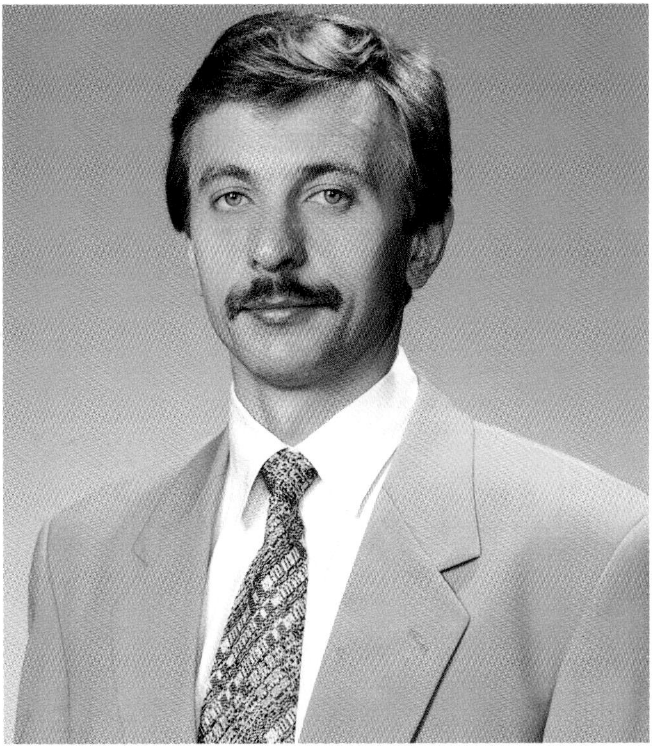

Aleksandr "Sasha" Lazutkin: The rookie flight engineer hoped his first mission to Mir would lead to a promotion, a hefty bonus and a new apartment.

Bad omens: Tsibliyev (*above, third from left*) and Lazutkin (*second from left*) share their worries about their upcoming flight with a famed psychic at an impromptu barbecue at Star City two weeks before the mission. *Below*, the two cosmonauts, accompanied by Reinhold Ewald, go through final ceremonies before their liftoff from the Russian base at Baikonur in Central Asia.

Mir: The eleven-year-old station, shown below soaring 250 miles above Cook Strait near New Zealand's South Island, was the crumbling Russian space program's last emblem of pride. American space shuttles—at right Atlantis docks with Mir in 1995—brought supplies that kept the station aloft, while an agreement hatched by the Clinton White House channeled $400 million in American money to keep the Russian program itself alive.

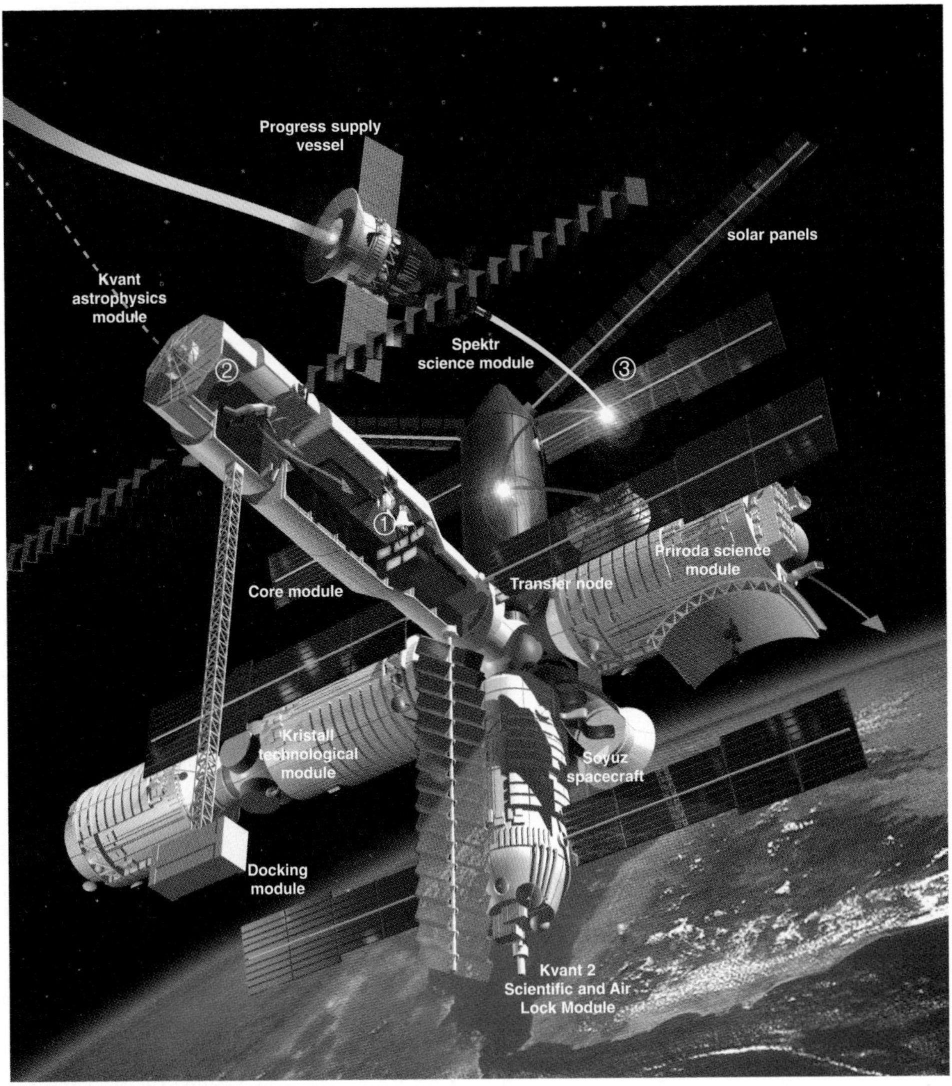

The Big Bang: The crash between *Mir* and a Progress unmanned supply vessel, June 25, 1997. Cosmonauts Tsibliyev and Lazutkin, *far right,* are in the core module, ① attempting to dock Progress by remote control. Astronaut Foale, ② *near right,* in the Kvant module, looks for the approaching supply ship, then rushes toward the Soyuz spacecraft for possible evacuation. He is only as far as the transfer node when he hears the crash ③, which punches a hole in the Spektr science module and damages a solar panel.

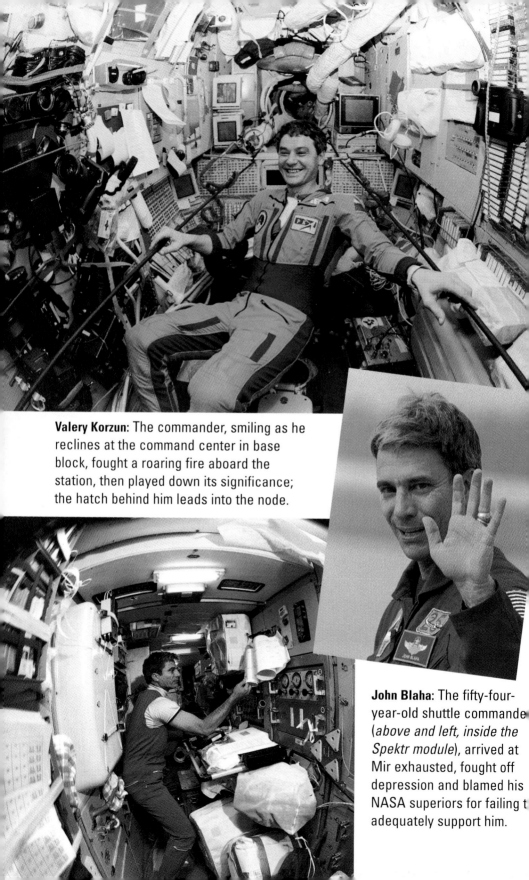

Valery Korzun: The commander, smiling as he reclines at the command center in base block, fought a roaring fire aboard the station, then played down its significance; the hatch behind him leads into the node.

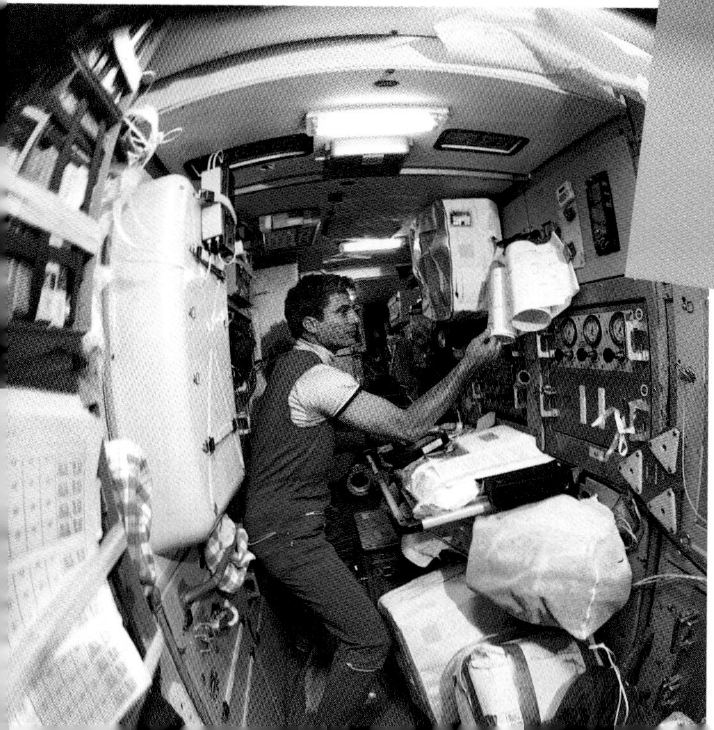

John Blaha: The fifty-four-year-old shuttle commande (*above and left, inside the Spektr module*), arrived at Mir exhausted, fought off depression and blamed his NASA superiors for failing t adequately support him.

Linenger in a gas mask: The astronaut feared his NASA superiors were conspiring with the Russians to cover up details of the fire.

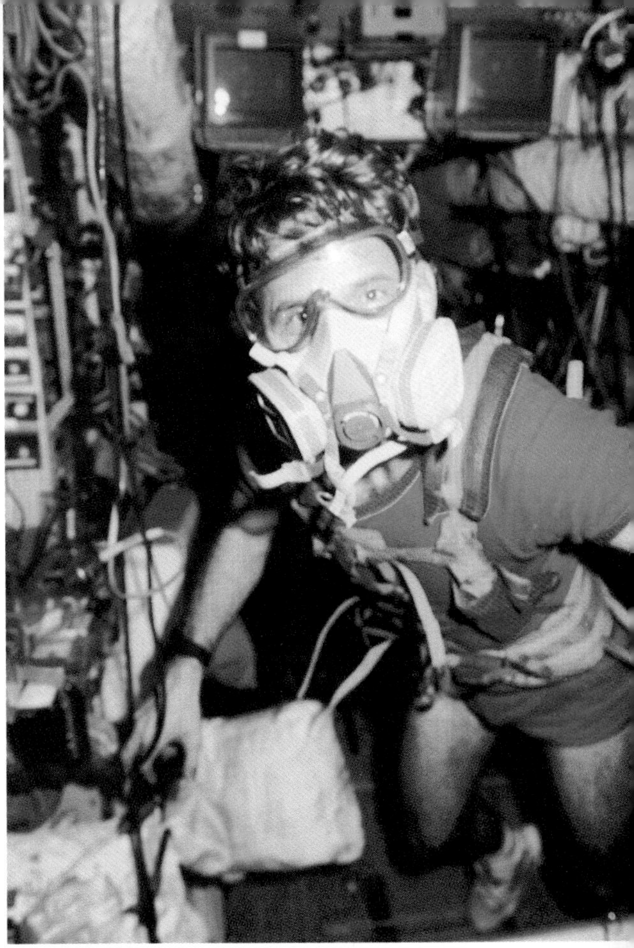

Korzun in the node: The station's central intersection, lined with the detritus of space flight as well as hatchways leading to the other modules, was to be the scene of a fateful drama.

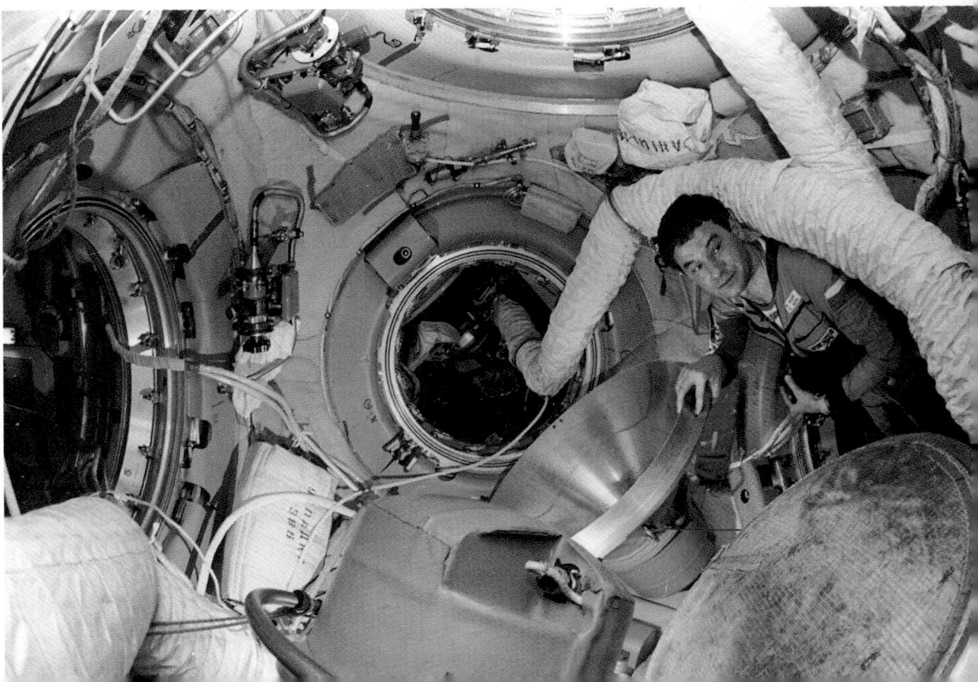

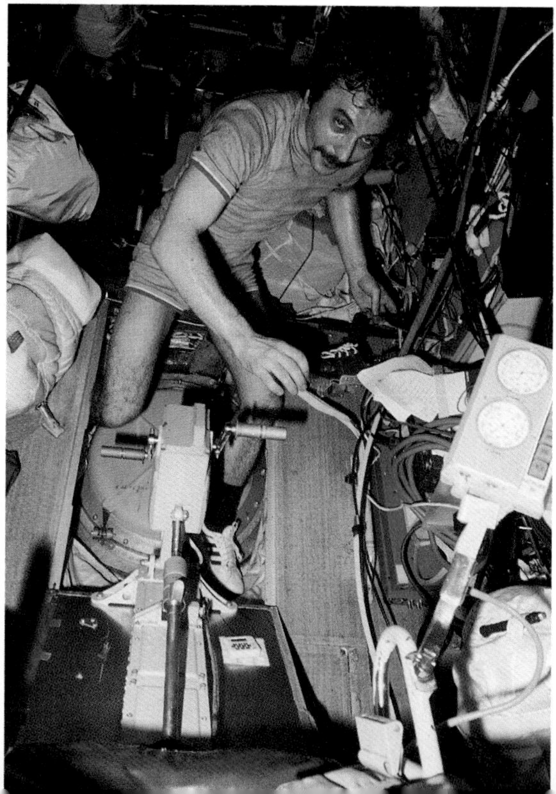

Mounting tensions: Lazutkin (*at left*), here exercising on a stationary bicycle, appeared overwhelmed by the cascading systems failures. Tsibliyev (*above*), grabbing a fast dinner with Lazutkin at the base block table, launched into frequent tirades at the ground; the hatch behind him leads into the small, darkened Kvant module, the site of the fire.

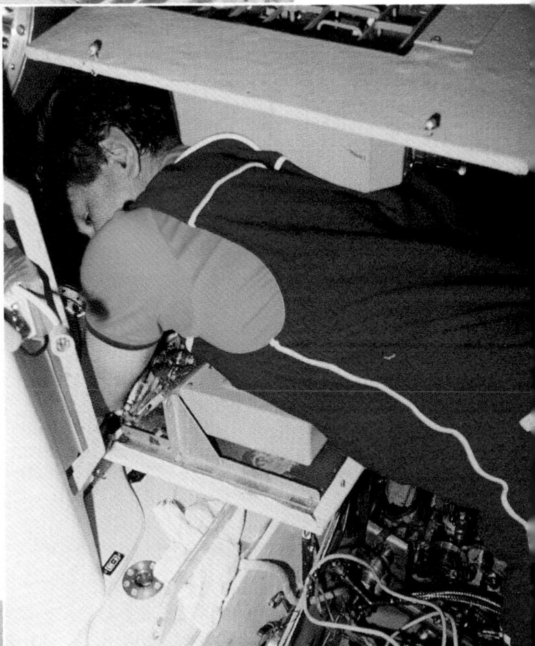

The commander (*above*), stayed up well past midnight most nights working with Lazutkin on repairs; at right, Tsibliyev slips behind a panel in search of coolant leaks. While the two Russians worked, Linenger (*below*), here outfitted for a sleep study, focused on his experiments.

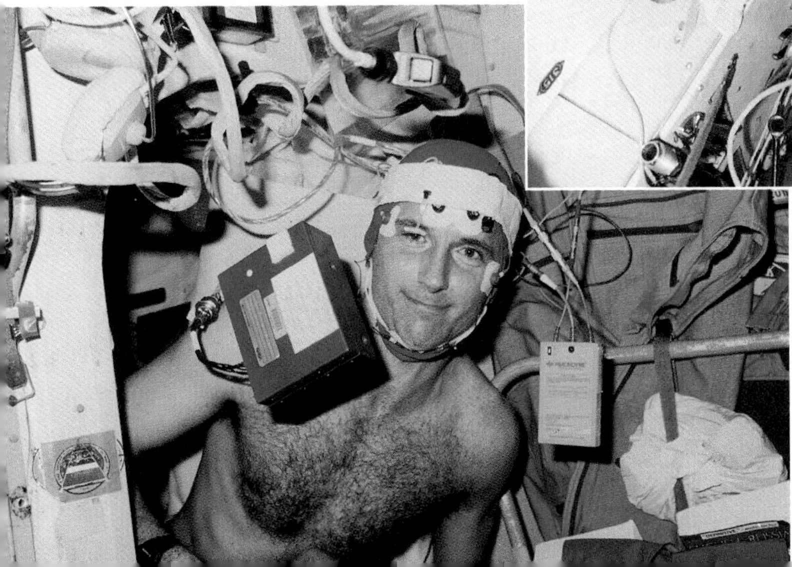

Bonnie Dunbar (*left*) and **Norm Thagard** (*above*): The first two astronauts to train at Star City wouldn't speak to each other.

Shannon Lucid: The veteran astronaut, here receiving an award from Boris Yeltsin in May 1997, overcame chaotic preparations to thrive aboard Mir.

Russian Mission Control: The TsUP, where ground controllers moonlighted as cab drivers and cats crept through the aisles in search of mice, controlled every move the cosmonauts made.

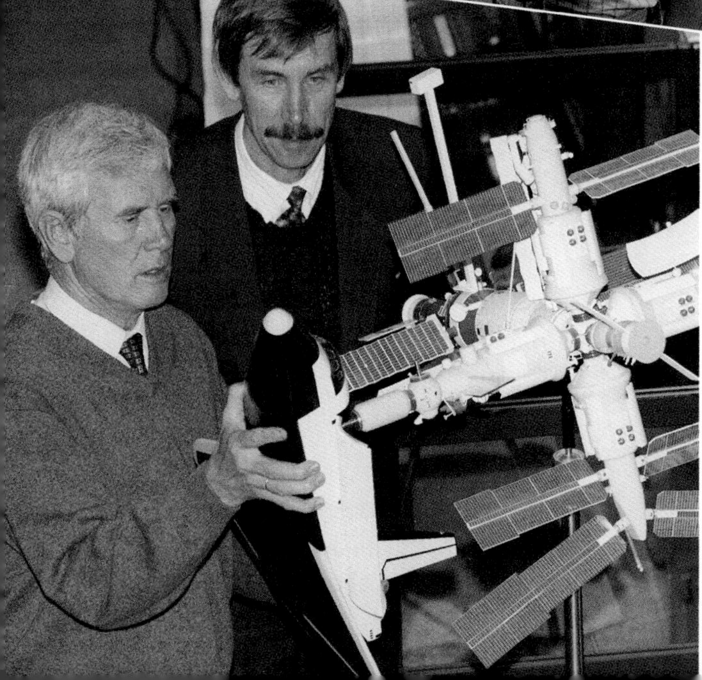

Viktor Blagov (*left*) and **Vladimir Solovyov** (*above*): The TsUP's two senior flight directors pushed crews toward the breaking point, then blamed them when things went wrong.

Mike Foale: The forty-year-old Britisher would prove the best-prepared of all the Mir astronauts, fully integrating himself into two Russian crews while surviving the first orbital decompression in the history of manned space flight. Below, Foale, Lazutkin, and Tsibliyev at a press conference after the collision.

A battered dragonfly:)amage to the Spektr module and its solar arrays crippled the station.

atoly Solovyov: The .eran commander o replaced Tsibliyev, re shown at the base ck table, was termined to make the rld proud of Mir ain.

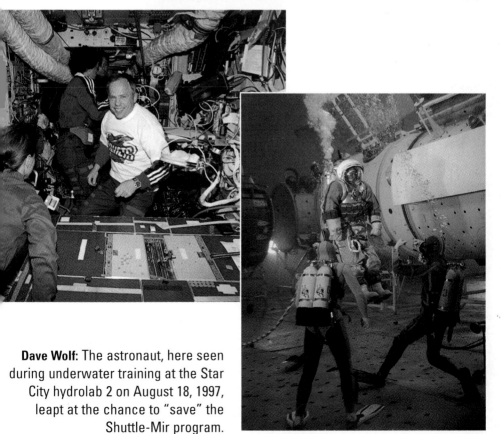

Dave Wolf: The astronaut, here seen during underwater training at the Star City hydrolab 2 on August 18, 1997, leapt at the chance to "save" the Shuttle-Mir program.

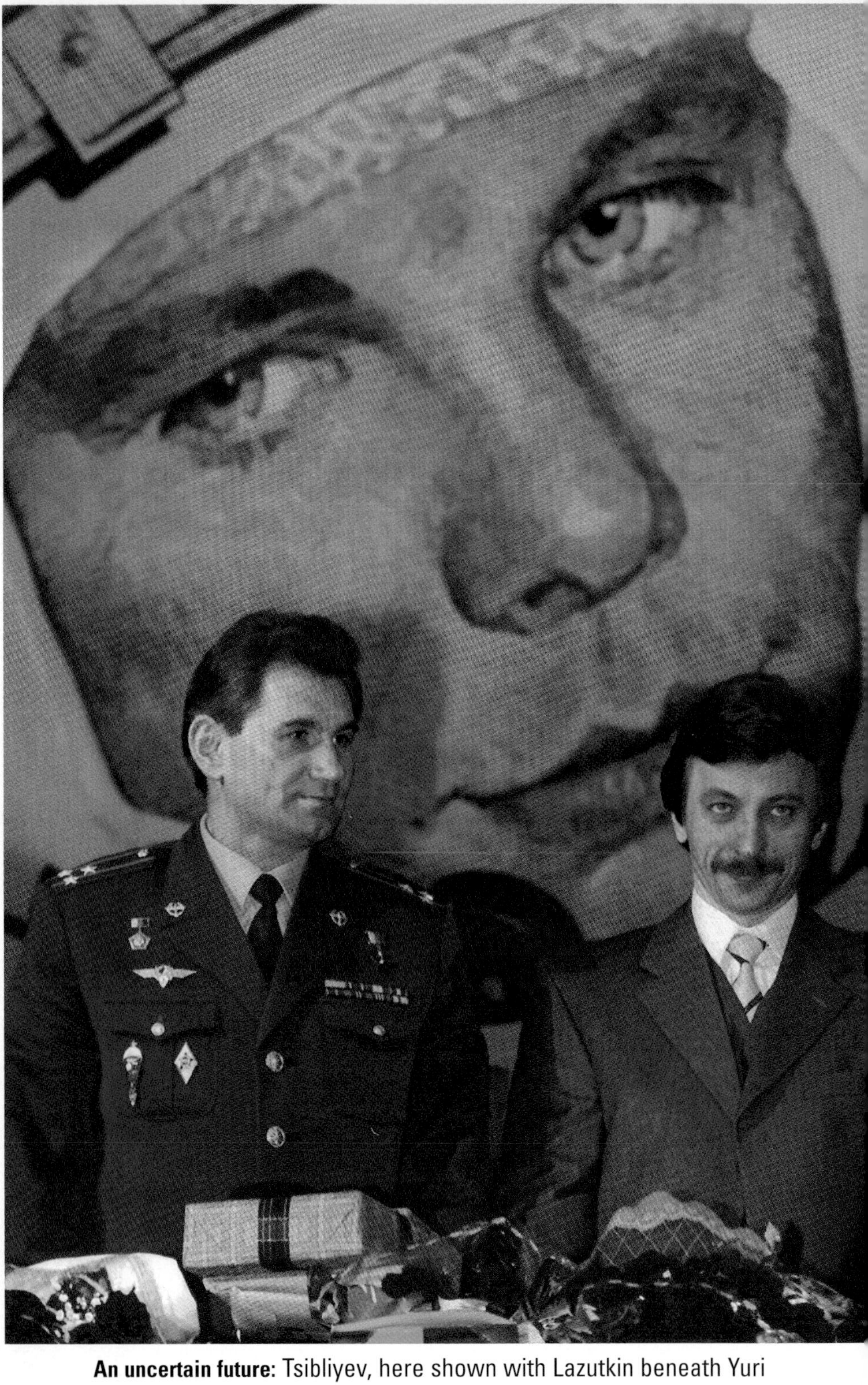

An uncertain future: Tsibliyev, here shown with Lazutkin beneath Yuri Gagarin's gaze at a Star City ceremony on September 5, 1997, lashed out at Russian officials upon his return to Earth; neither cosmonaut is expected to ever return to space.

13

November 1993
Monterey, Calif.

By November, Norm Thagard was in a black fury. As a Marine, as a doctor, he prided himself on his self-control, but this was too much for any man to bear. Here he was on the verge of graduating from the Russian-language program in Monterey, nearly sixteen months since the day Dan Brandenstein had asked him to be prime on the Mir mission, barely three months before he was scheduled to begin training in Star City, and still—still!—there had been no announcement that the Mir mission was his.

Every step of the way, it seemed NASA had thrown roadblocks across his path. When the Astronaut Office continued to insist there was no money to pay for the Monterey classes, Thagard volunteered to pay his own way. Only at the last minute had NASA agreed to help out, and then with a comically low per diem of ten dollars. Ten dollars a day. After gasoline and phone bills, it might be enough to buy a Big Mac. NASA wouldn't even fly him to California. He was forced to drive cross-country that summer in his family's 1983 Maxima station wagon, which broke down in the middle of Death Valley and had to be towed back to Las Vegas. There Thagard was forced to buy a new car—a necessity because NASA wouldn't pay for a rental car in Monterey. All told, Thagard figured he was out of pocket nearly two thousand dollars.

He blamed George Abbey. For a solid year the rumors had flown more or less constantly that Abbey was dead set on finding a way to give Bill Shepherd the Mir mission. Even after Thagard had flown to Star City in the fall of 1992 to help negotiate the terms of his living arrangements, the Shepherd rumors continued. He had returned to Russia in April and shown his wife, Kirby, the kind of apartment they would live in. Still the

Shepherd rumors continued. He sat through four months of Russian classes and still there had been no formal announcement of his appointment. The irony was, Thagard knew Shepherd didn't want his mission. The two astronauts had shared an office for a brief period that year, and in an awkward situation Thagard was obliged to listen to Shepherd's phone calls to Abbey. Incredibly, neither Thagard nor Shepherd ever brought up the matter, but Thagard got the clear impression Shepherd had better things to do than spend the next two years in Russia.

Then, just before Thanksgiving, Thagard got an alarming message from Bill Readdy, the shuttle commander who had surprised his colleagues by volunteering to be Thagard's backup. NASA hadn't planned on naming a backup for Thagard's mission, but when the Russians insisted, Readdy, a forty-one-year-old Navy pilot, had stepped forward. Readdy told Thagard of a phone conversation he had just had with Mike Mott, a diminutive aide of Abbey's in Washington. The astronauts called him "Mini-Mott."

"Mini-Mott," Readdy reported, "says George came out of a meeting and said that you, Norm Thagard, would never fly the Mir mission."

This was too much for Thagard. "I'm tired of this stuff!" he called and railed at Hoot Gibson, then head of the Astronaut Office. "I want to know: I'm either flying this mission, or I want out!"

Gibson tried to placate Thagard, but there was little to say. Dave Leestma had taken Thagard's name to Abbey more than a year before and nothing had happened, nothing at all. Gibson could never decipher Abbey's exact objection to Thagard, but it was clear he didn't want him on the mission. "George really didn't think Norm was the one that should be representing NASA in that kind of position," Gibson recalls.

Not for another two months, till the end of January 1994, would Thagard finally get the word: The Mir mission was his. Dan Goldin was set to announce it in a press conference at The Cape in early February. "We finally just beat [Abbey] down, which doesn't happen very often," remembers Gibson. But Thagard's jubilation was immediately undercut by confusion, and then worry. The new backup NASA named to replace Bill Readdy—who backed out after the Mir "18-B" mission failed to materialize—raised eyebrows throughout the astronaut corps: It was Bonnie Dunbar.

"Oh, no," said Thagard.

* * *

It had been tough enough finding someone to fly the Mir mission. When Readdy resigned as backup, Hoot Gibson despaired that he would ever find anyone to go to Star City for a solid year, train alongside Thagard, and then *not* fly. "Who in their right mind would want to do that?" Gibson remembers thinking. "Backup was just a crummy deal. I tried offering people all kinds of incentives. I tried to sweeten the pot. I said the backup could do the next increment. Or: 'we'll put you on STS-71 [the first shuttle flight to dock with Mir] and you can go to Mir aboard the shuttle.' That's a pretty sweet deal. People were fighting to get on that docking flight. But we still couldn't get anyone to go as backup."

Gibson assembled a list of twenty-five astronauts he thought could be arm-twisted into taking the assignment. Every single one who responded refused. He managed to find exactly one volunteer: Jerry Linenger. But Linenger hadn't been in space yet, and the Russians demanded that every astronaut flown aboard Mir have at least one mission's experience. To get him space experience, Linenger was rushed onto the crew of STS-64 in September 1994, then hustled to Star City to take the fourth Mir increment.

Running out of candidates, Dave Leestma mentioned Dunbar, who was then on an extended assignment at headquarters, helping to map out scientific experiments the astronauts would perform on Mir. Gibson wasn't sure. "Bonnie is extremely smart; interpersonally, Bonnie has had difficulties," recalls Gibson. "In fact Bonnie has had difficulties with many people at NASA. It's just her personality."

It was always difficult to put a finger on why so many astronauts disliked Bonnie Jeanne Dunbar. Born in 1949, the oldest of four children, Dunbar had grown up on an isolated forty-acre homestead in the Yakima Valley area of central Washington. The Dunbars didn't have indoor plumbing until Bonnie was a teenager. She always said she could ride a horse before she could walk, and she spent many long days herding cattle and repairing fences. As a child with no access to television, she read Jules Verne and H. G. Wells, and at the age of nine, in 1958, she scanned the night sky for Sputnik, which helped convince her she wanted to go into space someday. Boundlessly confident—in high school she was a cheerleader and homecoming royalty—she ran into her first real challenge at the University of Washington, where a professor told her engineering wasn't a suitable job for a woman since most engineering plants didn't have women's rest rooms. "So where do the secretaries go?" Dunbar shot back.

That was Dunbar in a nutshell. Ambitious, acerbic, and easily insulted, she rose to the top of her profession without completely shedding her insecurities. She was an attractive woman, with a lean jaw and hard, challenging eyes, whose natural reserve was reinforced by the stern blue suits she wore, with blouses buttoned tightly around her neck. Few astronauts ever complimented Dunbar twice; to do so invariably earned a sharp response: "What do my looks have to do with anything?" She was right, of course, but it did nothing to earn her goodwill.

After obtaining her master's degree in ceramic engineering at the University of Washington, Dunbar had gone to work at Rockwell International's space division in southern California. When NASA announced it would be taking women in its first class of shuttle astronauts in 1978, she applied but failed to gain acceptance; instead she took George Abbey's offer of a job as a flight controller at JSC, where she worked on the team that finally deorbited the Skylab space station to a watery death in the South Pacific. A year later, along with John Blaha, Bryan O'Connor, and Dave Leestma, she easily gained entrance to the next astronaut class. By December 1993, when Leestma summoned her to his office in Houston, she had flown aboard the shuttle three times. Despite her reputation for being prickly, she was an acknowledged favorite of several top NASA administrators, including Abbey and the woman who briefly preceded him as head of JSC, Caroline Huntoon—a fact that Dunbar didn't mind making clear to other astronauts.

"Bonnie," Leestma said when she arrived in his office that day in December, "I *need* you to go to Russia."

Dunbar's first reaction, like almost everyone else's, was no. Leestma pushed: She was their only hope. Everyone else had said no. Finally, Dunbar said, "Give me a couple of days to think about it."

Leestma gave her a look that said: Don't take long. "Bonnie," he said, "I need you to go *soon*."

Several days later Dunbar got back to Leestma with her answer: She would go. And yes, she would take a spot on the coveted STS-71 as well.

News of Dunbar's appointment surprised many in the astronaut corps, prompting predictions of immediate warfare between Dunbar and Thagard. Richard Jennings, JSC's director of flight medicine, pulled aside David Ward, a young flight surgeon who was to accompany the two astronauts to Star City. "You're going to have your hands full," Jennings warned. "Not that I feel sorry for you, but you're going to be in the middle of a hell of a situation."

Everyone, in fact, saw the pairing of Thagard and Dunbar as a recipe for trouble. "I'll never understand why we put those two together," sighs Frank Culbertson.

"Matter and antimatter," clucks Mike Barratt, a second flight surgeon detailed to Star City.

"Biggest mistake we made in the program," says Bill Readdy, the shuttle commander.

Within days of the internal announcement that Dunbar had been named his backup, Thagard says, he received at least four calls from other male astronauts warning him about her: Watch your back; he was warned she'll try to take advantage of you. Thagard's first thought was that Dunbar's selection sent the wrong message to others inside the agency. "You had to wonder how serious NASA was about the Russian missions when they sent someone like Bonnie," says Thagard. "She didn't know anything. She had no language skills. Why do it? It was definitely political. . . . The Astronaut Office had so many smart, gifted people who didn't have the psychological baggage. And yet they picked her."

In fact, the more Thagard stewed, the more he convinced himself that someone so clearly favored by Abbey and Huntoon could have been chosen for only one reason: to replace him as the first American aboard Mir. It was hardly an auspicious beginning for the new program.

February 3, 1994
Kennedy Space Center, Florida

The Phase One program began officially on February 3, 1994, when the dashing young cosmonaut Sergei Krikalev soared into space as a member of the six-person crew of a routine flight of the Space Shuttle Discovery. Goldin and Koptev were on hand at the launch to make the requisite rosy speeches. In Moscow, Boris Yeltsin released a statement calling the launch "a vivid manifestation of ever-growing cooperation and partnership between our countries and peoples."

After the launch Goldin stood before a bank of microphones to announce the names of the two American crew members who would travel to Star City later that month to begin training for the Mir 18 mission in March 1995. Thagard watched on television in his office in Houston; after eighteen months of waiting, his moment was finally at hand. And then Goldin announced the names: Bonnie Dunbar as backup, Bill Readdy as prime.

Readdy, standing in an office at The Cape with a crowd of NASA people, watched Goldin's lips in amazement. As people began pounding him on the back, congratulating him, he didn't know what to say. Thagard, in Houston, was dumbstruck. In a flash he jumped out of his seat and headed for Hoot Gibson's office.

When a reporter asked Goldin what had happened to Thagard, who was widely rumored to be the prime, the NASA Administrator said, "We selected the two best people for that flight. We think we have the absolute best people and we stand behind them." Not for another two minutes, when a NASA PR official crept onto the dais and slipped Goldin a note, did the administrator realize his mistake. "I gave you the wrong names," Goldin said. "I've been corrected. It's Thagard and Dunbar. I was talking to Readdy yesterday evening, and his name locked into my brain."

Thagard arrived at Gibson's office and was told of the mistake. Yes, he was told. Yes, he was going to Russia. Really. For real.

February 24, 1994
Moscow

The first NASA team landed on a snow-dusted runway at Moscow's Sheremetyvevo Airport in late February and, after a fitful night's sleep at a downtown hotel, was taken to the American Embassy. Inside, following a formal welcome by Ambassador Thomas Pickering, Thagard, Dunbar, and their half-dozen support staffers filed into an insulated cube inside a large room, where an embassy security man informed them they would be under constant surveillance by Russian authorities. Nothing that was secure in the U.S.—phones, e-mail, bedrooms—would be secure in Russia.

Even conversations on the street outside the embassy were recorded with the aid of directional microphones mounted on a church across the street. "Our Lady of Surveillance," the Americans called it.

The next morning a bus filled with Russian soldiers arrived to take them on the hour-long drive to Star City. At the depth of winter the base lay beneath enormous drifts of snow, its buildings darkened to conserve energy, the streets deserted, the air almost too cold to breathe. The team leader, or DOR (Director of Operations—Russia), a shuttle commander named Ken Cameron, craned his neck as the bus drove through the front gate. Everywhere he looked he saw walls and guards and guns—big guns; Kalashnikovs. "Well, guys, there's only a few of us cowboys," Cameron quipped, "and a hell of a lot of Indians."

With no fanfare the little group was driven through the pines and dropped off in front of the Prophylactorum, the marble-lobbied conference-center-cum-hotel the Russians had built twenty years before for the Apollo-Soyuz mission. Cameron and the team's flight surgeon, David Ward, took adjacent rooms, while Thagard and Dunbar were assigned spacious if drafty apartments in a run-down housing block packed with cosmonauts and their families. After a welcoming banquet that night, hosted by the base commander, Pyotr Klimuk, the Americans were left to their own devices. Unsure where to eat, Cameron and Ward began taking meals at the cosmonauts' cafeteria. Expecting to dine and mingle with the Russians there, they were instead ushered into a small back room with a single round table. If they tried to sit in the main dining room, the servers ignored them and the older cosmonauts, including Anatoli Solovyov, who was to head Thagard's backup crew, eyed them with undisguised disapproval. The Americans quickly abandoned the cafeteria when NASA sent word they would not be reimbursed for meals there. Making forays out onto the base, Ward found stores that sold vodka, fruit, and vegetables, but NASA doctors had made clear that Russian foodstuffs were to be avoided if possible. Hepatitis was a concern, as was lingering fallout from the Chernobyl nuclear disaster. Cameron began organizing Saturday trips into Moscow, where everyone bought food at the embassy commissary. For weeks Ward and Cameron subsisted on rice and beans warmed on a hot plate in their rooms.

The NASA team's sense of isolation was total, a situation made all the more unnerving by Star City's military trappings. The Russian officers in their royal-blue uniforms, who mustered in the main square every Monday morning, watched them nervously as the Americans wandered the base's wooded pathways. Every Tuesday the air crackled with the pop-pop-pop of troops taking target practice. There was no reliable telephone, so Cameron, to the Russians' dismay, rigged an umbrella-size satellite dish for his Inmarsat satellite phone on the balcony of his room and opened a spotty line of communication to Houston; later he did the same for Dunbar, who held the team's second Inmarsat phone. While the two astronauts sat in Russian-language classes, Cameron opened talks with Klimuk's people over just what privileges the NASA team could expect. Klimuk initially refused almost everything the Americans wanted. His people seemed amazed that Cameron expected to be given office space, despite a promise given by the Russian Space Agency. The problem, Cameron realized, was that NASA had negotiated its arrangements with the Russian Space Agency, not the Russian Ministry of Defense, and Star City's military brass saw no reason to honor commitments it hadn't agreed to. Cameron spent days negotiating with the American Embassy on an escape plan the NASA team could use in the event of a coup.

An early victory was won when Thagard and Dunbar each received a car and driver for their Saturday grocery runs. The trips into Moscow were notoriously hair-raising; the Russian chauffeurs were dubbed "Double O" drivers because, the Americans joked, they had a license to kill. The team's food worries ebbed with the arrival in April of Thagard's wife, Kirby, who quickly emerged as the team's de facto den mother, taking over the grocery runs and serving up sumptuous meals. On Saturday nights all the Americans began gathering at the Thagards' fifth-floor apartment, where after dinner they would watch *Blade Runner* or *Batman* or a Mel Brooks movie. Thagard drove Ken Cameron crazy by reciting every line of dialogue in advance.

The two flight surgeons, Dave Ward and Mike Barratt, who were to take rotating tours at Star City overseeing the astronauts' training, had the toughest job. While the Russian doctors and trainers were friendly enough, they were far from open. Both Ward and Barratt experienced episodes in which, while they were quizzing a Russian counterpart about Mir, a uniformed Russian officer barged into the room and barked, "You are forbidden to discuss that!" For the most part, details of Thagard and Dunbar's training remained a mystery. Training schedules were issued every Friday, and it was only then that the two doctors learned what the Russians had in store for the coming week.

Some of what they saw worried them. Barratt was surprised to see Star City still using a device known as the "barani chair," a rotating seat that is put through an escalating series of gyrating swirls and rapid up-and-down motions, causing a cosmonaut's head to swing and bob violently, like a broken dashboard ornament. The idea is to build tolerance to extreme motion; most cosmonauts vomit. But the chair keeps moving long after that, testing the occupant's ability to put up with extreme nausea. NASA long ago abandoned similar types of training, and Barratt wasn't about to let his astronauts go through it now. This sparked long negotiations with the Russian doctors, who gave in only after weeks of talks.

It was a routine Barratt and Ward were to engage in again and again. Some arguments they lost; many they won. Thagard and Dunbar were thus spared a turn in the high-temperature chamber, in which a cosmonaut is simply baked at temperatures that can top 120 degrees Fahrenheit. Nor did the Americans partake in the sleep-deprivation tests, which lasted up to five days. Nor did Thagard and Dunbar go through Star City's unique twist on parachute training, in which cosmonauts were dumped from planes and forced to do mathematics problems and map-reading exercises while floating to Earth.

But the flight docs' most exasperating disagreement was the most basic: whether Russian physicians would be allowed to rule on an astronaut's medical suitability to fly. The prospect of some poorly schooled Moscow doctor arbitrarily ruling him ineligible for flight was one of Thagard's greatest fears. Before arriving in Star City, Barratt, Ward, and their NASA superiors had negotiated a lengthy protocol with the Institute for Biomedical Problems that pointedly excluded astronauts from Star City's quarterly medical reviews. If NASA said its astronauts were healthy enough to fly, everyone had agreed, they would fly. But during the Americans' first week in Star City, the Russian doctors demanded blood and urine samples from Thagard and Dunbar. Barratt and Ward argued that this violated the protocols, but the Star City doctors, claiming they had no knowledge of any such agreements, insisted the tests were required under Russian law. And they were. If the astronauts weren't tested, the Russians pleaded, the doctors could be sent to prison.

Though Barratt and Ward won the first go-round that March, the situation flared anew every three months. Ward promised both Thagard and Dunbar that their missions would not be held hostage by Russian doctors, but Valery Morgun, Star City's senior physician, pressed hard for Russian medical oversight. The dispute eventually went all the way up to Dan Goldin and Yuri Koptev, and to Ward's amazement, NASA caved in. "None of us could believe it," Ward remembers. "We were now at the whims of

1950s medicine." It was a decision that caused several Mir astronauts, including Jerry Linenger, more than a few sleepless nights.

In those first weeks in Star City, Thagard and Dunbar managed an uneasy truce. Their days were heavily regimented: up at dawn to study, an initial language class at eight, a tea break at eleven, a fast lunch, more classes, exercise or yoga in the afternoon, followed by an evening of study before bedtime. Lectures on Mir and Soyuz systems wouldn't start until late spring; actual training with their Russian crewmates wouldn't begin until summer. Tension between the two astronauts flared almost immediately. Because Dunbar understood so little Russian, Thagard was forced to act as her interpreter during their classes. He stayed up late at night painstakingly translating Russian training manuals into English for Dunbar. It was an agonizingly slow and difficult process, and Thagard detested it. "This is impossible," he told Dave Ward.

Dunbar, in turn, was livid at the Russians' initial refusal to let her use the cosmonaut gymnasium, because, they explained, there was no dressing room for women. Eventually Dunbar's complaints to Ward won her a tiny changing closet in an upstairs room. "Bonnie," Ward recalls, "perceived everything like that as a slight against women."

And with good reason. Star City was a bastion of the Russian space program's famously blatant male chauvinism. In an oft-quoted passage in his book *Diary of a Cosmonaut,* Valentin Lebedev has written of his crew's treatment of Svetlana Savitskaya, one of the few female cosmonauts, during a 1982 mission aboard Salyut 7. Savitskaya, Lebedev observed, was forever "primping" and neatening her hair. "After a communication session we invited [her] to the heavily laden [dinner] table," Lebedev wrote. "We gave Sveta a blue floral print apron and told her, 'Look, Sveta, even though you are a pilot and a cosmonaut, you are still a woman first. Would you please do us the honor of being our hostess tonight?'"

To Dunbar's surprise, cosmonauts' attitudes had advanced little in the intervening years, a fact that was to startle NASA officials throughout the Phase One program. In a press conference preceding Shannon Lucid's flight in 1996, General Yuri Glaskov noted that it was a good idea to send a woman up to the cluttered Mir. "We know women love to clean," he said. In 1997, Valery Ryumin told reporters he opposed the decision of his wife, Elena Kondakova, to become a cosmonaut: "It's my opinion that a wife should stay at home for the most part, not at work and not in spaceflight. That's my opinion. There's nothing new in that because I think the majority of men will support me, because the majority of us would prefer that everything in our home is taken care of and everything is quiet and okay."

As for Dunbar, the Russians were forever making similar comments, sometimes in her presence. At first her Russian was so poor she didn't understand what they were saying, and the Russian interpreters, Ward noticed, would never translate the remarks. But as the weeks went by and her language skills improved, Dunbar began to pick up on the little comments, and she rarely let one pass without a tart response. If a Russian put his arm around her and said, "My, you look beautiful today, Dr. Dunbar," Dunbar would snap, "It doesn't matter what I look like. I've got a Ph.D." These exchanges left the Russian men scratching their heads. "Her degrees matter more than her appearance?" asked the cosmonaut Valery Korzun.

What really galled Dunbar was Thagard's attitude in these situations. "Norm never defended me, not once," Dunbar says today. "He was laughing at the jokes. He was doing things that, if he did them in [the U.S.], he'd probably have lost his job. He helped them laugh at me."

"It's not my job to defend her," responds Thagard. "You've got to understand, this is not the United States. It's Russia. What am I supposed to do, take on the whole Russian culture? It isn't my job to change this culture or to take offense, because they're going to act like they're going to act. It would have been a hopeless thing to defend her. When they made their little jokes, basically I ignored it. I never laughed or smiled or joined in."

The chauvinism extended to their classwork. In one memorable lecture, a Russian doctor patiently explained that in the unlikely event that Dunbar made it to Mir, she would be unable to use the station's toilet. The reason, he said, was that female urine had a different chemistry than male urine, and the Mir toilet could not process it. Dunbar immediately rose from her seat to challenge this assertion. "I started asking, 'What are the differences?' " Dunbar remembers. "He would not tell me what the goddamn differences were. I stayed at him: 'What is the basis for this?' And finally he said, 'Politics. It's what management says.' I just stared. There was no rationale for this at all."

To Thagard, a mannered southerner, Dunbar's willingness to confront the Russians "was just plain rude. I thought: This is the ultimate ugly American. She wants to impose her own attitudes on the Russians, rather than understand theirs. Bonnie just brought over to Russia all this feminist baggage that no one could understand."

These exchanges underscored a profound difference in the two Americans' approach to their training. Thagard's goal was to *become* a cosmonaut: He intended to train in Russian, read Russian, speak Russian, act like a Russian. Dunbar, with her minimal language skills, didn't have that option, even if she had wanted it. Where Thagard went along with everything the Russians demanded, Dunbar questioned everything: Why was there so much rote memorization? Why all the medical tests? She often interrupted the trainers to explain how NASA did things. Dunbar was no easier on the Americans. Barratt and Ward never knew whether to offer her help in class. If they didn't, she would complain she was being ignored. If they did, Barratt discovered, Dunbar was prone to snap, "Don't patronize me."

To Dunbar, the Russian chauvinism was a painful reminder of the hurdles men had set before her during her career. She found it exasperating that after thirty years she was still having to prove herself. "You just get to a point in life where you say, 'Why am I having to deal with this?'" she says. "I just lose patience with it now." But as a result of her snippiness, within her first weeks at Star City it was an open secret that few of the Russians wanted her to fly to Mir. The only way she would, they joked, was "with a red hand," meaning that every time she reached to flip a switch in the Soyuz or on the station, her commander would slap her hand until it reddened.

For two months Thagard and Dunbar managed to remain civil. Then, one day in May, they held a small press conference in the Building Three tearoom for a group of visiting American journalists. Everything went smoothly until one of the reporters asked Thagard about his contact with friends back home. Thagard mentioned that the communications situation needed to be improved. He didn't have an Inmarsat satellite phone and wasn't exactly sure why. At that point Dunbar interrupted, volunteering that the Inmarsat phones were expensive. Obviously, she said, not everyone could have one.

Thagard wanted to throttle her. It was easy for Dunbar to say not everyone could have an Inmarsat phone; after all, *she* was the one who had the phone Thagard thought should have been *his.* "It really ticked me off," he recalls. "She was saying that to make my stay over there more pleasant, I was more than willing to squander the taxpayers' money on this expensive phone." For the moment Thagard kept his tongue.

That night Dunbar received an alarming phone call from a secretary in Houston: Her father had suffered a heart attack and been admitted to a hospital.

"Which hospital?" she asked.

"I don't know."

"Well, who told you?"

"I heard it in the hall."

Dunbar stayed up all night working the Inmarsat phone and a ground line in a vain attempt to locate her father back in Washington, or to find someone in Houston who could. She was still on the phone at dawn, worried sick, unable to reach anyone who had a clue whether her father was dead or alive. In the morning, haggard and exhausted, she decided to go to her eight o'clock class, then check in with Ken Cameron during the tea break. Maybe Cameron could help. She left her fifth-floor apartment and trudged down the stairs, where, as usual, Thagard was waiting for her. He was angry.

"I would appreciate it if, when we have these press things, you not make me look bad," he began.

Dunbar put up her hands: She couldn't deal with this, not now.

"Norm," she said, "I just don't want to talk about it. This is not a good time." She walked off and, according to Thagard, shouted some angry comments over her shoulder; Dunbar denies this.

Whatever was said, the two astronauts made their way to class separately and took seats by each other in stony silence. When the lecture began, Thagard, to Dunbar's surprise, refused to translate for her. When the lecture concluded, he rose and left the room without speaking to her. The next morning he was not at the bottom of the stairs, waiting. At class that day he continued to ignore her. Afterward he told Dave Ward to tell Dunbar he would no longer translate training manuals for her.

Dunbar realized he was freezing her out. "What I did that morning he found unforgivable," Dunbar recalls. "I didn't know; I just thought in the interest of my own sanity I had to get out of there. [But] Norm has a line, and once you cross that line, you cross it forever. There's no making up."

Thagard freely admits freezing Dunbar out. "I didn't feel comfortable being around Bonnie after that," he says. "[That morning] she was shouting things out, in front of the Russians. People were turning their heads." He felt bad about her father but, "I got the impression it had happened several days before that," he says. "I thought she seized on it as an excuse to gain leverage over me."

From that day on, Thagard said as little as possible to Dunbar, avoided walking with her, and stopped helping her with her studies. Dunbar found the experience deeply humiliating. She saw the Russians, especially the women, casting glances at her, as if she had done something to earn Thagard's contempt. She thought she was getting fewer invitations to the get-togethers the Americans had every few nights. And the worst thing was, there was no one at Star City in whom she could confide. Her second husband, the astronaut Ron Sega, was six thousand miles away.

"I tried just to ignore it," remembers Dunbar. "Norm was stressed; what could I do? Go to my management and say, 'Normie doesn't like me'? This was not something I wanted anyone to know about. I was personally very embarrassed. If it got out, I felt they would blame me." In Dunbar's experience, that was the way it always worked. When a man and a woman couldn't get along, the woman was blamed. Besides, Thagard was the prime. She was the backup and, as far as the Russians were concerned, not a popular one. Dunbar was keenly aware that if she forced NASA to make a choice between the two of them, she was the expendable one.

Still, after several weeks of Thagard's silent treatment, Dunbar felt she had to talk to someone. When Dave Leestma visited Star City, she told him everything. She also confided in Phase One's logistics specialist, Travis Brice, although she curbed her criticism of Thagard because the two men were close.

"We don't have to be friends," she told Brice. "But he has to acknowledge that I exist. He has to recognize that I am his backup."

Brice returned to Houston deeply worried. He went to Leestma and suggested they find a way to get Dunbar's husband transferred to Star City as fast as possible.

"I think Bonnie's about to crack," Brice said.

By that summer the Americans began easing into the routines of Star City. Before rotating back to Houston in July, Ken Cameron managed to open a NASA office on the second floor of the Prophy, and he hired several Russian women to work in it.* The Thagards began swapping home-and-

*When Cameron rotated back to Houston from Star City, a Russian newspaper reported he was a spy who had been thrown out of the country.

home meals with Vladimir Dezhurov, the thirty-two-year-old rookie commander of his mission, and his wife; Kirby Thagard took a job teaching English at a Star City school. The warm weather seemed to bring out a playful side in the Russians. When Thagard, Dunbar, and Valery Korzun practiced emergency water landings in a Soyuz capsule at a local lake, Dunbar was surprised to see five little Russian boys grab one of the cosmonauts' rubber dinghies, swim up to their floating capsule, and begin rocking it. On sultry days the cosmonauts and their families lay out in microscopic Speedos. "One of my [Russian acquaintances] has had the same piece of food stuck between her two front teeth for the last three days," Mike Barratt wrote in his diary in July. "I have had to resist the temptation to pick it out."

Friendships sprouted like wildflowers. John McBrine, the garrulous exercise physiologist who would later room with Dave Wolf in Houston, arrived in April when the snow was still piled high; after a blowup with a former boss, his career options had boiled down to leaving Krug Life Sciences, the NASA training subcontractor, to work at a rehab clinic in his native Massachusetts or going to Russia. He chose Russia. One frigid evening that spring, McBrine and another Krug man, Matt Muller, accepted an invitation to an after-work birthday party for one of the Russian trainers in Building Three. There the two Americans were enticed into a vodka-drinking contest. The Russians, mostly senior men, took their shots in shot glasses. They poured the Americans' shots into teacups.

"We are older than you!" one of them teased McBrine. "You are young! Drink!" In barely a half hour Muller managed to down five cupfuls of vodka. McBrine made it to six before breaking into a convivial argument with one of the Russians over the relative beauty of Russian and American women. Both Americans ignored the impromptu buffet of pickled tomatoes, salted pork fat, and fish aspic.

"John, I gotta get out of here," Muller said after a bit. "I don't feel too good."

"Two minutes," McBrine said, continuing his spirited argument.

"John, man, I really need to go."

"Two minutes."

Muller's face had turned the sickly color of the buffet fish. "No, John, now," he said.

McBrine reluctantly tore himself away from his debate and led his wobbly friend to the door. They had barely passed the threshold when Muller began an impressive display of vomiting. By and by McBrine helped Muller down to the lobby, where they found their Russian hosts already assembled.

"John, you can't go home," one of the Russian trainers, Vladimir Dronov, said. He moved to take Muller and help him home.

"No, I'm fine."

Dronov insisted.

"No, I'm okay."

"No, we'll take him," said Dronov, and with that two of the Russians hoisted Muller between them. McBrine finally realized what was happening. In America friends don't let friends drive drunk. In Russia friends don't let friends pass out and die in snowdrifts. The next thing McBrine knew, he was being led along a snowy path by a woman he knew only as the "tea lady"; she served tea in Building Three's break room. The last thing he remembered was staggering toward the Prophy's front steps, sagging against the tea lady. He awoke in his bed at 3 A.M., to the sounds of a party outside. Rising slowly to his feet, he grabbed a beer and joined a crowd of his NASA colleagues, who immediately pelted him with questions about the evening's events. McBrine repeated his story about the tea lady.

"John," someone finally said, "you were brought home by two military police."

Apparently the police had found McBrine lying in a pool of vomit in the Prophy lobby. Escorted to his room, he had gone, wretching, to the bathroom sink, where he smashed a glass shelf with his forehead. Mike Barratt fixed him up and put him to bed. Muller had gotten home okay as well, but only after the NASA team sent three military patrols out in search of him. "We had a few nights like that," sighs McBrine.

September 1994
Moscow

By that fall, after two years of planning and discussion, the American science program for Thagard's mission was finally taking shape. It had been a nightmarish job for many of those involved. Phase One, everyone understood, was a political program; even though experiments in microgravity would be the focus of the astronauts' work aboard Mir, science remained an afterthought. Alex McPherson, a crystallography expert at the University of California–Riverside, remembers attending a hastily called meeting of top American space scientists at NASA headquarters shortly after the initial announcement of the astronaut exchange in 1992. "Once they signed this thing, they went to the scientific community and said, 'Now let's draft a document to justify what we've done.' " McPherson sighs. "That's generally how NASA operates." The attitude among the agency's midlevel bureaucrats was best summed up in an address Harry Holloway, NASA's associate administrator for life sciences and microgravity, gave at Woods Hole. "Gentlemen, I have good news and I have bad news," Holloway said. "The good news is we have just avoided nuclear war in South Asia. The bad news is we have this Mir mission."

The planning and scheduling of experiments on a shuttle mission is a process that typically takes three or four years to finalize. The bulk of the American science aboard Mir was to be performed in two new modules—Spektre, to arrive at the station in early 1995, and Priroda, in 1996—that the Americans had paid to outfit. Once these agreements were struck, NASA had barely a year to arrange the science Thagard would perform. "It was a mess; I've never seen anything like it," says an astronaut involved in the planning of Thagard's mission. "Somebody in Washington decided we were going to do this with the Russians, and the science part of it was just thrown together. This was insanity; you know, 'Let's do some science and do it real quick. Pull together some equipment we already have, get some scientists who will put their names on it, and quick, quick, quick.' "

Peggy Whitson, a NASA scientist, was named to head the team that would oversee Thagard's science program. Whitson worked with engineers from Krug Life Sciences, the NASA medical sciences contractor, to assemble a series of "off-the-shelf" experiments, most of which had been run on shuttle missions years before. "They came to us," recalls Mark Bowman, the senior Krug man on the project, "because we had a suite of experiments that were quick and dirty." Though Thagard was to be NASA's first long-duration crew member since the last Skylab mission, in 1975, many scientists outside the agency were still surprised to learn that most of his experiments would involve basic measurements of how Thagard and his Russian crewmates' bodies changed over the course of their three-month mission; bone deterioration remains one of the major challenges to long-duration space flight. "Everyone said, 'Well, why are you doing life science experiments on Norm? The Russians have been doing all these things for twenty years," recalls Whitson. "Well, actually no, the Russians haven't been keeping data for twenty years, at least none we could find, and we looked. They quote [data from] Skylab."

Whitson's team, armed with bulging binders that laid out dozens of experiments they planned to run aboard the station, arrived in Moscow for its first sit-down with the Russians in 1993. The meetings began badly, with a pronouncement from an Energia executive named Eduard Grigorev—the NASA team nicknamed him "Fat Eddie"—that the Russians would ferry to the station only about three hundred kilograms of American scientific equipment. The Americans exchanged astonished glances; their plans called for experiments requiring two thousand kilograms. "We can't do this!" several of the Americans began whispering to each other.

The Americans' consternation turned to confusion when they began slide presentations laying out their experiments. The Russians yawned. Their eyes glazed over. The Energia men, it was clear, had absolutely no interest in what the Americans wanted. At first Whitson's team couldn't decipher the Russians' attitude. Not until months later, when some of its members took a class in Russian culture, did Whitson's people understand what had gone wrong. In America's free market society, the customer, in this case NASA, has power over the vendor, in this case Energia. In Russia, still steeped in totalitarian mores, the vendor dictates to the customer. It was an attitude glimpsed every morning in Russian grocery stores, where butchers would tell women what steaks they could buy.

"We thought we were in a strong bargaining position. Boy, were we wrong," remembers Mark Bowman. "Their attitude was, 'Thank you for your $400 million, it's our space station, and we may serve you or we may not.' That was an eye-opener."

The Russian attitude was on full display the next day, when the two sides broke down into technical teams. "That was when the lectures began," recalls Bowman. The Russians, many of whom had worked in the space program since the 1960s, struck a sharply condescending tone. Before NASA could hope to fly a single experiment to Mir, they told the Americans, they had to develop an appreciation for Mir's pristine environment. No new American microbes could be introduced to the station, for fear they would grow out of control. Several of the Americans had to suppress laughter; as the Russians spoke, cockroaches could be seen scurrying down the walls. "If you wish to get the long-term experience in space that we have," a Russian engineer named Sasha Ermac told Bowman's group, "you will have to shed the tears and the blood that we have."

By the final days of its two-week visit, Whitson's team realized the truth of advice given every Westerner who attempts to do business in Moscow: To get anything done, it is necessary to befriend your Russian partners. In practice that means one thing: Get drunk fast. On their last night in Moscow, Whitson's team threw a banquet for their Energia counterparts at a restaurant across from the KGB's Lubyanka prison. Each table setting featured a bottle of vodka, a bottle of wine, and a bottle of champagne. The toasts began early, continued back-to-back all evening, and by midnight had grown steadily more sloppy and emotional. A Krug man named James Breeding, whom everyone called "Gumby," vowed to drink his Russian counterpart under the table and succeeded; when the Russian passed out facedown into his dinner plate, Breeding was photographed standing over him, striking triumphant bodybuilder poses. One of the Lockheed men, who was married, began necking passionately with one of the Russian women engineers. At the end of the evening it took nearly an hour to gather the drunken revelers back onto the NASA bus; every time it seemed everyone was aboard, someone new would stumble out and begin vomiting into the gutter.

Shipping the equipment Thagard would use was a nightmare. Russian customs snafus delayed shipments for weeks, sometimes months. The delayed shipments often sat in subfreezing temperatures in unheated warehouses, ruining a wide variety of laboratory paraphernalia. Petri dishes froze. IV bags burst. NASA bureaucrats were dismayed by Russian demands that all American equipment be tested to withstand gravity loads as high as 100 Gs. Where aboard Mir, they wondered, was there even the remotest possibility of gravity that high? The Progress launchings pulled nothing like 100 Gs. No, no, no, came the Russians' reply: The requirement had nothing to do with space. It had to do with the stresses the American equipment would endure on the *ground.* Progress ships were routinely placed aboard an unheated railcar for the trip to Baikonur; the G-force requirement was necessary to ensure that the experiment survived Russia's pothole-pitted roads. After a tortured set of negotiations, NASA arranged for its experiments to be flown to Baikonur and placed inside the new Spektr module.

* * *

By September 1994, seven months before Thagard was to begin his mission, much still needed to be done. Peggy Whitson finally forced a meeting with IBMP officials to work out the precise time lines and schedules for Thagard's experiments. But when she strode into the meeting at IBMP headquarters in Moscow, she quickly realized that the men scattered around the conference table weren't the senior officials she needed to see. They were scientists, and low-level ones at that. Her direct counterpart, Oleg Lebedev, was nowhere to be seen. "He can't make it today," one of the scientists began, explaining that it wouldn't be possible to discuss Thagard's mission. "But if we can't talk about 1A today, then we can talk about 1B." Phase 1A was Thagard's mission. Phase 1B was Lucid's.

Whitson could see what was happening. The Russians always seemed to want to talk about 1B and beyond; these were the missions whose success would bring them the $400 million NASA had agreed to pay. Thagard's 1A mission, which wasn't technically part of the NASA contract, was a goodwill flight, a freebie, and the Russians were far keener to map out money-raising missions than free ones.

Whitson began to get angry. "Well, fine. Maybe I need to go back to my hotel and change my plane tickets and go home," she told the room. "Because if you can't talk to me about 1A, then I can't talk to you about the IB program." The scientists pleaded and cajoled, but Whitson knew it was time to make a stand. "No, no," she said. "Bring the van back. I'm going home."

And with that Whitson walked out of the meeting. It was the first time she had ever tried such a thing, and she was a little frightened it would backfire. It didn't. Later that day Oleg Lebedev telephoned her hotel room and promised to come by the next morning to talk.

"There's a reason I didn't want to come talk to you," Lebedev told her the next day, as the two sat outside the Penta Hotel's second-floor business center. Lebedev seemed uncomfortable. Whitson could smell bad news in the air.

"Unofficially," Lebedev said, "the Spektr launch is going to slip."

Whitson nearly fainted. Almost all of Thagard's science program was on Spektr; the Russians had promised the new module would be ready and waiting when Thagard arrived aboard Mir in March.

"How long?" she asked.

"It looks like June." The last month of Thagard's mission.

This was catastrophic news. Whitson and her team had spent a solid year shuttling between Moscow and Houston, eating hotel food and missing their families. Now, in one fell swoop, everything they had worked for was melting away. Without his science program, what would Thagard do for four months in space? Knit? Lebedev held out the prospect of sending up some of the American experiments aboard two Progress flights. Thagard's equipment weighed almost four hundred kilograms. With luck, Lebedev said, the Progress ships could carry one hundred kilograms of American hardware.

Whitson and her team worked through the following weekend, mapping out a drastically reduced science program for Thagard to perform until Spektr arrived. They canceled some experiments, delayed others until Shannon Lucid's mission in 1996, and looked for ways to shrink the rest. Eventually they whittled the list of fifty or so experiments to twenty-eight. With luck Thagard would have enough to keep him busy.

October 1994
Star City

Relations between Thagard and Dunbar deteriorated through the fall. Because Thagard refused to translate for her—an awkward proposition to begin with—Dunbar was obliged to hire an interpreter to sit by her in class and whisper into her ear. This infuriated Thagard, who complained that with Dunbar's translator chattering away, he couldn't concentrate. "It was a distraction in class, and frankly it was undermining my ability to learn," recalls Thagard. "People had warned me about Bonnie. They said, 'Bonnie will always try to change the situation to optimize herself,' and that is what I saw. I was the prime and she was the backup. Normally you would optimize the situation for the prime, and that's not what happened."

Feeding Thagard's anger was his fear that George Abbey and the new JSC director, Dunbar's mentor Caroline Huntoon, would somehow find a way to replace him with Dunbar. His worries manifested themselves in an obsession with the Inmarsat phone that Dunbar seemed constantly to be using to call people in Houston. (According to several NASA officials, the bill for Dunbar's phone use during her yearlong stay at Star City came to just under $100,000.) "I don't know what she's saying when she's calling back to Houston, but it can't be anything good about me," Thagard told Dave Ward. "She's just trying to get this flight, you know she is."

"Look, Norm, there is no way anybody in their right mind is possibly going to say you can't fly this flight," Ward counseled. "Because there is no way Bonnie is going to be ready to fly. She's always going to be behind you."

Bill Readdy, the shuttle commander who rotated to Star City to replace Ken Cameron in July, discussed the "Bonnie Situation" with Hoot Gibson and Dave Leestma that fall. All agreed with Thagard that NASA management would use any excuse to replace him with Dunbar. "That was pretty obvious to all of us," says Readdy. "I tried to defuse Norm's paranoia on this. I'm not sure whether he believed me or not." Readdy had long talks with Yuri Glaskov, Star City's No. 2, and the round little general admitted that the Russians would never let Dunbar fly. "Privately, the Russians were very distressed about her progress and the fact she was still using a translator," Readdy recalls. "Most of them didn't think Bonnie should fly. They said it all the time. They used to joke about Bonnie right in front of her. We'd all be talking Russian, and she's just standing there. They'd say things like, 'She'll never make it. She doesn't understand us, does she?' And she didn't."

The announcement that Readdy would be replaced in November by Dunbar's husband, Ron Sega, sent Thagard into new spasms of fear. "The Ron Sega thing, anything that could be construed as favoring Bonnie, Norm saw as a plot to subvert him," remembers Dave Ward. "It just drove him deeper into the Russian world. He didn't have many allies back at home. Hoot Gibson was one, and he was about to leave office. In Norm's mind, all those things become harbingers of doom."

And then, as the swirl of political intrigues and interpersonal conflicts intensified, everything suddenly changed. It happened in October, when Thagard and Dunbar returned to Houston for a series of preflight medical tests. Most were for charting what NASA called "baseline data collection"—that is, determining the chemical and physical characteristics of the two astronauts' bodies to compare against data from the same tests taken while in orbit and after return to Earth. October 16 was "metabolic day," when both astronauts were scheduled for chemical tests. One of the experiments, to examine a theory that kidney stones formed faster in space, involved an injection of inulin, a plant derivative commonly used as a tracer chemical in kidney tests. The research project was most notable because it was headed by Caroline Huntoon.

The medical experiments NASA performed in space have long been a sensitive subject for the astronauts. Every astronaut had to sign a written agreement before an experiment could be performed on his or her body. The rare refusals, NASA researchers sometimes complained, invariably occurred in the final days before launch, at which point it was unlikely an astronaut would be yanked off a shuttle crew. Thagard, for one, had complained for years that astronauts signed away their rights when they agreed to fly. In 1984, he had attempted to back out of an experiment involving exactly the kind of rotating chair that the Russians still used. "It's making me sick," he told a NASA attorney named Hank Flagg one day at the astronaut gym, "and I want to get out of it."

"Norm, you can be fired for good cause, bad cause, or no cause," Flagg said. "You are required to participate in this experiment as a condition of employment."

Thagard pointed out that even federal prisoners were no longer required to participate in medical experiments.

"Prisoners are different—they have no choice," Flagg said. "You, on the other hand, can always quit."

The exchange stunned Thagard. "It was a blatant attempt to intimidate me," he remembers. "But what can you do? Unless you're willing to start a major court case, they've basically got you over a barrel."

That day in October, Thagard took Huntoon's test first. As a fighter pilot, he didn't like doctors one bit. He knew that he could only come out even with a doctor: You go into an examination room a member of a mission; with luck you leave the same way. Still, he was a Marine, and Marines did what they were told. He took the shot. After a few minutes he headed to the next examination room, for still more shots.

Dunbar went next. As Mike Barratt closed the door behind her, she lay down on a gurney. Dunbar's attitude toward the test, and toward NASA's medical community in general, was sharply different from Thagard's; her background was in life sciences, and she considered Huntoon a valued mentor. An hour after Dunbar entered the room, as Peggy Whitson, John McBrine, and Valery Morgun waited outside, one of the nurses burst through the door, obviously upset. "Bonnie's having a reaction, and Mike wants an ambulance," she said, disappearing down the hall to find a phone.

A minute later a second nurse emerged; she asked Whitson to call an ambulance. Both ambulances arrived shortly after, and Dunbar, awake and alert, was wheeled out and taken to St. John's Hospital in nearby Clear Lake. Initial reports indicated that Dunbar's heart had stopped and that she was suffering from anaphylactic shock; her condition was later amended to a severe anaphylactic reaction. Whitson and the others were left in a state of shock themselves. No one outside the room knew what had happened, and under NASA's strict medical privacy guidelines, no one other than the doctors was even allowed to discuss it.

Even today the incident remains intensely controversial at NASA. The general outlines of what happened leaked to the press and led to an internal investigation. During this review, astronaut Shannon Lucid came forward and volunteered that she had suffered a similar, though far less severe, reaction to the inulin when she underwent the same test during STS-58 in the fall of 1993; her warnings about the chemical, Lucid charged, had apparently been ignored. Even so, Dunbar downplayed the severity of her reaction. In the wake of NASA's investigation, several senior astronauts, including Thagard, were sharply critical of her behavior.

"That episode didn't endear me to Bonnie," says Thagard. "Whether she will admit it or not, she nearly died. She did no service to the Astronaut Office by downplaying what had happened. She put the office at risk, and other astronauts who had to take the test. And her reason was clear. She was protecting Huntoon. Bonnie is never going to do anything to hurt the [life sciences] group, because she considered it a part of her power base."

What happened in Building 266 that day is not disputed; what caused it is. Dunbar lay on a gurney and a NASA aide named Nelda Huber administered two injections. The first was a chemical called indocyanine green; the second was inulin. Both are considered "tracers" for the study of kidney function. Moments after the shots were given, both Barratt and Dunbar realized that she was experiencing a reaction. She began to sneeze. A rash appeared on her stomach. Later the two would differ sharply on how serious Dunbar's initial reaction actually was. Barratt believed Dunbar was going into shock. "It was a life-threatening reaction," says Barratt. "It was cookbook anaphylactic shock, just like the medical books say."

According to Barratt, the prescribed antidote was a dose of epinephrine, a clear liquid commonly known as adrenaline. NASA medical guidelines call for epinephrine to be delivered intramuscularly—that is, with an injection into a muscle. But the medical cart in the examination room that day—the "crash cart"—did not contain an intramuscular syringe. "That was a mistake," Barratt acknowledges. "We should have had one." Barratt elected to administer the epinephrine directly into the IV tube hooked up to Dunbar's arm. He says the solution was mild enough that there should have been no ill effects. It was a procedure Barratt says he had done several times.

But the moment Barratt began emptying the syringe into the IV, Dunbar experienced a searing pain in the back of her head—"like someone had stuck a spear right above my neck," she recalls. Her heart rate, normally about 55 beats per minute, soared to 220 beats per minute.

"Stop, stop it!" Dunbar remembers saying aloud. It was the last thing she said before momentarily losing consciousness.*

Dunbar remained in the hospital for a week. Interviewed by the investigating medical board, Barratt said there was no apparent reason for her reaction to the epinephrine shot. He speculated that the inulin had somehow altered Dunbar's body chemistry. "I've given many epinephrine shots to many people," Barratt says, "and Bonnie had an anomalous response"—a response that could not have been predicted.

Dunbar insists the incident is far simpler to explain. She dismisses the idea that the inulin injection caused anything other than a mild reaction. What caused her heart to race—and what caused her to lose consciousness—was the epinephrine injection into her IV. "It was like a slug of adrenaline delivered right to my heart," she says. "You simply don't give IV epinephrine to people who are conscious. That's what you do to people who are on their last legs." A NASA review board, however, cleared Barratt of any wrongdoing; it deemed his actions "appropriate and potentially life-saving."

Whatever the precise chemical cause for Dunbar's reaction, the incident had an immediate impact on the Shuttle-Mir program: By the time Dunbar was released from the hospital, Russian doctors in Star City had suspended her from further training. "They were done with her at that point; she was not going to fly, period," Bill Readdy recalls. "[They said] 'This person obviously has a medical problem that disqualifies her from

*Barratt says this version of events is incomplete. However, citing Dunbar's rights to medical privacy, he declines to provide further details.

all training. Case closed.' " There ensued a rigorous debate between the American and Russian doctors about Dunbar's fitness for flight. "The NASA doctors, their opinions were quite superficial," says Valery Morgun, the senior Russian doctor. "They told us, 'In five days, Bonnie can go and continue training, according to our opinion.' [But] what happened to her had to be researched and studied. Only after that, we said, after all the tests have been done, could we revise her fitness for flight."

A week after the incident, a JSC medical board agreed: Dunbar was to be disqualified from training until her condition could be studied further. Barratt was given the unpleasant job of walking down to Dunbar's office in Building Four and telling her the bad news. It was an intensely awkward moment for the young flight surgeon. "Well, the board has met," he said, "and the decision has been made that you not return to Star City until further tests are made."

Dunbar didn't take the news well. According to Barratt, she exploded in tears of rage, castigating Barratt and every other doctor at NASA.

"I know my body!" Dunbar railed. "I know I can fly!" All Barratt could do was listen. (Dunbar says she doesn't remember the incident, but denies she is prone to temper tantrums. "I talk hard," she says, "but I don't get angry.")

Years later Barratt would call Bonnie Dunbar the single most determined person he had ever met, and in those first days after her release from the hospital Dunbar showed why. Nothing would stop her from getting back to Russia. She felt no aftereffects from the incident and wouldn't listen to anyone who suggested she might not be fully recovered. Bob Cabana, who had just replaced Hoot Gibson as head of the Astronaut Office, offered her an easy way out of the Phase One program. "You can leave now and no one will think less of you," he said.

"Yes they will," Dunbar argued. "I am not a quitter. I will not give up."

There was just one problem: A mysterious spike had appeared in Dunbar's EKG readouts, indicating an abnormality in her heartbeat. Barratt checked the literature and found only one similar case in the last century, a twenty-year-old British woman who had shown similar cardiac irregularities after a bee sting. Both he and Dunbar tried in vain to track down the woman or one of her doctors. Barratt and other NASA doctors checked with cardiac experts around the country and encountered a bewildering variety of opinions. "The opinions ranged from 'She should never solo in a Piper Cub' to 'She's fine, let her go,' to everything in between," he recalls.

The lack of unanimity put the NASA medical board in a bind. Faced with Dunbar's ironclad resolve to return to Russia, they had no clear medical reason to stop her. And so a week after its first vote, the board reversed itself and cleared Dunbar to return to limited training; a second vote would be required to return her to full training. Barratt, as a member of the board, voted against the move. He had been to Russia; he had seen the country's antiquated hospitals. If Dunbar had any sort of relapse, he warned the board, he had no confidence a Russian doctor could revive her. Faced with Dunbar's resolve, he did the only thing he could: He picked up a cardiac defibrillator and an oxygen tank and stuffed the equipment into a backpack for their flight to Moscow.

When Dunbar and Barratt arrived back at Star City in the first week of November, Colonel Yuri Kargopolov, the training chief, announced that Dunbar would be limited to classroom training. She was barred from anything involving physical activity, even something as innocuous as lying on her back in a darkened Soyuz simulator. According to Barratt, Dunbar rose and engaged Kargopolov in a spirited debate about her fitness for training. According to Dunbar, she remained silent and whispered to Barratt, "Mike, this is unacceptable."

The immediate problem was the Soyuz training. Without it, Dunbar knew, there was no way the Russians would ever clear her to fly aboard Mir. For a week she tried in vain to get approval for the training. Finally, the morning her crew was scheduled for the simulator, she showed a supportive letter from NASA doctors to her commander, the veteran cosmonaut Anatoli Solovyov. Solovyov read it in silence. "You will stay with me," he said as they walked into the simulator building. "You are on my crew." Inside, Solovyov led Dunbar directly into the Soyuz simulator, telling the instructor, "Bonnie's on my crew; she's getting in the simulator with me." As she lay on her back and strapped herself in, Dunbar thanked Solovyov. "I will never forget Anatoli for what he did," she says.

But the overall question of Dunbar's fitness remained. Her husband, Ron Sega, who had taken over as DOR, had expected the American side to press hard for her reinstatement; instead, he says, he was dismayed by the yawning silence that emanated from the Phase One office back in Houston. Barratt came by the couple's apartment every few nights to take Dunbar's EKG, which showed steady signs of improvement; still, no one other than Dunbar and Sega seemed to be pushing to get her recertified. Late one night in the third week of November, Sega realized they couldn't wait forever. He asked Barratt to help him learn the answers to two questions. First, the American doctors in Houston were asked: What was the earliest date NASA could recertify Dunbar to fly? The week of December 12, Barratt was told. Second, General Glaskov was asked: What was the latest date Dunbar could reenter the program and still complete her flight training? Glaskov's office came back with a date, December 23. It gave them a window of eleven days.

On December 9 Dunbar returned to Houston with Valery Morgun and underwent a series of treadmill tests at Baylor Medical Center. She returned to Star City several days later with a letter from the JSC medical board, cleared to resume full training. The letter, however, meant nothing to the Russian medical establishment. If she was to resume training by year end, it was clear, the Russians would have to be pushed, and hard, Sega and Dunbar decided to take a gamble. On the evening of December 14 they joined the rest of the NASA contingent at a reception at the American ambassador's residence in Moscow for Al Gore and Viktor Chernomyrdin. Standing to one side, Thagard watched in amazement as Sega strode up to Gore, shook his hand and leaned in for what was obviously an intense conversation. He assumed Sega was taking his wife's case straight to the Vice President. "I was just flabbergasted," recalls Thagard. "It was clear to me that [the question of] Bonnie's medical status went way beyond the medical people. It was going to be settled at the diplomatic level."

In fact, according to Sega, he and Gore were discussing a baseball cap he and the crew of STS-60 had given the Vice President the year before. The man he sought out that night to discuss Dunbar's problem, Sega says, was Dan Goldin, who was at the reception with Yuri Koptev. When Sega took Goldin aside, he was under the impression that the NASA Administrator had been fully briefed on Dunbar's health. But from the blank look Goldin gave him, it was clear he knew little about the case.

"I think you need to read this," Sega said. He handed Goldin a sealed envelope, which Goldin opened and read. In it was the letter the American medical board had written clearing Dunbar to fly aboard Mir. "We need to get the Russian medical folks going," Sega said, "so that we can resume the training, because we are running at the limit of when you operationally can do all the activities needed to fly."

Sega watched as Goldin walked over to where Koptev stood. The two men talked for several minutes, and while Sega joined the NASA group in a photograph with Gore, he saw Koptev nodding his head. Afterward Goldin assured Sega that the matter would be taken care of. Everything would come down to the Russians' Grand Medical Commission, which was scheduled to ponder Dunbar's status two days before Christmas.

December 23, 1994
Moscow

Friday morning, December 23, Barratt nervously donned his only tie and sports coat, gathered Dunbar's medical reports into a briefcase, and, along with Dunbar, Morgun, and a pair of translators, boarded a bus that took them to the Seventh Central Aviation Hospital, which had been described to the Americans as Russia's premier aviation-medicine facility. While the Grand Medical Commission met behind closed doors on other matters, Barratt and Dunbar were taken on a tour of the hospital's new Alternative Medical Center, which seemed to prescribe a bewildering mix of herbal, chiropractic, and Zen-like cures for all manner of ailments. To Barratt, it seemed medieval.

By and by the commission was ready, and while Dunbar sat in the corridor outside, Barratt and his translator walked into a broad auditorium and looked out upon the doctors. After a young doctor read aloud a description of the October 16 events in Houston, the room erupted in a heated debate, which Barratt was hard put to follow. Many of the doctors clearly opposed Dunbar's reinstatement; others seemed to waver. If any of them favored it, they stayed quiet.

After an hour of debate, a vote was taken, and to Barratt's surprise, the vote was for Dunbar to be reinstated, with one caveat: Her status would be reviewed at launch-minus-thirty-days. Dunbar was led into the auditorium and given the news. She smiled politely and made a brief statement of gratitude. As the meeting adjourned and several Russian doctors came down to shake his hand and pat his shoulder, Barratt was left with the clear impression that the entire session had somehow been stage-managed. Behind the scenes, he guessed, the powers that be had already made their decision.

He was right. Valery Morgun acknowledges that political considerations overrode his staff's fears about Dunbar's health. There was no avoiding the simple fact that without a backup for Thagard, the entire mission—and the future of Russian-American space cooperation—was at risk. "We just didn't have any choice, you know?" says Morgun. "Both the Americans and the Russians understood that there was no way we could cancel the flight—these contracts had been signed by Gore and Chernomyrdin themselves. We knew that if this flight didn't happen . . . well, we wouldn't be having this conversation today."

While the commission's approval cleared Dunbar to return to full training status, signs remained that the Russians would never let her go to Mir. At a Star City dinner not long after, Valery Ryumin rose and gave an elegant toast to Thagard. When he finished, someone at a nearby table piped up, "Let's hope he doesn't slip on any ice!"

There were some nervous chuckles around the room. "I don't care if he has a broken leg," Ryumin announced. "He's flying anyway!"

In the last months before NASA's first visit to Mir, many within the agency remained worried about what they would find there. There were rumors the station was infested with cockroaches and fast-growing fungi, even a superbacteria that ate certain kinds of plastic. Most of all the Americans worried about the air. There were reports that air samples from Mir showed signs of benzene, a toxic hydrocarbon associated with motor fuels. That November, Peggy Whitson got word that NASA headquarters wanted a new air sample from Mir before Thagard's mission would receive final approval. To do so meant asking the Russians to ship a small piece of sampling equipment aboard their next Progress, have a cosmonaut take the measurement, then send it down on the subsequent Progress. Whitson cringed. It was bad enough that NASA should ask the Russians to lug a piece of equipment at the last minute aboard the tightly packed Progress. Worse was the message implicit in the request: NASA didn't trust the Russians' measurements.

But what made Whitson dread the task was that it required a presentation before Ryumin, whose thundering tirades had achieved near-legendary status among the Americans. Whitson's pitch came in the Russian Space Agency's cavernous main conference room in Moscow, with U.S. and Russian delegations facing each other across an oval table surrounded by stadium seating. Whitson was nervous. "We were asking for a big favor," she recalls, "and I knew it."

But nothing prepared her for Ryumin's reaction when she finished her presentation.

"You Americans worry too much about crew health issues," the big Russian said. "We've had no one die up there. You send up these huge medical kits! We have almost nothing! Your astronauts must not be as strong as our cosmonauts." As Ryumin launched into a discourse on the weaknesses of American astronauts, Whitson glanced covertly at Frank Culbertson. "I wanted to see how Frank was reacting to having the astronauts called a bunch of wusses," she recalls. Culbertson remained impassive. Whitson's own reaction to Ryumin was visceral. "I really did want to crawl across the table and rip his face off," she remembers.

"You know this is a very big inconvenience," Ryumin concluded. He turned to Tommy Holloway, who at that point was still Culbertson's boss in the Phase One office.* "What do you think I should do?"

Holloway remained diplomatic. "Well, I know it's a big inconvenience," he ventured, "but if you told me it was important to your program that I fly a 1.8 kilogram piece of equipment aboard the shuttle, I'd fly it."

Ryumin struck down the proposal anyway. The Americans would not get their air sample. Holloway didn't have much better luck with questions about the Russian food eaten aboard Mir. NASA nutritionists wanted a full breakdown of what Thagard would be eating, how many calories and nutrients each food item carried. Ryumin, all but shouting at Holloway, would have none of it.

"Why are you worrying about the food?" he asked, rising from the table. "I was up there for an entire year, and I ate it. You Americans worry too much."

And so it went. Along with studying how Thagard's body adapted to the rigors of long-duration space flight, the highest priority for NASA scientists was studying Mir's environment. In addition to experiments measuring the quality of the station's air and drinking water, NASA was

*Not all the early negotiations involved American requests of the Russians. Under the terms of its contract with the Russian Space Agency, NASA had agreed to ferry to Mir a half ton or more of potable water on each of its Shuttle missions. Ryumin objected to the water NASA proposed to deliver. It was purified with chlorine, which Ryumin insisted could not be processed by the station's electrolysis system. He demanded that NASA deliver "Moscow water"—that is, water that had been purified Russian style, with silver nitrate. The Russians, Ryumin proposed, would gladly do this for NASA, for a price of $8 million. Holloway balked, electing to purify the water with an American system that cost about $500,000.

especially curious about the level of vibrations aboard Mir. Almost anything—the rumbling of engines, ventilation systems, even a cosmonaut running on one of the treadmills—could cause the station to vibrate, and scientists working with precise measuring instruments needed to know how bad the vibrations would be. Only reluctantly had Energia executives, who marketed Mir to Western countries as a laboratory, agreed to let NASA string accelerometers throughout the station to measure vibrations. "The Russians were very sensitive on that; they didn't want our investigators to publish how bad the microgravity environment was on Mir," recalls Whitson. "The Russians make a lot of money on Mir, and they were afraid it would ruin their profits if the Americans published that data." In the end, NASA discovered that vibrations justified its sending to Mir a specially insulated piece of equipment called a Microgravity Isolation Mount, which was designed to run its most sensitive experiments.

Ryumin scoffed at NASA's concerns, as well as its proposed science program. "We couldn't understand why the Americans kept raising issues that were of no importance to us," he recalls. "The reason for that, we thought, is they were considering an American astronaut to be a different person from a Russian cosmonaut. They were so worried about conditions. They felt the existing conditions were not what they would like for NASA astronauts. So of course we didn't see the need for any of this. Everything they wanted to do had been done already. I can't say any of it was very helpful or made much sense."

14

February 3, 1995
TsUP

The Space Shuttle Discovery, commanded by Abbey's favorite son, Jim Wetherbee, erupted off Pad 39B at the Kennedy Space Center in a ball of orange flame just after midnight that Thursday morning. In the first Russian-American space rendezvous in twenty years, STS-63, as the mission was called, was scheduled to reach Mir three days hence, on the evening of February 6. Wetherbee and his crew, which included NASA's first woman pilot, Eileen Collins, as well as Vladimir Titov and Mike Foale, were to circle the station at a distance of thirty or forty feet in a dress rehearsal for STS-71, the first shuttle-Mir docking, scheduled for June.

Not long after Discovery reached orbit, a problem arose. NASA controllers discovered fuel leaking from one of the shuttle's aft thrusters. It was a common glitch, thought to be caused by bits of fuel residue propping open the thruster's outer valve. On other missions shuttle commanders had managed to clear stuck thrusters by warming them, so Wetherbee maneuvered Discovery to face the sun: It didn't work. The leak persisted, a glittering trail of white snowflakes gushing out miles into space. Ordinarily NASA would ignore the leak, but the rendezvous with Mir changed everything. The leaking fuel was nitrogen tetroxide, which is highly caustic. There was the potential, however slight, that if it came in contact with the station's outer hull it could damage sensitive instrumentation.

The unpleasant job of explaining this to the Russians fell to Bill Reeves, one of MOD's senior shuttle flight directors, a folksy, laid-back, fifty-year-old Arkansan who was one of only three flight directors left from the Apollo era. Before visiting Moscow a year earlier to oversee the installation of communications links between the TsUP and Mission Control in Houston, Reeves had never been to Russia. But he had been named STS-63's liaison to the TsUP after establishing a close working relationship with Viktor Blagov and his Russian ground team—a bond largely forged with the help of an interpreter named Boris Goncharov.

Goncharov, whose advice and guidance were to prove invaluable to a number of the initial Americans to work at the TsUP, was the first Russian Reeves ever met, standing outside customs at Sheremetyvevo Airport on Reeves's first trip to Russia, in 1994. Goncharov was a big man, about fifty, with a full head of brown hair and a booming baritone voice. He always wore a dark suit with a dark shirt; to Reeves, he looked like a gangster. By training, Goncharov was a research scientist who worked in computer systems at the TsUP; to make money he began moonlighting as an interpreter. Theirs was an instant friendship; after barely two days working with Goncharov, Reeves felt he had known him his entire life.

On follow-up trips to Moscow, Reeves began bringing Goncharov American books, usually something by Stephen King or one of Jeff Foxworthy's "You Know You're a Redneck" books of humor. Between meetings Goncharov would sit in a corner of the NASA suite reading Foxworthy's jokes and laugh so hard tears leaked down his broad Slavic face. Reeves would be in the middle of some crucial analytical work, and Goncharov would amble over, his brow furrowed, and ask him to explain one of Foxworthy's southern anecdotes. When Reeves explained, Goncharov shook with laughter.

Reeves was watching before the monitors in the NASA suite at the TsUP when American ground controllers first identified the leak. Before he had time to tell anyone what had happened, Blagov entered the room and said something in Russian to Goncharov. "Bill, these gentlemen want you to come downstairs to the Blue Room for a press conference," Goncharov said.

"Uh, well, sure," said Reeves, who hadn't known he would be required to speak to reporters.

The group walked downstairs, past the cavernous break area with its giant bust of Lenin, and entered the big formal conference room where the Russians talked with the press. CNN and television crews from around the world were already set up and waiting. As he sat on a dais beside Blagov, Reeves was uncomfortably aware that he would probably have to disclose news of the thruster leak before telling Blagov. "They asked me right off the bat to talk about the launch, [so] I told them all the flowery words—you know: 'Successful launch,' 'Embarked on a new historic joint mission,' all this kind of crap," Reeves recalls. "Then I told them, 'We did have one problem: with a thruster leak,' and I explained it." Afterward, when Reeves took Blagov aside to explain the matter in detail, Goncharov commandeered the conversation, making certain Blagov understood exactly what had happened. "Blagov got all excited about it," Reeves remembers. "Immediately he began saying, 'We don't want the shuttle coming thirty feet from Mir and having this fuel getting on the Mir and causing damage.' They were very, very worried."

All that afternoon Reeves shuttled between meetings with Blagov and the NASA suite, where he kept the Houston control team updated. Again and again he explained to the Russians that the leak was not a concern. What little fuel was escaping the shuttle would be long gone by the time Wetherbee reached Mir. Blagov listened but remained adamant. Until the Russians were satisfied the shuttle was no longer a danger, the Americans did not have permission to close within a thousand feet of Mir.

The next morning Reeves was waiting in the NASA suite when Blagov came in. "Bill, come here," he said. "I want to talk to you."

Reeves and Goncharov followed Blagov downstairs to Vladimir Solovyov's office. Inside, Ryumin, Solovyov, and a half-dozen specialists sat around a table crowned with a massive, five-foot-tall model of Mir. "Bill, let me tell you what our problem is," Blagov began. He pointed to a little Soyuz jutting out from the Mir model. "You see this little thing on the Soyuz? This is a horizon sensor. When the Soyuz separates from the Mir for reentry, this sensor is critical to reentry. I cannot afford to get anything on this sensor. I could lose the crew. This is my problem. You have got to convince me that I will not get anything on this sensor, or we will not let you get close to Mir."

Reeves spent the better part of an hour explaining to the group that there would be no leak by the time the shuttle arrived. Eventually Blagov appeared to accept the argument.

"But if you stop this jet from leaking, you've lost some performance, haven't you?" he said. "How can you convince us this is a safe operation if you lose redundancy in the jets? Is there danger of crashing into Mir if you have another failure?"

It took all day to persuade Blagov and his superiors that the shuttle had sufficient redundancy to avoid losing control. Discovery was set to rendezvous with Mir the following evening, and the Russians set a deadline of ten o'clock the next morning to iron out their concerns. Otherwise, Blagov made clear, there would be no fly-around.

At nine the next morning, Reeves grabbed Goncharov and headed to Solovyov's office. Again Reeves made his points: There would be no leak, and there was no chance the shuttle could lose control and ram the station.

"Okay, Bill, I understand everything you are saying." Then Blagov paused, as if for dramatic effect. "But what about the one-hundred-eighty-gram snowball?"

Reeves stared. "What damn snowball? What are you talking about?"

"The one-hundred-eighty-gram snowball you mentioned in your fax."

"Viktor, I don't know what the hell you're talking about."

Blagov handed Reeves a copy of a fax the Houston control team had sent over the day before. Reeves had never seen it, but he instantly realized what it was: a NASA engineer "worst-casing" the situation, imagining a scenario so unlikely it would never occur. In this case, the fax laid out the possibility that if fuel backed up inside the manifold and froze, it could form a ball of ice weighing about one hundred eighty grams, not quite half a pound. In a worst-case basis, the snowball could be forcibly ejected by the shuttle's thrusters toward Mir.

"Viktor, I'm seeing this for the first time," Reeves said. "But my first reaction is I think there's a problem here that's really not a problem. This is a nonproblem. There is no one-hundred-eighty-gram snowball. There is not going to be a one-hundred-eighty-gram snowball. Let me go back and talk to Houston, and I can explain it to you better."

Reeves returned upstairs, called Houston and heard exactly what he expected to hear: There was no snowball. In the meantime the Russians pushed back the deadline one hour. Reeves headed back downstairs to Blagov's tiny office and knocked.

"Come in."

Reeves led Goncharov into the room and shut the door firmly behind him. "Viktor," he said, "we are not leaving this room until you are convinced there is no one-hundred-eighty-gram snowball."

Blagov grinned. Reeves sensed an opening. A moment later there was a knock on the door, and Ryumin stuck his head in.

"Oh, good, I've got both of you here together," Reeves said, pulling Ryumin into the office. "Now, none of us is leaving until you guys are convinced there is no one-hundred-eighty-gram snowball."

Reeves made his case. "This is just so far out, it has no bearing on reality whatsoever," he told the two Russians. "The leak has stopped.

There is no ice in the jet. There is no ice. There isn't going to be any ice. There is no problem with redundancy. This is a nonproblem."

Blagov and Ryumin exchanged looks. "We believe you, Bill," said Blagov. They were cleared to forty feet.

At 9:20 P.M. Moscow time, as Discovery hovered just 40 feet below the Kristall docking port, Jim Wetherbee hailed Mir in English. As historic moments in space history go, it wasn't exactly the equal of Neil Armstrong's speech on the moon. But Wetherbee was determined to say something eloquent for the occasion. He had written his short speech the night before and had Titov translate it into Russian. But now, as Foale and the rest of crew remained transfixed on Discovery's position, determined not to let the shuttle creep toward the station, Wetherbee was unsure when to begin talking. "You think they're ready for me?" he asked aloud.

No one said anything.

"All right, well, I'm starting."

Wetherbee keyed the mike and began. *"Mir, Discovery. As we are bringing—"*

Suddenly Story Musgrave, the astronaut acting as capcom for the mission, broke in. Much like intercontinental telephone calls, there is a small delay in communication between Discovery and the ground, and Musgrave had begun to speak before Wetherbee's first words came through. Realizing his mistake, Musgrave immediately signed off.

"As we are bringing our spaceships together," Wetherbee continued, shaking off the interruption, *"we are bringing our nations closer together."*

It was at this point, with Wetherbee in midsentence, that Foale realized Discovery was beginning to inch toward Mir.

"Pull it out! Pull it out!" Foale said urgently.

Wetherbee grabbed the stick and did so, even as he finished his message. *"The next time we approach, we will shake your hand, and together we will lead the world into the next millennium."*

Wetherbee then repeated his greeting in broken Russian.

Russian commander Aleksandr Viktorenko's reply was ebullient. *"We are all one! We are all human! . . . This is almost like a fairy tale. It's too good to be true."*

As he pulled away from the station for good, Wetherbee breathed a sigh of relief. As first dates go, it wasn't perfect, but it was a start.

March 14, 1995
Baikonur

Thagard stood out on the frigid Kazakh steppe in his white Russian flight suit and pretended to urinate. To one side his crewmates Vladimir Dezhurov and Gennadi Strekalov were merrily watering the bus tires, but Thagard couldn't bring himself to join them. It was broad daylight, after all, and there was a second bus behind them. Off in the distance, he could see spectators already milling around the rocket. For all Thagard knew, there were photographers nearby. That was just what he needed, a picture in *Time* magazine with some smarmy caption: "Norman E. Thagard, the first American to fly into space aboard a Russian rocket, prepares for flight by urinating outdoors."

This was the least of Thagard's worries, of course. Even as he walked the final steps to the Russian rocket, Thagard still feared George Abbey would find a way to replace him with Bonnie Dunbar. "I'm not going to believe Abbey is going to let me go until I actually get up there," Thagard had told Dave Ward on a preflight trip to Baikonur.

In fact, Thagard never learned how close he actually came to being bumped from his historic flight, and it had to do not with Dunbar but with Strekalov, the squat, curmudgeonly old cosmonaut who liked his vodka and had the long pink nose to prove it. Two weeks before the flight, when the crew had come down to Baikonur for a series of preflight tests, Strekalov felt a twinge in his lower back as he jammed himself into a model of the Soyuz. By the time he got to his feet, the twinge had become a throbbing pain. Today Strekalov, in a diagnosis few Western doctors could agree with, describes the injury as a pinched nerve he suffered because of a flow of cold air beneath his back as he lay in the capsule.

Whatever the precise cause of the injury, Strekalov zealously kept his secret from both his crewmates and the Russian physicians. Everyone knew the rules: If one member of a crew is disqualified, the entire crew is disqualified, and the backup crew flies. "I stopped having pain after twenty injections," Strekalov remembers with a sly grin. "I had a very good doctor. *My* doctor. I can't name him. Only me, the doctor, and the doctor's wife know who he is. I was afraid the [other] doctors would find out my secret and cancel my flight. The risk then was that the whole crew wouldn't fly. Bonnie Dunbar would have been the first American aboard Mir. But I told no one."

In the event, the launch of Mir 18 went smoothly, and two days later, after an uneventful flight, Thagard crawled through the Kvant docking port to become the first American to board the Mir station. He didn't know it at the time, but he was destined to have one of the more frustrating missions in the history of the American space program.

His first impression of Mir, Thagard liked to say, reminded him of a neighbor's cluttered utility room. It smelled faintly of sweat. The biggest surprise, in fact, was not the station itself but his commander, Dezhurov, a thirty-two-year-old rookie. In Star City, Thagard had formed what he thought was a close friendship with Dezhurov. But in space, much as John Blaha would discover with Valery Korzun a year later, Dezhurov underwent a transformation. Gone was the sweet, encouraging little brother Thagard had come to know. In his place was a bossy boy commander. Thagard first noticed the personality change during the changeover period, when, over dinner one evening, Dezhurov suddenly ordered Strekalov to fetch him something. No please, no thank-you. Just get it.

"Well, you're awfully bossy, aren't you?" Elena Kondakova, the Mir 17 flight engineer, observed. Kondakova was Valery Ryumin's wife; she could say anything she pleased. "Why don't you get it yourself?"

"I thought: Whoa, this is different!" remembers Thagard. It didn't take him long to figure out Dezhurov's metamorphosis. The young commander was obviously unsure how to deal with subordinates who were far more experienced; issuing commands seemed to be his way of establishing order and hierarchy. But understanding Dezhurov didn't make him any easier to take. On several occasions when the commander barked an order at him, Thagard wanted to tell him off. Instead, ever mindful that he was a guest aboard Mir, Thagard stayed quiet. He began watching Strekalov, whose demeanor and body language indicated he wasn't going to stomach Dezhurov's newfound bossiness much longer. Finally, a week or two into the flight, Strekalov blew up at Dezhurov. Thagard wouldn't remember what set off the veteran engineer, but his brief outburst clearly had its intended effect. Dezhurov began treating Strekalov a bit more gently.

Thagard wasted no time launching the American science program. The gear for all twenty-eight of his experiments was waiting for him, stowed in the unused base block toilet. On his second day aboard, Thagard took body-mass measurements of himself, Dezhurov, and Strekalov, a routine he would repeat regularly during the flight. He dug out the SAMS accelerometers—the ones the Russians had been so sensitive about—and began sticking the little units to walls throughout the station and stringing cables to connect them.

The key to Thagard's entire science program, though no one dwelled on its significance beforehand, was a single yellowish two-and-a-half-foot-square freezer that the European Space Agency had sent up for the German astronaut Ulf Merbold to use on a mission in late 1994. The ESA freezer, as it was called, was the one repository aboard Mir for the dozens of small green test tubes containing blood, urine, and saliva samples Thagard began collecting in his first days aboard the station. One of the first things Thagard did upon arriving on Mir was locate the little freezer inside a locker and find a spot to plug it in in Kvant. It hadn't been easy. The Mir 17 crew, like most crews, used Kvant for trash stowage. When Thagard stuck his head inside the darkened module, he found a floating junkyard of loose bags and containers, even an extra pair of blocky air-conditioning units, clogging its open spaces. Strekalov remedied the situation with a bit of typical cosmonaut ingenuity, taking a tennis net—Thagard never did discover what a tennis net was doing aboard Mir—and spreading it across the "floor" of the module. They then stowed all the loose equipment beneath the net.

Problems with the freezer began almost at once. The science procedures called for Thagard to note its internal temperature every few days, and he did so. The initial readings hovered around −15 degrees centigrade, right where they should be. But barely a week into the mission, Thagard noticed that the temperature was slowly rising. He asked the American ground team for guidance and was surprised when they couldn't offer any. "No one understood the operation of the device," he remembers. "They weren't prepared." At the TsUP, John McBrine asked a Lockheed engineer what they should do. "It's a European device," the engineer said. "What am I supposed to do about it?" McBrine then asked several other NASA people for advice, and all threw up their hands; he realized that no one was ready to take responsibility for something that was going wrong. Despite knowing nothing at all about the freezer, McBrine began trying to track down ESA engineers in France, Germany, and Spain, in search of a solution. The Lockheed man was dispatched to Barcelona to meet them.

Aboard Mir, Thagard was growing concerned. During his first weeks aboard the station, as temperatures inside the freezer inched upward, he worried that his samples would spoil. Finally, McBrine tracked down an ESA engineer who explained the problem: The freezer needed to be defrosted. Defrosting the unit, however, proved to be a far more problematic process than anyone envisioned. The freezer was divided into two compartments: the large lower cooling vault where the test tubes rested, and a smaller upper compartment that held a heat exchanger and fans for air circulation. Apparently ice was forming in the upper compartment, blocking the heat exchanger and preventing the unit from cooling properly. The ice, however, could not simply be chipped away; there was no door or access to the closed upper compartment. "The instructions to defrost the thing were ridiculous," remembers McBrine. "You were supposed to turn up the temperature enough to defrost, but not kill the samples, and open the [freezer] door."

Thagard judged this approach too risky. Instead he and Strekalov assembled a kind of cooler using pieces of Styrofoam wrapped in thermal blankets; they carefully lifted out the test tubes and stored them in this makeshift cooler, which kept the samples cold for several hours. Only then did Thagard open the freezer door and crank up the temperature in an effort to melt the ice. He tried it once, then twice. But there was no indication the ice was melting. Peering up through tiny gaps in the cooling compartment roof, he could still see slivers of ice. Temperatures inside the freezer, meanwhile, continued to rise. On the ground, McBrine placed urgent calls to anyone he could think of at ESA or NASA who might offer a solution. "I wasn't getting much support, from pretty much the whole world," he remembers.

The cosmonauts lost patience long before Thagard did. "The freezer we had on Mir was not supposed to be there," Strekalov remembers. "The freezer had not undergone enough tests and experiments on the ground from the specialists to make sure it would do okay in space. It seems that they were in too much of a hurry to get it into space."

The job of troubleshooting the freezer was complicated by the primitive communications links between Thagard and the ground. Before the mission, Al Holland had negotiated with IBMP officials for a package of what the Russians called psychological support items. Thagard was to receive two comm passes a day with his flight surgeons, Barratt and Ward, regular private family conferences with his wife, Kirby, and daily uplinks of world news. But once Thagard reached the station, Viktor Blagov's TsUP controllers refused to go along; Holland's agreement, they pointed out, hadn't been negotiated with the TsUP, and Blagov felt no obligation to recognize an agreement with IBMP. Ward and Barratt were reduced to literally begging for comm time from the various flight controllers. Some of the Russians, like Nikolai Nikiforov, relented after a bit of wheedling; others, like the disagreeable Ekrim Koneev, refused to let Thagard talk to the ground unless there was an urgent reason to do so. Ward discovered that everything on the TsUP floor revolved around personal relationships. Those flight directors he knew and befriended would help out; others, like Koneev, ignored him. "When I first got to the TsUP I thought: We're a third of the crew, we'll get a third of the comm time," Ward remembers. "Boy, was I wrong."

As a result, especially in the opening weeks of the mission, Barratt and Ward received only sporadic opportunities to talk to Thagard. Two and three days sometimes went by during which Thagard, much to his irritation, did not hear an American voice. For the first six weeks he received no news updates from NASA. Into the breach stepped an American ham-radio operator named David Larsen. Larsen, a forty-five-year-old psychologist who lived high on a ridge in northern California, had been speaking intermittently with the cosmonauts aboard Mir since the first ham radio was installed aboard the station, in about 1990. Using the "packet" system, he had begun exchanging e-mail messages with the more talkative Russians, who surprised him by insisting they communicate in English, even though Larsen spoke Russian. Early in the Mir 18 mission, Thagard seized on Larsen as a reliable channel to slip messages to his wife and others. On Mother's Day, it was Larsen whom Thagard relied upon to send Kirby flowers. And it was Larsen, not NASA or the TsUP, who on his own sent Thagard regular summaries of world news. "The news you send, for now, is the only news I get," Thagard wrote Larsen in an e-mail message on May 18. "I haven't even talked to our capcoms in over two days."*

*Larsen was to have varying levels of success chatting with the Mir astronauts who followed Thagard. Shannon Lucid was not an avid ham-radio user. John Blaha was; Larsen sent him regular updates on the Dow Jones Industrial Average. Jerry Linenger was the least cooperative. On February 12, a month into his mission, Linenger sent a message to his ground crew complaining that several ham-radio operators, including Larsen, were sending him unsolicited e-mail messages. "Somebody back there [in Houston] needs to talk to these guys ASAP because they're asking me questions and I don't like to respond to them," Linenger said, according to the daily NASA logs.

* * *

As temperatures inside the freezer continued to rise, John McBrine realized they were staring at the potential loss of Thagard's entire science program—at least everything until the Spektr module arrived in June, near the end of the mission. McBrine racked his brains but couldn't come up with any ideas on how to defrost the freezer. Finally, he called Peggy Whitson and said, "We're fucked."

Thagard brooded over the situation for days. NASA, he felt, had rushed this mission for political reasons and didn't seem to care that he was now paying the price. The more Thagard mulled the situation, the more he realized it was useless to blame his ground team. Blame, he came to believe, lay further up the ladder.

"There is no doubt in my mind: If I had been the [astronaut] George Abbey wanted, I would have got the backing and support I needed," Thagard said in a February 1998 interview for this book. "But I wasn't George's boy. And so the structure just wasn't put in place to help me. If I had been Bill Shepherd, you can bet the situation would have been completely different."

The scope of the developing disaster was slow to dawn back in Houston. Hired out of JSC's space station division by Tommy Holloway, Jim Van Laak joined the Phase One office on April 1 and was appalled by the chaotic state of Thagard's ground crew. No one seemed to be in charge. They had flight surgeons—flight surgeons!—acting as capcoms. Holloway and his people would sit in daily telecons with Moscow and endlessly debate what to do, yet nothing seemed to be getting done.

"Jeepers, Tommy," Van Laak warned Holloway, "you don't have anyone over there who knows how to run an operation. You might limp through Thagard's increment this way, but God forbid things go like this in the future." Van Laak recalls: "Tommy wanted the team to grow its own skills—you know: 'Let the children grow.' And it wasn't happening. By my judgment, they didn't have a clue. The fact was, we grossly underestimated the requirements of the job. We really thought the Russians would do all this for us."

What was needed, it was clear, was someone with operations experience to take control of the Moscow team. That meant MOD. But when Holloway approached MOD officials, he found they had no interest whatsoever in helping out. "MOD, bless their hearts, had their head up their ass and didn't want to get involved," Holloway recalls.

"We didn't want the job because of the manpower requirements," remembers MOD's Phil Engelauf. "We saw Phase One as an interesting but not particularly useful diversion of resources. . . . Our first reaction to Phase One was: It's not very effective. You know, we gathered a bunch of experiments, gave 'em to Norm, and pretty much forgot him. There was no infrastructure in place to support him. Zippo." For the time being, Phase One was on its own.

The drama over the ESA freezer dragged on for several weeks. Both Dezhurov and Strekalov made yeomanlike efforts to fix the unit, but neither man made headway. At one point, a Russian engineer suggested chopping through the grillwork that separated the freezer's two compartments, to get directly at the ice; the grillwork was lightweight aluminum, Thagard was told, and could be easily cut. Strekalov examined the metal closely and agreed with Thagard: The grillwork was heavy metal, too heavy to cut.

"They don't know what they're doing down there!" Strekalov exclaimed. "They're not up here! They don't have a clue!"

He turned to Thagard. "In the old days it wouldn't be like this," he said. "In the old days we had experts who knew their jobs."

Barely a month into his mission, Thagard gave up. He unplugged the little freezer, which was now too warm to properly cool his test tubes, and flung it into a compartment. Of the one hundred or so samples of blood, urine, and saliva he had taken, he managed to save maybe twenty, inserting them in niches in Mir's main refrigerator and a pair of tiny secondary freezers. Until the Spektr module arrived in early June, the American science program for Mir 18, the first-ever joint Russian-American space station mission, was all but over.

The only experiments Thagard could still perform were a handful of basic metabolic tests, mostly measuring things like bone loss. It wouldn't be enough to keep him busy for six weeks. But then, around the first of May, Thagard lost the ability to perform even these rudimentary experiments. The news came in a private medical conference abruptly convened by Russian doctors at the TsUP. Every three days since the beginning of the mission Thagard had taken a series of measurements of himself and his crewmates. It's impossible to weigh yourself in orbit, but using Thagard's measurements, the ground could extrapolate the crew's weights. During the medical conference a Russian doctor asked Thagard: Did he realize he had lost seventeen and a half pounds?

No, Thagard said, worried. It was true: Thagard's weight had fallen to 141 pounds from 158 pounds. "You've lost lean body mass," a doctor said.

Thagard immediately understood the significance of his weight loss. It had nothing to do with his comfort or even his health. It had to do with his fitness as a scientific subject: The data from any experiment performed on Thagard would be tainted. Scientists wouldn't be able to tell whether changes in his body—bone loss, for example—were due to time spent in space or weight loss. As a guinea pig, Thagard had been rendered all but useless.

"My first thought was: Where have you people been for the last six weeks?" Thagard recalls. Why hadn't his weight loss been discovered earlier? But his next thought made him even angrier. Thagard had warned NASA scientists that something like this might happen.

It was his diet. Upon reaching the station, Thagard and his crewmates had learned that their food intake was to be monitored. NASA scientists had applied bar codes to many of the foods they considered to be staples of orbital meals, including canned fish, borscht, potatoes, and other relatively bland edibles. To his irritation, Thagard was instructed to eat only the bar-coded food. An entire second class of food, called the supplemental food supply, including almost all the "fun foods," like almonds and yogurts and candy, were not bar coded and were thus considered off-limits. Thagard was surprised to see Dezhurov and Strekalov happily ignore the dietary restrictions and plunge right into the supplemental food supply. Not Thagard. Thagard was a Marine, and Marines followed orders. But even Marines don't like canned fish. Thagard, in fact, couldn't stomach nearly a third of the Russian bar-coded food. Even so, he successfully resisted the temptation to dive into the treats.

"What percentage of your food is coming from non-bar-coded sources?" Thagard asked Dezhurov one evening over dinner.

"About twenty-five percent."

"And you?" Thagard asked Strekalov.

"About fifty percent."

Early in the mission, Thagard warned Ward and Barratt that the dietary restrictions could lead him to lose weight. The two flight surgeons, after checking with Houston, replied that it was all right for Thagard to eat from the supplementary food supply. Whatever he ate, though, he had to write down. This Thagard decided he simply couldn't do, for there was no such thing as scratch paper aboard Mir, no unused program on the laptop. When Strekalov went to sketch a schematic drawing for an EVA later in the mission, he was forced to use a felt marker on the back of an aluminum food-can lid. And so Thagard decided to stick to the bar-coded food. And so he lost seventeen pounds.

Few at NASA bought his explanation about scratch paper. Peggy Whitson and others felt Thagard would rather be a martyr than take the time to write out what he was eating. He had done exactly the same thing at Star City, when Whitson was forced to give him a stern lecture on his diet. The problems had arisen when Whitson's people scheduled weeklong "data takes" during Thagard's training, in which the astronaut was supposed to measure everything he ate and drank and collect his urine in receptacles. At one point Whitson discovered that Thagard's food intake dropped drastically during the data-gathering periods. "We noticed it when his urine volume was way down," she recalls. "[It turned out] he didn't want to pee in the bottle, so he would not drink. He didn't want to eat because he was tired of asking Kirby to measure out his food. I told him, 'Norm, you know better than anyone else that this affects this data.' " Thagard said he would do better next time.

"Norm could have written things down," recalls Barratt. "This was Norm's way of making a point that this [experiment] wasn't planned out. There wasn't a good enough understanding for that first flight of what food was available and how this diet experiment would go. It just wasn't thought through well enough."

Now it was too late to begin eating responsibly. "No one took responsibility for the whole food fiasco," Thagard recalls. "But then no one would take responsibility for anything." With his health in mind, the Russian doctors ordered him to begin eating snack foods. "You can eat anything except your crewmates," one of them said. Thagard didn't appreciate the joke. It was the first of May, and with the demise of the American science program, he had nothing to do. Nothing at all.

May was hell. Every morning, Thagard wriggled out of his sleeping bag and shaved—carefully making sure to retrieve any errant whiskers—then combed his hair and ate breakfast at the base block table. And then he would stay there, floating around the table, for most or all of the day. As Dezhurov and Strekalov busied themselves with EVA preparations and repairs, all Thagard could do was watch. He spent hours staring out the windows at Earth, or at the blackness of space. He had brought with him exactly one book, a volume of *New York Times* crossword puzzles, but he couldn't bring himself to do puzzles during the day. The irony was crushing: Here he was, the first American aboard a Russian space station, and all he had to do was crossword puzzles. It was humiliating. He talked a little on the ham radio. He ate lunch. He ate dinner. At night he watched movies on a little monitor in base block. He found to his surprise that the Russians had stocked Mir with a large supply of French and Italian soft-core pornographic films. They were almost embarrassing to watch, but he watched, as did Dezhurov. Strekalov usually sat with them, reading a book.

"Norm, how are things going up there?" John Blaha asked Thagard in a rare video comm one day in May.

"Well, John, see where I'm floating right now?" Thagard responded. Blaha could see he was located just outside his *kayutka*, by the base block table. "I spend ninety percent of my time right here."

Back in Houston, Al Holland and the NASA psychological group were growing worried about their underworked astronaut. "Norm had to twiddle his thumbs for a month, which is difficult for anyone to do, especially a high-achieving astronaut," says Holland. "Work underload is a terrible thing. It can be very devastating. You have to be careful with it. But there wasn't much else he could do. [The Russians] trained him on systems, but they wouldn't let him touch the systems. They absolutely would not let him touch the switches."

Thagard hinted to the Russians that he was willing to help out on the cleaning and maintenance chores that kept both men busy.

"Don't be sad, my friend," Dezhurov told Thagard at one point.

"I'm not sad; I'm bored," said Thagard. He had trained on Mir systems. He knew how to fix things. But Dezhurov didn't seem at all inclined to give him responsibility. They were friends and crewmates, but both Russians clearly still considered Thagard a foreigner who couldn't be trusted with the hardware.

The one thing that kept Thagard busy was a pair of unexpected medical emergencies, the only incidents of their kind during the seven-mission Phase One program. Neither was disclosed publicly at the time, but the more serious of the two injuries can be glimpsed on a scrap of NASA archival videotape. Strekalov's left arm, from wrist to shoulder, has turned a dull shade of blue, as if the whole limb were covered in a single gaudy tattoo. According to Strekalov, he cut his arm, and the cut had become infected. "I scratched it trying to reach something behind a panel," Strekalov says today. "It was blue and swollen to twice its normal size. Norm cured me. He gave me his medical kit, with all his medicine. There was a medicine called prednisolone; I was taking about ten pills a day. It's a lethal dose for a normal person. But not for a Russian cosmonaut!" Strekalov recovered, but not before his injury nearly caused the Russians to cancel a series of EVAs they needed to prepare the station for the arrival of the new Spektr module.

On May 18 the doctor suddenly found himself the patient. Thagard was exercising with a set of isometric cables the Russians call "expanders." He had inserted his feet into loops at the end of the cables and stretched the cables themselves across the back of his neck, allowing him to do deep knee bends. Somehow one of his feet slipped out of its loop, and the cable sprang upward, striking him in the right eye. Thagard saw stars; the pain was excruciating. When he forced himself to open the eye, his vision was badly blurred. Base block looked as if it had been draped in gauze. Worse, the eye was badly photophobic. Any direct light caused him searing pain. Even the tiny red LED indicator on one of the consoles hurt him to look at. Thagard quickly found an American medical kit, applied a local anesthetic, and taped a patch over the eye. He was just beginning to ponder what effects the injury might have on his mission, when Strekalov floated into the module. Thagard explained what had happened. "Oh, yeah, those things are dangerous," Strekalov said of the expanders. "That's why I never use them."

Thagard forced a smile. "Thanks for telling me that *now*."

A Russian doctor, put on comm from the TsUP, diagnosed the injury as a possible corneal abrasion and prescribed antibiotics. When the next day the pain remained the doctor suggested steroid drops. This time the medicine worked, and the eye began slowly to heal. A week or so later it was back to normal, but the incident spooked the NASA flight surgeons. "We were really worried about that," Mike Barratt recalls. "That was actually the worst injury Dave and I had ever worked."

While Thagard sat by and watched, Dezhurov and Strekalov busied themselves with the EVAs to prepare the station for the arrival of Spektr and, later, the Space Shuttle Atlantis, which was scheduled to perform NASA's first docking with Mir at the end of June. In a trio of spacewalks on May 12, 17, and 22, the two Russians dismantled a set of solar arrays outside the Kristall module, reinstalled them on the hull of Kvant, and stretched cables across the outer hull to connect them. When Spektr finally arrived at the station on June 1, the two Russians performed two more EVAs to install the module and open its four solar arrays. One of the arrays would not deploy, and the TsUP announced plans to perform an impromptu EVA on June 15 to fix it.

To the amazement of everyone, including Thagard, Strekalov refused to carry out the final EVA. Huge arguments broke out, both between Strekalov and Dezhurov, and between Strekalov and the ground. The old cosmonaut felt the EVA was too dangerous, and nothing would persuade him to venture outside the station.

"I was really yelling at them," Strekalov acknowledges. "This was my sixth flight, and I had every right to yell. I knew it was my last flight. I didn't have anything to lose. Volodya Dezhurov was young. It was his first flight. He didn't want to spoil his relationship with TsUP and the big bosses and big bananas there.

"There were three reasons for disagreement," Strekalov continues. "Problem number one was we didn't have a special tool for cutting the tube that was preventing the solar array from unfolding. This tool had been made specially and [was supposed to be] delivered by the shuttle. We call it a guillotine. It looks like a big garden shears. Second reason: The pressure in the station was only 590. You know we made five EVAs. And for each EVA you lost at least thirty millimeters of pressure. For the sixth EVA we would have lost another thirty. We would have been down to 560, and 550 is an emergency situation. At 550 cosmonauts are supposed to leave the station.

"The situation looked like we were going to be hostages to these last ten millimeters. We also had Norm Thagard. We had to pick up and pack Norm. I told TsUP, 'What do you mean? Why can't Norm [pack] himself?' But no. It took a lot of time. Plus, we were trained for these emergency situations and Norm was not, and we had to take care of him. The third reason was actually purely diplomatic. I didn't want to take any risks on the first international flight. And this was risky."

Spektr's arrival gave Thagard something to do at last. He spent his last weeks aboard the station unpacking Spektr and performing what few experiments he had time to complete. But even then the mission's drama was not over. As the EVA debates flared, the crew received startling news. Dezhurov's mother had died. The young commander was famously close with his mother, who had met and dined with the crew. Dezhurov was stricken. In tears, he disappeared into the Kvant 2 module, where he refused to speak with the ground for days. Down at the TsUP, Barratt and Ward were stunned; nothing like this had ever happened aboard a shuttle.

"Honestly speaking, he was down there for five days; you should probably write only three," says Strekalov. "We had to leave him alone for some time. He really, really suffered. We did all his work for him." Thagard was unsure how to treat Dezhurov. His instinct was to comfort him. But for all he knew of Russian customs, that was the wrong thing to do. He watched Strekalov. The elder cosmonaut dispensed with sympathy, left Dezhurov to his thoughts, and tended to the commander's basic needs. "You still need to eat," Strekalov told him after a day or two.

On June 27, the Space Shuttle Atlantis, carrying Hoot Gibson, Charlie Precourt, Bonnie Dunbar, two other astronauts, and Mir's replacement crew, Anatoli Solovyov and Nikolai Budarin, roared off into the heavens to begin its voyage to Mir. Two days later, on the morning of June 29, from a distance of sixty-five kilometers, the shuttle crew got its first glimpse of Mir in the distance. Ironically, it was Dunbar who radioed Thagard aboard the station: "We have you in sight." Just after lunchtime Gibson maneuvered Atlantis to a seamless mating with the station's Kristall docking port. Down at the TsUP, Dan Goldin leaped into the air and bear-hugged Yuri Koptev, as applause broke out in the auditorium.

Atlantis remained docked to the station for five days, during which a wide array of experiments were completed by both crews. Just before ten on the morning of July 4, as Hoot Gibson inched the shuttle away from the Kristall docking port, Solovyov and Budarin undocked their Soyuz from Mir and flew out to a position about one hundred meters from the departing shuttle. On direct orders from Energia's chairman, Yuri Semenov, the cosmonauts intended to film the shuttle's undocking; Semenov excitedly called it the "picture of the decade." Tommy Holloway had strenuously objected to this. "We argued it's too much risk; what happens if the station should lose power?" Holloway remembers. "How would they redock?"

"But it's our risk," Valery Ryumin countered. "Not yours."

For once, the Americans were right. Even as Gibson piloted the shuttle away from the station, Mir suffered a sudden power outage, sending it into a slow spin. Solovyov, in the Soyuz, was forced to perform an immediate emergency redocking, which, to the Americans' hushed relief, he achieved without further incident.

When the shuttle landed back at the Cape on July 7, Thagard emerged standing and waved. Peggy Whitson and the assembled scientists groaned: Another macho astronaut who insisted on walking out of the shuttle. This would prove to be a sore subject with the Mir astronauts. All were supposed to be taken off the shuttle on a stretcher, but every one, with the exception of John Blaha, walked off. As the first American in twenty years to endure a long-duration space flight, Thagard's greatest value to science was as a human guinea pig, especially a guinea pig who hadn't yet reacclimated to Earth. The act of standing prematurely ignited changes in Thagard's body that Whitson's team of scientists, ready to run a battery of tests, were eager to delay.

Thagard was whisked from the shuttle to an examination room, where the NASA team began putting him through his paces, taking blood and urine samples and watching him balance himself on one foot. As Whitson and her assistants hurried through their tests, Tommy Holloway entered the room. "Dan wants to see Norm," he told Whitson. Dan Goldin.

"Dan can come see him in here," Whitson said, too busy to focus.

"No; there's going to be a press conference," said Holloway. "All three crew members need to be out there."

Whitson sighed. She had a limited amount of time with Thagard, a matter of hours, before his body began to recover. All their science depended on getting experiments done quickly. Of course, Goldin's request was nothing new; she had faced the same problem on shuttle flights. Politics and public relations always ranked far ahead of scientific considerations.

But then Whitson heard the kicker.

"Dan wants to give him an ice cream sundae," someone said. "And a hot dog."

This was ridiculous. The food would sharply skew the neurosensory data. Eating ice cream would cause Thagard's glucose levels to skyrocket. Still, Whitson knew there was no use arguing. "Dan didn't care; he was doing this for the press," she recalls. "What am I supposed to say [to Norm], 'Dan is going to give you this ice cream but don't eat it'? It was just unfortunate we couldn't have finished up this stuff first."

Whitson sent for the sundae and the hot dog. "Okay, weigh it," she told one of her assistants, gritting her teeth. "Before and after."

That night Thagard, the two Russians, and Hoot Gibson's shuttle crew all flew back to Houston.* While Gibson and the shuttle crew were greeted

*Thagard's was the only mission in which cosmonauts returned to Earth aboard an American space shuttle.

by celebratory crowds and mobs of press, a NASA van quietly took Thagard directly to his home in suburban Friendswood. The NASA doctors had engaged in long debates about the wisdom of allowing Thagard to go home mere hours after returning from the longest space mission in NASA history; normally, he would stay in crew quarters under medical observation. But Thagard was adamant: He wanted to go home. Dave Ward grabbed a medical bag and stayed that night at the Thagards'. Thagard walked in, his wife and three sons in tow, and quickly drank a beer. His legs were wobbly, and Ward held his breath every time Thagard walked down a set of stairs or veered too close to the edge of a coffee table. By and by, Kirby helped him walk to his "sound room," where Thagard turned on his beloved stereo and cranked the volume of a classical CD as high as he could. "Norm was a happy man that night," remembers Ward. "He was a happy, happy camper."

The Russians were not happy. Strekalov and Dezhurov were taken to JSC in an ambulance. For the two melancholy Russians, who faced stiff fines and charges of insubordination for their refusal to perform the mission's final EVA, there were none of the vodka toasts and raucous celebrations cosmonauts were accustomed to upon their return to Star City. Already rumors were rife that Dezhurov would never fly again; Strekalov was retiring. Mike Barratt walked the two men, alone and unnoticed, to their darkened crew quarters in Building 259. The door was locked; Barratt had to fish out a key. Both men were clearly disappointed that Thagard was not staying with them. "Why should he get special treatment?" the Russian flight surgeon, Gene Kobsev, asked Barratt. Barratt had no explanation. "It was a real letdown for the Russians," he remembers. "They were kind of ignored."

Thagard's medical tests the next day went smoothly: He gave his blood, performed his balancing tests, and ran on treadmills with little difficulty. But to the surprise of researchers in both the Russian and the American camps, Kobsev refused to allow the two cosmonauts to undergo any tests involving physical exercise on their first two days back in Houston. Instead he ordered them to rest. Strekalov, who hadn't kept to his exercise regimen while in space, was a wreck, out of shape, dizzy, and exhausted. "This was a big deal," recalls Dave Ward. "You were losing what you flew for. You were losing the data. You just couldn't believe they were just kind of blowing off one of the most important days."

Ice cream and hot dogs weren't the only things Thagard's crew was ingesting that gave the scientists fits. Despite rules explicitly preventing it, Dezhurov and Strekalov began drinking alcohol almost immediately after their return. Whitson discovered they were getting it from the Russian doctors. Asking why, she was unable to get what she considered a straight answer. At one point, one of her people grabbed the cup from which a cosmonaut was drinking and had the last few drops of liquid inside analyzed. It was seventy-proof alcohol. All Whitson could do was sigh.

"It was a cultural difference," she says. "The Russians believe a lot more things are medicinal than we do. It probably did [affect the experiments] to some degree. In general, alcohol is a diuretic; it's going to bias the fluid status of a person. You go to all this trouble to measure their fluids, and then they go and drink a diuretic."

For Thagard, the first American mission to Mir was in many ways the most profound disappointment of his long and distinguished career. He was the first American to fly aboard a Russian spaceship, and the first to go into space after training in the Russian space program. In the process he set a new American record for days spent in space, surpassing the record established by the third Skylab crew. But public reaction to his historic flight was at best muted. The newsmagazines and television networks all but ignored him. At the press conference he gave after the flight, wire-service reporters homed in on his mild complaints about the station's food, missing his family, and the normal cultural isolation he experienced. He felt their subsequent stories portrayed him as a whiner. His bitterness escalated after a desultory visit to the White House, and climaxed a year later with the public adulation that greeted Shannon Lucid's return to Earth. "If Clinton knew who I was when I visited the White House," he recalls, "he gave no indication. But he was at the landing for Shannon. He just didn't have much use for a white Anglo-Saxon male. But for Shannon? A woman? It was night and day."

The Russians, meanwhile, were underwhelmed by Thagard's abilities in space. "His performance was adequate; he did okay," says Valery Ryumin. "We only realized later that in the U.S. it is not a good thing to show initiative. He did what he was told to do. He did not go the extra mile. He didn't try to go further, to go above and beyond the call of duty. That is the way of NASA astronauts. There's always things to do on the station. If you want to do something, you will find something to do. I always assumed he would find more things to do. He didn't."

* * *

Thagard's debriefings pointed up the obvious deficiencies in Phase One's approach to the Mir missions, especially in the structure and performance of the ground team in Moscow. "We weren't in sync at all," acknowledges Culbertson. "This was a whole new program for all of us, and we were learning as we went." What they needed at the TsUP, Holloway and Culbertson saw, were veteran flight controllers who had experience running space missions. That meant MOD, which had repeatedly declined to get involved in Phase One. Still, Holloway met with Phil Engelauf and other MOD officials; as expected, he found them reluctant to free up people for extended stays in far-off Moscow.

"They wanted first-level people, okay?" recalls Engelauf. "They said, 'We want volunteers to go out to Star City for eighteen months.' There was a deafening silence. Most of us have two- and three-year-old kids and don't want to live in Russia; that wasn't in the oath of office when we joined NASA. Not only couldn't we find volunteers, we couldn't spare people of the caliber they wanted. [But] we were basically told by center management, 'Phase One is broken. Go fix it.' "

Urged on by George Abbey, MOD did put out a tepid call for volunteers to work with Phase One. Holloway, justifiably wary of MOD's commitment to securing him quality people, personally approached a capable MOD console operator named Bill Gerstenmaier. A laconic Ohioan with a drooping mustache, Gerstenmaier had just returned from a sabbatical to complete his doctoral studies at Purdue and felt he needed a high-profile assignment to regain the attention of management. He signed on and was immediately dispatched to the TsUP to negotiate new and improved arrangements for Lucid's mission to Mir, which was scheduled to begin in March 1996. MOD's call for volunteers, meanwhile, snared a half-dozen relatively junior people, including Caasi Moore, the training supervisor who was assigned to John Blaha's mission; Tony Sang, who was given Linenger's; and Keith Zimmerman, a twenty-six-year-old, who was put on Mike Foale's increment. "It was basically Bill and three or four kids," says Engelauf, rolling his eyes. "You should have seen the people we dismissed."

MOD's behavior left Jim Van Laak boiling. "If I can say this tactfully, MOD was not interested in giving us their best talent for this," says Van Laak. "In MOD there is a mass psychology, not a philosophy, really: [And that is] if MOD could offer to let you go for nine months, you were obviously not very good. And they said that to people. They were too blind to see the opportunity, they felt no responsibility to the agency, and they polluted the environment for others. They poisoned the atmosphere for Phase One. They really did."

Summer 1995
Star City

A strange quiet settled over the American contingent at Star City as Thagard's mission came to a close. The second NASA mission to Mir, Shannon Lucid's, was to be the first official increment of the Phase One program. With Thagard and Dunbar gone, Lucid and John Blaha, who had arrived at Star City that January and spent their first months in language classes, expected to find themselves the new focus of NASA attention. To their surprise, they weren't. "I remember during Norm's training there were all these science people around, and so during his flight John and I said, 'Okay, they're going to help us now,' " Lucid recalls. "Then Norm landed, and they all disappeared. We were alone. John and I had a lot of conversations, you know, where we said we were actually worse off than Norm [had been]."

Much later, after she had graced the cover of *Newsweek*, after she had been named to the Space Hall of Fame, after she had appeared on *Nightline* and before congressional committees, Shannon Lucid's experience in the Phase One program would be glorified as the brightest moment of the new Russian-American partnership, a triumph of strategic planning, technology, science, and diplomacy. In fact, although Lucid never acknowledged this publicly, it was chaos. All but abandoned by NASA in Star City, uninformed about even the most rudimentary details of her mission, she was wholly unprepared for her stay aboard Mir, yet managed to thrive in space thanks to a personality blessed with large doses of patience and flexibility.

Born in 1943 in China to missionary parents, Lucid spent the first year of her life in a Japanese prison camp. After four more years in postwar China, Lucid's family settled in tiny Bethany, Oklahoma, where as a girl she stumbled on one of Robert Goddard's books on rocketry and vowed to make it to space one day. After years working as a scientist, she had been one of six women selected in the first class of shuttle astronauts, in 1978. In the ensuing years she had flown aboard the shuttle four times, twice with her friend John Blaha. She had a reputation as a cheerful, diligent, and quietly intense astronaut.

In August 1994, Lucid agreed to accompany Blaha to Monterey to begin Russian classes, even though she didn't yet have a flight assignment. Not until a few days before Thanksgiving, as she prepared to return to Houston for the holiday, did Lucid realize she was actually going to Russia. A secretary called to ask for some personal information for a visa.

"What visa?" Lucid asked.

"It's a visa for Russia."

"Oh," Lucid said with a chuckle. "Am I going to Russia?"

"I guess."

She and Blaha arrived in Star City in January 1995 and immediately attended all-day Russian classes in a darkened classroom—the overhead lights buzzed too loudly, they were informed. Their teacher spoke no English at all. The two old friends knew of the tensions between Dunbar and Thagard and made an informal pact to avoid similar pitfalls: No backdoor conversations with NASA management, no complaining. Lucid's apartment was freezing cold, while the Blahas' was so sweltering they kept a window open in February. Still, Lucid felt bad; their flats, she knew, were palatial compared to those of ordinary Russians.

Lucid's husband, Mike, had stayed behind in Houston, and in those first few months in Star City her stiffest challenge was finding ways to communicate with him. Until Dunbar left in March, Lucid walked over to her apartment many nights to use the Inmarsat phone. She felt guilty about imposing, however, and after a while began making nightly visits to the NASA office in the Prophy in an attempt to reach her family via e-mail. The walkways were coated with ice, and some evenings it took Lucid forty-five minutes of slipping and sliding across the base, her IBM ThinkPad hugged tightly to her chest, to reach the office. She was, by her own admission, computer illiterate and never managed to get the e-mail to work. Finally, she appealed to one of her daughters, who sent her a diskette that, with the help of a modem Dave Ward picked up at a Moscow Radio Shack, enabled her to sign on to the Compuserve on-line service. "And I had to pay for it all myself," she remembers. "NASA would not pay for it."

Still, news from the outside world made it to Star City slowly, if at all. One night in April, Lucid received a strange e-mail from her husband; it said only, "Don't worry. Your brother's okay." It wasn't until she sent him a reply asking what he meant that Mike Lucid sent her a second note to explain: Someone had set off a bomb demolishing the federal building in Oklahoma City. Her brother, a local lawyer who often visited the building, hadn't been in the area that day.

The isolation was numbing. The two astronauts debated how best to handle Dunbar's Inmarsat phone, only to be dismayed when the phone disappeared along with Dunbar. The only voice contact with the United States was the all-but-useless Russian phone system, which Lucid and Blaha ignored. "There was no contact between NASA and us in Star City—nothing—so we never heard anything," Lucid recalls. "I didn't even feel like I worked for NASA. I felt like I was working for myself. I never heard anything from the Phase One office. I wasn't even sure how it was organized." Not until November was a satellite phone system installed that allowed the two astronauts to begin talking to Houston. Even then they were obliged to pay for personal calls themselves. "Don't even talk to my husband about those bills," Lucid says. "We're *still* in the hole."

For months, until she began Soyuz training that summer, Lucid rarely if ever saw cosmonauts. Through the end of summer and into the fall she and Blaha slogged through fourteen-hour days learning the various Mir and Soyuz systems. While Blaha studied seven days a week, Lucid took Sundays off, riding her bicycle through the woods or taking the train into Moscow to sightsee. She assumed that at some point someone at NASA would begin teaching her the experiments she would perform in space. One day shortly after Thanksgiving Lucid took a call from a NASA public relations person, Debbie Rahn. Rahn said she had an ABC News correspondent who wanted to interview Lucid about the experiments she would be doing in space. "That'd be fine," Lucid said, "but I don't know what science I will be doing."

"You're kidding," Rahn replied. "That can't be."

"Tell you what," Lucid said. "You find out what I'm doing, you let me know, and I'll talk to the reporter."

Lucid in fact fell victim to a political struggle within NASA over which division would conduct her training. Thagard had been trained by Peggy Whitson's group of workers from Krug Life Sciences, the NASA contractor. After Thagard's mission, NASA decided to replace Krug with a new group of trainers from Krug's archrival, Lockheed Martin. The ensuing transition was far from smooth, and in the resulting confusion no one was assigned to train Lucid in the proper working of the experiments she would be doing in space; the new team seemed to assume the Russians would handle Lucid's training. To Lucid, who had spent eighteen solid months in detailed science training for her last shuttle mission, it was a shocking and disorienting experience. "There was no one," she remembers. "I didn't even have a clue what was going to be *on* the flight."

By Christmas, three months before her mission was to begin, Lucid was growing desperate. She sent pleading e-mails back to Houston, trying to find someone who knew just what she was supposed to do aboard Mir. It was no use. "I was asking them, 'Does anyone have a list of experiments for the flight?' " she remembers. "[The answer was] No. No one had a clue."

"It was just absolute chaos, even worse than what Norm went through," remembers John McBrine, who spoke with Lucid that December. Matt Muller, another Krug worker who talked to Lucid, begged NASA's life sciences group to send training equipment to Star City. Says Muller: "I told them, more than once, don't even bother sending it to the station if Shannon doesn't train on it." Some of Lucid's training equipment remained stuck in customs. Other items, including several cameras Lucid would have to use and maintain in space, made it to Star City but languished in warehouses. "I really wanted to use those cameras, to learn how to change the film in a microgravity environment, that sort of thing," says Lucid. "The cosmonauts had access [to them], but we couldn't use them. I found that bizarre. I never understood why."

In space, Lucid's stoicism would serve her well. At Star City it worked against her. "Shannon's personality is not to complain at all," says Bill Gerstenmaier, her ops lead. "So nobody really knew how badly her lack of training and procedures was hurting her. I was too busy. No one really knew until near the end."

Not until January 1996, in fact, after McBrine spoke to Lucid, did anyone in Houston seem to understand that she was in serious trouble. A rescue effort spearheaded by Matt Muller was hastily put together. But when Muller phoned Lucid to tell her he was coming to Star City to give her a three-day crash course on her science program, he was surprised when she told him not to bother. "Shannon said, 'I don't need it,' even though we all knew she needed it big time," Muller recalls. "What she was saying was, 'Don't go to too much trouble, we'll get through it, blah blah blah.' What we were hearing privately was, she is just clueless, there are experiments she hasn't heard of, she's in a lot of trouble. But Shannon did not want to call a lot of attention to the fact that this hadn't been the best training flow. It was very typical Shannon, not wanting to advertise her troubles."

Muller went anyway. In late January he arrived at the Prophy and, along with Bill Gerstenmaier, met with their frustrated astronaut. The meeting left everyone shaken. Because the new ops lead concept hadn't been explained to her, Lucid didn't even understand who Gerstenmaier was. When Muller offered to begin explaining her experiments, Lucid would have none of it. "Why are you telling me this stuff now?" Lucid demanded. "I could have used this *months* ago." Muller handed her a rough outline of her mission plan, which laid out which experiments she would do and when, and Lucid flipped through the pages in disgust. "Well, I don't know anything about *this,*" she said, pointing her finger at one experiment. She glanced at another page. "And I've never even heard of *this.*"

On the verge of tears, Lucid launched into an uncharacteristic tirade at the way NASA had mishandled her training. "She was saying, 'There's too much for me to do this in the last couple of weeks,' " Muller remembers. " 'Where the hell were you nine months ago? I can't take the time this close to flight and sit down for three days and work with you on this.' It was scary. It was the only time I've ever seen Shannon like that."

Gerstenmaier sat by silently, stunned by Lucid's outburst. "It was so unlike Shannon, I don't even want to talk about it," he says quietly. "She was not happy." That day Lucid came across as bitter, unprepared, and thoroughly disillusioned with her mission; emotionally, Gerstenmaier thought, she was at the end of her rope. "I left that meeting scared to death," he remembers. "She hadn't been trained, she didn't know her procedures. She didn't know anything about the flight plan."

Lucid soldiered on. Rather than panicking, she told herself she could live with a ruinous mission. "I had no idea how things were going to go in space," she recalls. "If things all fell apart, and your career's over, that was fine. I've had four good [shuttle] flights. I was going to have a good time. That was it. I figured I would take care of myself." With Thagard's downtime in mind, she got her daughters to put together a library of paperback books, mostly eighteenth- and nineteenth-century novels, such as *David Copperfield.* In February, just as she began preparing to return to Houston, she noticed a distinct change in her Russian trainers' attitudes. "A few weeks before launch, [the Russians] were like, 'Oh, NASA's really going to launch this person,' " she recalls. "They got really nervous. You could see it. They were very, very nervous about a female going up." Her trainers began nitpicking her in classes, pestering her with little questions she had answered dozens of times. For some reason they seemed obsessed with her use of the Mir toilet. "It was 'Do you think you'll know how to use the bathroom?' just over and over and over," she recalls.

Culbertson and Van Laak came to Russia for her final flight certification. At the ceremony, held at the Star City headquarters building in a copse of fragrant pines, Valery Ryumin rose and sternly criticized Lucid's ability to speak Russian. Her language skills were not adequate for the mission, Ryumin announced, but they would accept her anyway. "Oh, Shannon, don't worry," her crewmate Yuri Onufriyenko whispered to Lucid as Ryumin spoke. "Everything will be fine."

For all the chaos Lucid endured, her mission to Mir turned out to be by far the least eventful of the first five NASA increments. Arriving at the station via the Shuttle Atlantis on March 28, 1996, she became great friends with her crewmates, Yuri Onufriyenko and Yuri Usachov, who called her "Miss Shannon." After all the uncertainty surrounding her science program, she and her ops lead, Bill Gerstenmaier, managed to complete all her experiments, including those sent to Mir aboard the new Priroda module, which arrived at the station that summer. Their method was simple: Because Lucid knew so little about the experiments, Gerstenmaier learned them first, then wrote out instructions that he e-mailed her. Gerstenmaier estimates he uplinked more than nine hundred pages of procedures to Lucid during her six-month stay aboard the station. By August, in fact, Lucid had finished them all and, like Thagard, was left with nothing to do. She relied on her library to keep herself occupied.

Although it received little publicity, Lucid's mission was plagued with the same kind of coolant leaks and Elektron breakdowns that later beset Linenger, Tsibliyev, and Lazutkin. She was surprised that the Americans paid little attention to Mir's intermittent problems. "I remember asking Gaylen [Johnson, the flight surgeon] at one point, 'Does anyone at NASA even care?' " she recalls. "When I got back, it was clear NASA really wasn't aware what was going on."

One cog in the Lucid-Gerstenmaier team's smooth-running machine was Boris Goncharov, the joke-loving Russian physicist-cum-translator who had helped Bill Reeves the year before. Because he spoke so little Russian, Gerstenmaier, like Reeves, depended on Goncharov for almost all his communications needs. It was through Goncharov that Gerstenmaier negotiated the comm time he needed to work with Lucid.

One morning that April, Gerstenmaier ran into Goncharov in the hallway outside the NASA suite. The big Russian put an arm around the American's shoulder.

"I have something to tell you," he said. "I am going to die in seven weeks."

Gerstenmaier made a face. He thought Goncharov was joking. "Boris, you're not gonna die."

"No," Goncharov said. "I have a cancer in my kidney. It's the size of a baseball."

Gerstenmaier blanched. "Boris, you *can't* die. You're too young to die. You're going to fight this." He went on like this for a bit, encouraging and supporting his friend. Later he would regard it as a typical upbeat American response: Fight, fight, fight. Win one for the home team.

"No, Bill, it's okay," Goncharov finally said, shaking his head. "My daughter is married now. My work is done here. I am ready. It is time."

Soon after, Goncharov disappeared from the TsUP. When Gerstenmaier asked where he was, he was told Goncharov had entered a cancer institute, purportedly the best in Russia. Gerstenmaier immediately asked to visit. Vladimir Solovyov denied his request, apparently on grounds that the American might say something that discouraged Goncharov. "I decided I didn't care, I was going to go anyway," he remembers. Solovyov relented, saying Gerstenmaier could go, but only with an escort.

Gerstenmaier remembers the hospital as "a really ratty place," dirt and grit encrusted in the walls, old women mopping the floors with rags. In an upstairs room they found Goncharov and his wife. Goncharov looked like a man on the verge of death, ashen-faced and heavily sedated. Nevertheless he insisted on leaving the room to sit with Gerstenmaier in the pale spring sunlight, Goncharov telling all his favorite dirty jokes.

When it came time to leave, Goncharov asked to walk Gerstenmaier down the stairs to the outside door. His wife objected, but the big Russian insisted. Gerstenmaier could see it was important to him. Downstairs they embraced and said their good-byes.

A week later Goncharov was dead. An open-casket memorial service was held in the central courtyard of the TsUP, and Gerstenmaier, to his surprise, was asked to speak. He laid flowers on the casket, an even number of stems, as dictated by Russian tradition. Unsure what to say, he listened to the Russian speakers for clues. As he sat there, all he could think of was how much he needed Goncharov. Boris would tell him what to say and how to say it. Boris would know.

Eventually it was his turn to speak. He decided to keep it simple. "Boris was my friend, but he was more than a friend," he told the little crowd, explaining how valuable Goncharov had been to the American team. "We will miss him."

There was one sad postscript to Lucid's mission to Mir. Her friendship with John Blaha did not survive. While Blaha blamed many of the troubles he experienced aboard the station on Culbertson and NASA management, in time he also came to blame Lucid. In Blaha's eyes her sin was an unwillingness to prepare him for life aboard Mir during their weeklong changeover period that September. Blaha estimates that Lucid spent about an hour briefing him on the scientific procedures she had used during her mission; he would spend ten hours or more with Jerry Linenger. Blaha is not a person given to bitterness, but his feelings toward Lucid run deep. Still, it is almost physically painful for him to discuss his former friend. When the subject is brought up, he cannot bring himself to refer to Lucid by name.

"You can't believe the way I went out of my way for over a year, supporting this person, how many times I defended this person, saying how good this person was, how this person will do well in space," says Blaha. "My negativism toward her only began two weeks after the shuttle left. She used to tell people she didn't have enough to do on the station. That's why I had to spend ten hours putting the cameras together. The longer I was there, I was like, 'Why would somebody do that? To a friend?' I thought: Why didn't my best friend on STS-43 and STS-58, who I had helped in Star City, why didn't she help me? And I could never understand that. No one, including someone who used to be my best buddy, could ever answer that question for me. And she knows that."

Blaha's simmering resentment of Lucid only intensified when he returned to Houston. His angry debriefings, with their personal attacks on Culbertson, contrasted sharply with Lucid's generally upbeat tales of her own time aboard Mir. The problem, of course, was that Lucid's version totally undercut Blaha's: If all the awful things Blaha was saying were true, NASA officials justifiably wondered, why had Lucid thrived?

"That is the greatest dichotomy of my life," says Blaha. "I will never understand it. I know what that person really felt. I know what she told me when we were alone together on the station. I would say that what she told me then was the truth. Why she would tell people things differently when she got back to the ground, I'll never know. She did a disservice to a lot of people that followed behind us. I don't understand why she did what she did. When you don't provide accurate information, then unfortunately those that follow are going to get hurt by it. That occurred.

"I never want to talk to her about it," Blaha continues. "I will never communicate with her again. That is my feeling. I can't look at her and communicate with her, because, I mean, she could never explain to me the questions I would ask her. She could never explain even if she tried."

"I know John didn't talk to me when he was back," Lucid says today, acknowledging the split. "I've heard from someone that John is really, really angry. I'm really sorry about it. I don't know what I would have done differently. I talked to John for hours during the changeover. I tried in every way I knew how to help John." If she spent more time in the shuttle during the changeover, Lucid says, it was only to allow Blaha to forge his own relationship with Korzun and Kaleri. As for her debriefings, she insists she was as truthful as she could be.

"I have to agree with Shannon on this," says Valery Korzun, the Russian commander who observed the Lucid-Blaha changeover firsthand. "The American astronauts didn't understand the importance of the changeover. During the changeover we discussed this with Shannon. Shannon was willing to tell him all these things [he needed to know]. I believe John didn't pay much attention to the details he was supposed to hear from her."

PART THREE

May 18 to September 25, 1997

15

Tuesday
June 24, 1997
TsUP

"Don't worry, it's been real quiet," her predecessor was telling Kate Maliga. "You can go for days without a single reporter calling."

Maliga was NASA's new press contact at the TsUP, a stylish thirty-one-year-old Syracuse University graduate who had just been detailed from headquarters in Washington for a two-year stint in Moscow. Monday had been her first day on the job, and the woman she was replacing, Cathy Watson, had spent the afternoon escorting her through the TsUP's dim hallways, showing her the Baskin-Robbins around the corner, and briefing her on her responsibilities. Things were quiet, Watson said, almost dead. Everything was going smoothly aboard the station, just the normal minor breakdowns and foul-ups—an Elektron stoppage here, a coolant leak there. The American press had all but forgotten about Mir and the new astronaut on board, Mike Foale. In the previous two weeks, Watson said, she had received exactly one phone call of note, from a CBS producer thinking about a piece on the training out at Star City.

Maliga was introduced to the new faces who had replaced Jerry Linenger's ground team a month before. The new ops lead was a moon-faced kid named Keith Zimmerman, a twenty-nine-year-old Texas A&M graduate who had grown up in a Houston suburb near the Johnson Space Center and begun a work-study program there at the age of nineteen. Zimmerman's systems expert, one of three rotating specialists who filled the new position Jim Van Laak had helped create after the fire, was James Medford, a quiet, graying thirty-four-year-old who spent his days cataloging everything that went wrong aboard the station. The new flight surgeon was Terry Taddeo, a thirty-seven-year-old doctor from Cleveland.

There were changes afoot back in Houston too. Though it hadn't yet been announced, Frank Culbertson was preparing to step down as head of the Phase One program. Culbertson had his eye on commanding the last shuttle voyage to Mir in the spring of 1998 and was scheduled to enter training for the flight that fall. Charlie Precourt, who commanded the mission that returned Linenger to Earth a month before, was the leading candidate to replace him.

Tuesday, Maliga's first full day on the job, she woke at her new apartment in downtown Moscow, dressed, and strode to the door to catch the NASA bus to the TsUP. The door was jammed; it wouldn't open. It took her fifteen minutes to force it, by which time she had missed her van. She got the next one, but when she finally arrived at the TsUP, she realized she had forgotten how to find the NASA suite in the building's labyrinthine corridors. She wandered aimlessly until a Russian worker took pity on her and, in broken English, directed her to her office. Hurrying down a set of darkened stairs, Maliga tripped and fell, badly bruising her knee. She finally made it to her office, limping, late, and thoroughly humbled. She told herself it didn't matter. Things were quiet. They were almost dead.

Jerry Linenger returned to Houston an angry man. Like Thagard and Blaha before him, he was a mass of conflicted feelings, jubilant at seeing his family but furious at the elements within NASA he felt had abandoned and ignored him on the station. Unlike Thagard and Blaha, Linenger was also intensely critical of the Russians; the Mir station he had experienced was falling apart around him, and he had been repeatedly surprised at both Tsibliyev's and the TsUP's ignorance of the station's primary systems. The Russians more than returned the feeling, leaking several critical comments about Linenger's performance to Moscow newspapers. "Linenger had a nervous breakdown," one anonymous TsUP official sniped to a Russian reporter. "He stopped talking with our guys. He locked himself up in the Spektr module . . . He said it openly: 'The station is dangerous. Get me out of here. And don't bring Foale until the Russians put everything in order.' "

Thagard had channeled his anger at George Abbey; Blaha wanted to throw Culbertson in jail. As he swam long, slow laps in the backyard pool at his house on Armand Bayou—a part of his physical therapy—Linenger developed far more elaborate systems of blame. He began muttering darkly about conspiracies involving Russian and American politicians, suggesting that someone in one—or maybe *both*—of the space programs had attempted to cover up his fire report. He knew very little about the Indian missile compromise, but what he heard led him to draw comparisons to past political scandals. "It's almost scary," he told one visitor. "It's like an Iran-Contra type thing." He felt he and Tsibliyev and Lazutkin had been guinea pigs the politicians were prepared to sacrifice in the name of world peace, and while he was careful never to say this kind of thing around reporters, word soon spread through JSC that Linenger was shaping up to be a problem.

Jim Van Laak was among the first to sense it. As Culbertson's self-described "wing man," Van Laak was capable of savaging anyone who criticized the Phase One program. There had been few signs of Linenger's feelings in his initial debriefing, a private session with Culbertson and Van Laak in Culbertson's fifth-floor corner office in Building One. Linenger stuck to the facts and kept his opinions mostly to himself. But Van Laak made a point of attending the astronaut's first major debriefing, on Friday, June 6, when Linenger stood to address a group of fifty or so NASA managers in a Building Four conference room on the subject of his spacewalk.

Astronaut debriefs are typically intensely civil, highly technical affairs. But from the moment he walked before the crowd that day, Linenger's presentation was sharply different in tone. Mir's exterior hatch, he announced at the outset, was "jury-rigged," held together by a pair of C-clamps. Russian hydrolab procedures were "ridiculous." The entire EVA process was "winging it." Support documents were "terrible . . . almost unusable." Mir solar arrays had "razor-sharp" edges. He characterized Tsibliyev as all but useless in matters of repair.

"I'd say, 'Well, what—why are we turning this valve today, and what line are we trying to isolate?' " Linenger said. "[He would say] 'I don't know.'

" 'Is this the leak that affects Elektron systems?'

" 'Who knows? The ground—they don't tell me anything.' "

Linenger went on. "They'd call for commands of 'Alpha One-One.' I'd say, 'What was that?'

"[He would say] 'I don't know, some kind of orientation.'

"I'd say, 'I'm not sure what they're doing.'

"It was really surprising to me that really no one on the station has a concept of what the ground's trying to do. And the ground doesn't know what they're doing a lot of times is my general impression."

"And that's very different from the U.S. experience?" someone asked.

"Oh, a hundred percent," Linenger said. "You could at least chastise the commander if [you] knew what was going on. Or somebody on board knew what was going on. But they didn't really care if the people on board knew what was going on. And again, like with cooling loop leaks, I don't think they knew the system, basically. They'd tell you, 'Well, go look over there. Is there a pipe there?' "

Sitting in the audience, Van Laak was growing madder by the moment. This was a Jerry Linenger he hadn't seen before; this was a man with an audience, pacing and gesturing theatrically, making it sound as if he had survived a descent to the ninth circle of Hell. This was the Linenger that his classmates had tagged "Hollywood." "Jerry was in his acting mode," Van Laak recalls. "He was being very melodramatic. Intuitively, I had great confidence he was blowing things way out of proportion. I briefly considered standing up and shouting, 'Jerry, cut the bullshit!' " But as Van Laak well knew, the worst thing a NASA staffer could do was to be seen stifling an astronaut's freedom of expression.

He took a deep breath and simmered. Van Laak simply didn't believe there were shaky clamps on the hatch. It sounded too far-fetched, even for the Russians. Linenger's use of terms like "razor-sharp" was infuriating. No one who had ever been outside Mir had used words like that. Van Laak wasn't the only one surprised by Linenger's performance. "Jerry stepped over the line," remembers astronaut Rich Clifford, who had EVA'd outside Mir on STS-76 and who sat in the audience that day. "If he had given those concerns to Phase One in a private forum, it would've been okay. But this was practically a public forum. Turning it into a bitch-and-moan session is not something you do. It's like in the military, where the first rule is, 'Don't embarrass the boss.' He embarrassed the boss."

When Linenger finished the question-and-answer session, there was a lot of backslapping and "way to go"s and good feeling. But not from Van Laak, who returned to his office and faced a decision regarding how to deal with what Linenger had said. In the following days Van Laak made halfhearted attempts to learn more about the hatch situation. He put in a call to the Russian liaison officer at JSC, Vladimir Dezhurov—the same Dezhurov who had commanded Thagard's mission—but Dezhurov's English wasn't very good, and he seemed unable to fathom what Van Laak was saying. The conversation went nowhere. Finally, Van Laak decided to go with his gut feeling and simply ignore Linenger's warnings.

"We didn't believe what Jerry said, those of us with exposure to EVA stuff," Van Laak recalls. "We weren't planning to do another EVA, so this was no big deal." Van Laak then took a step to make sure it wouldn't become a big deal. He directed that no written transcripts or notes of Linenger's debriefing session be circulated. That way he could be reasonably certain the contents of the debriefing wouldn't leak to the press. "I knew there was a kernel of truth in what he was saying," says Van Laak. "I knew there were gross violations of our standards. But in terms of safety, I thought we were still okay."

At the same time, Van Laak began keeping track of the handful of interviews Linenger was starting to give—to CNN, to *Reader's Digest,* to *Vanity Fair,* to *People* magazine. The public affairs office was excited at the prospect of Linenger's dramatic, first-person account of "surviving" the fire making *People*'s cover; instead it was buried on page 107. Later that summer, when events in space put Linenger in greater demand, Van Laak actively discouraged reporters from approaching him. He suggested that Linenger hyped and changed his stories to make them more titillating. Linenger's motive, Van Laak said, was to build up his public image in order to retire from NASA and make a run for Congress in his native Michigan.

"Now, you talk to Jerry, he tells a story about Lazutkin's eyes swelling up in his head like golf balls," Van Laak scoffed to one journalist.* "The Russians will tell you the guy had allergy problems." In fact, both Lazutkin and his Russian superiors attributed the eye incident to contact with ethylene glycol. If a reporter pressed, Van Laak was prepared to go for the jugular. "In my opinion, Jerry Linenger does not have the right stuff," Van Laak said. "The fire was no big deal. I've been in three burning airplanes in my life. It's a serious incident, sure. But you go on. That's just the way life is."

These were cracks that by the end of the summer would grow into a chasm dividing Linenger and the Phase One program—a chasm that seemed to widen every time Linenger opened his mouth. On Monday morning, June 16, ten days after his EVA debriefing, Linenger stood up after the Astronaut Office's weekly meeting and gave his Mir debrief to the corps. Several astronauts who were in the crowd say it was as caustic and withering a presentation as they had ever witnessed. Linenger's stories seemed to grow more vivid with each retelling. To his description of the "winged" spacewalk he now added dramatic accounts of the Near Miss and the fire. In no uncertain terms he said Mir was unsafe.

Sitting in the audience, Blaine Hammond was appalled. The man his

*The author.

"Maggot" classmates had long ridiculed as *Blaine Hammond, World-Famous Astronaut,* was winding down his days at NASA in preparation for retirement, probably at year end. Like others at NASA, Hammond didn't much care for Linenger, especially having flown with him on STS-64 and glimpsing firsthand his willful ways. But as the Astronaut Office's safety officer, he was not concerned with the style of Linenger's presentation. He paid attention only to the facts, and Hammond found the facts alarming. *C-clamps on the hatch? An out-of-control Progress? Fire extinguishers locked to the wall?* Why hadn't they heard this before? In thirteen years at NASA, Hammond had never listened to an astronaut speak more candidly about how frightened he had been in space, or how dangerous a spacecraft had become.

Like other astronauts, Hammond had harbored reservations about the Mir program from the beginning. As far as safety went, he felt it was an impossible situation. NASA was being asked to guarantee safety on a station it knew next to nothing about. Astronaut training at Star City was rudimentary at best. Hammond, like so many other astronauts, tended to roll his eyes at the very idea of the program. It was a Washington boondoggle, a political thing. A George Abbey project. And that, of course, meant that no one—*no one*—would speak up against it. At least not publicly.

The next day Hammond took his notes from Linenger's debriefing to a Preflight Assessment Review for STS-94, a shuttle mission scheduled to launch on July 1. A dozen or so people were there from JSC; others were conferenced in from Washington, Kennedy, and the Marshall Space Flight Center in Huntsville, Alabama. Fred Gregory, the former astronaut who headed the safety office at headquarters, opened the meeting with a side issue, asking people how they felt about what they had learned of the fire.

When it was his turn to speak, Hammond reviewed the substance of Linenger's debriefing the day before. "The things Jerry is saying are very telling," Hammond said. "Jerry is very, very upset. He doesn't think this is a worthwhile program. He says the Russians will never abandon Mir. He said he talked to the commander, and he said they would never abandon Mir. Never, under any circumstances. We can keep saying it's safe because they have the Soyuz up there, but if they won't use the thing, it might as well not be there. My biggest concern is that I don't think the Russians will ever leave, no matter what happens."

Gregory politely cut him off. "Well, that's a policy matter," he said. "That's not for us to decide here."

"Well, I disagree," said Hammond. "I think it is our concern. If we're basing our thoughts on whether it's safe up there, we have to believe they'll abandon it. And I don't think they will. Jerry says they won't."

Gregory wouldn't pursue it. This was a policy matter, he insisted, and therefore out of their purview.

For the moment, Hammond let the matter drop. He had a feeling, though, that it wouldn't be the last time the subject would come up.

When Mike Foale crawled through the Kristall docking port into Mir on the morning of May 17, one of the first things he did was peer intently into Tsibliyev's and Lazutkin's faces. He had heard the rumors about bulging eyeballs, nasal irritation, and skin burns. But looking into the cosmonauts' faces, he saw nothing out of the ordinary. Both men seemed tired, but other than that they were the same men Foale had known in Star City.

Technically and emotionally, forty-year-old C. Michael Foale was destined to be the best-prepared of all the American astronauts who journeyed to Mir. His conversational Russian was excellent, far better than that of Thagard or Blaha. His easy laugh, gregarious charm, and overall flexibility were rivaled only by Shannon Lucid's. He was, in short, everything Linenger wasn't—open, friendly, easy to know. Foale had grown up the son of a Royal Air Force pilot on air bases all around England. His mother, Mary, was American, a Minneapolis girl who had met Foale's father, Colin Foale, after crossing the Atlantic on the *Queen Mary* during a family vacation in the 1950s. She emigrated to marry him, and when their son, Michael, was born, in 1957, he was registered as a U.S. citizen at the American Embassy in London. When Foale was three his father was briefly transferred to a British air base in Cyprus, and the pilot of the military transport jet spiriting the family across Europe briefly gave him the controls over the Alps; from that day on, all the little boy could talk about was becoming a jet pilot. A few years later Foale refined the dream during one of the family's biannual trips to visit his mother's family in Minnesota. There at a state fair he saw the blackened space capsule in which John Glenn had returned to Earth. By the time Mike Foale was nine, his dream was set: He would become a fighter pilot, then an astronaut.

He read everything he could on space: newspapers, magazines, a book by Wernher von Braun. But when Foale, at sixteen, was a student at the prestigious King's School in Canterbury, his dreams received a crippling blow. Undergoing an RAF physical with thoughts of winning a military scholarship to Cambridge, he failed the eye test. A tiny defect in his vision, though it did not require his use of glasses, prevented his qualifying as a fighter pilot. For years afterward Foale's ambitions wandered. At Cambridge he majored in physics, figuring he would do something in geology or geophysics. He didn't know what, and for the first two years of university he didn't really care, neglecting his schoolwork in favor of parties and rowing races and fooling around. But then, in his final year of undergraduate studies, Foale retook the RAF eye test on a whim and, to his surprise, passed. The defect was gone, and a recruitment officer urged him to sign up immediately.

Foale thought long and hard before declining. By that point he had read that NASA was creating a new class of astronauts called mission specialists, who didn't have to be pilots. The RAF thus seemed a roundabout route to space. His dream resurrected, Foale bore down on his classes, earning a master's in physics, and forged ahead into doctoral studies, specializing in laboratory astrophysics. He was an adventure lover, signing on for two expeditions that surveyed underwater antiquities, one off the Greek coast, the other off the Isle of Wight. But his old dream remained, and when a NASA delegation passed through Cambridge in 1981 Foale arranged to meet some of the visiting Americans at a London pub. He struck up a conversation with a JSC man named Andy Hoboken, and when Hoboken casually mentioned that he should come to Houston someday, Foale leaped at the chance, jumping on a Braniff flight straight to Texas. Hoboken set up an interview with one of George Abbey's men, and Foale was impressive enough to earn an appointment to see Abbey himself. Foale had no idea who or how powerful Abbey was, but he listened when Abbey urged him to give NASA a call after finishing his doctorate.

Foale appeared in Houston a year later, only to find NASA in the midst of a Reagan-era hiring freeze. Disappointed, he took a job in shuttle payload operations at McDonnell Douglas, which got him one step closer to NASA. In June 1983 he wangled his way into a payload officer job at MOD; working out of Mission Control, he supervised payload operations on four shuttle missions. But he still wasn't an astronaut. He interviewed for the astronaut corps in 1983 and was turned down. Too young, they said. Two years later, he was again rejected.

By 1986, the year Challenger exploded, throwing NASA into turmoil,

Foale's boyhood dream was slowly dying. He had time on his hands, which he used to become a top-notch Windsurfer, racing in national and international competitions. At twenty-nine he had the lean, tight body and boyish, sunburned look of a surfer, and his warm blue eyes and British accent earned him a reputation as a bit of a playboy. His skirt-chasing days ended shortly after he spotted the lead singer of a local party band called the Kitty Projectiles. She was a Louisville girl named Rhonda Butler, she was dressed like Madonna, and she wouldn't give him the time of day. A few weeks later he ran into her in a crowd at the Houston Music Festival, and this time they clicked. He taught her Windsurfing and helped get her a job at NASA. They married soon after.

By early 1987 Foale was preparing to abandon NASA. Before resigning, he went in for his third and final astronaut interview, telling anyone who asked that he was leaving NASA for good and was no longer all that enthralled by the notion of becoming an astronaut. He didn't know it at the time, but it was exactly the kind of talk NASA recruiters liked to hear; the agency learned long ago that starstruck candidates often made poor astronauts. Much to Foale's surprise, NASA wrecked his plans by accepting him as an astronaut candidate with Group 12 on June 5, 1987.

He was soon identified as one of the Astronaut Office's up-and-comers, a capable, brainy spacewalk enthusiast. Between 1992 and 1995 he flew three shuttle missions in quick succession; between flights he served stints as deputy chief of mission development and head of the office's Science Support Group. He began hearing the calls for Mir volunteers in 1993 and ignored them. "In all honesty, I just thought long-duration flight was going to be too hellish," he recalls. "It didn't appeal to me at all." When Bob Cabana or anyone from the Astronaut Office hinted that he sign up, Foale made clear he had no interest. Even if he wanted to stay aboard Mir, family considerations all but prevented it: Rhonda had quit her NASA job to be a full-time mother, their two children were toddlers, and Foale had never been away from his family for more than two weeks. Bonnie Dunbar had warned everyone that Star City was no place for young children.

By the summer of 1995 Foale had his sights focused squarely on STS-86, which was to feature a pair of joint Russian-American EVAs to install a power module on Mir's hull. Foale was already working with the Russians supervising STS-86's EVA planning and expected to be named to its crew. Then the astronaut Scott Parazynski, who was found to be too tall to fit comfortably into the Russian Orlan space suits, was removed from his

Phase One increment. Foale, already deeply involved with the Russians in the planning for STS-86, knew he was an obvious candidate to replace Parazynski. "I could feel the searchlight going around," he recalls, "and I could feel it settling on me."

By the time rumors began to circulate that he would be asked, Foale had all but resigned himself to accepting the assignment, telling Rhonda it would be necessary to take a bullet for the program. Then, on a trip to Star City in October 1995 to work on the STS-86 EVAs, one of the Russian secretaries made an offhand remark that spun his head around.

"Oh, Michael," the woman said, "we're so looking forward to seeing you next month when you begin training."

Foale was astounded. "What?" he said.

Even before the conversation ended, Foale knew the searchlight had found him. According to the secretary, he was already scheduled for reassignment to Star City, and no one—not Culbertson, not Cabana, not George Abbey—had even bothered to ask him. "I was *pissed!*" Foale recalls. But he knew it was useless to object. Before talking to anyone at NASA, he called his wife from Russia and said, "Rhonda, it looks like we've been sold down the river." The Foales were going to Star City, like it or not. They arrived on November 22 and moved into the Prophy. The long-delayed American condominiums still weren't ready, and the two Foale children ran rampant through the Prophy for the next several months. Foale got the distinct impression that General Klimuk ordered the condos completed in large part to rid the Prophy of his two energetic toddlers.

During Linenger's mission Foale closely followed the difficulties the American had with his two crewmates. Having spent many hours with Linenger during training, Foale wasn't surprised by the situation. "I knew Jerry was doing the same thing the Russians were doing to us, putting his hands up and pushing them away," Foale recalls. He felt Linenger's travails were simply the worst of a line of dysfunctional crew relationships. None of the Americans—not Thagard, Lucid, or Blaha—had managed to become fully integrated into a Russian crew. Foale was determined to change that. Science was important, yes. He would tend the greenhouse and bake his alloys and take his snapshots of Earth. But the more Foale thought about it, the more he wanted his top priority to be amazingly simple: He wanted to make friends.

Language was the key. Blaha and Lucid had both emphasized that the biggest barrier to forging solid relationships with their crewmates was poor language skills. Early in his training at Star City, Foale noticed how at the end of a long day the other Americans returned to speaking English with their friends and family. He soon decided to stop speaking English whenever he could, even at home. He spent every last minute of spare time reading and speaking Russian. He bought Russian children's books and read them aloud. His kids hated it, but it helped him.

What he really needed, Foale saw, was total immersion, which was difficult with his family around. Ironically he got the chance in March 1996, when he joined a Russian delegation for a twenty-one-day visit to Houston. Tsibliyev and Lazutkin and dozens of other Russians took over the Residence Inn on Bay Area Boulevard, and Foale, his family left behind in Russia, stayed with them, sitting up till all hours chatting and telling jokes and drinking massive quantities of vodka. For three weeks Foale drank and drank and drank, speaking nothing but Russian. By the time he returned to Star City his social Russian was better than he'd ever expected. Where Thagard had avoided telling jokes for fear of offending the Russians, Foale plunged right in, keeping up with his growing list of Russian friends joke for joke, vodka for vodka.

Back at Star City, he began cultivating Sergei Krikalev and his wife, Yelena. Alone among the active cosmonauts, Krikalev invited the Americans into their flat. The Krikalevs were the first of his Russian friends to begin addressing Foale in Russian, and when Rhonda and the children returned to Houston that summer, Foale began staying with the Krikalevs at their dacha on weekends, speaking only Russian. Much as Valery Korzun was perceived as the most Westernized of the Russians, Foale soon established a reputation as the most Russianized of the Americans.

It wasn't true. The more Foale learned of the Russian culture, in fact, the more he loathed it, at least the enormous segments that hadn't changed since the fall of the Soviet Union. "I hate—absolutely hate—the Russian Communist culture of male power and control," he says today. "This doesn't go back eighty years, to the Revolution. This goes back thousands of years. It's the same culture, of a Russian male oligarchy. Their culture hasn't changed. It's just evolved. It's a handful of powerful men controlling everything." Foale found hopelessly naive NASA's vision of establishing with the Russians a joint International Space Station organization that would be entrepreneurial and responsive to the needs of its workers. "The Russians are *never* going to do that," he says. "They are total masters, in maneuvering, in bargaining. All they want is to be masters. It's almost a game to them."

For hundreds of years the czars had overseen a vast feudal system, with landed barons controlling the lives of millions of poverty-stricken serfs. Foale thought he glimpsed the legacy of this system at work in the space program. At the TsUP, the flight controllers at their brightly lit consoles were the unchallenged masters. Their serfs—their *slaves*—were in space. That was the word Foale began using: slaves. The cosmonauts were slaves.

Foale traced his thinking to his first trip to the TsUP in 1993, when he had seen in Viktor Blagov's office a poster that portrayed Mir as a puppet controlled by the strings of Mission Control. He had been standing with shuttle commander Jim Wetherbee.

"Did you see that, Jim?" Foale said when they left Blagov's office.

"Yeah." Wetherbee smiled.

Foale turned to Blagov, who was walking with them. "That's an interesting poster," he observed.

Blagov, too, smiled—"a supercilious little smile," as Foale remembered it.

"I mean, that's not really kind to the cosmonauts," Foale went on. "Surely the cosmonauts would be offended by this."

"Well, no," Blagov said, uncomprehending. "That's just the way things are."

Later Foale mentioned the poster to Krikalev.

"Don't you find that offensive?" he asked.

"No," Krikalev replied. "That's the way it is."

By the time he arrived aboard Mir, Foale had expanded his goals of befriending and integrating himself with the cosmonauts to include a bit of subversion. He wanted to see whether he could get Tsibliyev and Lazutkin to think for themselves, to act independently of their TsUP controllers. He began his campaign while the shuttle was still docked to Mir. When he had to use the bathroom, he made sure to use the one on the station, in the Kvant 2 module. It was a small thing, but he wanted to convey to the two Russians his eagerness to become a member of their crew. When the shuttle undocked on May 21 and he formally joined the Mir 23 crew, he made sure he ate every meal with Tsibliyev and Lazutkin and, more important, attended every comm pass. Linenger hadn't, and thus had cut himself off from Mir's daily routine.

"We noticed these little things," remembers Lazutkin. "What was different about Mike was he started studying everything right away. He said, 'Show me how everything works.' And we gave him things to do. He asked all the right questions. Mike amazed us by mastering [systems] theory sometimes better than we knew it. He was like a child, and he was growing, and he grew into a real cosmonaut engineer."

"When Michael first came on the station," Tsibliyev recalls, "the three of us were sitting around the table, and Michael said, 'I came here to make you happy with me.' He didn't say he came to make himself happy. He came to make *us* happy. The fact that he integrated into the crew was because of these qualities." Like Korzun before him, Foale noticed that Tsibliyev was far more stern in orbit than he had been in Star City. Foale wasn't surprised. "I'm used to the two-facedness, I'm used to this schizophrenia," he recalls. "I saw it and I expected it. And I knew what it was caused by. That poster. When you're forced to be a slave, that slave generally looks around for someone to be a master to."

Foale was willing to become Tsibliyev's slave, if that was the best way to win the commander's confidence. From the outset, he saw it wouldn't be easy. Idle chitchat, usually over dinner or in the slower hours before bedtime, provided Foale with his best opportunity to work himself into Tsibliyev's favor. He tried engaging him in conversation about all manner of subjects, but the two men had little in common. They couldn't talk about Windsurfing or physics or politics. Somehow Foale had to find a way to draw him out. He decided to bore in on the one subject everyone had in common, childhood and family life. Lazutkin opened up quickly, but then the gentle engineer was an easy mark, friendly with most of the Americans. Tsibliyev was a tougher nut to crack, even when it came to talking about his family. The commander harbored memories of his Crimean childhood that he wasn't eager to share.

In the end Foale realized there was only one topic with which he could engage the commander. "There's almost *nothing* in common between us," Foale recalls, "except the master-slave problem."

Foale decided to give Tsibliyev what he came to think of as the Poster Test. "Don't you find that offensive?" he asked.

Not really, Tsibliyev said, not taking the bait.

"I find it incredibly offensive," Foale said, pushing a bit. "But the fact that you don't is what really bothers me."

It didn't bother Lazutkin at all. Lazutkin in fact remained the loyal company man, the rookie who refused to believe his bosses at the TsUP were anything other than thorough, flexible professionals. Later, after the mission, Foale and Tsibliyev would disagree on the extent to which the commander had bought into Foale's "subversive" line of thinking. "I remember this talk," says Tsibliyev. "I remember I told Michael we do have enough independence. But when we do repair jobs, and other important matters, we have to discuss it with the TsUP, because you have to understand that we are not working on our private automobiles, we are dealing with expensive equipment, important to the whole country. It is an enormous responsibility."

Foale tells a different story. "Vasily would respond to these issues about the TsUP, and I would play on that," he recalls. "He seemed to really understand what I was saying. Later, when Vasily began to vent at the ground, to me it looked as if his rebelliousness was an expression of our conversations. I saw that Vasily's relationship with the TsUP was in trouble. I realized he would suffer after that flight because of it. I then got worried I had gone too far."

Tsibliyev and Lazutkin were thrilled to be rid of Linenger. But as nice as Foale seemed, the fact was they didn't have time to dwell on it. Although the press hadn't picked up on it, the coolant leaks had renewed themselves with a vengeance, and by late May the two weary Russians were staying up past midnight most nights, nosing around behind the panels in base block and Kvant in search of their origins.

"We were tired," recalls Lazutkin. "We were like little hamsters that had to be doing everything. We were tired psychologically. We were exhausted. We couldn't find the leaks. And the fact that we couldn't find the leaks put constant psychological pressure on us. That's why Vasily was so irritated all the time."

Tsibliyev's mood, which had improved only slightly with Linenger's departure, darkened progressively as June wore on, especially after something that happened on June 5. It had been another grueling day spent in search of leaks. That night Tsibliyev and Lazutkin once again stayed up late, running a pump to force the air out of one of the base block coolant pipes. The ground had estimated it would take forty-five minutes to an hour to pump all the air out of the system, but after an hour and a half the two exhausted Russians were still pumping.

"Where are we getting all this air?" an exasperated Tsibliyev finally asked. "This is impossible. We've got to be sucking in air from somewhere. Where is it?" He motioned toward the dimly lit Kvant module on the other side of the dinner table. "Sasha, will you go look in there?"

Lazutkin began to float toward Kvant, but Tsibliyev stopped him.

"Sasha, wait—you stay here. Someone's got to watch this while the pump's going. I'll go check."

During their first weeks aboard Mir, most astronauts and cosmonauts tend to fly with their arms extended before them, like Superman. After a while they begin to get the hang of weightlessness, and almost all take to flying headfirst. Tsibliyev had done so early on, much to his regret that night. He pushed off from where he was working with Lazutkin and managed a graceful arc to the dinner table. Reaching the table, he braced his feet on the far side and pushed hard, setting a trajectory that would take him straight into the mouth of Kvant, headfirst. It was a maneuver he had managed literally thousands of times, and he did it without thinking. But then, just as he pushed off, he saw something he never expected to see. It was like a scene from a science fiction movie.

There, hanging in midair like some greasy toxic balloon, was the largest single drop of ethylene glycol he had ever seen. It was the size of an overinflated basketball, and it floated directly in his flight path. There was nothing Tsibliyev could do to avoid it. Frog-kicking his elbows and legs in a frantic but futile attempt to brake his speed, he hit the giant drop head-on. Its oily edges slathered his face and hair in a noxious embrace, coating his nose, mouth, cheeks and ears. For the longest moment he hung there in midair, clawing at his face.

"Sasha! Please, the towel!" he sputtered.

Lazutkin was at his side in no time with a wet towel and a dry one, and Tsibliyev spent the next half hour washing and wiping his features.

"Thank God," he breathed, "I at least managed to get my eyes closed."

The next morning Tsibliyev awoke in his *kayutka* feeling as if he had eaten a six-course dinner of antifreeze. His stomach felt heavy and queasy; just the sight of food made him nauseous. His eyes were worse. Even though he thought he had closed them in time, his vision was so blurry he could barely see. "I sat at the computer and I couldn't see anything," he remembered months later.

All that day he barely spoke. He swam about lethargically, as if the station's atmosphere had been replaced with chocolate pudding. "He wouldn't eat," remembers Lazutkin. "It was difficult for him to speak. Psychologically, he was just beat. Withdrawn. He didn't talk."

The next day was no better. It was as if the accumulated stresses of the last four months—the fire, the Near Miss, the leaks, the all-nighters spent repairing pipes, the EVA, the tensions with Linenger—had all crested at once in a tidal wave of exhaustion that simply inundated him.

He was a vegetable. Lazutkin watched him all day and worried. Once again they queried the ground about the effects of ethylene glycol. The answer they received, to no one's surprise, was not to worry. Linenger had been tested for ethylene glycol poisoning upon return to Houston. *"Jerry showed perfect results,"* a TsUP controller told Tsibliyev on June 9.

"That's no surprise to us," the commander replied. *"He wasn't exposed to anything, because he didn't do any systems work in contaminated areas and exercised like hell all the time."*

Tsibliyev was slow to recover physically, even slower emotionally. On the third day he managed to keep down some milk. The good news was that in the meantime Lazutkin found the leak, a crack in a Kvant module pipe. He donned a surgical mask and succeeded in cutting out the corroded pipe and replacing it with a rubber hose.

"That's great news," the TsUP radioed up. *"What you have found is the crack that has given so many headaches in the last three months. So it looks like this is the end!"*

The repairs left Kvant a mess, and it took days to clean, a period in which Tsibliyev's mood deteriorated further. Then, on June 12, with the last repairs complete, the schedule dictated that they begin a new experiment, a demanding twelve-night sleep study. Tsibliyev dreaded it. According to the guidelines laid down by the Russian doctors, he and Lazutkin were to sleep every night from 11 P.M. to 8 A.M. wearing a sky-blue cloth cap fitted with various sensors, one of which jutted down over the wearer's left eyelid and looked like a tiny golf putter; it was intended to measure the cosmonauts' eye movements during sleep, but it was so intrusive it interrupted their sleep, an experimental anomaly NASA's Al Holland had ruefully noted. For two of the twelve nights Tsibliyev and Lazutkin were to add more gadgets beneath the cap, slathering gel onto their hair in various spots and affixing electronic sensors to their scalps. Both men had accomplished this unruly setup in run-throughs on the ground, and ended up looking like a pair of spastic rock singers.

The experiment was to climax on one terrible night. While wearing the cap, the hair gel, and the entire inventory of sensors, the two Russians were to sleep with a blood-drawing catheter inserted into veins in their arms; they were to awaken every thirty minutes and draw blood. Tsibliyev drew the line at this. After all they had endured, it was just too much.

"What if something happens, some emergency situation, what am I supposed to do with people in caps and catheters?" he told the doctors. *"How could I evacuate or take care of them quickly?"*

To his surprise, the doctors backed down—but only after appeals from "Steve" Bogdashevsky, the Star City psychologist. "Don't you understand, you can't have them do these experiments now," Bogdashevsky remembers pleading with the doctors. "They can't. They're just too tired." The doctors told Tsibliyev that he and Lazutkin would still wear the catheters but take only four blood draws during the night. It could have been much worse, Tsibliyev knew. For another set of studies, cosmonauts are ordered to sleep—or, more accurately, attempt to sleep—wearing what is called a "core temperature monitor." The penny-size probe is inserted into the anus.

By his own reckoning, Tsibliyev hadn't had a full night's sleep in weeks, and once the sleep study began he got even less. *"The sleep study only cause[s] insomnia and totally screws up the sleep cycle, which has been shaky enough even without any procedures,"* he complained eight days into the experiment, according to the NASA logs. *"Twelve days of sleep study is just way too much, way too much."* At one point, Tsibliyev all but begged the ground to cancel the last several days of the experiment, griping that he was getting no sleep at all. The shift flight director, in tried-and-true TsUP form, said he would think about it.

It was all too much for the bone-weary commander—the lingering antifreeze hangover, the exhaustion, the sleepless nights. By the week of June 16, Tsibliyev had become a growling Russian bear, snapping at every technician, doctor, and specialist the TsUP put on comm. If before he had been irritable, he now became combative and argumentative. On Saturday, June 14, he angrily told the TsUP that he "will not write down his overtime any more, because this is becoming ridiculous and we work *all the time*" (emphasis added in NASA log). By his count he and Lazutkin had gotten exactly two days off in the last four months.

A typical exchange came the following Thursday morning, when the TsUP drew Tsibliyev's ire by asking for the serial numbers of some missing items. At 0841 the NASA logs report: "Crew is saying they do not have the habit of discarding stuff without ground's prior approval, therefore the possibility that they could have trashed something useful should not even be a consideration."

0843: Silence on comm. "Crew is going through their notes . . ."

0847: More silence. "Crew promises to brush through their records at their leisure, 'cause they 'don't have anything else to do anyway.' "

Friday morning, June 20, Tsibliyev again awakened in a foul mood, spending most of a late-morning comm pass barking at the ground. "Vasily is pissed," the NASA logs note at 10:59. "[He] tells ground that the 'had to be done yesterday' philosophy doesn't work in [an] environment where you cannot find anything. The situation is seriously aggravated by things like tomorrow's MK-108 monitored exercise [experiment], 'cause they are definitely not in the best shape in the world. Exercise was not done yesterday—crew was in no mood for that, were way too frustrated to even step on the treadmill."

By noon Vladimir Solovyov had had enough. The TsUP's senior flight director summoned the general who ran Star City, the former cosmonaut Pyotr Klimuk, and asked him to stand by on a telephone for the two o'clock comm pass. According to the NASA logs, Tsibliyev was clearly surprised to hear Klimuk come on the line but began complaining about the sleep study anyway.

Klimuk abruptly cut off his commander.

"A job is a job," the general told Tsibliyev, according to the NASA logs. "You went up there to work, not to relax and have fun. It wasn't a vacation when I was up there in the '70s, you know [it was tough]. Be tough and hang in there. Difficulties come with the job. Don't take it personally. That's the way it works in our business."

"Okay," a chastened Tsibliyev replied.

Klimuk signed off.

When Tsibliyev was certain the general was gone, he blasted the ground once again. "Vasily sounds pissed," the NASA logs note. "Vasily promises he'll never ever complain about anything again and says he did not appreciate ground folks setting him up in front of General Klimuk."

Mark Severance, the young MOD engineer who had orchestrated the Dryden-Wallops upgrades, was at the NASA console for the episode. When the comm pass ended, he slipped out without saying anything. "I was a little shocked," he remembers. "Basically Vasily got a royal ass-chewing. We really don't talk with the Russians about that kind of stuff. It's considered their business. Militarily it was probably the right thing to do, but when you have somebody in the isolation of a space station, it's probably not." Keith Zimmerman, who was also there, agrees: "It was a very Russian way to deal with the problem."

The Americans, too, were becoming alarmed by Tsibliyev's mental state. Back in Houston, Al Holland grew worried after talking with Linenger. "Vasily is right on the edge," Linenger told Holland. "These guys need a break. They're just exhausted. They need to come home." Holland put in calls to his IBMP counterparts in Moscow, but to a man the Russians said everything aboard Mir was *normalno*—normal, typical. Frank Culbertson was worried enough to call Foale's flight surgeon, Terry Taddeo, and ask him to get the Russian doctors to intervene. Maybe, Culbertson suggested, the doctors could get on comm and pet Tsibliyev a little, tell him what a great job he was doing.

Taddeo tried. "I talked to them twice, and I swear to God, I got blank stares," Taddeo remembers. "They just don't know how to convey encouragement. They don't do that. Finally I told Frank I wasn't getting anywhere. He suggested *I* do it. So I did. I got on comm and told the guys what a great job they were doing. I don't think it made much of an impression either way."

"When is the end of the sleep experiment?" Tsibliyev asked the ground Saturday afternoon, June 21.

"Morning of the 24th is the last day."

"What do you think about this [*last few days*]. *Maybe we can cancel it."*

"Let's not be in a hurry," the ground replied, putting off the weary commander.

It was at this point, with Tsibliyev so exhausted and quarrelsome that his commanding general was obliged to intervene, that the TsUP began preparations for a second attempt at the hair-raising manual docking exercise that had gone awry on March 4. It was to prove a fateful decision. The factors underlying the test hadn't changed. The Russians were still running low on the Kurs automated docking units, with no prospect of solving the impasse with the Ukrainians. They simply had to find a way for the cosmonauts to dock incoming ships themselves.

The decision to proceed with the test, however, was far from unanimous. In the wide hallways that led from the TsUP auditorium to the building's two-story break area, technicians and doctors openly questioned whether Tsibliyev was still capable of managing the difficult test. One of the few to voice his opposition formally was Sergei Krikalev, the popular young cosmonaut who had been the first Russian to fly aboard the shuttle, in 1994, and who was working as Blagov's deputy for the Mir 23 mission.

"I was against it, but I was not heard," recalls Krikalev. "The crew had made more mistakes than average, and my feeling was they were not comfortable with this operation, this test. Just listening to the crew, you could *feel* they were uncomfortable. In my experience, if a pilot doesn't feel comfortable to fly, it's better not to fly. But this was on the level of human feeling, of intuition, it was nothing concrete. Many specialists were against the test, but they would not speak out. I told this to Blagov. His answer was, 'We don't have any official or formal reason to change the flight plan.' You see, it was all on paper [that we would do this]. If anything is on paper, it is very difficult to change. If something is on paper, you need something concrete to change the flight plan."

Blagov and other TsUP controllers were not blind. They were fully aware of Tsibliyev's mental state, and it worried them. They decided to put the matter before the IBMP doctors. Bogdashevsky says he had grave doubts about the docking test but withheld them. "I did not give any recommendations at that time, because I knew that nothing would prevent them from doing [the test]," the psychologist recalls. "You see, if the system has a goal, the system will try to achieve that goal. We as doctors try to take into account that concern. And this was a political goal. For the people who are directing these politics, this is sacred, this test. Nothing stops them." Bogdashevsky pauses in his recollection, then looks aside. "We did not fulfill our professional duty to the full," he says. "We just didn't pay that much attention to it. We thought the cosmonauts would be able to manage it."

"This decision was approved by the doctors," Blagov recalls. "They had the right not to sign this decision, to make the ruling that Tsibliyev was not in a good state of mind, that he was not sound enough psychologically to do this kind of test. They did sign off. But frankly speaking, the doctors were not in the best situation. They were told this experiment was going to be carried out, an experiment that when it proves out will be very profitable for future flights. So there was pressure on them to approve this."

Amazingly, given everything Linenger had said in his debriefs about the Near Miss, no one at NASA seems to have grown alarmed about the upcoming test. "We knew very little of the technical detail," admits Van Laak. "The words that were used to describe it were part of the problem. It was called a TORU test. Well, they had used TORU around the station millions of times. So at that point we didn't know much, and they told us less. I guess now that sounds pretty stupid."

Aboard Mir, Foale knew nothing of the upcoming test until the days immediately before they were to attempt it. No one at NASA in Houston or Moscow briefed or warned him about it, or told him what had happened in March. But that weekend, when Tsibliyev explained why they were setting up the TORU manual docking equipment, Foale could see how much it worried the commander, and he peppered Tsibliyev with questions, and the commander reluctantly discussed the test, describing what they would be attempting and how the camera had failed three months before. When he pressed for more, Foale found the commander resistant. He sensed that both his Russian crewmates were uneasy with his curiosity. They kidded Foale that he was a spy and joked that he came from "Langley," the Virginia home of the CIA.

"I asked a lot of technical questions, and that was clearly off-putting to them," Foale remembers. "They didn't know how much they were authorized to tell me. The whole assumption in Russia is if you haven't been authorized to tell someone something, you say nothing at all, and that is how they acted."

Tuesday morning, the day before the test, Tsibliyev managed a rare bit of good humor.

"The sleep study is over!" he rejoiced during the morning's first comm pass. *"It is big holiday for us!"*

But as the day wore on, Foale could see how bothered Tsibliyev was. Over dinner that night the two Russians spoke openly of their worries.

"This is a bad business, Sasha," Tsibliyev said.

"It's all right, Vasily," replied Lazutkin, trying to sound upbeat. "The engineers need this, and you can do it."

Tsibliyev shook his head no. "It's bad," he repeated. "It's a dangerous thing to do."

The two cosmonauts, in fact, were among the few to understand that this test was exponentially more dangerous than the test in March. After weeks of analysis the TsUP had come to the conclusion that the most likely reason for the camera malfunction in March had not been Tsibliyev's failure to push a button. It was almost certainly the Kurs radar signal's interference with the camera signal. And while that finding gave Tsibliyev a mild sense of redemption, the solution the TsUP came up with scared him to death. This time the Kurs signal would be turned off completely, meaning that Tsibliyev would have no telemetry data, no information about the incoming ship's speed and range, to rely upon. Instead—and this would floor NASA engineers when they learned of it later—Tsibliyev was to estimate the speed and range himself, using a handheld laser rangefinder and a stopwatch. It was a recipe for disaster.

Preparations for the test continued all day Tuesday. On Monday the Progress cargo ship had been undocked from the Kvant docking port and placed in an orbit about forty-five kilometers above the station. In the hours before the docking, the TsUP planned to bring it to a position roughly seven kilometers out. Tsibliyev's job, as in March, was manually to maneuver the ship from the seven-kilometer point to a position about fifty meters from the docking port, a task he was to accomplish while out of communication with the ground. Once communication was reestablished, Tsibliyev was to steer the Progress the final meters to a soft docking.

At one point, Foale managed to coax Tsibliyev into discussing what they would do in the event the camera aboard Progress failed again. The two men discussed a "breakout," the aviation term for aborting a test in midflight. Tsibliyev told Foale that the TsUP had emphasized that he could break out at any point up to four hundred meters by simply taking his hands off the TORU controls. Outside four hundred meters, by the TsUP's calculations, the Progress would fly harmlessly by the station. Abort inside four hundred meters, and it was anyone's guess where the cargo ship would go.

"At 1830 do physical exercise, please," the ground informed Tsibliyev in late afternoon. *"This is very important. Then try to get some rest. Tomorrow is a tough day."*

16

June 25
9:09 A.M.
Aboard Mir

"Pressure is at 780. Progress is in position. Elektron is on."

Tsibliyev has donned his navy-blue dress uniform for the docking test. As ground technicians voice up the last of the pretest instructions, he jokes a little about the water puddles that have been condensing inside the dank Kvant module.

"[Kvant] has dried up quite a bit," he says. *"It still stinks."*

"That's what was happening for the last one, two days?" a capcom asks.

"Yes, a little bit. It has dried up after the undocking, after the unit has heated up. The damp has dried up a little bit, but now we have a mushrooming mold."

"That's a lot of fun!"

"With mushrooms . . ." Tsibliyev thinks aloud. *"Hm, too bad we don't have any oil to fry them up."* He laughs.

Someone shouts from the top hall of the TsUP. *"How is your mood?"*

"Great!" says Tsibliyev. This is not the time for griping.

"Happy to hear," responds the shift flight director. *"Do you have any questions regarding the operation?"*

"I don't think so."

10:47 A.M.

"We have switched on the heat at 10:10 A.M.," Tsibliyev says as he goes over the last of the pretest preparations with ground technicians. *"Indicator and Astra are on."*

"We are getting the data on the Progress," the TsUP reports. *"The distance right now is 45 kilometers. The end of the shadow is at five kilometers from you. The docking velocity is at 2.9 meters per second. Precise braking impulse is at 12:01 at one degree. If this corresponds with what you have, that means that everything is going according to the plan. Correction of the orbit to the Progress is at 11:41. We recommend you not to switch anything on till you get out of the shadow."*

Tsibliyev rogers. Everything is on schedule. The Progress is approaching. In one hour he will take over the controls and begin the manual approach.

11:43 A.M.
Aboard Mir

"I am switching on UKV2D, A17, A18," Tsibliyev is saying. "Tape recorder is on." He glances at the instrument panel before him. "Time is 11:43. Switching on Communications Line Three. Transmission."

Tsibliyev is back where he was three and a half months before, sitting before the TORU controls in base block, hands resting by the black joysticks to maneuver the incoming Progress. He and Lazutkin pass between them a copy of the instructional memo the TsUP has sent up, laying out in detail what they should see if everything goes as planned.

Lazutkin, floating to one side of the commander, is nervous. Foale hovers behind them, ready to help as needed.

Tsibliyev punches a second series of buttons.

"Switching on transmitter radios A77, A78."

He stabs another button, activating the television camera aboard the Progress. It's a tense moment. Lazutkin, peering over Tsibliyev's shoulder, remembers all too well what happened in March. He has personally attached the camera cables this morning, double- and triple-checking them to make sure the connections are tight.

"Switching on LIV . . . ," Tsibliyev intones. He makes sure the button is firmly pressed.

Together all three men stare at the screen. After a moment it flickers to life. A backdrop of billowing clouds swirls across the Sony monitor, moving from bottom to top at a rapid clip. Overlaying the picture is a white checkerboard grid. Somewhere among the clouds is the station. Right now it is too small to be seen. Tsibliyev is pleased. He is happy to see anything at all.

"The picture can be seen quite well," he continues. "The picture is shaking a bit, but on the whole, it's not bad."

For a moment the screen jitters and blurs. "A line has appeared," Tsibliyev says. "Let's wait a bit."

They wait a few seconds.

"The picture is not bad now." After a moment the horizontal line on the screen disappears.

"Testing TORU," Tsibliyev says, pushing a new series of buttons. As he does he consults the radio memo the TsUP has sent up.

"Page twenty-four, paragraph four," he tells Lazutkin, who reads along. "It's paragraph two of the main radio memo." The TORU assembly seems to be in working order.

"Switching on 'BPS Original.'" This button, once lit, signals that the Progress is ready to fire its thrusters.

The button doesn't light up.

"There is no receipt yet," says Tsibliyev. "Switching on 'BPS' again. The light should come on."

The button still doesn't light up.

"Pressing till the receipt comes on . . . Pressing again, still no receipt."

He turns to Lazutkin. "Why not? Sasha, write it down: 'BPS Original is not on when pressed.'"

Tsibliyev tries it again. Suddenly the button begins to flicker. "It's blinking," says Tsibliyev. Then it goes out. He pushes it again.

"The receipt is not coming through!" he says, beginning to grow frustrated. "It doesn't come on!"

Lazutkin checks the memo.

"Format once again," says Tsibliyev.

A warning light appears. "The sign 'Moving Toward Each Other Forbidden' is on, for some reason. It's not clear why the 'Moving Toward Each Other Forbidden' sign is on."

Perplexed, Tsibliyev takes a few moments and repeats the entire procedure. This time the necessary lights come on.

"It has come on, everything is fine," he says. "Once again [pushing] 'BPS Original.' There is a receipt!"

Lazutkin breathes a sigh of relief. Everything seems to be in working order.

It is 11:51. If everything goes more or less on schedule, the Progress should be at the fifty-meter point, ready for final docking, at about 12:08.

Seventeen minutes.

"The test of equipment is over," says Tsibliyev. "Ready to do manual docking. Going to page twenty-eight, paragraph four-point-one."

He reads down the instructions.

"Switching on 'Work.'"

A message appears on his screen: "Taking Away Allowed."

"Taking away allowed," Tsibliyev intones.

He can assume control of the Progress at any time. The station remains a faint blip on his screen, a dot he can barely see against the undulating clouds. He and Lazutkin are peering down at the screen when the dot suddenly disappears. Both men search the clouds for any sign of the station.

"Well, here," says Tsibliyev, pointing. "Maybe . . ."

Lazutkin shakes his head.

"No, it's not us," the commander says.

"I can't see it in the window to be able to check," says Lazutkin.

Neither man is especially bothered. In time, they know, both the Progress and the Mir will come into view, the station on the screen and the cargo ship outside their windows.

Tsibliyev, wasting no time, decides to begin. He leans over and flips a switch.

"Switching to 'Manual Docking.'"

The TORU joysticks in front of Tsibliyev would be familiar to any American teenager who plays video games. Much of the commander's control rests in the left joystick, the "translation controller." Moving the knob to the left moves the Progress to the left; moving it down moves the ship down, and so on. Directly beside this left joystick is a small lever Tsibliyev controls with his left pinkie finger. Pushing the lever forward accelerates the ship; nudging it back fires braking thrusters. The right joystick, the "rotation controller," controls the incoming ship's orientation, its roll, pitch, and yaw.

Now, taking control of the ship, Tsibliyev uses his left pinkie to push the acceleration lever. Seven kilometers out in space, thrusters fire aboard the Progress, and the hardy cargo ship begins moving on a slow arc toward the station.

"Here it is," says Tsibliyev, gesturing toward the screen. The dot that is Mir seems to reappear among the clouds. Tsibliyev can't be certain it is the right dot.

"I'm watching the picture," he says to Lazutkin. "If it's us, then we are in the center of [the screen]." Right where they should be.

"It's difficult for me to tell," says Lazutkin, now peering out the base block window. "Nothing can be seen out there."

Several minutes go by with no obvious change on Tsibliyev's screen. Mir remains a dim dot in the center square of the overlay, winking in and out among the clouds. Lazutkin and Foale, glued to the base block windows, search the blackness outside for the first sign of the giant birdlike Progress. By and by the dot on Tsibliyev's screen appears to be growing larger.

"The target is moving up," says Tsibliyev. His voice sounds firm, confident.

"So do you have control or not?" asks Lazutkin.

"Control is going on fine." Tsibliyev turns to Foale.

"Michael, can you hear me?" he asks. "If you could try and measure the distance, it would be great." He glances at the clock. "The time now is 11:54."

Fourteen minutes.

Foale takes the laser rangefinder and swims into the Kvant module behind the two Russians. The device, which looks like a set of binoculars, bounces a red thread of laser light off a target, in this case the Progress, to estimate its distance from the station. Foale passes the scorched area where they fought the fire in February and quickly comes to rest at the far wall of Kvant, where a small porthole affords a crimped view outside the station. Pressing his head to the tiny window, he looks left and right, then again. There is no sign of the approaching ship—and thus no way to gauge its speed and distance. The rangefinder is useless if he can't see the ship.

Tsibliyev is peering intensely at the screen. The dot—the station—has disappeared again.

"The target cannot be seen, for some reason, Sasha. . . . Let's see."

Lazutkin returns to Tsibliyev's shoulder, looks at the screen, then points out a smudge that might be the station.

"If that's it . . . ," Tsibliyev says, staring at the little smudge. "That's it! It's approaching the center of [the screen]."

The commander's relief is palpable. "Such an interesting star!" he exults. "It takes up half a square." Half a square on the screen's checkerboard overlay. The dot is growing.

"I am turning down the right side a bit," Tsibliyev says, adjusting the cargo ship's side speed. "Maybe left is good."

"Probably so," says Lazutkin. "I will go look. Do we have a minute and a half? At least I can try and check the distance." At this point Lazutkin reads in the TsUP memo, they should be able to spot the Progress from at least one of the windows on the station, if not from Foale's limited-sight post in Kvant. He guesses it should be about five kilometers out.

"Go ahead," says Tsibliyev.

Lazutkin takes the rangefinder from Foale and swims off into the node, turning in to Kvant 2. Quickly he checks each of the windows in the module. Nothing. The Progress is nowhere to be seen. Lazutkin reverses course and returns to the node, then turns abruptly and dives into Kristall. Moving with a sense of urgency now, he ricochets to the near window.

Nothing.

He turns and checks the next window.

Nothing.

By the time he reaches the third window, Lazutkin is not surprised he cannot find the Progress. Outside he sees the outer hull, space and stars. Nothing but stars.

"So now it's half a square," Tsibliyev is saying as Foale watches. "It's somewhere a little far out, about 2.5 kilometers." He takes a moment to refigure his math. "The distance must be 2.9, about three kilometers. It's three kilometers, and the speed is about three meters [per second] . . . It's gradually moving up and it's practically standing still one degree up and 0.5 degrees to the right from the cross. . . . Up a bit . . . Well, the station can be seen more distinctly, with all the panels already . . ." The smudge on the screen is now identifiable as the station. He can almost make out the individual solar arrays.

"There is control on all channels," Tsibliyev says. "Good. . . . Stabilization to the right and to the left is working. . . . The station is practically in the center of the cross. . . . There is no movement on the angle lines. . . ."

Everything is on schedule. The commander turns to Foale.

"And where is Sasha?"

"Sasha I think went to go measure the distance."

A moment later Lazutkin returns.

"I can't see it," he says.

"Sasha, the speed is probably not that high," says Tsibliyev. "I have a feeling that it's like it's standing still."

Tsibliyev takes a moment to recalculate the cargo ship's speed. Lazutkin, floating over and past, returns to his post beside him. When he is finished with the calculation, Tsibliyev shows it to Lazutkin.

"Like they promised," Lazutkin says. "Two-point-five meters [per second]. It's practically okay."

"Yes, it's practically okay."

But it's not, and both men know it. Tsibliyev motions to the screen. At this point, according to their instructional memo, Mir should fill an entire square on the checkerboard grid overlay.

"But now it's only half a square," says Tsibliyev. The Progress is approaching too slowly.

The two men stare at the screen, unsure what to do. It is at this point, Lazutkin would recall months later, that he began to grow worried.

"Maybe we need to push it?" Lazutkin wonders, suggesting that Tsibliyev fire a forward thruster to increase the ship's speed.

"Well, it's like . . ."

Tsibliyev doesn't finish his thought. He says nothing for a long moment. "Well, if we push it," he finally says, "then we may have to start braking. But it will, of course, take a long time if we sit and wait like this."

They continue staring at the screen. Lazutkin takes a moment to reread parts of the memo.

"You see?" says Tsibliyev, pointing toward the screen. "It looks like it's getting bigger very slowly. I feel like it's standing still."

Lazutkin looks again at the radio memo. He hasn't a clue what they should do now.

"Well, should we increase the speed a meter per second?" Tsibliyev wonders.

"Uh . . . I think so."

"Otherwise we are hanging here and it practically is not approaching."

They continue staring at the screen, like two confused schoolchildren pondering how to fix their balky physics experiment.

Seconds tick by.

Tsibliyev peers closer. The station appears to be slowly growing bigger.

"Look!" he says excitedly. "It looks like it's approaching!"

Lazutkin nods. The station does appear to be growing.

"The size is about 0.7 squares," Tsibliyev says. "Let's not be in a hurry."

Lazutkin gives the commander a quizzical look.

"Sasha, why hurry?" says Tsibliyev.

Lazutkin nods. "There's no sense in giving a braking impulse now [either]," he says after a moment.

"No, I will not apply the brakes yet."

As the seconds tick by, the station continues growing bit by bit. It is 12:03. Five minutes to go.

"It's 0.9," says Tsibliyev. "It's about 0.8 or 0.9 squares." He refigures his math. "That means the distance is a little over five kilometers, right?"

"Yes," Lazutkin says, nodding. "Over five."

"Maybe 5.5."

Lazutkin manages a nervous smile. "It would be good to see a little bit of it," he says. Lazutkin is beginning to experience an uncomfortable sense of déjà vu. Every few seconds he is glancing out the base block window, but there's still no sign of the approaching ship. The solar arrays block his view. Foale, checking the tiny window behind the two Russians, hasn't managed to catch a glimpse of the ship either.

"You can't see it at all, right?" Tsibliyev asks.

"It can't be seen!" Lazutkin blurts.

"It should be behind us. Maybe the speed will increase."

Tsibliyev remains perplexed. "If we brake, we don't need to do it by two meters here. Half a meter will be enough. Otherwise it's too much." He is still worried that the Progress is approaching too slowly; if he brakes hard, he fears, it might even creep to a halt.

Tsibliyev wants to check the speed again.

"Sasha, give me the stopwatch, please. I forgot to take it."

Lazutkin looks around.

"It's on the wall," says Tsibliyev.

Lazutkin retrieves the watch and hands it to the commander.

"The time now is 12:04. I will brake now."

Tsibliyev nudges the braking lever with his left pinkie to fire the ship's reverse thrusters. He holds it for fifty-three seconds, a period that should slow the oncoming Progress by about one meter per second.

"Now everything is working fine," he says. "There is braking . . . Okay . . ." The station is steadily growing larger on his screen. "One-point-five squares if we count all these deviations."

After the required fifty-three seconds, he releases the braking lever.

"I stopped the braking," he announces. "Time . . . 12:05:50 . . ." Two minutes to go.

Lazutkin looks out the window, then at Foale. Neither man says anything. Lazutkin is beginning to sweat. They should see something by now.

Tsibliyev still sounds confident.

"Okay . . . ," the commander says. "Let's check out how it's moving. . . . It can't be seen too well with the Earth in the background. . . . Okay. . . . Another braking impulse should be given. . . . Two meters . . . Good."

Tsibliyev nudges the braking lever and holds it again, this time for fifty-two seconds.

"I am switching on the stopwatch," he says. "I am giving another braking impulse and turning down the side. . . ."

No one says anything.

At 12:06:51, with Lazutkin and Foale floating silently behind him, looking out their windows, Tsibliyev releases the braking lever. According to the instructional memo, the Progress should be just a kilometer or slightly less above them, moving down toward the docking port. Once the ship arrives at a point about four hundred meters away from the station, Tsibliyev will slow its speed to a crawl and begin inching it forward to the fifty-meter point, where it will be readied for a soft docking at the Kvant docking port.

Ninety seconds to go: According to the TsUP's plan, the Progress should be approaching the four-hundred-meter mark. But neither Lazutkin, peering out window No. 9, nor Foale can see anything. Both men know the ship has to be out there somewhere, just beyond their view; on the screen, the station now fills four entire squares on the checkerboard overlay. An eerie feeling washes over Lazutkin. Looking at Tsibliyev's screen, he feels as if he is being watched. But no matter what they do, they cannot find the onrushing watcher.

Tsibliyev nudges the braking lever one final time.

"It's moving down . . . ," he says.

Suddenly Lazutkin spots the oncoming Progress, emerging from behind a solar array that until that moment had blocked his view. The ship appears huge—bigger than he could have imagined.

It is heading right for them.

"My God, here it is already!" Lazutkin yelps.

Tsibliyev can't believe it. "What?"

"It's already close!"

"Where is it?"

Now everything begins to happen fast. As Lazutkin looks on through the window, the brightly sunlit Progress appears to be heading straight for a collision with base block, its twin solar arrays making it appear like some shiny white bird of prey swooping down on them.

"The distance is one hundred fifty meters!" he shouts.

Tsibliyev thinks Lazutkin must be mistaken. His left pinkie remains clamped on the braking lever. The Progress should be moving at a crawl.

"It's moving closer!" Lazutkin says. He looks outside again and sees the big ship coming on inexorably. "It shouldn't be coming in so fast!"

"It's close, Sasha, I know; I already put it down!"

Tsibliyev is holding the controls tightly, his left pinkie clamped on the braking lever. The ship should be slowing. It doesn't seem to be responding.

To his horror, Lazutkin sees the Progress pass over the Kvant docking port and begin moving down the length of base block.

Tsibliyev sees it on the screen.

"We are moving past!" he shouts.

Lazutkin remains glued to the window.

"It's moving past! Sasha, it's moving past!"

Lazutkin, watching the Progress come on, turns to Foale.

"Get into the ship, fast!" he tells Foale, directing him to the Soyuz. "Come on, fast!"

Foale, who has still not seen the Progress, acts quickly, pushing off the wall, shooting across the dinner table, and hurtling over Tsibliyev's head toward the Soyuz, which rests at its customary docking port on the far side of the node. Then, just as Foale passes over the commander, something happens that may or may not have a profound effect on all their lives: One of Foale's feet whacks Tsibliyev's left arm. Later, everyone on board will disagree on the effect this accidental bump may have had on the path of the onrushing Progress.

As Foale passes, Tsibliyev sits frozen at the controls, his face a mask of concentration. He is convinced he can keep the Progress out away from the station, that if he holds tightly enough to its current course it will still miss them. Not until the last possible second, when the hull of the station ominously fills his entire screen, does the commander realize there is no avoiding a collision.

"Oh, hell!" Tsibliyev yells.

As the black shadow of the Progress soars by his window, Lazutkin closes one eye and turns his head.

* * *

The impact sends a deep shudder through the station. To Lazutkin, still glued to the base block window, it feels like a sharp, sudden tremor, a small earthquake. Foale, swimming through the node toward the open mouth of the Soyuz, feels the violent vibration when his hand brushes the side of the darkened chamber.

"Oh!" Tsibliyev shouts, as if in pain. He stares at his screen, barely comprehending what has happened. He says aloud, "Can you imagine?"

The master alarm sounds, eliminating all but shouted conversation.

"We have decompression!" Tsibliyev yells. "It looks like it hit the solar panel! Hell! Sasha, that's it!"

Confusion breaks out as Lazutkin turns and begins to swim toward the node, intent on readying the Soyuz for immediate evacuation.

"Wait, come back, Sasha!" Tsibliyev barks.

It is the first decompression aboard an orbiting spacecraft in the history of manned space travel. As Lazutkin hovers beside him, waiting for an order, Tsibliyev remains at his post, staring dumbfounded at the screen, looking for all the world like the captain of some stricken celestial *Titanic*.

"How can this be?" he asks. "How can this be?"

After that his words are drowned out in the manic din of the master alarm.

Floating alone in the node, Foale pauses. After a moment he realizes he is still alive. His ears pop, just a bit, telling him that whatever hole has been punctured in the hull, it is probably a small one. The station's wounds, whatever they are, are not immediately fatal. They should have enough time to evacuate.

He turns and faces the entrance to the Soyuz, where a tangle of cables, a mass of gray-white spaghetti, spills out of the escape craft's open mouth. Executing a deft little flip, he turns backward and enters the Soyuz feet-first, extending his legs behind him, his head and shoulders protruding from the capsule.

As he turns to look back toward base block, Foale fully expects Lazutkin and Tsibliyev to come charging into the node after him to begin the evacuation. They don't. Foale waits five, ten, then twenty seconds. There is no sign of the Russians. They remain somewhere back in base block, out of his sight.

After roughly a minute of waiting, Foale begins to worry. He is certain the Progress struck the station either in base block or in Kvant. These are considered "nonisolatable" areas—that is, a hull breach in either area cannot be sealed off. In emergency drills simulating a meteorite strike against the hull of either module, the crew is given no option but to abandon ship. Foale can't understand why Lazutkin and Tsibliyev aren't evacuating.

Tsibliyev swivels out of his seat and crouches by the floor window behind him. There, barely thirty feet away, so close he feels he can reach out and touch it, he sees the Progress sagging against the base of one of Spektr's solar arrays. It looks as if the long needle on the leading edge of the cargo ship's hull has pierced a jagged hole in the array's winglike expanse. He can't be certain, but the Progress appears lodged against the hull. Lazutkin crouches by the window and looks down. He sees it too.

The commander turns, thinking he will fire one of Progress's forward thrusters to, as he later put it, "kick it" off the station. But just as he begins to leave the window, he sees the cargo ship shift and move forward once again, striking and denting a boxy gray radiator on the side of Spektr's hull. Then it keeps moving forward and, after a long moment, floats free again.

Tsibliyev holds his breath, hoping that the Progress will now fly free of the station without hitting any more of its outer structures.

Where are they?

Foale can't understand what Tsibliyev and Lazutkin are doing. Emergency procedures mandate that they immediately evacuate the station, but the two Russians are nowhere to be seen. It occurs to Foale that his two crewmates are doing something to try and save the station, when in fact they should be evacuating. This kind of going-down-with-the-ship mentality wouldn't be unusual among the pride-soaked cosmonaut corps, he knows; it is precisely the reckless kind of behavior Linenger has been warning everyone about. Foale crawls out of the Soyuz and begins to fly back toward base block, intent on finding out what is going on.

But the moment Foale emerges from the Soyuz, Lazutkin hurtles out of base block into the node. In a flash he is at the little ship's entrance. Foale, realizing that Lazutkin is now prepared to begin the evacuation, is unsure of his role.

"Sasha, what can I do?" he asks.

Lazutkin ignores or doesn't hear the question; the alarm remains so loud it is difficult to hear anything. Moving with the fury of a man in hand-to-hand combat, Lazutkin grabs the giant, wormlike ventilation tube and tears it in half. Wordlessly he seizes cable after cable, furiously rending each one at its connection point. Foale watches in silence.

It takes Lazutkin barely a minute to disconnect all the cables. At last, only one remains. It is the PVC tube, which channels condensate water from the Soyuz into the station's main water tanks. Lazutkin cannot separate it with his hands. He needs a tool.

A wrench. They need a wrench. Lazutkin looks frantically for one all around the node, which is lined with spare hatches and tools and equipment. He and Foale spend nearly a minute in search of a wrench before Lazutkin finds one, floating by a blue thread. He hands it to Foale and shows him how to unfasten the PVC tube. Foale retreats into the Soyuz, applies the wrench, and begins turning as fast as he can.

When he is certain Foale knows how to unfasten the PVC tube, Lazutkin turns toward the entrance to the Spektr module. Foale, while saying nothing aloud, remains convinced the leak is in base block or Kvant. Lazutkin doesn't have to guess. He has seen the Progress lodged against the Spektr module's solar array. He assumes that whatever breach the hull has suffered, it almost certainly occurred in Spektr. Lazutkin pushes off from the Soyuz entrance, arcs across the node, and shimmies into Spektr.

Diving headfirst into the module, he immediately hears an angry hissing noise from somewhere below and to his left. It is, he knows, the sound of air escaping into space. His heart sinks. At this moment, Sasha Lazutkin is certain they are all about to die.

12:14
TsUP

Keith Zimmerman, Foale's young ops lead, trots down to the floor and takes his post at the NASA console a few minutes before the beginning of the comm pass. With him stands Terry Taddeo and their translator, a man he knows only as Aleksandr. Zimmerman, like the other Americans, has not spent much time worrying about today's docking test. He has heard little of Linenger's debriefings, and what he has heard wasn't especially alarming. "A lot of [Linenger's] debriefings were focused on the fire, and the Near Miss was just seen as a miss: big deal, it happened, no big deal," recalls Zimmerman. "Missing on a docking attempt was not exactly a new thing."

"I didn't know anything about [the test]," recalls Taddeo. "I had no sense this was something they had not done before. I thought: Okay, they're going to fly it in, no big deal. There was no sense that this was something unique, or a potential problem. So I had no idea there was any danger at all."

Zimmerman and Taddeo's attitudes underscore the near-total lack of understanding among NASA personnel at the TsUP about the basic nature of today's test. Zimmerman, like other Americans following events that day, has utterly failed to distinguish between a cosmonaut's manual docking of a Soyuz capsule from a distance of ten meters and the remote-control manual docking of an unmanned Progress ship from a distance of seven kilometers. The Russians haven't volunteered much information about the test, and Zimmerman's team hasn't asked many questions.

Zimmerman and the other Americans, in fact, have no idea how close to a collision Tsibliyev came three months before, they have no idea how exhausted the commander has grown over the last two months, and they have no idea how dangerous today's test is. And as Zimmerman dons his headphones in preparation for the comm pass, they have no idea whatsoever that at that very moment, two hundred fifty miles above, the three men they are pledged to support are fighting for their lives.

12:15
Aboard Mir

Lazutkin realizes immediately that in order to save the station, he has to somehow seal off Spektr. Like all the other hatchways, it is lined with wrapped packets of thin white and gray cables, eighteen cables in all, plus a giant wormlike ventilation tube.

A knife, Lazutkin thinks: I've got to find a knife to cut the cables. While Foale remains inside the Soyuz, finishing off the PVC tube, Lazutkin soars back through the node and dives headfirst into base block, where he sees Tsibliyev poised to begin talking to the ground. Vaulting over the commander's head, Lazutkin shoots down the length of base block, past the dinner table, and into the mouth of Kvant. He remembers a large pair of scissors he has stowed alongside one of the panels, but when he reaches the panel, he is heartsick: The scissors aren't there. Then he sees it: a tiny, four-inch knife—"better to cut butter with than cables," as Lazutkin remembers it. Normally he uses the blade to peel the insulation off cables that need to be rewired.

Lazutkin grabs the knife and flies furiously back down to the node. Sticking his upper body into Spektr, he grabs a bundle of cables and instantly realizes his plan won't work: The cables are too thick to be cut with his little blade. Each of the bundles is fitted into one of dozens of connectors that line the inside of the hatch. Frantically Lazutkin begins grabbing the cable bundles one after another, unscrewing their connections and tossing the loose ends aside, to float in the air.

After a moment Foale emerges from the Soyuz, where he has finally disconnected the PVC tube, just as Lazutkin finishes ripping apart the first few cables. Foale is immediately surprised to see Lazutkin working at the mouth of Spektr. Still believing the leak is somewhere back in base block or Kvant, he is convinced that Lazutkin is isolating the wrong part of the station. If Foale is correct, sealing off Spektr will be a disastrous move. It will actually reduce the station's air supply, thereby causing Mir's remaining atmosphere to rush out of the breach even faster.

"I was still very concerned we were isolating the wrong place," Foale remembers. "I was not going to stop him physically—yet. But that was my next thought: Should I try and stop him."

Instead, intimidated by the sheer fury with which Lazutkin is tearing at the cables, Foale floats by and watches. As Lazutkin rends each line, its loose end floats out into the node—"eighteen snakes floating around, like the head of Medusa," Foale recalls. Foale begins grabbing the loose lines and binding them with rubber bands he finds in the node.

Finally, he says something.

"Why are we closing off Spektr?" he asks, pressing in close to make sure he is heard over the alarm. "It's the wrong module to close off. If we're gonna do a leak-isolation thing, we have to start with Kvant 2."

Foale is about to say more, when Lazutkin cuts him off.

"Michael," he says, "I saw it hit Spektr."

And with that Foale at last fathoms Lazutkin's urgency: Seal Spektr, and they save the station.

It takes almost three minutes for Lazutkin to tear apart fifteen of the eighteen cables. The remaining three cables don't have any visible connection points. They are solid and unbreakable. Lazutkin thinks of the knife. He retrieves it from his pocket and slashes a thin data cable for one of the NASA experiments. The next moment he slashes a leftover French data cable from one of the Euro-Mir missions. One cable remains. One cable whose removal will allow them to seal the hatch and save the station.

But Lazutkin receives a rude shock when he begins sawing into the last and thickest of the three cables. Sparks fly up into his face.

It is a power cable.

Foale sees a frightened look cross the Russian's face.

"Sasha, go ahead!" Foale urges. "Cut it!" A beat. "Cut it!"

But he won't. He won't cut it.

17

12:18

At the moment the Progress strikes the station, Mir is streaking two hundred fifty miles above the mud huts of Agadez, a forlorn Sahara oasis town in the central African country of Niger. In the minutes that whiz by as Lazutkin and Foale attempt to close the Spektr module's inner hatch, the station flies a northeasterly course across Egypt, then over the eastern Mediterranean near Cyprus, until reaching the limits of Russian communications range over the mountainous interior of Turkey.

It is at this point that Tsibliyev, still floating anxiously at the command center in base block, hears the TsUP hailing him. Nikolai Nikiforov, the shift flight director, is at the command console in the TsUP. Vladimir Solovyov stands out of sight in a separate control room, used for Progress docking operations. There is static, and for a moment Tsibliyev's words cannot be heard.

"Siriuses!" Nikiforov shouts. *"Siriuses!"**

Suddenly Tsibliyev's voice breaks through the static. *"Yes, yes, we copy! There was no braking. There was no braking. It just stalled. I didn't manage to turn the ship away. Everything was going on fine, but then, God knows why, it started to accelerate and run into module O, damaged the solar panel. It started to* [*accelerate*], *then the station got depressurized. Right now the pressure inside the station is at 700."* Module O is Spektr.

Solovyov, immediately realizing what has happened, gets on comm. *"Guys, where are you now?"*

"We are getting into the [*Soyuz*]. . . .*"*

Chaos breaks out for several moments on the floor of the TsUP as Solovyov and the other controllers try to determine exactly what is happening aboard the station. The comm breaks up for a moment.

*"Sirius" is the Mir 23 crew's code name.

"Copy, damn it!" Solovyov barks.

"Oh, hell," Tsibliyev blurts out. *"We don't know where the leak is."*

"Can you close any hatch?" interjects Nikiforov.

"We can't close anything," Tsibliyev says hurriedly. *"Here everything is so screwed up that we can't close anything."*

As Tsibliyev's words crackle over the auditorium loudspeakers, Keith Zimmerman can't understand anything the commander is saying; he speaks very little Russian. Then, suddenly, his interpreter says, "They hit something."

Zimmerman wrinkles his brow. "What do you mean?" he says. From the interpreter's even tone of voice, he guesses that maybe one of the cosmonauts has hit his thumb with a hammer.

"The Progress," Aleksandr says quietly. "It hit the station. The pressure is going down."

Zimmerman goes numb. This is not something a twenty-nine-year-old MOD assistant is accustomed to hearing.

"W*ait, Vasya, what are you doing now?"* Nikiforov asks.

"We are getting ready to leave. The pressure is already at 690. It continues to drop."

"Can you switch on any blowers?"

"I think we can."

"Open all existing oxygen tanks."

"Sasha," Tsibliyev hollers to Lazutkin, "have you closed the hatch?"

Lazutkin's reply is drowned out as the station's master alarm continues braying.

"Vasya," says Solovyov, *"what are you doing now?"*

"DSD has turned on. We managed to close the hatch to module O." This is wishful thinking; as Tsibliyev speaks, Lazutkin still hasn't cleared the last cable from the hatch. DSD is a depressurization sensor.

"Module O. Has [the Progress] run into module O?"

"Yes; it hit module O."

"Is the hatch closed right now?"

"Sasha is closing it right now."

"What's happening with the pressure?"

"DSD turned on when pressure dropped down to 690."

Solovyov interjects. *"Can you pass through [the node] right now? We should have extra [oxygen] tanks somewhere in TSO."*

"I know that," says Tsibliyev. TSO refers to the air lock at the end of the Kvant 2 module.

Solovyov's call for Tsibliyev to retrieve one of the station's pumpkin-size oxygen cylinders is a standard response to depressurization scenarios in both Russian and American simulations; until this moment it has never been tried in an actual crisis. Releasing oxygen into the Mir's atmosphere, Zimmerman realizes, means Solovyov has decided to begin "feeding the leak"—that is, replacing air that has already begun to whistle through whatever hole the Progress has poked in the hull of the Spektr module. Feeding the leak won't save the station, but it should give the crew precious extra minutes. How many depends on how fast the station is losing air.

"So open them up," Solovyov orders.

"I [*will start*] *doing that right now. I am taking off the ears"*—the headphones—*"and am taking off to do that."*

"But someone has to stay here to maintain the connection!" Solovyov pleads. *"Then I can't make it."*

Tsibliyev does it anyway. Ripping off his headphones, he leaves his post, turns, and swims out over the command console and into the node.

"Guys?" Solovyov asks. *"Someone pick up!"*

There is no answer.

"Sasha?"

No answer.

"They have left. . . . Guys? . . . Someone respond."

James Medford returns from lunch and plops down in a chair in the NASA operations room behind Michael Malyshev, the interpreter, who sits at the long table topped by four monitors. Medford, as one of the three new rotating Mir systems specialists, is in his sixth week in his job and has developed only the most rudimentary understanding of how the station works. He spends his free time poring over Russian training manuals. Medford knows the Russians have some sort of manual docking exercise coming up, but like Zimmerman, he understands little else about it. None of the Americans have been involved in the test's planning, of course. And despite all the information Linenger gave during his debriefings, Medford has only the dimmest understanding of what transpired in March.

Staring out the window at a warm, lazy Moscow afternoon, Medford barely registers the fast-paced Russian chatter as it comes over the speakers. Malyshev is an excellent interpreter, and Medford figures he will read over the comm when it finishes.

Then, with no warning, Malyshev springs from his chair, standing upright.

"Holy shit!" he exclaims. "Progress hit Mir, and the pressure's dropping inside the station!"

Medford feels a massive surge of adrenaline through his upper body. "If I'd had an EKG on at that point," he remembered months later, "it would have been scary."

Medford stands and watches over Malyshev's shoulder as the interpreter hurriedly scribbles down the exchanges between Tsibliyev and the ground. He remains silent, not wanting to interrupt Malyshev's concentration. Every few moments the interpreter quickly updates him.

"They're trying to close the Spektr hatch," he says.

Medford glances at the monitor at the far right-hand side of the table. It carries real-time updates of Mir's temperature, pressure, carbon dioxide levels, and other key life-support indicators. The internal pressure, which normally hangs steady around 780 millimeters of mercury, is at 690. And dropping.

Medford is frightened. It isn't the pressure level itself. Engineers generally don't worry about the pressure unless it somehow drops below 600 or so; at 550, Russian regulations call for immediate emergency evacuation. It is the rate the pressure is dropping that scares Medford. At this rate, he guesses, the station has barely thirty minutes of oxygen left. It's a good guess: NASA estimates later put the figure at twenty-eight minutes.

As Medford watches in amazement, the white digital reading on the NASA monitor continues to drop.

685.

680.

"Have they got the hatch closed?" he asks Malyshev.

"Hold on, hold on."

Kate Maliga and Cathy Watson walk into the room. From their stunned expressions, Medford can tell they know what is happening.

"Are you hearing this?" one of them asks.

Medford shushes the two women and turns back to the telemetry screen. The pressure is still dropping.

675.

670.

668.

* * *

Lazutkin won't cut the power cable. Again he and Foale plunge down into the darkened morass of loose cords and equipment and lids and seals that line the node walls. Somewhere in the chamber's dim recesses, Lazutkin believes, there must be a plug for the power cable. Foale rips aside cable bundles and runs his hands over the walls. Lazutkin cranes his head, looking, looking.

There. Lazutkin pulls at the power cable and follows it to a plug inside Spektr. With one furious yank, he rips it from the wall.

Immediately Foale and Lazutkin turn to confront Spektr's inner hatch. Lazutkin reaches into the module and pulls on the hatch to close it.

It won't budge.

Both men instantly see the problem. With the pressure dropping inside Spektr, all the air inside the station is rushing past them, seeking to escape through the unseen breach into open space. It is as if they are trying to close an open door while an invisible river surges through it. Lazutkin realizes he could slip into Spektr and push from the inside, but then he would be trapped within the sealed module. He would die quickly, a hero of the motherland, but Sasha Lazutkin isn't ready to die yet.

Again he and Foale tug at the hatch, straining to pull it closed.

It won't budge. Nothing they do will make it move a single inch.

Tsibliyev streaks through the node, hurtling past Foale and Lazutkin as the two men continue struggling with Spektr's open hatch. Diving headfirst into Kvant 2, he flies swiftly to the far end and pulls down hard on the wheel-like hatch cover to the air lock.

In seconds the door swings open. Sticking his upper body inside, Tsibliyev grabs one of the air tanks, hugging it to his body. Slamming the hatch shut, he propels himself back up the length of Kvant 2, carrying the pale-yellow air tank like a football. Crashing back through the node, he turns and flies into base block, swings down into his seat, and slaps the headset onto his head.

"Emergency TSM X light went on," he says, scanning the console. *"The pressure is at 670. I have brought this* [*oxygen tank*] *into the main module."*

"Did you manage to seal the hatch?" asks Nikiforov.

"Is the hatch closed?" asks Solovyov.

"[We are] closing it right now. Not yet. We have not yet turned on the [oxygen tank]."

"Start it up! Start it up!" urges Solovyov. *"Open up all the valves."*

"I am opening it."

The moment Tsibliyev turns the valve on the air tank, a screeching hiss fills the station and echoes through the TsUP auditorium.

"Watch the mano-vacuumeter while you are opening it up," instructs Solovyov. This is a handheld pressure gauge.

"I have opened it."

"Do you have a mano-vacuumeter on you?"

"Of course."

Feeding the leak should increase the pressure.

"Get it up to 730," Solovyov instructs.

"All right."

"What volume are you trying to reach?"

Tsibliyev seems to misunderstand the question. There is a moment of confusion. *"I am [back] in the main module."*

"Everything is working except module O?"

"Yes, everything except module O."

"Great. What is Sasha Lazutkin saying?"

Tsibliyev glances toward the node, where Lazutkin is trying to make himself heard over the din of the master alarm.

"I can't hear him. He and Michael are trying to put away the cords in order to close module O." A beat. *"The pressure is at 668."*

The commander begins to try and explain how the accident happened.

"Wait," Solovyov orders. *"We'll do that later."*

"Okay."

"Right now, take care of the pressure."

"The pressure is going up."

"What is it at now?"

"Wait a second, I'll tell you."

"How much?"

"670."

"670. Vasya, leave [the tank] right now," urges Solovyov. *"Go back to help the guys close the hatch."*

"If we don't do it now, what do we do?" says Tsibliyev, ignoring the order for the moment. *"The comm pass is almost over. . . ."*

"Don't worry. This will be enough time to send all the instructions."

"We are trying to pump up the pressure, right?" he asks.

"What is it at the moment?"

"675."

"Can you use anything else to help the pressure?"

"We have [another] cylinder in the [Soyuz]."

Tsibliyev, who by now has switched off the master alarm, turns and yells at Lazutkin: "Sasha, take it out!"

The commander then returns to Solovyov. *"Right now [the pressure's] at 675, but the air is still coming out [of the tank]."*

"Let it go," says Solovyov. *"But do not switch on the second [tank] until we tell you to do so."*

"All right." Tsibliyev glances down at the control panel. *"The Emergency ZMX light came on, the DSD has started, and the light indicating depressurization turned on."*

"Received."

"It's so bizarre," says Tsibliyev, again bringing up how the Progress hit the station. *"Everything was so seamless. I had a feeling that there was a braking, although we went through all the commands."*

The hatch won't shut.

Its outer surface is smooth, with no easy handholds. Neither Foale nor Lazutkin can risk slipping his hands around the hatch's outer edge, for fear of losing a finger.

"The lid! Let's get a lid!" Lazutkin urges.

Foale realizes that with the inside hatch unable to close, they will have to find a hatch cover to push onto the module's open mouth from the outside.

Each of the four modules attached to the node originally came with a circular lid, vaguely resembling a garbage can lid, which sealed the hatch from the outside. All four of the lids are now strapped to spots on the node walls. They come in two sizes, heavy and light. Lazutkin reaches for a heavy lid, but it is tied down by a half-dozen cloth strips, each of which, he realizes, he will have to slash to free the hatch cover underneath. He simply doesn't have enough time to cut all the strips.

Instead Lazutkin reaches for one of the lighter covers. It is secured to the node wall by a pair of cloth straps, both of which the slim Russian quickly severs with the knife.

Together both men lift the lid and set it over the open hatch. The lid was originally held in place by a series of hooks spaced evenly around the hatch's outer edges, and Lazutkin thinks they will have to work this mechanism to seal the hatch. But the moment the two men affix the cover to the open hatch, the pressure differential that foiled their earlier efforts now works in their favor. The lid is sucked tightly into place.

Lazutkin isn't satisfied. He tells Foale to support the hatch cover while he finds the tool he needs to work the closing mechanism.

"*Vasya,*" Solovyov says, "*what hatch are they closing in module O? The one that needs to be pushed out or pulled in?*"

"Which one are you closing?" Tsibliyev yells over at Lazutkin.

Lazutkin says something inaudible.

"*The one that will be pushed toward the module,*" says Tsibliyev.

"*You mean the one that is part of the main module.*"

"*It's like a lid that will be pressed on.*"

"*Understood. So you are putting on the lid? Do you have some knife? Can you unplug the cables?*"

"*Yes, we have closed it and with that the light indicating depressurization has turned off.*"

At the NASA console Keith Zimmerman breathes a tiny sign of relief. This is the first good news he has heard. He scans the telemetry on his screen, paying close attention to the pressure levels. If the damage is limited to Spektr, and Spektr's hatch has been firmly sealed, the pressure should hold steady.

It is 12:21.*

Up in the balcony of the TsUP, Mark Severance has joined a group of visiting NASA officials to watch the docking. Entering the high-ceilinged space just as the comm pass begins, he gazes out at the giant display and notices that two lights are still lit, signifying two separate ships in space. If the docking has been successful, there should be a single light. That's funny, Severance thinks. Why haven't they docked by now?

The visiting Americans, none of whom can understand much Russian, are chatting away, oblivious to the drama below them, when Severance begins registering snippets of comm being broadcast from the overhead

*Foale's subsequent report on the collision indicates the hatch was closed at 12:21. NASA logs for the day indicate the time was 12:28; the 12:28 time may have been when the ground learned of the closure.

speakers. He hears the number "675" and realizes the Russians are taking pressure readings. *Why are they doing leak checks?* he wonders.

There is another pressure number, then another. *That's weird.*

Suddenly Severance registers the unmistakable bleating of the station's master alarm. Another NASA man, Jack Duce, cocks an ear and says, "Did you hear something about Progress hitting the station?"

"Oh, shit," Severance breathes. Suddenly everything makes sense. In a flash he leaves the NASA group and races out of the auditorium, charging down the carpeted hallways to the NASA suite, where James Medford and his small group remain frozen in silence, listening to Mike Malyshev's intermittent updates.

"Is this true about a collision?" Severance asks.

Medford's eyes are wide as poker chips.

"Yeah," he mumbles.

Severance reverses course and leaves the room, bounding down the internal stairwell to join Zimmerman at the NASA console. Neither says anything he will remember later. Severance looks over toward the flight director's desk, and for the first time he notices a pair of veteran cosomonauts, Nikolai Budarin and Sergei Krikalev, standing there. Budarin is a favorite of the Americans, a gung-ho, fun-loving extrovert always ready with a joke or an easy laugh. Severance is amazed by the glazed-over expression on Budarin's face. It takes the American a moment to realize it is fear. Then he looks at Krikalev, whose visage registers nothing of the sort. Krikalev is calm, focused, concentrated. Severance doesn't know it, but Sergei Krikalev is already thinking of ways to save the station.

"*If this light went off, that means DUZ is no longer functioning,*" says Solovyov. "*That's already good.*" DUZ is an alarm triggered by a sudden fall in pressure.

"*Yeah, good is good,*" replies Tsibliyev. "*But nothing is good up here.*"

"*What is the pressure right now?*" asks Nikiforov. "*How is it functioning?*"

"*It's slowly going up. Right now it's at 677.*"

"*Received.*"

"*I managed to switch off the TOR and the television. . . . Emergency ZMX light is for some reason still on. What does that tell you?*"

"*This means that your computer system is down,*" answers Solovyov. "*Leave it like that for a moment.*"

"*Right now the pressure is at 678 and is starting to grow.*"

"All right, Vasya," says Solovyov. *"Is the pressure in module O at 490?"*

"Right! There, it's falling. You can see that—if you look through the ninth window, you can see that one solar panel is damaged in module O."

"All right," says Solovyov. *"Let's first deal with the pressure. Is it still [growing]?"*

"Yes," says Tsibliyev. He mentions that he can still see the Progress outside the station. *"It's flying around us right now."*

For the first time it occurs to many on the floor of the TsUP that the Progress may be cartwheeling out of control nearby and could hit the station again. The thought has certainly occurred to Tsibliyev and Lazutkin.

"Where is the Progress now?" asks Solovyov.

"I can see it right now in the ninth porthole. It's about 150, 200 meters away. It's either getting away or flying around us."

"Okay," says Solovyov. *"We'll deal with the Progress later."*

"All right."

"Can you open a few more [oxygen] cylinders in the TSO?"

"Certainly we can."

"Okay. First we'll take everything out of the [oxygen tank] from which you are pumping now, and then when the pressure reaches 720, 730, we'll take it out of the TSO."

Lazutkin hollers that he can see the Progress.

"I have seen it, Sasha," says Tsibliyev. "But right now it's 250, 300 meters away."

"Less!"

"Less?"

"What is the pressure, Vasya?" asks Solovyov.

"Right now it's already 680."

"Is [the tank] still hissing?"

"Yes, it's hissing."

Lazutkin is trying to tell Tsibliyev something about the Progress. "Sasha, about the Progress later!" the commander shouts.

"Later about the Progress," seconds Solovyov.

"It's spinning crazily around after the collision."

"Vasya!" Solovyov barks. Then, an aside: "He doesn't hear."

The comm breaks up for a moment.

"I can't hear you," says Tsibliyev.

"Vasya, right now you have to retune DSD. What's your pressure at? 680?"

"I have it now set at 690."

"Set it down to 670. What is the next number?"

"We have to set it at 660."

"Set it at 660. Put DSD on control and switch On. In this way you'll make the [alarm] light go off."

Tsibliyev takes a moment to reset the DSD sensor. *"I have just done it, and the warning light indeed went off."*

"Great! Now pump [the air] out of the [tank] till it stops hissing. You have to close it and control the pressure with the mano-vacuumeter. We have to understand the nature of it, whether our station is hermetic or not."

"Got it."

"I am watching it again," Lazutkin hollers. The Progress.

"Is it far away? Where?"

"What's the pressure now?" asks Solovyov. It occurs to him now that they need to hurry; Mir will pass out of range of the Russian ground stations in just a few minutes. He murmurs an aside: "I am afraid that they will be out of our zone again and won't be able to hear us."

"683," says Tsibliyev.

"It's growing. Good.

"The Progress is moving ahead of us."

"Where is it?"

"It is rotating around the station."

"About one hundred meters away," says Lazutkin.

"Sasha, let me see."

"I think it went ahead. I will (go) look." Lazutkin pushes off and swims toward Kvant 2 to watch the Progress.

"Guys, let's do this the following way," says Solovyov. *"Right now just watch the Progress. The important thing is to take care of the pressure."*

"All right."

"What's happening with SUD right now? Is it in the 'indication' mode?" SUD refers to the station's motion-control system; if it has entered indication mode, Mir is in free drift and thus unable to keep the solar arrays toward the sun.

"Yes."

"Then we'll leave it."

With the station in free drift, its remaining solar arrays are unable to track the sun and thus generate power. With no new power coming into the system, the existing onboard systems will slowly begin to drain what power is left in the station's onboard batteries. It should take several hours for the batteries to drain altogether, longer if the crew shuts down most of the station's major systems. Solovyov, his eye already on the approaching end of the comm pass, begins instructing Tsibliyev which systems to shut down, and in what order, in the event power levels begin dropping while the station is out of contact with the ground.

"We are switching [*off*] *all that is not vitally important,"* Tsibliyev says.

"If you have real trouble with SEP"—SEP refers to power levels remaining in the batteries—*"the priorities will be the following: First switch off the Elektron, and only in the last moment* [*switch off*] *Vozdukh."* The Vozdukh carbon dioxide scrubber is the last thing Solovyov wants turned off, since the station is already running low on the replacement LiOH canisters.

"Elektron is switched off right now," says Tsibliyev.

"You should be fine with SEP. Just try to save it, but I don't think you will be anywhere near to switching off Vozdukh."

"We'll be watching the pressure gauge."

"What's the temperature right now?"

"It's quite chilly."

"You should give [*the pressure*] *time to stabilize."*

"I didn't get you."

"The pressure has to stabilize."

"Okay."

"What's the pressure right now?"

"689 and holding."

The enormity of what has happened overcomes Tsibliyev for a moment. *"It's so frustrating, Vladimir Aleexevich,"* he blurts out. *"It's a nightmare."*

"That's all right, Vasily," replies Solovyov, trying to keep his commander focused. Albertas Versekis, a docking specialist, joins Solovyov at his console. *"Now tell Albert chronologically what was happening with the Progress."*

"Everything was going as planned. We were thinking that we should give it some more space for acceleration. We ended up not doing it."

"All right."

"I started to put down the lateral velocity. It started to sink down. And then there was permanent braking—"

"Were you braking?"

"Yes, and I was trying to bring it down. I was holding it tightly with my hand to make sure that it passed away from the solar panel. It indeed passed on the side, but then it slightly bent to the left and punched the top solar panel of module O with the needle. Then it touched the attachable cold radiator with its top solar panel on the right side."

"Did it damage it?"

"Yes, a bit. However, it bounced back immediately. It seems the speed was not that great at that moment, and we probably did not have enough energy to brake [*the Progress*]*."*

"Got you."

And then the pass is over. It is 12:42.

Aboard Mir all three men reunite in base block. Lazutkin and Foale, still buoyed by the adrenaline rush of the hatch drama, actually force smiles. They have saved the station, and they are happy. Tsibliyev, though, looks anything but relieved. He sits at the command station, silent and dazed. No one says much of anything. There is nothing to do now but wait.

Solovyov strides onto the floor where a large group of technicians is huddled around the flight director's console beneath a giant model of the station. Viktor Blagov is there, a stunned look on his face. Valery Ryumin stands to one side, looking shell-shocked. Everyone is talking at once. Keith Zimmerman hovers at the edge of the Russian scrum, unnoticed. He has dozens of questions for which he needs immediate answers, but no one has the time to talk to him. Other than a direct hit on base block, Spektr is the worst place the station could have been hit. It is Foale's home and houses most of his private effects. It is also home to half the American experiments. Far more important to the Russians, the outer hull of Spektr holds the station's four newest solar arrays, which generate eleven kilowatts of the station's normal power output of twenty-six kilowatts. Without those arrays Mir is crippled.

Solovyov wastes no time with hand wringing or fingerpointing. Within minutes he and Sergei Krikalev and dozens of technicians begin brainstorming a recovery plan. If they are to have any hope of restoring the station's power, they must somehow get the Spektr arrays reattached to the main power supply. The obvious way to do that is by stretching cables directly from the Spektr solar arrays to some other part of the station. A Progress ship is already on the launchpad at Baikonur, scheduled for a liftoff on Friday, in two days. Solovyov's first thought is that cables could be loaded onto that Progress and be taken to Mir and mounted during a spacewalk.

Up in the NASA suite, James Medford is already dialing Frank Culbertson's home number in Houston when Zimmerman returns from the floor, having failed to get Solovyov, Blagov, or any of the crowd of busy Russians on the floor to answer his questions. The phone rings six times before Culbertson picks up.

"Hello?" he says. From the thickness of his voice, Medford realizes he has awakened his boss from a deep sleep. It is 4:45 A.M. in Houston.

"Frank, this is James Medford. Do you need a minute to get awake?"

Culbertson, realizing the call is from Moscow and grasping that it must be important, snaps awake.

"No, that's okay. What's up?"

"We've got kind of a situation here," Medford said, "and I need to put you on the speakerphone."

Zimmerman takes over and explains what has happened: the docking test, the collision, the depressurization, the race to close the Spektr hatch, the loss of half the station's power. When Zimmerman finishes, all Culbertson can say is: "Oh, wow."

There is not much else to say. Culbertson tells Zimmerman to keep him posted. Then he phones Van Laak, who is also asleep. "This one's bad," are Culbertson's first words. "The Progress hit the Spektr; the ship's decompressing. The crew's okay, and they're readying the Soyuz. I'll see you in the office."

At his home in suburban Friendswood, Charlie Precourt sees the news on CNN as he is preparing to leave for his office. Precourt's reaction, like that of many at NASA, is not so much shock as anger.

"My God," he says aloud, "what the hell have the goddamn Russians done now?"

Linenger hears the news when someone from JSC calls him at his house on Armand Bayou. Afterward he slams down the phone, livid. "What the hell is going on!" he seethes. "Does someone have to get killed before anyone around here will listen to me?"

Aboard Mir

When they are able to recover their senses, Tsibliyev, Lazutkin, and Foale begin sorting through what happened. "As soon as the comm pass was over and we realized we were staying on the station, naturally the question arose: Why did it happen?" Lazutkin remembers. "We constantly thought about it, and Vasily thought about it. We kept thinking: What did we do wrong? All our conversations centered on the incident. Vasily was already starting to worry what the ground would do to him when he returned, because it was he who had controlled the ship. I felt at that point that I knew he wasn't guilty of anything. But I couldn't find the facts to prove it. I felt he needed psychological support, and Michael felt the same. We told him, 'Vasily, it was not your fault.' "

Among the dozens of questions the three men debated was what effect—if any—Foale's foot had had when it hit Tsibliyev's arm in the moments immediately before the collision. Months later, each of the three would have sharply different impressions of what had occurred. Lazutkin, for instance, is convinced the brief contact between Foale and Tsibliyev actually lessened the damage from the collision.

"Before Michael hit Vasily with his foot, the Progress was flying straight toward Spektr, its back end pointing forward," he recalls. "[Vasily] took his hand off the controls, and the ship changed its position. As soon as Michael hit Vasily's hand, [the ship] moved, and it hit Spektr with its side. If the ship had continued flying the way it was flying, it might have been much worse. It would have hit with the sharp edge of the rear, rather than the blunt edge of the side."

Foale doesn't believe the bump had any impact on the flight of the Progress. To move the ship at all requires that one of the TORU levers be clamped down for at least several seconds; a slight jolt, Foale argues, would have had no effect. Asked later about the incident, Viktor Blagov would express ignorance. "This is new to me," he would say. "But afterward, when we looked at the videotapes, we didn't see any changes in the path of the Progress. It was very smooth."

Months later, sitting in a quiet tearoom off a colleague's office in Star City, Tsibliyev takes a deep breath when the question is put to him. "Okay," he says, "no lies. Just the way it was." He pauses and thinks, sticks his tongue into his cheek, then makes a little wave with his right hand, as if to ward off the question. "Michael is a good guy. I love him as a brother. Whatever he says is good [enough] for me. We should forget about this. What is important is we survived."

But, he is pressed, didn't the impact of Foale's foot have any effect at all? Tsibliyev smiles nervously. He is clearly uncomfortable with the question. "Only the three of us know what really happened there. There are some things that you can't write about, that have to remain secrets between cosmonauts. Michael is a very good guy. He's a very, very good guy."

Again he stops to think.

"But the answer is no, it didn't help. You can write this: After [the collision], Michael asked me, 'Was it I who did that? Who caused the collision?' I said, 'Yes.' He said, 'Sorry.' I said, 'That's okay.' The fact is, little things contributed to what happened, and we had a collision. That was one of the little things."

1:52 P.M.

"*Vasya,*" Solovyov says as Mir comes back into range. "*How are you?*"

"*Nothing has changed since one* P.M.," Tsibliyev replies. "*The pressure remains at 692.*"

"*Great. Let's leave it that way.*"

"*At 12:47 the Progress was 400 meters away. At 13:30 it was already 2,600 meters away. It is getting away but keeps revolving around the station.*"

"*Okay, we understand the problem with the Progress. If the* [Spektr] *module is pressurized properly, we'll leave it that way and keep with that pressure. Now we will have to deal with SEP* (the power). *There are a number of problems there.*"

The capcom begins to dictate a series of computer commands that should enable Tsibliyev to turn off several warning lights that have come on.

"*Which* [oxygen tank] *did you use for increasing the pressure?*"

"*The yellow one.*"

"*Just one of them? Have you touched the other one?*"

"*No, we left it for the moment.*"

To get the solar arrays pointed back toward the sun, they will eventually have to stop the station's slow roll. First Solovyov needs to know how bad the roll is. "*Could you assess the rotation velocity in relation to the Earth?*"

"*About a degree a second.*"

"*That's enough.*"

Foale chimes in. *"Yes, one degree a second."*

"Let's hold on for a minute." Before he can go much further, Solovyov needs to know how bad the damage to the solar arrays really is. *"Can you see the damage through the ninth porthole?"*

"The solar panel is intact. I believe [*the Progress*] *has just punched it through. The solar panel did not fall off; it was just slightly bent. And it looks like the Progress probably touched the radiator with its own solar panel. I think it must have punched a bit in that area."*

"If we get television for the next session, would you be able to show it to us?"

"Certainly."

"Do it with just your hand. Record it through the window."

"All right."

"Have you recorded the TOR mode?" This video from the Progress as it approached Mir should show the collision as it occurred.

"Yes. We haven't yet got time to look it over."

"Please watch it before the next session."

"We'll try to get it to you."

The pass ends a moment later, after Tsibliyev has given Solovyov more information on the rate of the station's roll.

3:27 P.M.

"H*ow are you doing?"* Solovyov asks the moment the station returns into communications range.

"692," says Tsibliyev, reading the pressure gauge. *"The Low Voltage light has just gone on. We are turned with the* [*solar*] *panels' sides turned to the sun. We can't manage to turn it the other way around."*

"Let's look at the angular velocity."

They are spinning slowly.

"That's good," interjects the capcom. *"We have two refrigerators in module I. They get their power supply from electrical outlets in the main module."* These are the American science refrigerators, filled with blood samples and other samples Foale needs.

"I have switched off one of them."

"That's good," says Solovyov.

"Michael," Tsibliyev tells Foale. "Go ahead and show the record."

Foale pushes a button that broadcasts images from the camera in Lazutkin's hand. Up on the big screen in the TsUP's auditorium, grainy black-and-white video begins to flicker. The floor falls silent as the camera pans across the damage.

"Yes, we can see it," says Solovyov.

"Wow," Keith Zimmerman murmurs to Terry Taddeo, the flight surgeon. The two men are standing at the NASA console. Zimmerman points up to the screen. "You can see where that panel is bent. Look." The camera widens, showing the entire array. "That array's obviously worthless at this point."

Zimmerman peers intently at the damage. Among the issues Culbertson has raised is whether the damaged solar array is loose or, worse, flapping, which could be a risk to future shuttle dockings. From the video, Zimmerman can see that the array is punctured and bent but appears to be stable.

"Can you get a close-up of the dent on the panel?" Solovyov asks.

"It's downward. On the left side."

"Is this the only damage on the solar panel?"

"Yes."

"Let's focus once again on the main body of the space station."

"It's difficult to film."

"Apart from this dent, are there any other places of collision?"

"No, I can't see any," says Tsibliyev. *"We looked at it out of every porthole and couldn't find anything else."*

"Vasya, have you watched your record [of the collision itself] on the VCR?"

"We didn't have time for that, as we already had a Low Voltage indicator."

"All right," says Solovyov. *"Let's switch off the television so that we don't waste your supply of energy. I would like you nevertheless to watch that tape this evening."*

"All right."

Tsibliyev takes a few moments to report on which systems remain functioning.

Valery Ryumin takes over the comm for a moment. *"Guys, when you were closing the hatch in module O, did you realize the damage was in O, or did you notice some other signs?"*

"I heard the noise when I was inside module O," answers Lazutkin.

"Was it the sound of hissing air?"

"Yes."

"In what area?"

"When you fly into the module, it's on your left. I forgot which panel in particular."

"Is it the one facing the floor?" Solovyov asks.

"I think so."

"The one that is in front of the table," says Tsibliyev. *"The blow was not too strong, it was more like a jitter, and when it started to move away, I noticed that the solar panel was damaged. I probably caught it on the solar cell and scratched that radiator, but there weren't any visible marks—even its antennae were intact. That's amazing."*

"Vasily, can you hear me?" Solovyov asks. *"By the noise of the leaking air, you should be able to judge whether it's a big hole or a small hole."*

"I haven't been in there."

"I was in the node," Foale interjects. *"I didn't hear anything. I just felt the pressure drop. I felt it in my ear. . . . Even when I was holding the hatch there, when I was helping Sasha . . ."*

"We are confronted with the following situation now," says Solovyov. *"We have just blown up the* [*oxygen*] *tanks. During the next session we'll try to get the* [*solar*] *panels in the right position. That means that we will have to manage the power till the next session."*

When the pass ends, Solovyov consents to a quick briefing with the journalists who are already scurrying into the mezzanine. "As a result of the accident, the systems of the station have lost a lot of power," Solovyov says. "The equipment, the scientific program, that comes second. We are trying to see what we can do to continue to fly. It appears it may be possible to do it. We already planned spacewalks for this crew, and if we work hard now, and keep the cargo ship, we can fix some cables, put them on the surface of the module, and use the power from them. The problem with the energy supply will probably be solved. The second problem is what to do with the decompressed module. We have to find the leak and seal it. Maybe during the spacewalk we will do an inspection. It looks like we will need to do two spacewalks. The Progress will stay [in Baikonur] until cables are made. We will be able to postpone the flight a maximum of ten days."

The reporters sated, Solovyov turns his attention back to the station. Now that they have assessed the damage, he needs to make certain they will not lose all power. With the station in free drift and the solar arrays askew, the station's onboard batteries are draining—how quickly they can't be sure. At the next comm pass he will need to work with Tsibliyev to stabilize the station's orientation, point the solar arrays back toward the sun, and begin recharging the batteries. With luck they will make it to the pass without losing all power.

4:59 P.M.
Aboard Mir

They don't make it. Four minutes before the pass begins, all lights go out in base block, plunging Tsibliyev, Lazutkin, and Foale into darkness. As the pass opens, the lights in Kvant 2 begin to flicker and dim. After a few moments, they, too, go out. Mir goes dark, the only illumination coming through the windows as the station nears the end of a daylight orbit. The three men can hear the whir of the gyrodynes as they start to power down.

"I want to report that the power has turned off in the main module, the gyrodynes started to slow, and the control board doesn't work," Tsibliyev tells the ground as the pass begins at 5:03. *"And we have two indicator lights on: Low Voltage and Failure of the Central Information System."*

"Turn off everything in the main module," orders Solovyov. *"In case of communication failure in the main module, turn on a transmitter and receiver* [*in the Soyuz*]*."*

"Got you."

"Switch off the air heater in [*the Soyuz*]. *You can switch off both refrigerators."*

"Received," answers Foale.

As the brief pass ends, the station enters a nighttime orbit, and without power Tsibliyev and the others float in total darkness. One by one the last lights have flickered and gone off. There is nothing but blackness. And when the gyrodynes finish powering down a few minutes later, there is silence. It is as if the station has died.

An awful thought crosses Tsibliyev's mind. *"Thank God,"* he tells the ground, *"the Progress is far away from us now."*

* * *

The sudden loss of power dismays the crew. "They blew it," Foale remembers. "The control center just blew it." Instead of watching video of the collision, Foale felt Solovyov should have immediately powered down all major systems, which might have allowed the station to conserve the power in its batteries. Still, at first Foale isn't especially worried. Months later, he likened the feeling to surviving a terrible auto accident, only to return home to find the lights in his bedroom won't work.

It is only when he realizes that Tsibliyev and Lazutkin have no idea how to *regain* power that Foale begins to worry. They seem to be caught in a catch-22. To generate power they must get the solar panels pointed toward the sun, which they can do with manual controls: It takes less than an hour to successfully reorient the Kvant 2 solar arrays. But as long as the station continues its slow spin, there is no way to keep the panels in the proper position. Somehow they have to stop the station's spin. Normally this could be attempted with the station's thrusters, but with power out, the thrusters will not fire.

"No one had any ideas," Foale remembers. "*No one.* Not the TsUP, not Tsibliyev. He was desperate for ideas. He was looking at me. I was really surprised he was looking at me. But there was no one else."

To everyone's surprise, not least his own, it is Foale who comes up with the plan. The Soyuz, he says. The Soyuz still has its auxiliary power. Why, Foale wonders, couldn't they use the thrusters on the Soyuz to reorient the station? By firing bursts against the prevailing roll, they ought to be able to slow or even stop the station's motion.

Even before he finishes the thought, Foale is uncomfortable with what he is saying. This is, after all, unprecedented. No American astronaut has ever been allowed to do significant systems work aboard Mir, much less chart a course of action. But Foale sees that Tsibliyev is all but paralyzed, unable to think straight. "I was nervous; to me this was purely experimental," Foale remembers. "It was the first time they had ever shown any dependence on my skills. That made them *very* nervous."

Lazutkin is nodding. Yes, he says. Theoretically, it could work. "We knew the idea to use the Soyuz engines appeared before our time," Lazutkin recalls. "Similar jobs had been conducted by one or two crews, [but] that was when the station was small, not as big as it is now. We were amazed that Michael, who didn't know that during training we had been told this was possible, that he came to the idea himself."

It is Tsibliyev who doesn't like Foale's idea. Even if the ground grants them approval to try it, how are they to proceed?

"It's fine for them to tell us to do that," he says, "but *how* do we do that? We have not been trained for this."

It's true, they haven't. It is a gaping hole in their training requirements, the cosmonauts will later admit. But the more Foale thinks about it, the more he is certain the Soyuz is their best shot. Still, Foale is uncomfortably aware of his own lack of experience with Russian systems. And there is one other thing that bothers him. It isn't the idea itself; it is the fact that it is his first idea. When it comes to technical fixes—whether in training, on the simulator, or on the shuttle—Foale has the impression that his initial ideas are usually wrong.

He doesn't say this aloud.

"Vasily, you know, I'm not the cosmonaut here," he tells Tsibliyev. "I haven't been trained on the Soyuz. You have. You've got to be comfortable with this."

And he's not. It is all too much for Tsibliyev to grasp. The Soyuz, after all, is their only escape route to Earth. If they begin fooling with the thrusters, they risk using up the Soyuz's precious propellant. Without propellant, they cannot return home.

"Vasily, how much fuel do we need?" Foale asks.

Tsibliyev shrugs. "I don't know."

"Well, how much do we have?"

The commander thinks for a moment. "About six hundred kilograms, I think."

"How much do we need for the deorbit burn and the undocking and everything else?"

Tsibliyev does some calculations. "Maybe two hundred kilograms."

"We've got plenty of fuel for this!" Foale says, even as he thinks to himself: Do we? He is convincing himself that they can do this. "Vasily, this is no big deal. How much could we possibly use? Half a kilogram? We've got the fuel. We can afford to make mistakes here."

But for the moment, Tsibliyev won't budge. It is too risky. "It was all a gut feel," Foale remembers. "It was all entirely at the commander's judgment. Vasily didn't want to do it."

Still, Foale persists, and Lazutkin thinks it is worth trying. Finally, Tsibliyev says he will ask the ground what to do.

"He had no choice," Foale remembers. "It was clear that if we didn't do it, we would lose the station."

6:36 P.M.

At half past six Mir comes into range of a German-owned ground station near Munich, which the Russians agreed to rent after the mothballing of their fleet. The German signal is weak, lasts barely ten minutes, and won't allow the transmission of telemetry data. It is better than nothing, but just barely.

"Is the Soyuz on autonomous power supply?" Solovyov asks.

"Not yet."

"Make sure it is right after this pass." A technician gets on comm and walks Tsibliyev through the procedure.

"You will need to switch off Vozdukh and switch on the American [*LiOH*] *absorbers,"* Solovyov says.

Solovyov reminds the crew that should they be working in Soyuz, they will need to return to base block for the next two comm passes, which are over the American Dryden and Wallops stations. Neither station is set up to receive a signal from the Soyuz.

"Okay."

Tsibliyev remembers—though it is not indicated on any written transcript of the discussion—that it is at this point that he asks Solovyov for permission to try firing the Soyuz thrusters to reorient the station. "They said they would get back to us on this," he recalls.

By the conclusion of the 7:52 pass they have permission from Solovyov to try Foale's plan. Everyone swims down to the Soyuz. Tsibliyev takes a seat behind the control panels. Foale and Lazutkin position themselves at the windows, watching to see whether the grouping of thruster firings the TsUP has given them will put the solar arrays in the sun's path. Tsibliyev pops the thruster button several times, as per the ground's instructions. After a minute both Foale and Lazutkin agree: The arrays remain in darkness. The station's roll continues.

"I [had asked the ground], 'Are you sure this is going to help the ship?' " Tsibliyev recalls. "They said, 'We are not. But try anyway.' So I tried, and it did not help. But TsUP had no chance to make the right recommendation, because they didn't have the right telemetry. [Without telemetry] TsUP could not determine the precise roll of the station. What they told us actually made the situation worse."

"These first actions were all wrong," confirms Lazutkin. "First of all, it was difficult to see what we were doing. The station was rotating as if it were in nature. I remember the ground asking us, 'Give us the specifics of your angles.' And we looked and we saw the Earth through one window, and then another window, and then it was rotating this way and then another way. It was very difficult to understand. That's why the ground couldn't give us correct recommendations."

Dejected, the crew returns to the base block table. Tsibliyev wants to wait for the next pass, at 9:25, to ask the TsUP what they should do. Foale is beginning to think they will have to do this all by themselves. Floating around the base block table, he and Lazutkin decide to construct a rudimentary mock-up. "We put on a flashlight and put it overhead and imagined it to be the sun," Tsibliyev remembers. "Earth was the table. We had a small model of the Mir station. We put this Mir model under the light and moved it this way and that, and looked at its relationship with the sun. We were like children playing."

"Even if we used the Soyuz engines, we had to determine which way to send the pulse," Lazutkin remembers. "On the ground, we were never trained to do this. We were only told we *could* do it. Even now the crew relies on the ground for answers to questions like this. There are specialists on the ground, and they are supposed to come up with these solutions. But in this situation, we realized the ground didn't know what to do. They couldn't teach us how to swim."

Foale gets out some paper and begins to sketch. By holding his thumb up to the window, he can estimate the speed of the station's roll. The difficulty comes in calculating the Soyuz's exact position in relation to the station. They float back down to the Soyuz, and Tsibliyev again takes his place behind the controls. Foale studies his position intently. He can't figure out which direction the Soyuz thrusters will be firing—left, right, up, down.

"Hold on," Foale says.

He floats back to base block and then retraces his movements into the Soyuz: If standing in base block points his head "north," then Tsibliyev seems to be sitting at a forty-five-degree angle to "north." Lazutkin tries it as well and is just as confused. "There were some comic moments there," Lazutkin recalls. "When I looked at Vasily, who was in the Soyuz, I couldn't figure out what position he was in relative to the position of the station. If you open the door [to the Soyuz] and you see a man in this position, you can tell him, 'Your right side is your left side.' But what if the man in front of you is sideways and his head is upside down? How do you tell him what to do? It was very confusing."

Finally, Foale takes a deep breath and calculates a new series of thruster firings. According to his math, if Tsibliyev fires the right thruster for exactly three seconds, it should just about halt the station's roll. Tsibliyev, though far from certain Foale's plan will work, agrees to try it, but only after the TsUP approves the plan.

"When we understood all this, and Michael had made his drawings, it turned out we had to make these very short impulse [firings]," he remembers. "We tried to explain it to the TsUP, but the [comm] passes were so short, we couldn't. So the TsUP said, 'Okay, guys, you try it, let's see what happens, because we have to do something.' "

They are on their own.

"We can *do* this, Vasily," Foale urges Tsibliyev. The commander looks skeptical.

They return to the Soyuz. "Okay, three seconds," says Foale. "Try it three seconds."

Tsibliyev presses the thruster lever three times, quickly.

It doesn't work. Foale, looking out the windows, sees that the solar arrays remain in darkness.

"Vasily, how long did you hold the thruster?" Foale asks.

"I didn't hold it. I just hit it." Pop. Pop. Pop.

Foale realizes Tsibliyev is being conservative in an effort to save propellant. "That won't work, I don't think," he says. "If you just hit it, that's not pressure enough. We need more than that. You have to actually hold it down for three seconds."

There are more calculations and another drawing or two before Tsibliyev finally sits and follows Foale's directions. He nudges the thruster lever for one . . . two . . . three seconds—and releases.

Foale and Lazutkin study the rotation and the solar arrays. After a moment they begin to smile.

"I think it worked," Foale says.

Months later, Viktor Blagov will roll his eyes at the difficulty the three-man crew had in reorienting the station. "The cosmonauts have been doing this [maneuver] for years," he insists. "These guys did not invent this. It's unpleasant for them to hear this, but it's true. The TsUP invented this. We have done this [maneuver] at least fifty times over the last twelve years at the station. It's easy to do. Just read the instructions. It's all in the instruction manuals. I guess I wasn't surprised that they didn't do that. In a period of stress like this, anything is possible."

TsUP
9:30 P.M.

By nine, Keith Zimmerman had sent most of the NASA contingent home with instructions to get a good night's sleep for what promised to be several days of hard work ahead. Mark Severance and Terry Taddeo volunteered to remain behind, bunking down on couches in the NASA suite. Taddeo had a travel bag packed with emergency medical supplies. If the crew was forced to evacuate the station during the night and return to Earth, Taddeo was ready to jump on a flight to meet them. When everyone else had left, Severance turned to Taddeo and said, "I wonder if we're still going to have a station in the morning."

June 26
12:40 A.M.
Aboard Mir

By one o'clock Foale's plan has allowed the crew to all but halt the station's slow roll. There is just one problem. The station's new orientation leaves the solar arrays outside Kvant 2 out of the sunlight, which means there is no power in the module. The toilet is in Kvant 2; it cannot work without power. Lazutkin remembers that one of Reinhold Ewald's experiments had the German astronaut urinating into a series of condoms and bags, and Ewald had left about fifty of the bags behind on Mir. Foale manages to dig them out and puts them in the toilet for use until full power can be restored. That at least covers their urination needs. No one is eager to talk about how to defecate into the bags. "Number two," Foale remembers, "was a lot harder. You just had to grin and bear it."

"Have you called our homes?" Tsibliyev asks the ground during the 12:40 pass. *"Are they panicking?"*

"Everything is okay," Solovyov assures him. They will begin the recovery plan in the morning.

2:18 A.M.

"The pressure is at 691.6," Tsibliyev is saying. *"CO_2 is starting to get absorbed. It's down to 5, although the dew is still at 22. The voltage of SEP is at 28.5. Oxygen is at 161."*

"Received," says Solovyov.

"We still have the Low Voltage light on in D. It's dark and quiet over there. The solar panel is turned away from the sun."

"So far everything looks fine."

"Are you guys still up in TsUP?"

"People have mostly left. My God, there were so many people here. I have never seen that many. . . ."

"I can imagine. They are probably blaming us left and right."

"No, we weren't talking about that."

"We were again talking about the situation here. It happened the same way as last time [*in March*]. *The only difference is that the last time it moved safely away, you see, and this time it hit the station because I started to direct it manually. I still can't get why. And how did its speed accelerate so much by the end? You know, at the beginning it took us a while to make it move at all. We were even thinking about pushing it a bit. And at the end it got so high that we weren't able to manage it."*

"Did you watch the video?" Solovyov asks.

"No. . . . We were trying to gauge the speed of the collision. We think it must have been something like one meter, because as soon as I noticed that it was getting

in the way of the station, it had already punched it and bounced back. If it had been moving at a greater speed, we would have crashed everything. . . ."

"Received."

"You see, that first time we managed to divert it. . . ."

Tsibliyev pauses.

"And this time . . ." He can't make himself say it. *"Sorry."*

He goes on. *"We should have stopped the whole damn process at 200 meters and left the bloody Progress altogether. But I was trying to pull it closer. I thought that we'd have time to bring down the speed."* He glances down at the computer. *"Right now my y-axis speed is 0.5 degrees"* per second.

"I have a feeling that the charge of the main module's panels is going fine," says Solovyov. *"I think we should be fine in regard to power. It's normal that it takes two or three* [*days*] *to get out of that pit. . . ."*

"We hope so," says Tsibliyev. *"We'll have a regular* A.M. [*comm*] *session, right? I have set my alarm clock to wake us up should we fall asleep. However, it's set in the system. If something goes wrong with* [*the station*], *we may not wake by the time the session starts. You should send us a signal if you notice we are late. We'll try not to fall asleep, but you never know."*

"Received," says the capcom. *"You could take a nap for one circuit."*

"By the way, Michael laid down to rest a little bit."

"You should take turns too."

"After such stress and frustration, I don't think we'll able to do that for a couple of days. We are unable to switch off for more than a couple of seconds."

"No, guys, you should try to rest at least for a couple of hours."

"I am afraid this won't happen. But we'll manage, and we'll have enough time to catch up with sleep later."

Twilight
Houston

At the end of the day, Van Laak and Culbertson are the last ones remaining in the Phase One offices. They spent the afternoon clustered around the circular table in Culbertson's corner office, culling information from Moscow and relaying it to Abbey, Goldin, and dozens of others. The mood is bleak. As if they don't have enough to worry about in space, Congressman Sensenbrenner has already been to Goldin's office, demanding a sweeping new safety review of Mir before another astronaut can replace Foale.

Van Laak pauses in Culbertson's doorway as they prepare to leave. For the first time that day he is beginning to feel the tiniest bit of optimism. "You know," he says, "there's a chance we can pull this all together."

Culbertson is quiet for a long moment.

"I don't think so," he says, looking up. "I think we're done."

18

Nowhere were the contrasts between the American and Russian manned-space programs more vivid than in their reactions to the collision. In Houston, Culbertson and Van Laak awoke Thursday morning believing there was a strong chance that the entire Phase One program was finished. The station was crippled. With Spektr sealed off and Mir's power supply cut in half, the American science program was all but over. They were ready to give up.

Not so the Russians. While the world press flocked to the TsUP's mezzanine to point their cameras down on Solovyov, Blagov, and their ground controllers, the Russians calmly went about the business of resurrecting the darkened station. "You have to remember, we've been operating this station for a long time, and believe it or not, we've had a lot of abnormal situations," observes Valery Ryumin. "Our guys are simply tougher than the Americans in these situations."

"Very often we had to calm the Americans down," recalls Viktor Blagov. "From the very beginning, Culbertson's reaction was very sharp, very worried. But then he saw our reaction was calm, and he trusts us—at least he says he does. He believes our actions are sometimes 'aggressive.' That's the word he uses. In Russian we translate that as 'hasty.' That was the meaning I took. He always tried to figure out how justified our actions were. You see, the Americans look at every problem from a philosophical viewpoint. First they start thinking about how to examine the problem. Then they research the problem, to find out how they want to look at it. Only then do they take action. We start in the middle of this process. We tackle big issues first. Only after that do we look at the philosophical questions, [such as] 'How did this go wrong?'"

The differences in the two programs' reactions to the collision reflected their experiences in space. Despite two years of work aboard Mir, the Americans remained largely ignorant of the practical realities of long-duration space flight. When there was a mechanical breakdown aboard a shuttle, the mission was simply ended, and repairs were performed later on the ground. The Russian space stations, of course, never had this luxury. When something went wrong aboard Mir, cosmonauts were forced to fix it in space, which had given the Russians twenty years' experience in the kind of seat-of-the-pants repairs the Americans had only read about in books. While NASA tended to study and diagram things to death, the Russians had developed the skills needed to fix things on the run.

The crisis aboard Mir, the Russians were at pains to point out, was hardly their first, or even their most challenging, rescue mission. Space aficionados still speak in awed tones of how cosmonauts revived the Salyut 7 station in June 1985, after the TsUP reversed a decision to abandon it. Frozen and empty, the station had drifted without power for eight months when two cosmonauts, Vladimir Dzhanibekov and Viktor Savinykh, approached in their Soyuz and managed a difficult manual docking with the station, which was in free drift. The temperature inside the station was well below zero when the two men, wearing fur-lined jumpsuits and oxygen masks, clambered inside with flashlights. Every surface was coated with frost and icicles. Working without ventilation to clear their carbon dioxide, the men grew cold and sleepy, repeatedly retreating to their Soyuz to rest and regain strength. As they worked to resuscitate the station over the next several days, patching burst pipes and thawing water supplies, headaches plagued them. Eventually they managed to revive several of the station's dead batteries and from there gradually switched on all the station's major systems. It took weeks to slowly resurrect Salyut 7, but by the time Dzhanibekov and Savinykh finished, they had shown that the Russians could almost literally bring a space station back from the dead.

June 26
6:44 A.M.
Aboard Mir

"Everything is the same," Lazutkin is saying. His voice is heavy and tired.

"Sasha, I hope that Michael and Vasya are resting," says the capcom. *"We hope—"*

"Yes."

"That's good. I think you should lie down, too."

"Later," says Lazutkin. *"Once I finish everything."*

"The thing is, we are not going to do anything during this session, as the chief staff will show up any minute, and they will be busy making decisions. You know we have a full plate of important decisions to make."

It has been a rough night on the station. All three men snacked a bit around four, but no one wanted to eat much. To do so means urinating into Ewald's bags, or something too messy to contemplate.

"I still have that picture in my eyes, with the Progress suddenly popping up in the ninth porthole," Lazutkin says. He manages a weary laugh.

"Where were you at the moment?"

"I stood next to it. Started to run from one porthole to the other, then stood by the ninth. It was really, really quiet."

"What part of the Progress hit?"

"God knows what it hit with. It was moving with its nose forward. When I saw it, I got out of there. It was flying straight at us."

8:16 A.M.

By eight o'clock the TsUP is already inundated by the press. Hundreds of reporters and cameramen throng the wide carpeted hallways, milling about outside the NASA suite, trying to grab anyone they can for a quick interview or a sound bite. For every comm pass that day, Mark Severance has to elbow his way out through the crowds, dodging the microphones and tape recorders thrust into his face, to make his way down to the floor. "Mr. Severance! Mr. Severance!" they yell. "Can you give us a statement!" On the floor, the NASA console is situated directly below the mezzanine, which is lined with television cameras and photographers. "Mark! Mark!" the more aggressive lensmen take to shouting. When Severance involuntarily turns, someone snaps his picture looking surprised. It gets so bad in coming days that the NASA men can't use the bathrooms in peace. Reporters accost them at the urinals. Severance finally has enough. "Could someone please move these damn people out of my face?" Severance erupts during a telecon with Houston. "We're trying to do a job here." A few days after that the press is briefly barred from the TsUP.

At a quarter past eight a capcom hails the station.

"Sirius, establish contact with TsUP."

"Received," says Tsibliyev. *"Good morning."*

"Did you at least take a nap, Vasya?"

"I was knocked out."

"We understand."

Vladimir Solovyov gets on comm. *"This is a very important moment. Is [Kvant 2] facing the sun's active surface?"*

"I will check it once again," says Tsibliyev.

"We'll build our strategy around that."

Tsibliyev checks the module out the window. *"It is lit, but is overcast by a shadow from the module that is crossing it in the middle."*

"Is its computer system on? What about the vents? Is it breathing?"

"I will go look," says Tsibliyev.

After a request from Solovyov, Lazutkin presses a button and the computer video of the collision begins downloading to the TsUP. The picture quality is not good.

"What can you do?" says Tsibliyev, sounding resigned to the picture quality.

"We did not manage to get any closer to it to get a better picture," says Lazutkin.

"That's all right," Solovyov interjects. *"We were trying to get at least something."*

When the video is downloaded, Solovyov explains their recovery strategy. It is what Viktor Blagov calls the "Coffin Scenario," a long-established plan to resuscitate the station from a total loss of power. The first step is for Tsibliyev and the crew to gather the best and newest batteries from throughout the station, lug them into base block, and allow them to begin charging off the solar arrays, which, now that the station's spin is stabilized, are able to point toward the sun. Once these batteries are fully charged, a process that should take twelve hours or so, they can be used to power up base block's guidance and control computers. Later, batteries from elsewhere in the station can be used to power up other modules. All told, the recovery should take about two days.

Following Solovyov's orders, Tsibliyev, Lazutkin, and Foale spend most of the day moving batteries around the darkened byways of the station. It is slow work, made more difficult by the rampant clutter clogging the modules. The trio's only illumination comes from the pen-size Maglite flashlights they keep clamped between their front teeth as they work. The putty-gray batteries, each the size of a countertop television, are located beneath floor panels that are blanketed by all manner of spent and full containers, flight bags and cable bundles. It falls to Foale to move much of the debris, after which, using a Maglite and a screwdriver, he locates and unscrews the panels. Tsibliyev follows in his wake, rapidly unplugging the ten thick connecting cables that hold each battery in place. Foale asks to unplug the cables as well, but the commander stops him. "They didn't want the American doing it, only the Russian, because screwing it up could affect their bonuses," Foale remembers with an ironic smile. "That makes the Russian more responsible, you see. So I did everything up to where any screwup wouldn't affect their bonuses."

Freeing the batteries one by one, Lazutkin and Foale take turns pushing them into base block, where they are hooked up to solar arrays to begin the recharging process. With any luck they should have power back the next day. Foale spends his free time minding the ham radio. He manages to get an encoded message for his wife written and sent down to Dave Larsen, the ham operator in California. Larsen forwards it to Rhonda Foale, who is visiting her family in Kentucky.

11:21 A.M.

By that morning it is already clear that the TsUP's initial ideas on reviving Spektr by draping cables on the station's outer hull will not work. The cables would weigh a ton or more, far more weight than a Progress can carry. Sergei Krikalev gathers a small team of cosmonauts and engineers to brainstorm an alternate solution. The only way they can see to save Spektr, and regain the station's full power, is for the cosmonauts to somehow enter the dead module in space suits and jury-rig a power supply. From the rate the station has depressurized, Krikalev believes the hole that has been punctured in Spektr's hull—wherever it is—is roughly three centimeters wide, about the size of a postage stamp. If Tsibliyev and Lazutkin can somehow wriggle into the dead module, they could conceivably find and patch the hole.

Before they can seriously entertain this idea, however, they have to make sure it is possible to enter Spektr in a space suit. A little after eleven, Krikalev gets on comm with Lazutkin. *"About your passage into the module—I heard you have a lot of stuff piled up there,"* says Krikalev.

"It's not because of the stuff," says Lazutkin. *"It's difficult to pass the corner because of the KFA"*—the massive camera at the back of Spektr.

"You mean in the area of the hatch? Is your hatch wide open . . . ?"

"Yes," says Lazutkin. *"The hatch is simply covered up with the lid."*

"That means you can take it away, right?"

"Yes."

"That means that if we make the decision that you enter the module wearing space suits, you should be fine, right? Will there be enough space?"

"Yeah . . . ," Lazutkin says, catching on. *"You could pass."*

"Do you think you could make your way up to the solar panel cables?"

"I don't think so. You have KFA and Balkan." Balkan is a large German camera anchored in the center of the module. *"I would have to squeeze in even if I were wearing just my underwear."*

"It'll be great if you could examine the cable of the damaged solar panel."

Lazutkin says he would be willing to try. *"If you will help me find it."*

"It's roughly in the center of the module. . . . In the space suits, you should be able to get close to the cable."

"Who knows . . . ," says Lazutkin, not at all certain they could pull off this kind of thing. *"I'm not sure we can get behind those panels."*

"Is there a lot of stuff behind the panels?"

"The whole area is stuffed."

While a skeptical Lazutkin talks with Krikalev, after several tries Foale finally gets his first chance to talk to Keith Zimmerman on the station's secondary channel.

"Michael, this is Keith."

"Okay," says Foale. *"What do you need to know?"*

"First, how are you?"

"Great. As good as one can be without your stuff."

"What personal/medical stuff do you need?"

"I propose that we send up a complete shuttle medical kit," says Foale. *"Things like aspirin."*

"Copy. What personal stuff do you need? Hygiene, et cetera."

"Exercise shoes. I will need those for sure. I don't have any on board that fit me. Also, my treadmill harness and the expanders."

"Copy all."

"I'd really like a shaver and toothpaste and a toothbrush. Three tubes of toothpaste."

Foale spends several more minutes telling Zimmerman what computer and scientific items he could use. At 11:44 he hands the comm back to Tsibliyev.

4:05 P.M.

Krikalev is on comm again, quizzing Lazutkin and Tsibliyev.

"Could you tell me objectively what is the current state of [Spektr]?" he asks. *"We need to understand whether it's feasible to get to the damaged solar panels' gear wearing a space suit."*

"It's hard to say," says Tsibliyev. *"You'd have to model that yourself."*

"We'll do it here. We have a rough idea that there will be two narrow spots. However, there should be another passage by the floor. My question is whether you have a lot of stuff piled on the floor."

"There's the cyclo-ergometer," says Lazutkin. The stationary bike.

"It's located in the center, right under the table," adds Lazutkin.

"All right," says Krikalev. *"All we need is to make a strategic decision whether we are going to go there or not. We'll certainly model that first down here. However, no one will be able to make a better reality check than you guys."*

"We should be able to bring the tabletop down," says Lazutkin.

"On the sides there are things that narrow the passage, right?"

"Yes," says Lazutkin.

The day ends with one disconcerting mix-up. Each of the solar batteries is unique, with its own individual charge level and age. The ground has emphasized how important it is to keep track of which battery has been moved where, so that all the batteries can get the highest possible charge. Tsibliyev asks Foale to write down the serial numbers and positions of all the batteries they have moved—a task the commander isn't entirely comfortable assigning Foale.

"He was our brain," Tsibliyev remembers with a grin. "We used to tell him, 'You are the brains, we are your hands.' We didn't want him to be involved in dirty work. It was awkward to ask him. But he did everything we asked." In honor of his prestigious university degrees and his new status as the station's "brain," Tsibliyev has given Foale a new nickname, "Cambridge."

But when it comes time to read the battery information to the ground, the brain fails them: Foale realizes he has lost the list somewhere in the station's murky depths. "So I said to the ground," Tsibliyev recalls, " 'You guys probably can wait for this list till later, right? When we have power.' And TsUP says, 'That's fine.' And Michael was really relieved."

"M*ike, just a couple quick things for you,"* one of Blagov's capcoms tells Foale shortly before midnight. *"Be advised that the press is getting all the air-to-ground between us and you, except the PFCs and the PMCs, of course."*

Foale laughs. *"Have I said something?"*

"Just to make sure you don't say anything!"

"Okay."

"No more comments about being caught with your pants down!"

Foale will prove as good as his word. For the next two months he speaks of little on open comm except the health of his two main experiments, the slender mustard plants growing in the greenhouse and the tiny beetles—used for a study into sleep patterns—stored in a locker down in Priroda. That night he makes a makeshift bed for himself in Priroda. There is no power in the module yet, and it is cold. He wakes several times during the night, shivering.

"Did my wife call?" Tsibliyev asks shortly before midnight.

"Yes, she called this morning."

"How is my family? They did not die of fear?"

"They are okay. I will call them."

"We are going to sleep with a clear conscience."

Friday, June 27
5:50 A.M.
Aboard Mir

"MCC answer Mir."

Tsibliyev is awake and on comm. *"Answering."*

"Vasily, how do you hear MCC?" The speaker is a controller named Igor Topol.

"Hello, Igor, we hear you loud and clear. At 5:18 the Central Onboard Computer failed when module D power went off three minutes prior to leaving the night pass. . . . Do you copy?"

"Yes. Go ahead, Vasily."

"Before that I got up to go to the bathroom at 1:43 and I saw that the power in module D went off one minute prior to entering the night pass. And then I woke up again at 3:12—three minutes prior to entering day pass—and I looked at the computer. I saw six set points for the gyrodynes, and then it went off again at 5:18."

To Tsibliyev's dismay, Mir has lost all power during the night, derailing the recovery plan. It will take Blagov and the TsUP controllers most of that day to understand why. What they find is this: Late the previous day, Russian controllers at the TsUP began sending up commands to Mir's central computer to test the gyrodynes. From all indications, power levels in the onboard batteries were now high enough to support the tests. In fact, they weren't. Large spikes in power tended to shoot down from the solar arrays to the batteries every time the station moved from night into a daytime orbit. Because of that, Blagov explained to the Americans, the batteries were protected by surge protectors. Somehow—no one would be completely sure why—the surge protectors prevented the batteries from charging properly. The Russians hadn't known this, Blagov explained, because they didn't have accurate telemetry data on the batteries. "That was a new one to us," remembers Keith Zimmerman.

Not realizing that the batteries hadn't fully charged, the ground controllers went ahead and ran gyrodyne tests. As a result, the batteries quickly drained, and a little after five that morning Mir's central computer crashed. All power was lost. "They pushed too hard," recalls Zimmerman. "They tried to recover too quickly."

"This is just too bad—we only had one orbit left," Tsibliyev says. *"We were so very close. I hope the attitude does not collapse now."*

"Probably it will, but not [too] fast," says Topol. *"The experts are coming in, so they can look at the problem, do an analysis. We are looking at the telemetry, but we don't see what else we had on that could cause this failure."*

"I looked at the display yesterday before going to bed, and the gyrodyne displays looked kind of funny."

"We are checking telemetry now. What time did you say you looked at the display?"

"A little after three. I thought we started spinning, and I went and checked, but we are not," says Tsibliyev. *"So now we have to conserve as much power as possible?"*

"Well, of course."

"My heart is breaking," Tsibliyev says. *"You wake up in the morning and look at all this, and it just looks so darn sad."*

10:29
Aboard Mir

All morning, as Blagov and his people scramble to find out why they lost power, the crew wrestles batteries back and forth between modules, starting the recovery process all over again. Solovyov, meanwhile, spends the morning caucusing with Krikalev on their plans for sending Tsibliyev and Lazutkin on an "internal spacewalk"; instead of an EVA, they are calling it an IVA. The goal would be to replace Spektr's hatch with a new cover through which electrical cables could be run to the base of the module's solar arrays.

"We are thinking about what to do next—it seems like you are going to have to do an IVA," Solovyov tells the crew at 10:30. *"They are planning on making a sealed adapter. For the hatch between [the node] and [Spektr] . . . At first you have to prepare the hatch cover, then change the hatch covers and connect the cables. This is the minimum you have to accomplish. As for the maximum, one of you will have to enter [Spektr] and assess the possibility of accessing the Balkan area, where the solar array drives are located."*

"It's going to be very tough," Tsibliyev says.

"We realize that it's going to be tough. That's why what we want you to do is to assess the possibility of passing into that area. In other words, if you tell us no, we will plan one way. If yes but obstacles are present, we will think how to overcome the obstacles. There are only design studies for now. This is just a preliminary idea, for you to understand what is in store for you. When you do this, Mike will have to be in the [Soyuz], you understand."

"We gave it some thought already. Let's assume we go there, into [Spektr]. How will we enter it? Will we do a fly-around?"

"So far we are not talking about a fly-around. We have [normal] procedures. . . . In the case of an abnormal situation you will have to enter the [Spektr] and close the hatch, then use a BMP [oxygen tank] to pressurize the Spektr. . . ."

"What can we use to pressurize [the node]?"

"You will pressurize [the node] per normal procedures, equalizing pressure with [base block]. This is also a normal procedure."

"We are going to lose a lot of air," Tsibliyev cautions.

"You will lose exactly the amount that will go from [the node]. It is a small pressure drop, a normal procedure. The only problem is that you were not trained to do it. When fit-checking the space suits, you will have to do the training. Then go in pressurized suits from [the node] into [Spektr]."

"I can't imagine how," says Tsibliyev. *"It's impossible to get there. We checked on it already, and we can't get in there with the sleeves that we have."*

"The training will be done for just that purpose. If you tell us it's impossible, we will plan our future work based on that knowledge. If you tell us it can be done, we will plan accordingly."

As he listens to the conversation, Foale, too, is skeptical. Spektr is his home, and it is one area of the station he knows better than either cosmonaut. He doesn't see any way a spacewalker can reach the rear of the module, and he is deeply worried about snagging or tearing a space suit when shimmying through the entrance. In his mind he begins to think of the IVA as cave diving, that is, scuba diving into a darkened cave. He hopes the Russians know what they are doing.

Houston

By Friday morning Culbertson and Van Laak, in concert with George Abbey, believe they have hashed through every possible scenario they may face in the coming weeks. Many revolve around whether—and how—to remove Foale from the station. Tsibliyev and Lazutkin are scheduled to return to Earth in mid-August, and there is a third seat on their Soyuz; Foale could take it. Or the next shuttle mission—STS-86, scheduled for liftoff in late September—could be moved up to retrieve him. No one, of course, *wants* to cut short Foale's mission. Everyone understands the repercussions of such a move: It would deeply humiliate the Russians, who could then be counted on to back out of the International Space Station. For the moment Culbertson & Co. have no choice but to follow the Russians' lead. Valery Ryumin, with whom Culbertson and Van Laak hold a teleconference each morning, insists they will be able to return the station to full power within weeks, probably by mid-July. For the moment, everyone agrees, all NASA can do is wait.

Both Linenger and Blaine Hammond attend a large meeting in the Phase One conference room Friday afternoon in which Culbertson reviews Mir's status. It is an exercise they will undergo on a daily basis for the next month, and it is all but totally useless. As Culbertson reminds himself again and again, Mir is a Russian station and they are guests aboard. Linenger makes a small fuss at one point by suggesting that Tsibliyev has violated Russian safety procedures by refusing to evacuate the station immediately after the collision. Linenger remembers from his Russian training that commanders have no choice in the matter: If they are in a decompression situation and there is less than forty-five minutes of air remaining, they must evacuate. Tsibliyev didn't. "The fact is," Linenger tells the meeting, "I don't think Tsibliyev would ever evacuate that station for any reason. A Russian commander, I am telling you, will go down with his ship before making the decision to return to Earth in shame."

To one side, Van Laak simmers. Linenger is getting to be a real pain. Van Laak doesn't think the Russian forty-five-minute rule is hard and fast. He feels certain it must give the commander some leeway for individual decision making. After the meeting he retreats to his office with Shannon Lucid to research the question. Culbertson, meanwhile, takes Linenger aside. The meeting, like many safety reviews at JSC, has been tape-recorded, and Linenger's conjecture about Tsibliyev is not the kind of thing Culbertson wants on a tape that could be reviewed by the NASA inspector general's office or a congressional committee. It could serve as ammunition for NASA's enemies seeking to shut down Phase One.

"Jerry," Culbertson says, "if you've got things like that to say, it's best to say them in private."

Linenger, in fact, has ruffled NASA's feathers all the way up the line.

On the afternoon of the collision, when he sat beside Culbertson at the first of what would become almost daily NASA press briefings, he had irked JSC's public affairs handlers by refusing to play down the incident's severity. Afterward one of the JSC PR men, Rob Navius, had phoned to feel him out. "Headquarters is a little curious, Jerry," Navius asked. "I mean, do you have an ax to grind here?"

"No ax to grind here," Linenger told him. "I just want people to know the truth."

Saturday, June 28
Aboard Mir

Saturday is a day off for the crew. By that afternoon the batteries have recharged sufficiently to return most power to base block. Tsibliyev is relieved; they have only three Maglite batteries left. The rest of the station remains silent and dark, and it will probably stay that way until they can somehow reconnect the four big solar arrays on Spektr. Foale has found an extra pair of running shoes, toothpaste, and a toothbrush.

"Life is getting back to normal for me, as far as everyday living," Foale tells Zimmerman in a late-afternoon pass.

Still, he and the Russians remain skeptical about the internal spacewalk the TsUP is planning. The Progress containing the tools they will need is due to launch from Baikonur in a week, on Saturday, July 5; if all goes well, it will reach them the following Monday. Solovyov is talking about an IVA into Spektr sometime after the middle of the month. Until those preparations begin, other than ongoing repairs and sopping up the pools of water that have begun to accumulate in the darkened modules, there is little for the crew to do. Foale's science program is in disarray. Half of it is lost inside Spektr. He hopes still to be able to harvest the plants in the greenhouse; the beetles he hopes to hook up to a power supply inside Kristall. In the meantime, all three men are forced to sit tight and anxiously ponder the rescue mission TsUP is planning.

"I've never done this before," Tsibliyev tells Blagov Saturday morning.

"You'll be given some training. This procedure is not easy, but possible."

"Still, I can't imagine how we can do that wearing space suits."

"We'll try to modify your suits."

"How? Do you want to make a miniskirt out of a space suit?"

"Don't worry about that."

By Saturday night Foale has moved his sleeping quarters into the backup air lock in Kvant 2. In his new quarters he shares a cramped space with the TORU docking apparatus, which has been disassembled and stored. He begins to joke that the TORU is his lover.

One of the ploys Foale had used to forge camaraderie with his two Russian crewmates was the construction of an impromptu theater down in the Spektr module. Every Saturday night beginning in his first weeks aboard the station, Foale had selected an American film for the three of them to watch together on a computer monitor he mounted on one wall. One Saturday they watched *2001: A Space Odyssey.* Another of Foale's selections was *Big Blue,* an ocean adventure. Lazutkin loved watching the star, Rosanna Arquette, swim. There were no Russian subtitles. Foale did a running translation, complete with voices.

Among its many other impacts on their lives, the collision destroyed Foale's movie theater and most of his movies. A few remained in other modules, however, and he and Lazutkin resolved to keep the tradition going. "We knew Vasily was very worried, very down; he thought he was guilty of causing the collision," Lazutkin recalls. "Michael and I tried to direct his mind away from these thoughts."

By Saturday night enough power has returned to the station that Tsibliyev decides it is time for all three men to take a much-needed rest. Foale decides to show a movie he thinks will capture everyone's interest: *Apollo 13,* the 1995 space drama starring Tom Hanks. The three men lie in a tight row in the backup air lock; Tsibliyev reclines with his arms propped behind his head and allows himself to enjoy Foale's narration. "It was so relaxing," he remembers, "like a village movie theater." *Apollo 13* becomes Tsibliyev's favorite movie. "That film, it is the best of the best," he recalls. "Tom Hanks as the leading actor is so good. Everything in the film is so realistic, so truthful, so dramatic, it's just perfect."

Afterward all three men agree that their circumstances are far less dangerous than those of the Apollo 13 astronauts. "We felt that, especially from a psychological point of view, their situation was much worse than ours," remembers Tsibliyev. "We at least had a spaceship that could get us home. With Apollo 13 they had to fly all the way around the moon in order to get back to Earth."*

Later in the mission Foale receives an e-mailed copy of an article Apollo 13's commander, Jim Lovell, has written comparing the two flights. "Jim said that 'I understand how these guys feel up there, because I've been there as well,' " Tsibliyev recalls. "[He said] 'I know their courage and bravery.' It was quite wonderful to hear that from Jim."

Sunday, June 29
Aboard Mir

Sunday, as gyrodyne tests continue, Solovyov schedules the crew for its first televised press conference since the collision. Lazutkin doesn't like the idea. He thinks there are better ways to use the station's fragile power supply than chitchatting with reporters. "We were told the whole world was watching," he recalls. "We said, 'Well, so what?' What does that have to do with us?" But Lazutkin, ever the company man, says nothing. It falls to Tsibliyev to question the TsUP's decision.

"I was also against [the interview]," the commander recalls. "When they told us we had to do this interview, I said, 'Why? We don't have enough power. We have to save it for something more important.' But TsUP said it was more a political question. We had to show up, so people of the world could see we weren't dead, that everything was fixed."

Despite their doubts, all three men expertly play the roles of confident, upbeat space fliers during their nine-minute video press conference. "Thank God, everything works now," says Tsibliyev, forcing a smile. Adds Foale: "I have vast experience now. Even when things go wrong, it is great to work with these wonderful guys."

When the interview ends, the crew learns someone else wants to speak to them: It is Yuri Semenov, the chief executive of Energia, the man who

*Lazutkin remembers *Apollo 13* being viewed on the first Saturday night following the collision; Foale says "that sounds right." However, it may well have been the following Saturday night, July 5; it seems more likely that the crew would wait until the station regained full power before using even a small amount of power to watch a videotape.

refers to the station as "my Mir." Semenov gets on comm and asks each of the men to say aloud whether he wishes to continue aboard the station.

"It was sort of a check-in call, so that he could say, 'I have spoken to the crew, and they want to go on,' " remembers Foale. "It was really a setup. I knew what he wanted the answer to be. It was our one opportunity to say we wanted out. Of course there was no doubt whatsoever what we would say." One by one, Tsibliyev, Lazutkin, and Foale pledge to stay on board for the duration. "We didn't have a single thought of going back home, of leaving the station," remembers Tsibliyev. "[On the other hand] only a crazy man would tell the chief designer he wanted to leave, even if he did."

Monday, June 30
Kennedy Space Center, Florida

The day before the launch of STS-94, a routine shuttle mission, Fred Gregory, the safety chief at NASA headquarters, convened an impromptu teleconference of NASA safety officials from Houston, Washington, Huntsville, and The Cape to discuss the Mir situation. The Kennedy leg of the meeting was held in a conference room at the headquarters building, and Blaine Hammond, who had flown to Florida for the shuttle launch, arrived early, eager to get a reading on what everyone was thinking about Mir. For the past week Hammond had been quietly canvassing his fellow astronauts and found they generally shared his mounting fears for Foale's safety. In informal hallway chats everyone tended to say the same thing: *It's awful! That damn station is falling apart! We don't have any business being up there in the first place. They gotta get Mike down.*

Of course, to a man, none of the other astronauts was willing to stand by Hammond and say this in an open forum. They were scared of retaliation. Hammond had tracked down Shannon Lucid and to his surprise found her among the most critical of the Russian program. Lucid said flatly that without a full science program there was no longer any reason for Foale to be aboard Mir.

Glynn Lunney was already in the conference room when Hammond arrived. Lunney was a legend at NASA, a longtime flight director who had worked the Apollo missions. "I think it's really dumb that we're leaving Foale up there," Hammond volunteered. "We need to let the Russians fix this thing and let Foale come down."

"We should just not send Wendy, and bring Mike home," Lunney said, mentioning Wendy Lawrence, the diminutive Navy flier who was scheduled to fly the next Mir mission in late September.

Hammond seized on the idea. When the teleconference convened a few minutes later, he brought it up.

"We ought to be able to take a time-out here and do it graciously," Hammond said. The next Russian crew, accompanied by a French astronaut, was due to arrive on Mir in early August. "They've already got the Soyuz going up," Hammond said. "What we ought to do is take the French guy off, put three Russians on, and bring back Mike on the Soyuz. The Russians could work more effectively with three Russians on board, better than with two Russians and an American. We keep all the infrastructure in place, keep the shuttles flying, keep the cooperation going. The cooperation would not and should not cease just because an astronaut is not on board. If it does, then we have the wrong kind of partnership anyway."

"It's not our job to recommend policy," interjected John Casper, an astronaut who was the safety chief at JSC.

"I think that's right," said Fred Gregory, who was patched in from Washington.

"But we can't talk about what's safe without recommending policy," said Hammond. Even as he said it, of course, Hammond knew there was little chance of NASA management signing on with the idea of a gracious time-out. "It was a perfect idea, absolutely perfect," he remembers. "But everyone was too afraid. It was pressure, political pressure. We knew Goldin and Abbey were not going to let this thing flounder."

Patched in from his office in Houston, John Casper didn't agree with Hammond's thinking. He listed the various Mir systems that were still working. "We've evaluated every system," he said, "and it's safe."

"One thing we haven't evaluated is the psychological component of the system," Hammond countered. "Mike is really left with nothing to do. I talked with Shannon Lucid, and Shannon told me it would be criminal to leave someone up there with nothing to do."

This was a harsh statement, especially by NASA's ultra-civil standards, and its utterance would be news throughout the corridors of the agency by the end of the week.

"It's a psychological nightmare for Mike," Hammond continued. "You really can't do much. They don't let you do much. What Shannon told me, and what I believe, is that we should not be sending people up there, or keeping people up there, in this condition. They have leaks all over the place, a sealed-off science module. Mike's now just an in-flight maintenance guy, doing Stepin Fetchit work, and he's not trained for that."

Glynn Lunney asked to make a statement. "I fear we've reached the point of diminishing returns," Lunney said, "where the science we're getting is outweighed by the risk. I think if we lose somebody, we may have a hard time justifying to the American people what our person is doing up there."

Fred Gregory spoke next, and Hammond listened closely to Gregory's tone. "I got the sense Fred wanted to agree," Hammond recalls. "But he knew what the answer was supposed to be, and he was too scared to say anything. I could see that Fred had his marching orders and was trying to get information that he could massage to let Goldin feel good, let him do what he really wanted, which was leave Foale up there. Fred likes his job in Washington. He likes it a lot. He was not going to say anything to endanger it."

Out in the hallway after the meeting, a number of people, including Lunney, took Hammond aside and quietly thanked him for his comments. "You said the right thing in there," Lunney told him. "It took courage to do that." Hammond wanted to feel good. He wanted to believe his little speech had made a difference. But what he heard in Fred Gregory's voice told him it hadn't. Gregory hadn't even bothered to take a vote of the room. "I walked out of there thinking it's clear nothing is going to happen," Hammond recalls. "They thought it was just the ravings of ol' Blaine Hammond, who nobody listens to anyway."

Hammond pondered the meeting for days. The more he thought about it, the more he was struck by the irony of the situation: While the safety of Mir was already a topic of world debate, a question being explored on the front pages of the *New York Times* and the *Washington Post*, the one place where the debate should be taken most seriously, inside NASA, seemed all but devoid of genuinely skeptical discussion.

"America was saying Mir was unsafe, everybody in the world was saying it, but nobody inside NASA was saying it, and for a simple reason," Hammond says. "Going against George Abbey was the kiss of death.

Speak up and you can just kiss away the chance of ever going into space again."

Aboard Mir

Tsibliyev and Lazutkin spend much of the day tracking down serial numbers of the eighteen individual cables that have been disconnected or severed from the Spektr module. The TsUP needs to know exactly which cables ran through the hatch, so the Energia engineers will be able to properly build the hermetic plate. Foale is stunned that the TsUP even has to ask.

"This was unbelievable, really," he remembers. "I mean, [the ground] should *know* what these cables are. It's really extraordinary that they didn't. In all honesty, they had lost track of the experiments they fly that were routed through that hatch."

"It's not that the ground lost track," Lazutkin counters. "That was normal. The station was launched a long time ago, and the people who knew which cable was where have retired. These new people didn't know this. They just looked at papers, and the guys sent up recommendations based on what they read. Well, there were discrepancies. For example, they couldn't tell us where a particular cable was. They would say, 'Well, we think it's there. Maybe.' And so we would have to start looking."

Foale, of course, was right: This was not the kind of record keeping one would ever see aboard a shuttle.

Moscow

By Monday morning the TsUP's plan for saving the station is firming up. Despite the crew's reservations, Tsibliyev and Lazutkin will be sent on a delicate five-hour internal spacewalk into Spektr at 4 o'clock on the afternoon of July 12, in eleven days. If they need additional time for training, the IVA, as they are calling it, will be delayed a day or two. The crew, wearing the bulky Orlan space suits, will seal off the node from the rest of the station and depressurize it, then open and enter Spektr; Foale will remain behind in the Soyuz. Once inside Spektr, their goal will be to hook up cables linking the module's solar arrays to the station's main power supply. They will also search for the quarter-size hole through which Lazutkin heard air hissing.

Hooking up the cables will be tricky. To do it, Tsibliyev and Lazutkin will first remove the lid that has been placed over Spektr's hatch and replace it with the module's original funnel-shaped docking "cone," which after Spektr's arrival at the station in 1995 had been removed and strapped inside the node. Once the cone is fitted onto the hatch, the two Russians will attempt to lay the power cables through it. The key to the entire IVA is a small circle of aluminum, forged by Energia engineers over the weekend, that Tsibliyev and Lazutkin will install at the tip of the docking cone. This "hermaplate," as Energia has decided to call it, contains twenty-three holes through which the power cables can be threaded and joined to the station's main power supply. If the plan works, the hermaplate will allow the cosmonauts to run power cables into the node while keeping Spektr sealed tight.

The job of plotting the IVA has fallen to a group of cosmonauts and Energia engineers led by a crinkled elf of a man named Oleg Tsygankov, Mir's maintenance chief. Quickly mobilizing his forces on the afternoon of the collision, Tsygankov, who had quietly toiled deep in the innards of the Russian program since the 1960s, had just seven days to assemble all the tools and procedures the crew would need to revive Spektr. His first questions involved the Orlan space suits the cosmonauts would use. Without constant cooling, anyone inside the suit quickly faces sauna-like conditions. In a vacuum, the Orlan is cooled with the help of a device the Russians called a "sublimator," which uses cold water to lower the suit's temperature. The sublimator, however, will not work effectively in a partial vacuum, and Tsygankov's group expected the cosmonauts to encounter at least some air still trapped inside Spektr. On the ground, extra water is simply funneled into the suits via hoses. The hoses, however, are only two and a half meters long. To enable the cosmonauts to get into Spektr Tsygankov's men first located a set of ten-meter cables. That raised questions about pumps. Could the standard Orlan pump send water all the way through ten meters of cables as effectively as it could through two and a half meters? To Tsygankov's relief, it could.

While his assistants oversaw the forging of the aluminum hermaplate, Tsygankov spent the weekend assembling the tools Tsibliyev and Lazutkin would use to manipulate the cables. The Orlan spacesuit gloves, while more tactile than the American suits, were still hardly ideal for working with small cables. Tsygankov designed a tool, a kind of inside-out celery stalk, that would slip over each cable and give the cosmonauts a better feel for what they were doing. By Monday, while that and other devices were being forged in the Energia shops, Tsygankov remained concerned Tsibliyev and Lazutkin wouldn't have sufficient tools for the job. He hopped in his car and made the rounds of Moscow hardware stores, where he bought wrenches, pliers, and flashlights. His favorite purchase was a large American-made police flashlight. The tools, the hermaplate, and the cables themselves would be sent to Baikonur later in the week for the launch of the Progress that Saturday.

While there were gnawing worries about what Tsibliyev and Lazutkin would find inside Spektr—floating broken glass was one concern—Frank Culbertson and other NASA officials were generally comfortable with this first stage of the Spektr rescue effort. It was the little-publicized second stage of which they were deeply skeptical. Theorizing that the collision with the Progress had wrenched the foundation of one of Spektr's four solar arrays, Russian engineers guessed that the most likely location for the tiny breach in Spektr's hull was at the base of an array. Tsibliyev and Lazutkin would look for this hole during the IVA. If they couldn't find it—and many engineers were skeptical that they could—a later crew would look for the hole during an EVA outside Spektr.

To fill any hole or gap they would find near the base of an array, Tsygankov's engineers built a device that looked a bit like a coffeepot with a long spout. The pot could be filled with a liquid resin that a spacewalking cosmonaut could squirt into the cuplike base of the damaged array; when the resin hardened, it would theoretically fill any holes in the hull. That, at least, was the Russians' hope. If it worked, this would serve as a temporary seal on the hull and allow Spektr to be repressurized. At that point, a cosmonaut working in shirtsleeves could enter the module and search the inner hull in an attempt to permanently repair the leak from inside the station. In the event that the temporary seal did not hold, Valery Ryumin told Culbertson, the cosmonaut inside Spektr would probably have two or three minutes to evacuate the module and seal the hatch from the rest of Mir.

The audacity—and obvious risks—of the Russian plan startled Culbertson. "We just didn't see this as anything we could support," recalls Van Laak. "The idea was that the station would not be put at risk, only the person inside Spektr. We simply didn't think that was a credible plan." Despite American opposition, the Russians planned to forge ahead with their two-stage recovery plan. All of Tsygankov's tools, from the police flashlight to the "coffeepot," would be on the Progress scheduled to blast off Saturday from Baikonur.

9:28 P.M.
Aboard Mir

"Larissa Ivanova called yesterday," a capcom is telling Tsibliyev, referring to his wife.

"Called where?" The commander's voice is heavy and tired. It has been a long day.

"At my place."

"I can imagine what state she is in."

"We had some champagne."

"Seriozha, please don't."

"Why not? She looked good and fresh. She is concerned, of course. We went through a small debrief with her, in which my wife took part for some reason. . . ."

"Thank you, Seriozha."

"[I heard] a very good joke on TV yesterday. This is something for you before you go to bed. A patient comes to see a doctor and tells him that he has trouble sleeping, that he is in a difficult situation because of it at work, and that his wife was leaving him. And the problem was that he had been having the same dream over and over again over the last two months. 'What dream?' [the doctor asks]. 'About cockroaches playing soccer. And that does not let me sleep. Help, please.' Having taken a long time to think about it, the doctor comes out with a pill."

The comm breaks up. *"Can you hear us? . . . Do you copy . . . Good night, guys."*

Tsibliyev breaks in. *"Finish the joke, for Pete's sake. The doctor comes out with a pill."*

"And says, 'Take the pill and you will sleep like a log for two weeks.' The patient says, 'Can I take it tomorrow?' 'Why?' 'They've got the finals tonight.' "

"Understand."

"Okay. Good night. Thanks for the good work."

Tuesday, July 1
Kennedy Space Center, Florida

The repercussions from Blaine Hammond's remarks the day before were swift. Booting up his laptop that morning, Hammond found an angry e-mail from Bob Cabana, the head of the Astronaut Office. Hammond noticed Cabana had cc'd both Culbertson and Dave Leestma.

"Blaine . . . I would like to talk with you," Cabana began. "I was told that you stood up at a meeting and said 'it would be criminal for us to send Wendy to Mir.' You represent CA [the Flight Crew Operations Directorate] and CB (the Astronaut Office) at the meetings you attend. We've been over this before, when you make a statement like this you had better have cleared it through Dave [Leestma] and I. Our primary goal right now is to help the Russians fix Mir and ensure that it's done correctly. We will evaluate whether Wendy goes when the time is right. Your job is to make sure the system is supporting, doing all the right things to fly safely, not to express personal opinions that may or may not coincide with policy. A better statement for you to have expressed at this meeting is 'what do we need to do to ensure Mir is fixed properly and we are confident that it is ok to send Wendy.' That's what I would have expected from my safety officer, not an emotional personal opinion."

Hammond was so angry he couldn't see straight. His e-mailed reply to Cabana, at times long-winded and self-serving, blasted the Astronaut Office for sacrificing safety to advance a political agenda. "Bob: It's too bad you obviously receive your information second or third hand. This is not exactly what I said nor how I said it," he began. ". . . [W]hat I did say was to repeat what Shannon told me a couple of days prior to the meeting,

[when] she was in the process of sending you an E-Mail note, i.e. '**Without meaningful work to do**, it would be criminal to send Wendy or any other astronaut to Mir.' According to both she and Jerry, [repair work] does not constitute meaningful work and with Spektr lost, most, if not all, of the doable American science is gone and therefore it would be a bad idea to send Wendy to replace Mike under those conditions. . . . I merely offered my strong support for a proposal which had extreme merit then (and now, given the current health, fatigue and stress situation of the crew), i.e. to offer the Russians an opportunity to help themselves to repair Mir and restore it to a more operational (and safer) station for everyone's benefit. . . .

"Common sense is whispering (no, shouting) to us that something's not right here and we should not be trying so hard to find contrived reasons and rationale to override that voice when a very sensible alternative approach is suggested. It can be argued (and was at this telecon) that perhaps we have achieved a plateau presently in marginal returns for having an American astronaut on board Mir. If there is another major incident . . . and we aren't so extremely lucky the next time as we have been so far, we will have a terrible time justifying to the American people and Congress, not to mention the families, just exactly why it was so vital to expose our Astronaut to such risks. . . .

"As far as being the (Astronaut Office) Safety Officer goes, I am not nor will I be just another mouthpiece to echo the official party line. We have far too many who do that already. My job is—or, at least, should be—to be the 'conscience' of CA/CB, the dissenting opinion which may not always be considered but should be. Done right, this job doesn't always win friends in high places but it does earn respect. My only agenda is to ensure that whatever is done is the right thing and is done safely—period. If that happens to coincide with NASA policy or politics, then great, because that is ***ALL*** that NASA should be concerned with. . . . I have ***nothing*** to lose by speaking out and only my honesty, common sense and integrity to protect. We ignored our collective common sense in January 1986 and if I have anything to do or say about it, we won't make that mistake again."

Hammond knew Cabana wouldn't miss the reference to January 1986. It was the month of the Challenger explosion.

19

Tuesday, July 1
2:00
Aboard Mir

Crumph-crumph . . .

Tsibliyev cocks an ear and listens.

Crumph-crumph . . .

There. He hears it again. The noise—it sounds like a muffled explosion—appears to be coming from inside the sealed Spektr module. Floating in base block, Tsibliyev turns and looks for the others. Lazutkin is asleep in his *kayutka*. He isn't sure where Foale is, probably down in Priroda. Moving to one of the windows, Tsibliyev peers down toward Spektr and is startled at what he sees: A snowstorm in space. Wrapped around Spektr is a floating blizzard of snowflakes, not one of them moving, each hanging eerily in stasis, as if watching him.

His first thought is that the flakes are frozen bits of fuel. They look just like the little ice crystals that form when the station's thrusters spit out drops of fuel. But no: There are too many. Tsibliyev points the camcorder out the window, but the flakes are too small to appear on film. They are almost invisible during daylight, and they disappear altogether during night orbits. Only in the first rays of sunlight do they appear, glistening and twinkling. By the next morning they are gone.

"This remains a puzzle for us and for TsUP," Tsibliyev recalled months later. "Everyone was just scratching their heads and could not come up with an explanation."

Star City

"All right," John Blaha is saying, "here she comes."

That Tuesday, Blaha sat in the darkened TORU simulator at Star City and tried to navigate the Progress toward a docking with Mir, just as Tsibliyev had. By that morning the Russian docking trainers were theorizing that the crew had limited the Progress's maneuverability by overloading it with garbage, which might have shifted the incoming ship's center of gravity. They had attempted to duplicate the cargo ship's shifted center of gravity—its "C.G."—on the flight computer, and Blaha was now attempting to duplicate Tsibliyev's challenge in docking it.

On its first approach on the simulator, the Progress came in too fast, and Blaha managed a last-second abort, sending the ship spiraling away from Mir. He had the trainers rerun the scenario. On his second docking attempt, he rammed the Progress smack into base block, killing everyone aboard the station. "Don't feel bad," one of the trainers told him. Nine out of ten cosmonaut commanders who had attempted what was being called the "C.G. scenario" had crashed.

Blaha had arrived in Russia on Monday. He hadn't wanted to come. He was preparing to retire from NASA and had a job interview scheduled, but a direct plea from Dave Leestma had won him over. NASA, Leestma told him, badly needed someone with experience on the ground in Moscow to monitor the Russian recovery efforts. Blaha was deeply skeptical he could do much of anything, and upon arriving at the TsUP his fears were confirmed. "I didn't need to be there," he recalls. "[Sending me] was too reactive on our part, as if we really thought we had a role here, as if we could control what the Russians did or didn't do. When I got back I told Frank and all his people, 'You guys aren't in control of this the way you think you are. I know you think you are. But you're really not.' "

Every morning at ten Yuri Semenov convened a crisis meeting in the large briefing room across from the NASA suite in the TsUP. Blaha spent the entire week sitting in on these and other meetings, listening as Russian officials debated the best way to revive Spektr. It was immediately clear to him that the Russians needed no help from NASA, a feeling he communicated back to Culbertson. He was also struck by the youth of the NASA ground team. "I frankly didn't understand what they were even doing there," he recalls. "I mean, they sat around and listened to what the Russians were doing, but they couldn't really do anything. That's not a knock on them; it's just true. They were basically just an information conduit."

Blaha returned to Houston at week's end, determined to wash his hands of the Phase One program. He still felt Mir was safe, and he still felt working with the Russians was the best thing to do in the long run. But if NASA thought it held an ounce of influence with the white-haired men in the TsUP, they were kidding themselves.

9:00 A.M.
Houston

News of Mir's travails quickly became a fixture of newspaper front pages around the world. Most of the coverage emphasized the age and general decrepitude of the station, which in America generated hundreds of bad jokes. "This just in," Jay Leno quipped one evening on *The Tonight Show*. "A meteorite has crashed into the Mir space station. Did over a million dollars in improvements." The jokes in turn created a new pop-culture archetype—the broken-down Russian space station—which in months to come would surface as a metaphor in popular speech and even a movie plot or two. The 1998 summer film *Armageddon* would feature a loony cosmonaut struggling to maintain his dirt-streaked space station.

The irony, of course, was that it wasn't the station that was in danger of breaking down. It was the people inside it—especially Tsibliyev. From the first hours after the collision, Al Holland had begun compiling everything he could on the commander's mental and emotional state. At his request, Culbertson's office had started translating every comm pass between Mir and the TsUP. From these transcripts Holland could clearly see how guilt-stricken Tsibliyev had become. From conversations with Linenger he knew the commander was exhausted. Now the question became: Could Tsibliyev's deterioration get Mike Foale killed? And if so, should NASA attempt to pull Foale off the station? In an effort to learn more, Holland threw out calls to everyone he could think of: Terry Taddeo and the NASA ground crew at the TsUP; the astronauts and flight surgeons still training at Star City; the Russian doctors at IBMP.

That Wednesday morning Holland listened in on the first of what would be weekly teleconferences with his Russian counterparts. Though medical ethics prevents Holland from publicly discussing Tsibliyev's mental state, his questions to the IBMP doctors clearly indicated NASA's areas of concern: How much sleep was the commander actually getting? Did he have a history of mental disorders or of erratic behavior under stress? Had he ever suffered from depression? How confident was the TsUP that he and Lazutkin could actually pull off the IVA? What worried Holland most was the idea that Tsibliyev would attempt something foolishly heroic in an attempt to regain the professional respect he had lost by wrecking the Progress. Psychologists call this an "undoing behavior," because it seeks to "undo" a previous mistake. Would he be prone to take unnecessary risks during the IVA?

That morning the Russian doctors characteristically downplayed the stresses weighing on Tsibliyev. He had no history of erratic behavior or depression, they said, and seemed properly focused on the IVA. In their words, everything aboard the station remained *normalno.* Holland wasn't so sure. A key indicator of the commander's mind-set, he decided, would be the opinions of his peers at Star City. Was he now a pariah? Or would there be a rallying effect to support a troubled comrade? From what Holland could learn, it appeared there was widespread sympathy for Tsibliyev among almost everyone at Star City. To a man, the cosmonauts blamed the collision on Energia and the TsUP and considered Tsibliyev a hapless victim of their incompetence.

In the end, Holland decided that Tsibliyev had no predisposition toward doing something irrational. He was a normal man undergoing normal reactions to abnormal conditions. "I think he held up pretty well," Holland recalls. "He had a lot of guilt, but then who wouldn't?"

4:52 P.M.
Aboard Mir

It has been another day of quiet hard work for the crew. Foale helps Lazutkin wipe up the pools of water condensing on the walls in Priroda.

With the station's power supply halved, all power remains out in both Priroda and Kristall, and without a working ventilator system, water gathers on the walls. The two men work for several hours trying to jury-rig a system of hoses to blow air down to the darkened modules.

That afternoon, during a break from his work, Foale is eight minutes into his weekly personal medical conference with Terry Taddeo when Lazutkin hears the five gyrodynes in the Kvant module suddenly begin to power down.

"We can hear them slowing as we speak," Tsibliyev tells the TsUP.

Neither Solovyov nor the cosmonauts have any idea why the gyrodynes are failing. A little after seven the TsUP decides to power down the station's remaining gyrodynes, just to be safe. Solovyov tells the crew to stand by, that they will examine the problem overnight and decide on a fix the next day. In the meantime Tsibliyev maintains attitude by firing the station's thrusters.

By the next morning the TsUP has identified the problem as a failed Omega sensor, which will take the crew several hours to replace. It is the same kind of sensor that failed in March when Linenger was aboard the station.

"Yeah, we understand," Tsibliyev says in a low voice when a capcom explains the situation. *"So this is going to be another sleepless night. This is becoming a normal way of operation. . . ."*

The commander's mood darkens during the day. By Thursday evening, as they begin the repairs, he is snapping at the ground regularly. When an EVA specialist gets on comm and asks if he has begun readying space suits for the IVA, Tsibliyev is unable to hide his impatience.

"Well, we were only able to assemble them—do you know how hard it is to do anything here when all this is happening? . . . ."

"Vasily, here is what we want you to do," the specialist says. *"For tomorrow morning we have planned for the second comm pass, at 9:27, space-suit checkout from the transfer compartment."*

"Oh, damn it. So what the hell do you want us to do?"

When the specialist tries to explain, Tsibliyev cuts him off.

"I don't understand squat," the commander says.

Friday, July 4
Aboard Mir

Once Tsibliyev and Lazutkin change out the failed Omega sensor, replacing it with a spare, they are able to look forward to a quiet weekend. For the first time since the collision, much of the world's attention is focused elsewhere: NASA's Pathfinder vehicle has landed safely on the surface of Mars, and for several days the hardy little robot rover all but monopolizes press attention with its vivid photos of the red planet. The Spektr IVA is now scheduled for July 17 or 18, and aside from making sure the gyrodynes power back up correctly, mopping water, and continuing repairs on the Elektron and other systems, there remains little for the two cosmonauts to do but wait. The Progress carrying the tools they will need for the IVA is to blast off from Baikonur on Saturday.

"So there is hope," Tsibliyev sighs to Blagov late on Friday.

"The Progress is okay and is being prepared for launch according to schedule."

"May God help it."

The next morning the Progress blasts off on schedule from Baikonur. If all goes well, it will arrive at Mir on Monday. *"Your order is on the way and we guarantee delivery in two days or it's free,"* Mark Severance jokes to Foale that morning.

Tsibliyev, worried about another docking failure, is not looking forward to the ship's arrival. *"What I think about this cargo vehicle,"* he tells the ground at one point, *"is that when it arrives we'll load it with water, which we have a lot of, and fly it away. I do not want to challenge crew life, my own, our families."*

By Sunday afternoon most of the gyrodynes have powered back up, returning the station to the steady position it will need for the Progress's arrival. *"Looks like everything is okay,"* Tsibliyev tells Solovyov that morning. *"All gyrodynes are in place."*

"Excellent," says Solovyov. *"We were told the Progress [initial] approach is complete. They have performed a single-burn maneuver and conducted the TORU test from the ground. Everything is normal."*

"Thank God."

Monday, July 7
8:55 A.M.

More than one hundred journalists and other observers cram the upper mezzanine of the main control room as the new Progress approaches Mir this morning. During the early-morning hours, the incoming ship has responded perfectly to a series of thruster firings radioed up by the TsUP. By a little before nine the ship has closed to within two hundred meters; if all goes well with the Kurs system, it will dock automatically. Tsibliyev, still edgy and irritable even after a weekend of rest, is sitting in base block at the TORU controls, which have been reinstalled for the first time since the collision. If the Kurs system fails, he stands ready to take control of the incoming ship. On his Sony monitor he can clearly view the station from the camera mounted aboard the Progress. Lazutkin and Foale hover at the windows.

"You are watching the TV, right?" Tsibliyev asks the ground as the Progress enters its final approach.

"Yes, we are," replies Solovyov.

"Range now is about 120 meters," Tsibliyev says as the ship approaches. *"115 meters, 100 meters. Rate is 0.9 [meters]. Range is 80, 75, rate is 0.8."*

So far there is no sign of any problems with the Kurs system.

"Okay," Tsibliyev says, his eyes still on the TORU screen, *"I'm keeping my hands ready."*

A snow squall crosses the screen for a moment. *"Interferences,"* the commander says. *"Nothing is seen on TORU [screen]. I do not see very well because of the disturbances."*

"We have a good view," says Solovyov. *"Everything is fine, Vasya. Trust me."*

"Shadow from the cross is showing one o'clock," Tsibliyev says. *"The target is practically in the middle of the docking cross. Range is 30 meters, rate is 0.25 or 0.3 meters per second."*

Down at the NASA console, Mark Severance watches as the Progress glides slowly toward the docking port.

"Oh, come on, get there, baby," he murmurs. "Get there, get there. . . ."

"Everything is okay so far," says Solovyov.

"Good," says Tsibliyev. *"Range is 18 meters, rate 0.20."*

The Progress inches forward. *"Now 16,"* says Tsibliyev. *"Twelve . . . Ten, moving right one degree . . . Five . . . Four . . ."*

Severance is holding his breath. It's almost there.

A moment later the Progress docking probe nudges the docking port.

"Now we have contact," says Tsibliyev.

"We have contact," confirms Solovyov. *"Congratulations."*

On the floor of the TsUP, Viktor Blagov breaks into a wide smile and turns to Severance. "Well, Mark," he says, "Mike has his new toothbrush." As flashbulbs pop above them on the mezzanine, the two men rise and shake hands.

That afternoon the senior IBMP doctor, Igor Goncharov, gets on comm to brief the crew on the upcoming week's activities, which are heavily oriented toward unloading the Progress and preparing for the IVA, now just ten days away.

"Continue taking vitamins and, Michael, you should take vitamins as well, as he doesn't have a [medical] kit," Goncharov directs. Foale's kit was lost inside Spektr. *"So, now, you have many things to do. Take care of yourselves and do not overwork, okay?"*

"We'll try," says Tsibliyev. *"We've got to accomplish what there is."*

"Yes, and do all the unloading step by step. Stay away from injuries."

"It's not the first time we've done this, you know," says Tsibliyev.

The unloading, in fact, takes most of the week. For much of the time the commander remains snippy.

"Medical experts would like to get a complete form 020," a capcom informs him at one point. *"In preparation for the EVA."*

"We still have to finish unloading," the commander replies. *"Do they need it later today?"*

"They needed it yesterday. Vasya, I understand everything, but they have to have it today."

"They'll have it. We don't have 100 hands. We work like bees here."

Tsibliyev does get one piece of good news on Thursday. Amazingly, there is some debate inside the TsUP whether to ask Tsibliyev to attempt another—a third—long-distance TORU test on the newly arrived Progress. Tsibliyev has argued strongly against doing so. On Thursday, Blagov tells him he can relax. *"Your argument was the most convincing one,"* says Blagov. *"There will be no TORU test on this vehicle."*

"Understand," says Tsibliyev. *"Yes, first we have to understand the whole scheme. We did not have much time for training [the first time]."*

Several times the cosmonaut Nikolai Budarin gets on comm to brief Tsibliyev on the procedures they will use for the IVA. Krikalev and other cosmonauts, closely observed by a newly arrived team of NASA astronauts as well as camera crews from Western television networks, have been rehearsing these plans in the Star City hydrolab since the previous Friday. Budarin assures Tsibliyev that threading the power cables through the hermaplate will be easy, even in the bulky space suits. Still, by Thursday, just a week before he and Lazutkin are scheduled to perform the IVA, Tsibliyev remains unclear about precisely what they will be doing.

"The EVA is only a few days away and we are not ready at all," he complains to Budarin Thursday afternoon. *"Don't know what we'll need, or where we'll be positioned. We think one thing and then you tell us another."*

"Vasya, if you have any objections, please let us know right away," Budarin tells him. *"We'll make corrections to our procedures. Agreed? Because you know better there."*

"We don't have time to study the documentation and see if there are errors there," Tsibliyev snaps. *"Anyway, [we] have to wait for three days to get an answer."*

Tuesday, July 8
Houston

In those first two turbulent weeks following the collision, many inside NASA came to believe—as Culbertson had warned Van Laak—that the Phase One program was doomed. "I felt certain the Americans would back out of Phase One at that point," acknowledges Andy Thomas, the astronaut who had volunteered to be Dave Wolf's nonflying backup for the final Mir mission in early 1998. "I just didn't think the Russians would be able to pull their program together." Adds Travis Brice, one of Culbertson's men, "Around here, I'd say people felt the chances were about forty/sixty that the program was over."

Sentiment in the astronaut corps ran strongly against continuing the program. "The question in most of our minds was, 'What are we still doing up there?' " remembers the former astronaut Rich Clifford. "Spektr is dead, so we really can't do science. Are we sending astronauts up there just to sit around?" Other than Linenger and Blaine Hammond, however, few if any astronauts or senior agency officials were willing to voice their opposition. "You want to support NASA in what it's doing," says Clifford. "If NASA wants to do it, and they tell you it's safe, questioning management is not something you go public with. You don't want to risk your career."

No one was hit harder by the collision than Culbertson. Almost every day now he was obliged to walk over to the briefing room in JSC's Building Two and subject himself to the barbed questions of journalists from around the world, including many piped in from Washington and The Cape. He was good at dealing with the press. He was, in fact, the perfect face for NASA to show the public: calm, professional, straightforward. Headquarters loved him. "Frank," Dan Goldin had called and told him after the collision, "for purposes of this mission, *I* work for *you* now."

But even as Culbertson fed the press the upbeat quotes the NASA hierarchy needed, privately he harbored doubts about the wisdom of continuing to work with the Russians. For the first time since joining the Phase One office three years before, he had serious reservations about the Russians' basic competency to run a space program. It wasn't just the collision; it was the fire, the Near Miss, the leaks, *and* the collision. "This whole program began with the premise that they knew what they were doing," Culbertson remembers thinking at the time. "Now we've seen a whole series of things where you begin to wonder whether they are capable or not. If they're not, then we have to reassess the whole program."

But was that even possible? At its inception Phase One was a political program, a construct of politicians, not engineers and pilots. Culbertson had no idea what would happen if he recommended canceling it. Backing out now, he knew, would almost certainly lead the Russians to back out of the International Space Station. Would the politicians listen? Would they even *let* him back out? Culbertson didn't know, and at some level he didn't want to know. In trusted NASA fashion, he decided to work the problem and hope the politics would sort out in time.

On a personal level, Culbertson felt betrayed. It was not a word he used, but when he examined his feelings there was no denying it. He had worked hard to earn the trust of his Russian counterpart, Valery Ryumin, and thought he had it. Ryumin had mellowed a bit over the months; his flashes of temper now seemed few and far between. Ever since the fire, Culbertson had been pestering Ryumin to begin treating the Americans as real partners aboard Mir, not as guests, and when Ryumin made reassuring noises, Culbertson started to believe it was true: They *were* partners now, real partners, with no secrets between them. But the collision had rudely put the lie to that. When all was said and done, the Americans hadn't had any idea what their "partners" were up to.

Every morning since the collision, Culbertson and Van Laak had been holding teleconferences with Ryumin. At first the two Americans simply used the calls to gather information to share with their superiors. But as the waves of international criticism began to buffet NASA, Van Laak realized it was time to stop listening and start making some demands. "Frank," Van Laak told Culbertson in the days immediately after the collision, "we simply can't continue this program without a role in the decision making, without real influence over what's happening."

Culbertson agreed, but he was skeptical. For three years he had tried off and on to change the Russians' ways, with a notable lack of success. Even if Ryumin agreed to open up and genuinely begin sharing technical information, how could they even be sure it was accurate? Would Ryumin lie? Culbertson just didn't know. Van Laak hectored him to get tough with Ryumin, yell a little, throw a verbal chair or two. But it wasn't in Culbertson's nature. Even while downing a dozen Diet Cokes a day, he remained achingly civil at all times. He asked Ryumin for daily updates on Mir's situation, and Ryumin promised: From now on, NASA would promptly receive all the latest technical data on the station's performance.

"There have to be concrete changes here," Culbertson told Ryumin. "We need absolute confidence that you in the Russian program will do nothing concerning safety or docking without briefing us thoroughly."

"Yes, yes," Ryumin said. And for the most part, he was as good as his word. In the two weeks after the collision there had been a few communications failures, but Culbertson was generally satisfied with the information flow. Timely technical updates, however, did not a true partnership make, as Culbertson was keenly aware. He and Van Laak needed to get involved in the decision-making process, and both men were uncomfortably aware that they weren't—a feeling that grew increasingly clear as the Russians' plans for resuscitating the station emerged in the week after the collision.

For one thing, Ryumin was insisting that the Russians intended to forge ahead with plans to send the French astronaut Leopold Eyharts along with two replacement cosmonauts to Mir in early August, a decision that would place six astronauts aboard the station for a period of three weeks. Culbertson and Van Laak adamantly opposed the move. This was no time for the station to be taking on unnecessary guests. They were worried that six men would place a serious strain on Mir's already overworked life-support systems, especially the carbon dioxide removal system. Culbertson had been urging Ryumin to cancel or at least delay the French mission, but so far Ryumin hadn't budged. The Russians, Culbertson knew, were anxious to get back into their normal routines and show the world they could quickly recover from even the showiest disasters.

But he and Van Laak knew there was another, just as important, reason for pushing forward with the French mission: money. Ryumin wouldn't say how much the French were paying for Eyharts's stay aboard Mir, but Van Laak believed a rumor he heard that it was in the $35 million range. He and Culbertson suspected that the driving force behind the TsUP's rush to revive Spektr in mid-July was to get money from the French in August.

"God damn it," Van Laak told Culbertson. "This is going to put a real stress on the life-support systems, but the Russians are going to go ahead anyway. All because of the money."

What galled Culbertson about Ryumin's intransigence was a little-known facet of the station's life-support system. Just as the Russians were forced to use a supplementary oxygen system when more than three people lived aboard the station, so, too, were they forced to rely on a supplemental carbon dioxide system; the backup system used canisters of lithium hydroxide—LiOH—to help scrub the station's atmosphere. Because the Russians had developed some kind of manufacturing problem, most of the LiOH canisters aboard Mir were supplied by NASA; thus the only reason Ryumin was even able to host the French mission was because of the Americans' largesse. What made matters worse was the fact that the station's supply of LiOH canisters was running low. To provide enough clean air to support Eyharts's three-week mission, the Russians would be forced to eat up fully half of the remaining canisters. This was a clear breach of an agreement Culbertson and Van Laak had wrung from the Russians after the fire that laid out minimum life-support levels aboard the station. If Mir ran below a certain number of oxygen or LiOH canisters, Ryumin had promised that the station would be evacuated. Now Ryumin wanted NASA to issue a waiver of those guidelines so the Russians could go ahead with the French mission—a mission Culbertson opposed.

"I have serious concerns that the life-support, power, and attitude-control systems will not support the load of six crew members for 21 days," Culbertson wrote Ryumin in an unusually stern (by NASA standards) letter the week of July 7, "so I need to understand the rationale for exceeding minimum crew [levels] at this critical point in the program. . . . In my opinion, consuming half of the LiOH reserve to support this higher than necessary crew load is not consistent with our agreement, so I would like to hear of an alternative approach that can leave the LiOH reserve intact. I am also concerned that U.S.-provided LiOH not be used to meet Russian contractual obligations with other nations."

The French mission, however, was nothing compared to what the Russians had in store for Spektr. Culbertson didn't need to sit through Al Holland's briefings to know that Tsibliyev and Lazutkin were dog tired. He listened to comm passes every day and could hear the stress and anger in Tsibliyev's voice. "Valery, this is coming from the heart; this is what I really feel," Culbertson told Ryumin early that week. "You really need to think carefully about doing this. This crew is very tired. They're going to be under a lot of pressure to succeed at all costs, and you know as well as I do, that's when crews make mistakes." But Ryumin held his ground: Tsibliyev and Lazutkin were healthy enough to do the IVA, and they would do so.

If all the hazard analysis cleared the IVA—safety experts in NASA were worried about broken glass or chemicals floating around inside the airless module—Culbertson told himself he could live with the two Russians carrying out the internal spacewalk. It was the second stage of the Spektr recovery—the idea of a shirtsleeved cosmonaut working in Spektr after the hull was temporarily sealed—that truly frightened him. "I have serious reservations about the plan to repair and repressurize the Spektr module, both in terms of the safety of the repair operations and the operational impact of the repair activity," Culbertson wrote Ryumin. "Our science program would not benefit significantly from the restoration if successful, and the expected impact to the crew timeline during [the next mission] would be great."

Ryumin listened to all Culbertson's objections but insisted on going ahead. Over and over he told Culbertson everything was okay, everything would be taken care of, don't worry, we have done this before. At first Culbertson assumed Ryumin was plying him with empty reassurances.

But the more he thought about it, the more he worried that Ryumin, in the midst of one of the worst crises in the history of manned space flight, really wasn't all that concerned. "You know, he's pretty cavalier about the hazards here," Van Laak told Culbertson at one point. Culbertson tried—unsuccessfully, he felt—to make Ryumin understand that this was no longer a technical challenge that could be overcome by the cosmonauts' Russian ingenuity. It was now a political situation.

"The confidence of our government in this program is definitely severely hurt by all this, Valery," Culbertson told Ryumin after the collision. "The only way they are going to let us continue this is if we can show we are active participants in what is going on."

He drove the point home in the letter. "[A] pattern seems to be developing," Culbertson wrote, "including the operational attitude control errors during STS-81, the failed Progress docking in March, and the collision in June. Regardless of my personal confidence in the integrity and professionalism of the Russian flight control team, I am forced to take action to improve my insight into and confidence in the operational decisions affecting the safety of our American crew on Mir." In the letter, Culbertson asked Ryumin to supply a list of things the Russians had never given anyone before: a six-month timetable of all maintenance, repair, EVA, resupply, and docking plans.

Ryumin grumbled a bit but ultimately agreed to Culbertson's demand. Culbertson's next request, however, was to prove far more difficult for the Russian to swallow. Everyone from Sensenbrenner to the *New York Times* wanted to know what had caused the collision. NASA had quietly formed its own internal committee, headed by the astronaut Mike Baker, to investigate the causes; Abbey, not wanting to alienate the Russians, had insisted that Baker's group operate in complete secrecy. "The future of the program," Culbertson remembers thinking, "would depend on the cause of the accident. Were there fundamental holes in the Russian system where we couldn't afford to work with them anymore? Was there a situation where they could no longer support a manned space program? That was the worst-case basis."

Culbertson didn't know the answers. But whatever the precise technical cause, he was confident Tsibliyev wasn't solely to blame, as Ryumin and the other Energia executives wanted the world to believe. The TsUP had planned the test. Star City had trained the commander. If the White House was to allow the program to continue, Culbertson knew that the Russians would need to come up with a credible explanation for what had happened. And that meant that every segment of the Russian space program—the cosmonauts, the TsUP, Energia, Star City—would have to accept its share of the blame. Culbertson tried to make this clear to Ryumin, but here the crusty old Russian drew the line. The collision, he stated emphatically, was Tsibliyev's fault. And Star City's.

Culbertson took a deep breath. This wouldn't be easy. Finger-pointing between Star City and the TsUP had broken out within hours of the collision. Blagov's people angrily blamed Tsibliyev for pilot error and Star City for failing to train him sufficiently. The generals at Star City lashed right back. In those first days after the collision, they hatched a theory and leaked it to Russian journalists: The collision had been caused by too much garbage having been loaded onto the Progress, throwing the ship's center of gravity out of whack. While on its face this put the blame on the cosmonauts, in reality it shifted the blame to the TsUP; everyone knew Blagov's controllers signed off on every ounce of trash that was loaded onto the ship. Star City trainers had even programmed this "center of gravity," or "C.G.," scenario into their simulators and let John Blaha crash the Progress a time or two to make their point. It took a few days before the Americans understood what was going on. "We thought that Star City would 'cook the books' with this C.G. scenario to blame Energia," recalls Van Laak, "and that's just what they did."

Somehow, Culbertson saw, Ryumin had to be convinced of the need to accept blame. Early that week, Culbertson called in the shuttle commander Charlie Precourt and asked him to accept a quick, quiet mission to Moscow. Precourt boarded a flight to Russia and arrived that Wednesday, July 9. He met with Ryumin at his Energia office the next day, then accepted an invitation to visit him at his dacha on Saturday. Over a long lunch there, Precourt tried to hammer home the need for every segment of the Russian program—Energia, the TsUP, Star City—to accept responsibility for the collision.

"You need to be completely open about this, because the American public is not going to let us keep flying with you if you're not," Precourt told Ryumin. "Don't blame the crew as the sole problem of this thing. Because it's not going to be a credible explanation. That's not credible in our eyes. And you know we can't take that explanation to Congress. This crew is fatigued, they've had no simulator training for six months, and then they're handed a situation where they are given no radar data. That's *never* been done before, even by a fresh crew."

Precourt left Ryumin's dacha with little hope his message had gotten through. Of Ryumin, he says, "He was going to blame the crew either way."

Sunday, July 13
9:39 P.M.
Aboard Mir

A gnawing worry has crept into Tsibliyev's mind even before he gets on the stationary bicycle and begins pedaling. There is something wrong in his chest. He can feel it. He first noticed it Friday night, as he tried to fall asleep in his *kayutka*. A delay, a little pause, in his heartbeat. If his heart had been an engine, it felt as if one of the pistons was sticking.

Thump-thump . . . thump.

Thump-thump . . . thump.

At first he ignored it. It was only when the new rhythm persisted Saturday that it dawned on him it had begun the night before. It frightened him. He knew what it would mean if there was something wrong with his heart and the doctors found out: They wouldn't let him participate in the IVA. "He felt something was wrong before the medical checkup, and he told us about it," Lazutkin recalls. "He said he didn't feel good, that something was wrong. Of course, we didn't know what until the checkup."

Tsibliyev attaches the electrodes to his chest, climbs on the bicycle, and begins pedaling. This, the first of two runs on the base block bike, is one of the physical routines the Russians go through in advance of every EVA. Down in the TsUP, a clutch of doctors has gathered on the floor to watch his readouts. As he pedals more rapidly, Tsibliyev keeps an eye on the small EKG monitor in the medical cabinet beside him. "I immediately noticed there were peaks on the heartbeat chart," he recalls. "I was afraid to start the second stage, in which I had to go faster."

But he does, and this time the spikes on the EKG grow even higher. "I knew the ground was watching it," he continues. "I started getting concerned that the ground will stop me from going on the EVA. [When I stopped] they immediately said, 'Vasily, there's a small problem. Let us think a little while, and we'll tell you [what it means] tomorrow.' "

"You could see the irregularities clearly on the chart," Lazutkin remembers. "I understood immediately that the reasons were the period after the collision. I remember distinctly that Vasily didn't have any sleep during this time, and if he did he had maybe one or two hours a day. All that time he worried about the collision, and he kept all these feelings inside."

"Finishing the exercise," one of the doctors notes as Tsibliyev pedals to a stop.

"Done," the commander confirms.

"Vasya, hold your breath as you exhale."

"Okay."

"You can breathe now."

"Thank you."

When Tsibliyev finishes the exercise, he shows the results to both Lazutkin and Foale. Worried, Foale goes down to Priroda and digs out an American medical book. Together the three men flip through the pages until they find an EKG that resembles what they have seen.

"That one," Tsibliyev finally says. The book shows the chart for a heart arrhythmia. It is not a heart attack but an irregularity sometimes prompted by stress.

Foale immediately realizes the significance of their findings. "I knew now I was really in the hot seat," he remembers. The Russians would ask him to take Tsibliyev's place in the IVA.

Monday, July 14
Aboard Mir

As expected, the Russian doctors confront a morose Tsibliyev about his irregular EKG Monday morning.

"When I tried to sleep on Friday night, my heart rhythm was a little bit unusual, and I have never had such a feeling before," the commander explains. *"I*

have felt no pain. I think the reason was the stress I felt following June 25, and it spilled over on Friday."

"You have to calm down," a doctor tells him. *"Healthy people often have this problem."* She orders him to rest, to get more sleep, and prescribes tranquilizer pills in the Russian medical kit.

"What a lousy time for this to happen," Tsibliyev moans. *"Will I be able to [do the work]?"* The IVA.

"Let us think and see," says another doctor.

All three crew members already know the answer to the commander's question: Tsibliyev won't be allowed to perform the IVA, a decision that is formally announced the next day. Foale, who is expecting the Russians to ask him to take Tsibliyev's place, is nevertheless surprised when Ryumin himself gets on comm that afternoon and requests him to do so. Foale says he is willing to do the work but will need to speak to Culbertson.

"Michael, can I say the initiative came from you?" Ryumin asks.

Again Foale says he will need to talk to Culbertson.

"We'll have to take it to the leadership [then]," Ryumin says.

Tsibliyev's heart condition startles Lazutkin. That night he takes Foale aside. "We saw Vasily's situation was getting worse," recalls Lazutkin. "With Michael we decided we should be ready if [he] got even worse. There were no doctors there, but we had a lot of medicine. Anything can happen to anyone there, and the ground wouldn't be able to help us. We had to be able to save him. We started looking at and reading different instructions and selecting drugs. We basically prepared the drugs, so in the worst case we wouldn't have to look for them. Of course, we didn't tell Vasily what we were doing."

Lazutkin readied the Russian medical kits, Foale the American. They pointedly decided not to tell the Russian doctors what they were doing. "The doctors had everything under control," Lazutkin recalls, smiling. "For me to tell the doctors, for me to express these sad thoughts to them, was useless. They are specialists, and I know they know the worst can happen, but nobody said it out loud. Besides, it would have been an extra psychological pressure on Vasily."

The news that he won't participate in the EVA is too much for Tsibliyev. That night in base block he finally breaks down. "Oh, what have I done?" the commander asks Lazutkin and Foale. "It's so shameful. How can I face my family?" Foale looks at Lazutkin; neither man knows what to say. Then Foale floats over to Tsibliyev and gives him a hug. Lazutkin immediately joins him, and the three men float there together for the longest time, each of them unsuccessfully fighting the tears that begin to dampen their faces.

Tuesday, July 15

In his regular morning call to Houston from Moscow, the timbre of Ryumin's voice was far different from the bullying tone he could adopt when angry. In its place was a soothing, solicitous tone Culbertson and Van Laak didn't often hear. "He wants to cut a deal," Van Laak whispered.

In fact, Ryumin wanted help. With Tsibliyev now sidelined, the Russians needed Foale for the IVA, and badly. That morning, in fact, they began pressing for Foale on all fronts. In another phone call from Moscow, Yuri Koptev of the Russian Space Agency assured Dan Goldin that the IVA was "easy," a no-brainer. Goldin wasn't so sure. "Murphy's Law," he quipped later that morning to Culbertson, "is still the law of the land."

Both Goldin and Culbertson realized that for the first time they had real leverage over the Russians. They had no doubt Foale would want to do the IVA, a hunch that was confirmed later that day when Foale talked to Houston on the ham radio. *"I can do this."* Foale told Culbertson. *"The task at hand is very simple. My role is very simple. I'm really not that concerned."* Foale said he could be ready to do the IVA by July 24, in nine days; Culbertson ordered a fast examination of Foale's EVA training just to make sure he was in fact technically qualified to do the work.

In return for Foale's services, however, Goldin and Culbertson were determined to get something back. They wanted the French mission canceled. Ryumin said he would think about it and get back to them.

Wednesday, July 16
Aboard Mir

By Wednesday afternoon the mood on the station has grown tense. After moping around for two days, Tsibliyev has grudgingly begun to accept the idea that he will be unable to perform his role in the IVA. But both Lazutkin and Foale get the clear sense that the commander hopes this is only a temporary change, that his heart arrhythmia will somehow abate and allow him to do the work. Foale is careful not to appear too forceful, too eager to replace him.

"I'll print by hand the steps for Mike," Tsibliyev tells the ground at one point. *"I will leave one copy in the node and one will be with me in the core, and [during the IVA] I will be prompting from there how and what buttons to push."*

"Of course," says the ground.

"Because, unfortunately, though I don't want to say that I let everybody down, I am sure we are going to come out from this situation with dignity. Mike is a smart guy, but there are abbreviations he doesn't understand yet and we need to work on those with him."

"You are absolutely right. I agree with you 120 percent. Your role will be like one of a guide. You will keep everything under control."

"Well, it looks like it's my cross to bear," says Tsibliyev. *"A guide!"*

"And what about us down here?"

"I know. But I did not mean this to happen, and God is my witness."

9:30 P.M.
Aboard Mir

That night Tsibliyev and Lazutkin stay up late, rearranging the bundles of cables that run through the node, which is to be depressurized during the IVA. Lazutkin is doing much of the work, working off a four-page radiogram the TsUP sent up. The memo divides the cables into three groups, those to be disconnected a week before, the day before, and the day of the IVA. Lazutkin is tired, and he makes special note of the group to be separated the last day. "I circled it with a red marker," he remembers. "The first two groups I did not mark."

He is busy detaching the group of "week before" cables when he grows momentarily confused. The heading that lists the group of cables to be detached the day before the IVA is at the bottom of a page, and he belatedly realizes he has disconnected one of the cables on the next page. "I missed the heading that said, 'One day before the walk,' " he remembers. "These directions were written on one page, and the list of cables [under that heading] was printed on the next page. The headline was at the bottom of the page, and I didn't see it. It just happened by chance. I turned the page and started reading the second page."

Lazutkin hastily reattaches the cable that he has mistakenly disconnected. Then, thinking he should tell Tsibliyev what he has done, he gets up and floats into base block, where Foale and the commander are working.

"What is it?" asks Tsibliyev, who is sitting at the command console.

"Uhh, I don't think I should have pulled one of these cables," Lazutkin says.

"What?" Tsibliyev asks.

"The cable. I reattached it."

Just then one of the station alarms sounds.

BOOOO . . . BOOOO . . .

Despite Lazutkin's quick reattachment, the damage has been done. The cable he momentarily disconnected furnishes power to the central computer. Without it, the computer has crashed. Tsibliyev looks at the computer screens and realizes the attitude control is beginning to break down. The station is entering free drift.

Tsibliyev is unsure what to do. A series of readings he has never seen before flash across his monitors. Fifteen minutes later Mir enters comm range, and he begins speaking to the evening's shift flight director, Ekrim Koneev.

"We have the following situation," Tsibliyev begins. *"At 21:33"*—a burst of static interrupts—*"lit up and immediately went out. We must have disconnected the wrong cable."*

"Just a sec," Koneev says. *"Let us look at the telemetry."*

"We need to think what to do. We connected the cables back."

Inexplicably, Koneev seems to be in no hurry. *"In 1.5 minutes we'll have telemetry."*

As they wait, Tsibliyev sees all sorts of numbers that he doesn't understand pop up on his computer screen. *"What is OSK-D?"* he asks. *"I do not recall such a mode. What is D? Thrusters?"*

"Just a sec," says Koneev. *"You are asking questions that cannot be answered right off."*

"This is for you to work on," says Tsibliyev, growing aggravated.

"Understand. Vasily, take it easy." Koneev says he will examine the telemetry and speak to him during the next comm pass.

Within minutes the station's onboard batteries begin to drain.

All night long Tsibliyev and Koneev struggle to understand what has happened. While they debate what to do, the batteries drain. For some reason NASA analysts would never fully understand, Koneev does not order the crew to turn off all major systems in an effort to conserve energy. By dawn the station has lost all power. They are back where they were the night after the collision.

It is a dramatic screwup, and Foale has no doubt what it portends for their future.

"You know what this means for the EVA," he tells Lazutkin at one point.

"I know," Lazutkin says, nodding. "They're not going to let us do it."

Neither man says anything to Tsibliyev, but several minutes later the commander comes to the same conclusion. "No way," he says. "There's no way they'll let us do it now."

When Vladimir Solovyov learned what had happened, he hung his head and rubbed his eyes for a long time. "This is like a kindergarten," reporters overheard him say.

Thursday, July 17
9:00 A.M.
TsUP

Phil Engelauf, the MOD flight director who had joined the growing American presence at the TsUP, had just walked into the NASA suite when one of the Russian flight directors, Viktor Chadrin, came in with an interpreter and asked to talk. The Russian managed a nervous smile.

"Well," Chadrin began, "the crew put us in another corner last night." He quickly briefed Engelauf on what had happened, adding that Koneev, the shift flight director on duty, "did not do a very good job of dealing with the situation." Engelauf knew Koneev well and understood; he was probably the most unpopular Russian in the TsUP, a vain and difficult man who was often defiantly uncooperative with the Americans. Engelauf was also struck, as he always was, by the Russian system's knee-jerk reaction, to always blame the crew when something went wrong.

That afternoon, as Blagov and the other controllers once more scrambled to restore power to the station, Engelauf made a point of obtaining a transcript of the air-to-ground communications from the night before. That evening he went over them in his hotel room. What he discovered startled him: Koneev appeared to have taken hours to identify and correct the situation, a delay that led directly to the station's total loss of power. Engelauf wrote up his hour-by-hour analysis of what had happened for a report he sent back to Houston.

Koneev, Engelauf discovered, did not seem to understand Tsibliyev's first report of a problem. "The crew continues to press, asking if they should not bring up thrusters, as a backup," Engelauf wrote. "The ground continues to tell the crew not to worry about it, at one time telling the crew to 'take it easy.' " There was no indication Koneev had the faintest idea what had occurred.

"If you read these transcripts, the crew calls down and says the vehicle is not performing correctly," Engelauf remembers. "It goes right by the ground. They just say, 'Oh, yeah, that's nice.' About four or five times Vasily calls down and says, 'Hey, the computer is spitting out garbage.' And the ground says, 'Well, we'll look at it next pass.' This goes on for four passes. It gets progressively worse, until they lose power altogether. These are classic symptoms of [what is called] cockpit resource mismanagement. It's a fairly classic case of [the ground] missing the first road sign and then driving right off the cliff."

Not until a pass at 2:29 that morning, nearly five hours after Lazutkin disconnected the cable, did it finally dawn on Koneev that the station was in crisis. "The crew finally tells the ground, 'Well, we have [disconnected those] cables,' " Engelauf wrote in his analysis. "The ground is totally surprised and asks what cable they are talking about. This yields a discussion in which TsUP finally grasps the situation onboard."

What alarmed Engelauf was not so much the incident itself, but rather its implications for the International Space Station. These, after all, were the same Russian ground controllers who would be working with NASA astronauts on ISS in two short years. Engelauf's conclusions were blistering. "There appears to be an inability on the part of TsUP, even when [telemetry] is available, to identify even major problems, like a loss of a major attitude sensor component . . . ," he wrote. "The ground does not appear to give credence to an evident state of concern on the part of the mission commander. The sense of team cohesiveness between the ground and onboard crew, to which we are accustomed, is absent. TsUP situational awareness is also lacking. Although they are advising the crew to power off equipment, they evidently didn't understand the severity of the power deficit nor pursue the cause [of] it."

"What the hell do these guys think they are doing?" an angry Jim Van Laak asks Culbertson that morning. "I mean, why the hell were they reconfiguring cables?"

The station's loss of power stuns everyone. Suddenly everything—the IVA, Foale's presence aboard the station, the future of Phase One itself—must be reassessed. Al Holland trots into Culbertson's office early that day and argues strongly against allowing Tsibliyev and Lazutkin to perform the IVA; if the two men are unable to disconnect a simple set of cables correctly, God knows how they could foul up a complex spacewalk. Culbertson has already reached the same conclusion. Somehow he isn't all that surprised to learn that Valery Ryumin hasn't.

"Well, we'll just have to postpone the EVA preps until we regain attitude control," Ryumin tells Culbertson, confirming the Russians' intention to forge ahead with the Spektr rescue mission. Culbertson does manage to wring one sop out of Ryumin. "Yes," the burly Russian acknowledges, "I guess we'll have to seriously consider canceling the [French] mission."

"Well, *duh!*" muttered Van Laak.

4:42 P.M.
Aboard Mir

"Vasily sleeping?" one of the Russian doctors asks Lazutkin.

"Yes."

The station remains quiet. While Foale and Lazutkin begin preparations to start moving the batteries into base block for the arduous two-day recovery process, Tsibliyev has been asleep in his *kayutka.* Mir lies in darkness except for the few systems that click on during the thirty-five-minute daytime orbits.

"Did you measure his pulse and blood pressure?" one doctor asks. *"Is he sleeping in his clothes? Do not disturb him. We will ask him to work at night. And you and Michael work with us now as long as you can."*

"Watch Vasily while he is sleeping," another doctor instructs. *"Check on him periodically."*

Just then Tsibliyev emerges sleepily into base block.

"He is flying up here now," says Lazutkin.

"Did you wake up by yourself," the doctor asks the commander, *"or did they wake you up?"*

"I woke up myself," Tsibliyev says.

"How do you feel?"

"I slept like a log."

"I understand. You had a sleepless night. Did you start feeling sleepy?"

"Yes."

"Vasily, how do you feel. Any weakness?"

"I just dropped asleep."

"By next session give us your pulse and pressure. And also we will take your EKG."

Solovyov comes on comm and lays out how they will proceed with the recovery process, which all three crew members are beginning to know by heart. *"This is a familiar operation,"* Solovyov says, *"and the sooner we do it, the sooner we will leave you alone."*

"In ten minutes we will enter the shadow," Tsibliyev sighs, noting the onset of a night orbit, *"and everything turns off. What a pity."*

Despite the tranquilizers he is taking, which put him to sleep for long periods, Tsibliyev insists on helping Foale and Lazutkin in their work that night. The doctors object. *"As long as Vasily is under treatment and [taking] pills,"* Igor Goncharov instructs the crew that evening, *"we want him to rest and Sasha to do the middle-of-the-night shift and let Sasha rest before, in the evening."*

"I will do this night by all means," Tsibliyev insists.

"Vasily, you should have a rest. Everything will be fine as planned."

"Roger. But we cannot give too much work to Sasha."

"We are government people and must comply," says Goncharov. *"Vasyl, as you are under treatment and medications you must rest."*

Tsibliyev reluctantly agrees. It is Lazutkin who pulls the midnight shift. Around three he becomes confused by a series of lights that pop up on the command console. Amazingly, he manages to keep his good humor.

"I still don't understand who keeps turning it all on," Lazutkin tells the ground.

"Who keeps turning it on?" the ground answers. *"There is a little bug sitting in a drawer, and when you push him, he turns on everything at random."*

"We are neighbors, but I haven't seen him yet."

"We hope you will be in attitude shortly, Sasha. This is the last communications pass for you. Thank you. Forgive us that it's been dragging for so long."

"Oh, God, I am myself to blame."

Friday, July 18

"Life keeps getting better and better," Ryumin announced sarcastically during the morning telecon with Houston. "We're going to rest the crew for three days. The gyrodynes should be back up and running by this weekend. Probably tomorrow." Culbertson pushed the obvious question: Has the TsUP decided to put off the IVA for the next cosmonaut crew? Ryumin said a decision should be made in the next several days.

In fact, it was all but made that morning. At a ten o'clock meeting at the TsUP, Yuri Semenov strongly hinted to his staff that Tsibliyev and Lazutkin would not perform the IVA. The work would instead be done in late August by the next crew, the veteran commander Anatoli Solovyov and the flight engineer Pavel Vinogradov, who would arrive at the station August 7. The new crew, perhaps with Foale's help, would then conduct a spacewalk outside Mir to examine the damaged hull. The French mission would be delayed until January. Semenov took the occasion to castigate IBMP doctors for having taken Tsibliyev's exhaustion too lightly. After weeks of downplaying the commander's obvious deterioration, the doctors have begun telling reporters that Tsibliyev was suffering from something they called "commander's syndrome," meaning that he obsessed too much about his responsibilities. "Everyone agreed that this was the right thing to do," recalls Blagov. "The doctors also agreed. It was not a strong recommendation they made, but we agreed anyway. [The crew] were making such simple mistakes. If they were to make a simple mistake during the IVA, it could be a very big thing."

That afternoon, without telling him of the pending decision, Solovyov attempts to prepare Tsibliyev gently for the bad news. The face-saving explanation the Russians are formulating for the crew—and will spring on them the next day—involves the risk of the station losing power during the IVA.

"How are you?" Solovyov asks the commander during a 3:30 comm pass.

"I want to work more but (the doctors) won't let me." Tsibliyev sighs. *"It feels like I have disappeared."*

"What are you talking about? You have been and are the commander."

Tsibliyev changes the subject. *"I've been busy correcting EVA procedures to adapt them for Mike to use."*

"Vasya, do not push correcting them," says Solovyov. *"I want to be frank with you. It looks like powerwise we overestimated our capabilities, because it appears that it is just one step from overall welfare to overall collapse. You can imagine what would happen if a crew member has his space suit on and there is a power loss."*

"We thought about it."

"We don't want to hurry [the IVA decision]. At this time we are weighing all pros and cons. There are several teams working here—I am just back from there—and we just don't know yet how to proceed. We need to think safety. It's a complicated situation. . . . We have abused you already and we need to get back to normal. We need to prepare the station; you understand many things better than we do."

The next morning, during a comm pass at 9:05, all the senior officials of the Russian space program gather on the floor of the TsUP to break the news to the crew. Klimuk and Glaskov come in from Star City. Semenov himself gets on comm and assures both cosmonauts that they are not being blamed for their recent mishaps; he pledges that they both can expect to be included on future crews. Neither Tsibliyev nor Lazutkin believes him. Nor is either man entirely surprised by the decision. Lazutkin had hoped for a delay in the IVA, but was ready to accept a cancellation.

"We just thought it would be postponed," he recalls. "We thought it would be a two-day delay. Then the ground said they were worried about the technical condition of the station. I just thought that was an excuse. Let me put it this way: I realized that they didn't trust us. But I didn't tell anyone. I tried to support the common point of view. But I myself understood the problem. I felt upset. But what could I do? Maybe if I were twenty years old I would announce how terrible this was. But now I'm a little older and I understand that we must report to management. We must abide by the ground's decisions."

Afterward everyone involved tries to console Tsibliyev, including several Moscow friends the TsUP puts on comm in an effort to boost his spirits. The commander doesn't want to hear any of it. *"I am sick of people trying to make me feel better,"* he snaps at the ground later that day. *"It seems like all we are doing here is sitting and crying all day long. I am not sure what to do, either wipe away sweat or tears."*

In Houston, Culbertson and Van Laak learn of the Russians' decision from a report on CNN. An aide to the White House science adviser, Jack Gibbons, calls Van Laak at home and browbeats him for failing to keep abreast of events aboard the station.

Sunday, July 20
Aboard Mir

Despite their obvious disappointment, Foale immediately notices, Tsibliyev and Lazutkin begin to relax a bit in the next few days. The pressure of saving the station in a complex IVA is gone, and for the first time the two Russians are able to concentrate on nothing more than going home in three weeks. Treadmill tests indicate Tsibliyev's heart arrhythmia is slowly easing, and the heavy load of tranquilizers the doctors have prescribed puts him to sleep for long periods.

"Vasily, you are making us happy," Igor Goncharov tells the commander Sunday night.

"Why?"

"You are doing better."

"I'm not sure how happy you are, but I am taking these pills and I am sleeping like a gopher, day and night, and cannot wake up."

"It's okay. You have to recover. How long can a man survive with no sleep?"

By Monday afternoon, after an official commission has ratified the decision to postpone the IVA until mid-August, the crew is actually able to joke a bit about restoring power to the cold, dank Kristall module.

"We need to make a draft in there," Foale says.

"Why don't you open the door and windows?" a capcom jokes.

"We already tried that, and you know, the draft was pretty bad, took all the air out," replies Tsibliyev, in a sly reference to the collision.

On Tuesday the commander returns to one of his favorite pastimes, making gibes at Jerry Linenger. Linenger has given a first-person account of surviving the fire for *People* magazine, and the commander has heard the rumors that Linenger is recommending that Foale be brought back to Earth immediately. *"What do you want to [take back to Earth with you in the Soyuz]?"* Viktor Blagov asks.

"Maybe take Michael?" Tsibliyev jokes. *"Jerry is advising him to do so."*

"[Someone] gave me an article to read about Jerry's reminiscences, in English," Blagov replies. *"Jerry writes that he was in charge of the fire-extinguishing work of the crew."*

"Excellent," Tsibliyev says. *"And what was [I], Vasily, doing there anyway?"*

"And during the EVA, when he was hanging on the boom, the commander was hauling him around on the boom, like a motor, wherever he wanted to go."

"Did he write about how we spent more than an hour at the [hatch] when he was afraid to take his hand off? I told him, 'Take your hand off and I will pull you aside.' And he said, 'No, I'm going to hold on.' And then when we were asked what we were doing there for so long, he said, 'We got lost.' I could not get him to take his hand off for an hour."

By the end of the week Tsibliyev has turned positively chatty, discussing everything from floods he can see on the Poland-Germany border to a haircut he has given Foale. *"Michael has gotten healthier looking. You won't recognize him when you see him with his haircut compared to before—big, powerful neck,"* Tsibliyev says at one point, laughing.

"You've made a soldier out of him!" a capcom exclaims.

"He was bugging me for a month, 'when are you going to cut my hair?' And I told him when the next cargo ship comes. It came [and] he said, 'Well, one has come and it hit us, so cut my hair.' And we did."

"It looks like he's been in the service for three months."

"Recently," Tsibliyev goes on, *"there was a magazine brought up on the cargo*

vehicle, and I forgot the name of that actress who was in the movie 'Striptease,' what's her name? Moore, yes, Demi Moore. She completely shaved her head, and every few hours she sits in a dark room and says things about herself."

"To whom?"

"To nobody. To extend her youth. And the most interesting thing is that she takes baths in the morning and in the evening with milk, and with something else—urine! And there was an interesting note that she and her husband live in separate bedrooms." He is laughing now. *"It's probably good for your health. Don't laugh. You've never tried it, you know."*

Saturday morning Tsibliyev actually manages to sit still and enjoy a completely pointless chat with a group of chirpy young women from the Krasny Octyabr chocolate factory in Moscow.

"Hi, dear Siriuses," one of the women begins. *"We came here to talk to you. We, the Krasny Octyabr representatives . . ."*

Tsibliyev is relieved. He tries not to think about what will happen when he returns to Earth. For now, he is entirely content to sleep, chat up beautiful young women, and discuss the finer points of Demi Moore's beauty secrets.

20

Wednesday, July 23
Denver

Now that the immediate crisis aboard Mir appeared to have passed, the question that loomed over NASA was simple: Should the Phase One program be continued? Should Wendy Lawrence, the astronaut scheduled for the next Mir increment, be sent to the station on her shuttle flight, STS-86, in late September? The ultimate call would be Dan Goldin's, but everyone involved knew the strongest influence on Goldin's decision would be General Thomas Stafford's independent commission. Composed of former astronauts, former NASA flight surgeons, and aerospace industry officials, the fourteen-person commission had been formed in early 1994 to act as an informal review board for Phase One, parceling out advice and guidance where needed and reviewing the program's operational readiness before each new mission. Stafford's group met two or three times a year with an identical commission the Russians had formed, headed by a senior scientist named Vladimir Utkin.

As an independent agency whose motives and budget requests were forever being questioned by Congress, NASA loved independent advisory committees whose opinions it could solicit to validate its own decisions. Many inside NASA, however, questioned just how "independent" the Stafford Commission actually was. The commission's executive secretary had an office at NASA headquarters. Stafford's right-hand man, the weathered former astronaut Joe Engle, had an office at JSC and sat in on many of Culbertson's meetings. A number of other commission members, including the former astronaut John Fabian, had long, strong ties to NASA or worked for companies that held NASA contracts. Stafford himself was a personal friend of Abbey and many senior Russian officials, and sat on the board of a company, Allied-Signal, that held NASA contracts. As the U.S. commander of the Apollo-Soyuz mission in 1975, a gold statuette of Stafford graced one of the glass cases at the cosmonaut museum in Star City.

Inside NASA there were few doubts as to what conclusion Stafford would reach when asked to review Mir's safety. "Tom Stafford is the equivalent of a lobbyist for the Russians," says Hoot Gibson, the former head of the Astronaut Office. "Every time we send an evaluation team over to Russia, it's a team that's headed by Stafford. And every time, Stafford always comes back with the answer [that] everything in Russia is great. [Sending Stafford] to look at Mir was a done deal. We knew that Stafford is going to come back and say, 'Yeah, it's safe.' Politically, it would have been a disaster if he hadn't. If Congress was told Mir isn't safe, why would they want to go do ISS?"

By late July the muttered allegations that Stafford's commission was a den of NASA good ol' boys determined to paper over Mir's problems had reached the ears of Roberta Gross, the NASA inspector general, who was quietly gearing up her own investigation of Mir's safety. That morning in Denver, as Stafford convened the commission, he was surprised to find two IG investigators waiting to enter the meeting. Joe Engle went out and met with the pair and patiently told them he couldn't allow them to attend the session. The problem, Engle explained, was Mike Baker's presentation on the collision. Baker's work was considered top secret; Abbey, not wanting to offend the Russians, had objected to Baker's even attending the meeting. But Engle had insisted, saying it was the only way Stafford's people could obtain hard data on the collision. Abbey had relented, though only after Engle gave his word that no outsiders would attend.

The IG investigators were clearly not pleased at being turned away. Not long after the meeting, Stafford was surprised to learn the IG was looking into the ties he and other commission members had to NASA. "The IG was convinced that this committee are good ol' boy networks [that would] just whitewash everything," says Craig Fisher, a former NASA flight surgeon who sat on the commission. "That really boils my blood. I submit to you that I've been in the space game for thirty-seven years. I'm not going to whitewash anything. Who else you gonna get? A bunch of schoolteachers or bankers? You go after people who have been there."

Joe Engle, who suddenly found himself the object of IG scrutiny, had a different take on the IG's investigation. "They were out to get me because I wouldn't let them into the meeting," he says.

Stuart, Florida

Already IG investigators were beginning to telephone many who had worked in the Phase One program—Linenger, Tony Sang, and others. A number of younger NASA workers were frightened by these approaches, which the combative Van Laak charged were akin to "Gestapo tactics." He and Culbertson simply didn't understand why the IG was prepared to criminalize the program's honest mistakes.

An IG investigator tracked down Blaine Hammond at his parents' home in Florida, where he was vacationing. Even though he remained safety chief in the Astronaut Office, Hammond agreed to cooperate quietly with the IG's investigation. He didn't think anyone else at JSC would dare speak out. Afterward he sat down and typed an eight-page memo laying out in detail his fears about Mir's safety.

"In my personal opinion, Mir is a disaster waiting to happen," Hammond wrote. "I admire the Russians for their adaptability and ingenuity in being able to solve the myriad problems that exist in this long outdate[d] station, but the fact remains, Mir is crumbling around them and, since the fire, I have had grave concerns for the safety of the American astronauts on board. . . .

"It is my strong belief that the Russians will never order Mir be abandoned for a failure until the situation is so dire as to be well beyond the point where it may not even be possible. First, the Mir is the last vestige of technical prowess and pride the Russians have remaining. To lose that would be to lose total international prestige and standing as a world power. Therefore, it is to their advantage to keep it going at *any and all possible costs*. . . . I also think the Russian view of life is somewhat different and a little less sacred than Western thought so that death as a hero is preferable to living in disgrace."

After finishing his memo, Hammond agreed to meet secretly with IG investigators at a Hilton hotel across from JSC.

July 28
Moscow

That Monday, after his trip to Denver to brief Stafford's commission, Culbertson walked off a long flight in Moscow, where he immediately disappeared into a series of meetings with Valery Ryumin and Yuri Semenov. Culbertson remained convinced that the key to persuading Congress not to cancel the program was to somehow get Energia to admit its complicity in the collision, as he tried to make clear to both men.

"Pilot error may well be a factor here," Culbertson told the two Russians. "But you don't get to that point alone. You and I as managers, and your people who put together the test programs, we all must share in the blame. The rest of the world is watching here, and if you use your old way of blaming one person, the old Soviet way, and not taking enough responsibility, people will feel like you haven't joined the Western world. You simply have to share the blame."

While polite, Semenov and Ryumin would not budge; they, like Blagov and others at the TsUP, remained convinced that Tsibliyev was solely to blame. But Culbertson's talks with the two Russians did lead to another agreement, one that Culbertson wasn't entirely comfortable with. The next day he was driven to Star City, where he sat down with Wendy Lawrence, who was scheduled to fly to Mir in late September. The thirty-eight-year-old Lawrence was a favorite of many in the Phase One program; unlike her friends Mike Foale and Dave Wolf, she had eagerly volunteered for Mir service in 1995. But just as she was preparing to leave for Star City, the Russians discovered she was too short for Orlan space suits. Lawrence was five feet three; the suits required an astronaut who was at least five feet four. Lawrence had gone to Star City anyway, filling one of the rotating slots as Director of Operations—Russia. Once in Russia, she had pushed hard for a Mir mission, and after she was remeasured, the Russians, to everyone's surprise, relented. Their only condition: Lawrence could not perform an EVA. She was just barely large enough to fit into the Orlan suit.

Some at Star City had noticed a change in Lawrence's gung-ho attitude since the collision. "Wendy was not as enthused as she was earlier," recalls John McBrine. "She was a little concerned that she would not be busy. She was concerned her mission was becoming more for political purposes than science. It obviously bothered her."

That day in Star City, Culbertson sat down with Lawrence and broke the difficult news. Since the next American aboard Mir—*if* there was another American aboard Mir—would need to be able to participate in the series of EVAs the Russians planned in an attempt to fully revive Spektr that fall, Lawrence was being replaced. Lawrence took the news like the pro she was. As a consolation prize she accepted a slot on the shuttle mission that would take her replacement to Mir. Then she rose and left the room and shook hands with the man standing outside, the astronaut who was to replace her.

Every astronaut who trained in Star City had his or her own way of coping with the stress and cultural isolation of the Russians' rigorous sixteen-hour days. Thagard had his Mel Brooks movies and a stray cat he adopted. Lucid had her Sunday bike rides, Blaha his notebooks, Linenger the cosmonaut gymnasium, Foale his family. Dave Wolf discovered The Hungry Duck. The Duck was a dingy, low-ceilinged bar-cum-disco in downtown Moscow where, in the absence of an actual dance floor, the patrons danced atop the bars and tables. By three or four in the morning the most drunken ones would begin falling off, often splitting foreheads and temples on the floor. Around five the bartenders took to spraying champagne all over the crowd, at which point sticky, hedonistic young Moscow girls could be counted on to peel off their shirts and dance topless. The young NASA crowd loved the Duck, and the bar's Canadian owner waved them in by the dozen for free.

But it was Wolf who was Lord of the Duck, so popular that the bartenders let him take shifts behind the bar. A typical Wolf weekend began Friday evening, when he and his former roommate John McBrine would down rounds of beer and burgers at the Starlite Diner, an American-style eatery popular with the expatriate crowd. By ten they were at the Duck, where they would stay till six the next morning. By seven they were back at the Starlite for breakfast. Then they would crash at the Penta Hotel during the afternoon, return to the Starlite for dinner, and head back to the Duck for another long night of dancing and drinking. Wolf, to no one's surprise, quickly became a favorite of the hard-drinking cosmonaut corps. He made a point of learning all the best Russian curse words and took pride in making sure the cosmonauts understood American ones. When the Kazakh cosmonaut Talgat Musabayev began hailing some of the NASA trainers as "you fucking goddamn bastards," they knew he had made Wolf's acquaintance.

Wolf had time for carousing in part because he had stumbled on a fast way to learn Russian, which allowed him to cut back on his language classes. At first he had labored in the classroom. The Russian technical manuals were impossible to understand. Then Wolf discovered a computer language program called "Context," which allowed him to translate the Russian manuals into English. Laying the Russian and English texts side by side, he found he not only could understand them, he was actually picking up the language faster than in his classes. Working with the Context program at night, he prevailed upon his Russian trainers to stop his language classes altogether, allowing him to begin Soyuz training ahead of schedule. He used his free time to explore Moscow's nightlife and in time even picked up a pair of Russian lady friends.

A wide smile crossed Wolf's face that day in Star City when Culbertson broke the news that he was to replace Lawrence as the next American to be sent to Mir.

"Think you're up to it?" Culbertson asked.

"You bet!" Wolf enthused, practically leaping out of his chair.

Wolf in fact was almost overwhelmed by the sense of redemption Culbertson's request instilled in him. All the shame and anger of the FBI sting operation, the embarrassment of his arrest in Indianapolis . . . it was all to be forgotten. Wolf wanted to scream: *They need me! They really, really need me!* "You have no idea how good that felt," he recalled months later. "NASA was finally putting their faith in me."

There remained one serious hurdle to Wolf's mission: EVA qualification. In order to be sent to Mir in September, Wolf would have less than a month to get the hang of the entire Russian spacewalk process, training in the Star City hydrolab and learning the ins and outs of the Orlan space suit. Officials in both the Russian and the American programs were far from convinced he could pick it all up in four scant weeks, even as he finished his other studies. "No one had ever accelerated a backup the way we were accelerating Dave," recalls Van Laak, "and not only were we pushing Dave, we were pushing him with a hell-for-leather EVA program. And Dave's reputation, for being reckless and for easy living, didn't exactly instill us with confidence in his ability to do this EVA."

Still, Wolf was all NASA had. The morning after meeting with Culbertson he began EVA training in the hydrolab. His days quickly became a blur. Mornings were spent underwater, learning how to use the Orlan suit. When he emerged from the hydrolab around noon, his flight surgeon, Chris Flynn, a tall, balding psychiatrist given to bow ties, would slip him a roast beef sandwich and a bag of Fritos as Wolf jogged to an afternoon packed with either more classes or sessions in the Russian pressure chambers, where he would lie still in an Orlan suit as Russian trainers lowered the pressure to simulate higher and higher altitudes. At the end of a long day of physical training, Wolf headed back to his apartment alone to study for his final examinations.

It was a lot to digest. And as Chris Flynn warned NASA officials, Wolf's training wasn't just difficult—it was dangerous. To finish his EVA certification by the end of August, it would be necessary for Wolf to spend long periods both underwater in the hydrolab and at simulated high altitudes in the hypobaric chamber. Normally NASA doctors spaced these two activities at least four days apart, fearing the effects of sudden changes in pressure; any less time, Flynn feared, drastically increased Wolf's susceptibility to the "bends," the disabling illness typically faced by scuba divers who surface too quickly. Yet to complete his training on time, Wolf would be forced to endure four- and five-hour diving sessions and altitude checks in the pressure chamber on alternating days.

As Wolf began his crash program that week, Flynn followed him, always on the lookout for the first symptoms of the bends: joint and muscle soreness, or even what Flynn termed a "neurological event"—that is, disorientation or mental confusion. Within a week Wolf began to complain about pains in his joints. Neither man, however, could be certain whether the pain represented the onset of the bends or simply the normal aches from a new set of physical activities. Flynn became worried enough that he persuaded the Russians to allow Wolf to spend an hour in an oxygen chamber, saturating his body with oxygen to offset any symptoms of the bends. "Every day—*every day*—I was concerned he was not going to make it," Flynn recalls. "The stresses on his body were just so intense. Dave was exhausted. Every night I left him he was exhausted, and I knew he was going back [to his apartment] to study."

For the first time since arriving in Star City a year earlier, Wolf stopped his trips to The Hungry Duck. Flynn and the other Americans watched him with a mixture of awe and dread. If anything happened to Wolf—a bout of the bends, even a twisted ankle—there was no one to replace him aboard Mir.* Any injury to Wolf, anything at all that derailed his crash training program, would almost certainly mean a "break" in the program—that is, a gap between Mike Foale's return to Earth and the arrival of a replacement American astronaut. "And if there was a break," recalls Flynn, "the entire [Phase One] program was over. Sensenbrenner would kill it. We all knew that. We all knew what was at stake."

Thursday, August 7
Aboard Mir

After a successful launch from Baikonur on Tuesday, August 5, the Soyuz carrying the next Mir crew—commander Anatoli Solovyov and flight engineer Pavel Vinogradov—arrives at Mir on Thursday. To Foale, "it was like the cavalry coming to the rescue." As is fast becoming custom aboard the station, the new crew's arrival is not without drama. Seconds before their Soyuz docks, Solovyov complains that he cannot get a clear image of the Kvant docking port; instead he switches on the manual docking system and guides the Soyuz in himself. Wags at the TsUP exchange amused glances. Solovyov's sudden "problem" earns him a thousand-dollar bonus for a successful manual docking.

There are hugs and handshakes all around as the two new men pile into the station. Vinogradov, whose nickname is Pasha, is a sharp-tongued rookie with the darkened visage of a coal miner. He and Lazutkin are close friends; they worked together in the same section at Energia. It is Solovyov who strikes Foale by his presence. The new commander is the Chuck Yeager of the Russian program, a squat, fifty-one-year-old Air Force colonel who holds the world record for number of spacewalks. There is a bit of the Old Soviet in Solovyov; a member of the Communist Party since 1971, he is a Hero of the Soviet Union and a member of the Order of Lenin. He is not especially friendly with the Americans at Star City, and

*The backup astronaut for Wolf's original increment was forty-five-year-old Andy Thomas, who with Lawrence's cancellation was now asked to take the last American mission to Mir, beginning in January 1998.

behind the scenes NASA officials suspect he is a stern critic of the two countries' collaboration.

Still, in the coming days Foale notices that Solovyov is utterly passive; Tsibliyev is to remain Mir's commander until his return to Earth the following week. Solovyov and Vinogradov in fact do everything they can to soothe Tsibliyev's worries about what lies in store for him back in Russia. "You could see the sadness in Vasily's face; he knew the collision had been his responsibility," recalls Vinogradov. "We tried to make him feel as good as we could. We all realized the entire system was to blame, but our bosses do not want to say that. All the cosmonauts knew the system was to blame."

Solovyov and Vinogradov are now scheduled to perform the IVA on August 20. Two weeks after that, Solovyov and Foale are to venture outside the station in an EVA to examine Spektr's hull. While they sketch out the EVA plans, Foale takes the opportunity to administer his Poster Test to Soloyov.

"You know that poster in Blagov's office?" he asks at one point.

"Da, da."

"Do you think that poster is an accurate depiction of life aboard the station?"

Solovyov shrugs. He is not going for the bait.

"Anatoli, are you a person that would do whatever the TsUP says, or would you do what needed to be done?"

Solovyov thinks for a moment. "I would do what needed to be done."

Foale feels good about his new commander. He thinks Solovyov may show initiative and break out of the master-slave relationship. "I had great hopes for Anatoli," Foale remembers. "But no. Anatoli would complain about the TsUP. But every time anything happened, he did exactly what the TsUP told him. He was exactly the situation depicted in that poster. He is the perfect Soviet cosmonaut."

Vinogradov is a bit freer with his opinions, as Foale, to his discomfort, is to note. On several occasions during the changeover period, Vinogradov remarks how "stupid" it is for the TsUP to be scheduling Foale for the second EVA; after all, Vinogradov points out, it is he, Vinogradov, who has trained on the ground with Solovyov. Why isn't he doing the EVA?

Foale squirms and says nothing. He knows exactly what the new Russian engineer is complaining about. Vinogradov, like all cosmonauts, stands to make an extra thousand dollars for every EVA in which he participates. The TsUP's decision to hand the EVA to Foale is taking money out of his pocket. To Foale's relief, both Tsibliyev and Lazutkin leap to his defense, telling Vinogradov how capable Foale has proven in space. Vinogradov eventually stops complaining, but his relations with Foale are to remain uneasy for several weeks.

Tsibliyev and Lazutkin spend their last week aboard the station packing for their return to Earth, briefing their two replacements on the station's status and steeling themselves for the storm they know awaits them back in Russia. On Wednesday evening, August 13, the night before they are to board their Soyuz for reentry, the two Russians stay up well past midnight, autographing and stamping dozens of posters, banners, and postcards to take back for their friends and family. Foale watches them work, worry etched into his face. With all the items they have left to stamp, the two men will probably get no more than two hours' sleep before their flight in the morning. "I thought it was the stupidest, most irresponsible thing I had ever seen them do," he remembers.

Foale urges Tsibliyev to get some sleep. The commander ignores him. Finally, Foale goes to Solovyov. "Anatoli, we have got to get these guys to go to bed now," he says. Solovyov agrees and speaks to Tsibliyev, who reluctantly retires to his *kayutka* shortly after.

Thursday August 14
11:53 A.M.
Aboard Mir

"Thank you for everything, you've been a great help to us," Tsibliyev radios the TsUP as he and Lazutkin prepare to undock the Soyuz for their return. *"The time flew fast. Many things remain unfinished. But the new shift will take over. And see you soon."*

"I hope the Rodniki [team] who are replacing us will do what they have to do," Lazutkin chimes in. *"Rodniki means 'sources,' so they are like sources of life and may God help them to breathe new life into the station."*

"What was bad about us will stay with us," says Tsibliyev. *"We are going out of comm and see you on the ground."*

"Thank you, guys," Vladimir Solovyov radios up. *"The entire TsUP personnel thanks you for your work and wishes you just one thing: Soft landing and see you back home. Good luck."* When the Soyuz finally pushes back, Tsibliyev mutters: *"Thank God."*

Lazutkin is worried whether Tsibliyev's weakened heart will hold up during their short but strenuous return voyage. It should take only about three hours to pierce Earth's atmosphere, open parachutes and float down to the steppe in a remote area of Kazakhstan; at its worst, they should pull about five Gs during the roughest part of the descent. "I just thought about it in the back of my mind; I didn't say anything," Lazutkin recalls. "Because of his heart problem, Vasily didn't exercise. Could he survive without having exercised? I realized that if he spends six months in space, and the last stage of preparation he doesn't do any exercise, I know that for a healthy body, descent is a problem, even with a healthy heart."

In the event, everything goes smoothly. At the last moment, the TsUP cancels the fly-around they planned to take more pictures of Spektr. No one has to explain why the ground is uneasy with the idea of Tsibliyev manipulating manual controls in the immediate vicinity of the station. Lazutkin watches the commander closely during the reentry, but Tsibliyev and his heart show no signs of stress. Not until the little capsule reenters the atmosphere, pops its chute, and begins floating down toward the ground does something go wrong. Later it will seem almost preordained, one last screwup in a mission that seemed cursed from the outset.

Suspended beneath the white parachute, the Soyuz is twenty yards from landing when one of its final thrusters fails to fire. The failure of this thruster, whose firing is supposed to soften the capsule's impact, causes the Soyuz to land on the steppe with a loud thud. Tsibliyev and Lazutkin are not injured, but the capsule is badly dented. The two men emerge wobbly and are carried from the Soyuz by ground technicians who have scrambled to the site. Russian reporters are already there, poking microphones in the two Russians' faces. "How do you feel?" one of them asks.

Lazutkin manages a weary smile. "What other feelings could we have?" he says. "We are just glad to be back."

Igor Goncharov, the IBMP doctor, is there and takes Tsibliyev's pulse. "Satisfactory," Goncharov says.

"Alive!" says Tsibliyev, forcing a smile. "I'm happy to be on Earth."

It is not until they board the small jet for the long flight back to Star City, having undergone all their medical checkups, that Tsibliyev begins to get angry. Just two days earlier Boris Yeltsin himself, in a rare public statement on Mir's troubles, blamed the collision on "human error"; a second statement, issued within hours of the cosmonauts' return thanking them for their "persistence, courage, and heroism," does nothing to mollify Tsibliyev. Even Viktor Blagov has chimed in, openly castigating the commander during an impromptu press conference at the TsUP. "We have found not a single fault in the systems on either Mir or Progress during the collision," Blagov says. "A logical conclusion follows—technology was not at fault here."

Aboard the jet, Tsibliyev warns a Reuters reporter that Russian officials will try to blame the whole thing on him. "It's very easy to find someone to shoot—pop, pop, pop, and he's gone," Tsibliyev says. "Whom does this help? Here you can't look for who is guilty."

When they reach Star City, Tsibliyev walks off the plane and accepts a warm hug from his wife, Larissa, and his daughter. He and Lazutkin step onto an old yellow bus, and as the bus leaves, a young soldier pounds his hand on a window and pumps his fist in the air. The two cosmonauts are whisked straight to the Prophylactorum, where they are greeted with bear hugs by a group of their peers. Aleksandr Serebrov, Tsibliyev's old friend who is now Yeltsin's senior adviser on space affairs, is there.

"Don't worry, Vasya, don't worry," he tells him. "Do not worry about the politicians. We will take care of you."

August 15
Aboard Mir

There is much to be done before the August 20 IVA. The new crew's first task together is a complex one, flying their Soyuz around from the Kvant docking port to the node, in order to make room for a Progress they are to dock in several days. Foale notices that Solovyov is tense. "All this put pressure on Anatoli more quickly than he thought," Foale recalls. In preparation for the fly-around, they turn off all the station's major systems, so that Mir will be maintainable in the unlikely event they are unable to redock and are forced to return to Earth. A few hours before the flight, the TsUP suddenly asks Foale to prepare a series of Russian cameras so that he can photograph the damage to Spektr. But the Russian cameras are dismantled and several of the pieces are missing. After a hurried conversation with the ground, Foale insists on using an American camera. He gets the sense that Vinogradov is irritated because he wasn't asked to take the pictures.

The fly-around goes smoothly, and forty-five minutes after leaving the station, Solovyov safely redocks the Soyuz at the node. When the three men reenter the station, however, they find that they have left the galley lights on. This failure constitutes a "black mark" on Solovyov's contract, for which he will almost certainly be docked in pay. Foale can see the commander is upset.

Saturday, August 16
Star City

Outside the window a military band was playing an incongruous rendition of "New York, New York" as Tsibliyev, attired in a natty new Reebok jogging suit, strode into the briefing room and addressed the assembled reporters. If his superiors wished that the commander would publicly take responsibility for his disastrous tour aboard Mir, they were disappointed. Tsibliyev came out swinging, and in a passionate self-defense blamed most of what had gone wrong on the dilapidated state of the Russian space program.

"The cause lies with problems on Earth," he told the reporters. "It's connected with the economy, with our affairs in general. Even the equipment needed to live aboard the station and that we requested to be sent—and we're not talking about coffee, tea, and milk—they just don't exist. The factories don't work, or have insufficient supplies, or they ask for, excuse me, crazy prices." This veiled reference to the dispute with the Ukrainians over the Kurs system went over the heads of most of the Western reporters.

"It has been a long tradition here in Russia to look for scapegoats," Tsibliyev fairly spat. "Of course it is easier to put all the blame on the crew. . . . Perhaps many want us to return as corpses, thinking that would have been great. Thank God everything turned out as it should have. . . ." This was a direct shot at the TsUP.

Lazutkin, standing at his side, was stunned by his commander's anger. "His behavior after the flight was a little bit unexpected to me," Lazutkin recalls. "I didn't think he would be that aggressive. The reasons are mostly psychological."

The press conference was over in barely thirty minutes. Months later Tsibliyev clearly regretted having been so combative. "After the flight you are so full of stars, so full of living, all you want to do is talk," he says today. "You'll tell anyone anything. But a lot of people have suffered from loose tongues like this. Norm Thagard was the first to talk about the same phenomenon. He only spent [four] months in space. Russian cosmonauts, they spend much longer. They pay a price for their openness."

Monday, August 18
Aboard Mir

This afternoon, after a one-day delay due to a computer problem, the TsUP plans to redock the cargo ship Progress-M 35, which has been circling the station while the two Soyuz craft filled both its docking ports. As usual the plan calls for the Progress to dock automatically via the Kurs system, but the ground asks Solovyov to ready the TORU docking system just in case. Foale brings out the system and sets it up in the middle of base block. He can see that Solovyov is nervous about the possibility of manually docking the incoming ship. "It put a lot of pressure on Anatoli at a time when he was still getting adjusted to space," Foale recalls. "The last person who had used [the TORU], after all, had caused a collision."

Foale, too, is nervous—"a little gun-shy," as he puts it. He crouches by the small window in Kvant as the TsUP brings the Progress in toward the Kvant docking port. To Foale the feeling of déjà vu is almost overwhelming. As he watches through the Kvant window, he tries to fight the fear building in his chest. From his vantage point, the ship seems to be coming in too fast. He tells himself he is imagining things—he is fully aware what tricks the mind can play—but eventually the fear overcomes him. The incoming ship is not slowing. "Anatoli, it's not braking!" Foale finally says. "I mean it, it's not braking."

"No, it's going to brake," the commander replies, his gaze never leaving the Sony monitor. "It's going to brake."

"No, no," Foale says, his eyes fixed on the Progress.

And then, just moments after Solovyov's reassurance, Foale can see the incoming ship's braking thrusters begin to fire, shooting out inverted cones of white exhaust. A feeling of relief washes over him.

His relief is premature. Everything goes smoothly until the Progress reaches a point 170 meters away from the station. Foale, who is videotaping the docking, can see the Progress clearly in space as it approaches.

Suddenly, with no warning, one of the station's alarms goes off.

BOOOO. . . . BOOOO . . .

A row of buttons light up on the central command console. Somehow the main computer has crashed. Within seconds the motion-control system fails and the station enters free drift. Poised before the TORU controls, Solovyov remains calm. He asks the ground for permission to take manual control of the Progress. Down at the TsUP, Vladimir Solovyov swiftly gives him the go-ahead.

Solovyov, his hands now gripping the black TORU joysticks, aims the incoming ship directly toward the Kvant docking port, and under his guidance the ship glides swiftly toward Mir. At fifty meters all is well. At forty meters, it remains under his control. Vinogradov hovers at his side, watching.

Thirty meters.

Twenty meters.

Fifteen meters.

Then, as the Progress passes the five-meter point, its nose pointed squarely at the docking port, Solovyov's screen suddenly goes blank. Somehow, flying completely blind, the commander keeps the Progress inching ahead. Moments later the ship's docking probe nudges the docking port.

"Congratulations!" the ground says. *"You guys are great!"*

Foale watches as Solovyov's shoulders sag in relief. Vinogradov wraps him in a hug. "That was a very big moment for Anatoli," Foale recalls. "He was only the second person to ever do that. Now he proved you could do it right. His confidence shot right up at that point."

Though the Progress is docked, the station's problems are not over. Minutes after the docking, Mir passes out of communications range with the TsUP. With the computer down, Mir has begun slowly spinning. The onboard batteries, whose charges are already low, start to drain.

"You know, Anatoli, I've been through this before," Foale volunteers. "What we need to do is spin the station in the opposite direction."

"Why is that?" Solovyov asks. Foale realizes the commander doesn't completely fathom the situation they are in.

"We're losing energy very fast. There's a chance we could lose all power." Foale explains how they had lost their power after the collision, when the TsUP failed to turn off the onboard systems fast enough. Solovyov seems to understand the argument, but he shows no inclination to take the initiative himself and halt the spinning station. Again Foale is hit with a sense of déjà vu. "[Anatoli] was *incredibly* uncomfortable with this idea," Foale recalls. "He didn't know what to do, and he didn't want to hear it from me. He wasn't about to do anything until the ground gave him an order."

"Anatoli," Foale says, "we may not ever get that order from the ground if we don't have power to receive it. You understand?"

"I understand, but we will wait to get the order from the ground."

There is no persuading Solovyov. For a half hour Foale waits anxiously for the station to reestablish contact with the ground. To his relief, it does.

"Guys, you are in a really bad energy situation; we want you to reorient the station," Vladimir Solovyov says as comm is reestablished. *"Anatoli, you ask Michael how to do this."*

Without a hint of chagrin, the commander complies. Foale can sense Anatoli Solovyov's respect for him growing.

The Spektr IVA is pushed back two days to give Solovyov and Vinogradov time to fix the computer—a piece of equipment called a data exchanger is quickly replaced—and make certain the station has stabilized. The crew uses the extra two days to finish securing their tools in the node, tying them to the walls with Velcro and bungee cords. During the IVA Foale is to remain in the Soyuz. In the event Solovyov and Vinogradov are somehow unable to repressurize the node after the IVA, they will crawl into the Soyuz and all three men will immediately return to Earth. Foale, who must seal off the Soyuz from the rest of the station, isn't at all comfortable with his role.

"Look, guys," he tells Solovyov and Vinogradov at one point, "I haven't trained on this stuff. You have to show me what to do."

"Da, da," Solovyov said, promising to show him. But as the day of the IVA approaches, the commander simply doesn't have time to tutor Foale. The urine-reclamation system is acting up again, and Solovyov is spending all his spare time repairing it. The night before the IVA, Foale still hasn't been shown the Soyuz controls, and he is forced to ask Solovyov for help once again.

"Anatoli, you've simply got to show me how to do all this stuff," he urges. Before the three men bed down, Solovyov finally complies, but in Foale's mind his lecture is rudimentary at best.

"He walked me through it, but it was very fast," Foale recalls. "I was hoping I would know what to do."

August 22
Aboard Mir

Vinogradov cannot sleep. He lies in his sleeping bag in his *kayutka*, twisting and turning, trying to get comfortable. It is hot inside the station, and he has butterflies. He is time-lined for six hours of sleep, but after barely four he gets up and makes something to eat. By and by Solovyov rolls out of his *kayutka* and joins him, then Foale. Everyone is nervous. No one will admit it.

By six the two Russians are busy performing final checkouts on their big white Orlan space suits, which are already in the node. They have gone through checkouts twice, but Solovyov insists they do it again, just to make sure. Then, just as they are finishing up, the commander finds a problem. His suit will not establish a voice link with the ground. The TsUP tells him to remove his communicator and switch it with the one on Vinogradov's suit to see whether it is the suit or the communicator that is malfunctioning. Solovyov objects. Changing equipment on a space suit at this point, after all the suit checks are completed, is considered a bad omen. The TsUP then checks its equipment on the ground and finds the problem lies in one of its computers. A fix is made, and Solovyov relaxes. The bad omen is avoided.

A little before ten, Solovyov, Vinogradov, and Foale gather in base block, where they observe the Russian custom of sitting for a minute of silence. It is Solovyov who breaks the quiet.

"Okay," he says. "Let's go."

In the node, Foale helps Solovyov and Vinogradov into their suits, which are difficult to close without help. Then he swims through the adjoining hatch into the Soyuz. Behind him he locks down the hatch between the Soyuz and the node, then retreats into the command capsule itself and locks down the hatch between the command capsule and the Soyuz's small living compartment. If for some reason the two cosmonauts are unable to repressurize the node after the EVA, they will be forced to depressurize the Soyuz living compartment, at which point they would enter the Soyuz and evacuate the station. Foale has his own reentry suit ready just in case.

Hovering inside the node in their space suits, the two Russians get the go-ahead from the ground to begin depressurizing. The node is barely seven feet in diameter, and it is crowded. Inside their suits, Solovyov and Vinogradov stand back-to-back and are unable to move much to either side. With his left hand Vinogradov turns a brown, star-shaped valve inside the node. Immediately he hears the loud hiss of air as the node's atmosphere begins escaping into space. They are to "depress" in three stages, first taking the pressure from 760 millimeters of mercury down to 540, then, with the TsUP's approval, dropping it to 280, then finally to vacuum. The entire process should take about forty minutes.

It takes only a few minutes for Solovyov, who is reading a handheld gauge, to realize that the pressure is not falling fast enough. Before the TsUP can figure out what is wrong, Solovyov suggests that a valve between the node and one of the modules is not closed. Within minutes the TsUP confirms the commander's guess; the pressure inside Kristall is falling, indicating that it is the valve between Kristall and the node that is leaking. Solovyov asks for and receives permission to fully repressurize the node, take off his space suit, crawl back into base block, and activate the necessary controls to close the valve. Foale volunteers to do it, fearing that Solovyov will be unable to close himself back up in his space suit, but the ground prefers the commander. Foale stays put.

It takes over an hour for Solovyov to close off the open valve and reenter his space suit. They are running behind schedule now, and everyone is eager to get going. Again, standing behind the commander inside the node, Vinogradov turns the depress valve; again they hear the loud hiss of air. This time the pressure declines as expected, reaching the 540 level in about ten minutes, then, fifteen minutes later, 280. All is clear. With one final twist by Vinogradov of the star-shaped valve, the pressure begins diving down toward vacuum.

Everything is finally going as planned. Then, suddenly, at 1:23, just as the pressure reaches 210 millimeters of mercury, Vinogradov feels something move inside the left arm of his suit, something feathery and light brushing against his inner forearm. With a start he realizes it is air. There is a leak in his suit.

"I started moving my hand," Vinogradov says on open comm, *"and the pressure is going down [in my suit]. When I move, I feel the air is moving."*

Foale is the first to react. *"Pavel, stop!"* he blurts through his headset. *"Stop moving your hand!"*

"Stop, stop moving your hand!" Anatoli Solovyov chimes in.

On the TsUP floor, Vladimir Solovyov and his ground controllers exchange worried looks. *"Pavel, don't move your hand,"* Solovyov says calmly. If Vinogradov moves his hand enough to loosen the glove, all the air in his suit could be sucked out in a matter of minutes. Vladimir Solovyov takes several moments to confer with Blagov and the others. As they talk, Foale is struck by Vinogradov's reaction. "He wasn't scared by it," Foale remembers, "but I sure was. I was as worried as I've ever been during spaceflight."

Vinogradov is in fact as frightened as he has been in his life, but he's struggling not to show it. He realizes the leak is somewhere in the fitting between his glove and his space suit. Fighting to remain calm, he takes his right hand and grasps his left wrist firmly, trying to stanch the flow of air from his suit. Behind him, Solovyov urges him to stay calm. Even with his glove clamped shut, Vinogradov believes air is still leaking from his suit. He estimates he has about fifteen minutes of air inside his suit.

"Pay attention!" Vladimir Solovyov barks after a moment. *"This is very serious."*

This Vinogradov does not need to hear. He knows it is serious.

"Pasha, don't worry," Solovyov goes on. *"Don't take any steps that are not well thought out. . . . We have time."*

Solovyov directs Vinogradov to close the star-shaped depress valve. The node is so small, the two men are pressed so tightly together, only Vinogradov can reach it. Awkwardly, he grabs the valve with his left hand, all the while keeping his right hand clamped around his wrist. Slowly, he manages to crank the valve closed. The ground estimates it will take seven minutes for the air to stop seeping out of the node. Vinogradov hopes the air in his suit will last that long.

Minutes tick by. The repressurization valve is on Anatoli Solovyov's side of the node, but the commander cannot begin repressurizing the node until Vinogradov's valve is completely closed. Vinogradov is quietly thankful his glove hasn't opened further. "If my glove had opened completely," he recalled months later, "I wouldn't have been able to close that valve at all. Anatoli could not get to it. Only I could close it. I don't even know what would have happened then." But he did know: If the commander couldn't find a way to close the valve, Vinogradov would suffocate.

After seven minutes the pressure inside the node stabilizes. Immediately the commander cranks the emergency repress valve, allowing air from base block to whistle into the node. With a loud hiss the pressure begins rising. It takes barely ninety seconds for it to climb back to 540.

If they are to have any hope of completing the IVA today, Vladimir Solovyov knows, they must quickly replace Vinogradov's leaking glove.

"Pasha, do you have a spare glove?" Solovyov asks.

"Da."

Both cosmonauts have brought bags containing two spare gloves into the node with them. When the pressure reaches 540, Vinogradov reaches for his bag, takes out his extra left glove, and quickly slides it on.

"You turn it, pull it with all your proletarian might, and break it out a little," Solovyov instructs. *"Minimize the amount of time when your hand is bare."*

It is 1:32. Vinogradov takes two minutes to make sure the glove's seal is tight. At 1:34 he announces that he is ready to continue. Vladimir Solovyov does a quick calculation to make sure that, with all the air they have used, there will be enough remaining to fully repressurize the node at the end of the IVA. According to his math, there is.

"Okay, then let's start again," Vladimir Solovyov says.

Again Vinogradov takes his left hand and turns the depress valve. The hiss of air fills their ears. In silence, the two cosmonauts wait as the atmosphere leaks back out into space. By 1:47 the pressure has fallen to 460. Five minutes later it is at 110. At 1:54 it has fallen to 50. Five minutes later, as the cosmonauts wait to enter vacuum, the station moves out of communications range with the ground. The men in the TsUP, with hundreds of reporters and cameramen clogging the mezzanine above them, must wait forty-five minutes for Mir to come back into range. In the meantime they can only hope there is not another unexpected problem.

* * *

"All right," Anatoli Solovyov says when the pressure gauge stabilizes at zero. *"Ready to go?"* The commander is anxious to get started. They are already two hours behind.

"I think so," says Vinogradov, fighting to retain his confidence after the glove incident.

Slowly, still inside the node, Vinogradov inches to his right to position himself in front of the Spektr hatch. Before the IVA, they had already attached the funnel-shaped docking cone with the hermaplate at its tip. It takes only a few minutes to remove the outer lid Lazutkin and Foale had thrown onto the hatch six weeks earlier.

At 2:14 Vinogradov sticks his head through the hatch into Spektr. Behind him Solovyov directs his flashlight into the module. Spektr is dark and silent. There is nothing floating in the air, no glass shards, no equipment, nothing. Then he spies a cluster of white flakes. They look like snowflakes.

"That's probably my shampoo," Foale pipes up. Out of sight in the Soyuz, he remains in communications with the two cosmonauts.

"That's my guess," Solovyov says.

Vinogradov crawls forward, sticking his shoulders, then his belly, then his legs, through the hatchway. To his relief, he slides in easily. Pulling himself erect inside the module, he is struck by the silence. The only sound is his own breathing inside his helmet. It is like a tomb.

Vinogradov turns around to face the module's inner hatch. Under his arm he has the bundle of eleven cables he is to string from the hermaplate into the connection points that encircle the inner hatch. In a separate bag he has brought the collection of tools Oleg Tsygankov's people have thrown together to help connect the cables; he didn't bring in the police flashlight. As Solovyov points his light onto the connecting points inside the hatch, Vinogradov immediately realizes something is wrong. At Star City, in the mock-up they used in the hydrolab, the cables' connecting points were positioned at the top of the hatch, which gave him plenty of space to maneuver the tools. What he is looking at now is completely different; in fact, it is exactly the opposite of the mock-up they practiced on. Inside Spektr the connectors are all on the bottom of the hatchway, hard by the wall of the module itself. One glance tells him there will be no room to use any of the tools.

"Oh, hell," he says.

A minute later the station reenters communications range.

"Everything okay?" Vladimir Solovyov asks.

"We entered and are working," Anatoli Solovyov replies.

"Did the hatchway open all right?"

"Opened easily. Here the module is in perfect order. Some white crystals flew out, looking like soap. Some case which carries the inscription 'Euroworld' is in there, stuck between some panels."

"Forget about the case. What are you doing now, Pasha?"

"Now I am trying to connect Joint 84. Unfortunately, everything here is vice versa. I cannot see where to connect to."

"Please explain what [you] mean."

Vinogradov explains that he is trying to insert the cables into the hermaplate using only his gloved hands. The gloves are heavy and awkward; his task is like threading a needle while wearing mittens. *"All the tools are with me, but I am trying to work with my hands. Everything turned out to be vice versa, 180 degrees turned over related to the model on the ground."*

Vinogradov is thankful that his trainers at Star City insisted on showing him ways to connect the cables without the tools. It is possible to do, but not easy. One by one he slips the first cables into their slots and jams them forward a bit. He has to be careful, lest the cable ends bend or break.

For several minutes he works in silence, his only words an occasional muffled *"Hell."*

"Please keep us in the know," Vladimir Solovyov asks. When Vinogradov remains silent for several more moments, he repeats himself. *"Pasha, tell us what you are doing."*

A few minutes later, just as the station enters a sunlit orbit, Vinogradov is startled by a series of sounds all around him. Suddenly the dead module seems to spring to life. Fans begin to whir. The ventilation systems kicks in with a low hum. Red LED readouts pop up on several pieces of equipment. He laughs out loud.

"Pasha, what is it?"

"The module works! Ventilators are producing noise. Everything else is buzzing too!"

Sunlight hitting Spektr's solar arrays has activated the module's emergency systems. Smiles break out on the TsUP floor. Someone says, "Russian technology!"

Vladimir Solovyov actually manages a joke. *"Do you mind taking a look around the module? Maybe some other members of the crew are still there?"*

The levity is short-lived. A moment later Vinogradov again begins to curse. *"I cannot install this damn nut. It does not fit into the joint."*

Vladimir Solovyov asks for more information.

"I do not like this at all," Vinogradov says.

By 2:57 Vinogradov has three of the eleven cables connected. On the TsUP floor the controllers eye the clock. The two cosmonauts have enough air to work till around 6:00 P.M.

As he works, Vinogradov's labored breathing fills the TsUP auditorium. The next morning the *New York Times* will compare listening to the operation to the sounds of a crack team of auto mechanics changing sets of spark plugs.

"Hell, what is this?" Vinogradov says a few moments later. *"Needs to be tightened a little. Maybe I should turn it with my legs?"*

"Pasha, maybe you should take some rest," Vladimir Solovyov suggests.

Sweat is streaming down the back of his neck, but Vinogradov vows to continue. *"When I start to prepare the next cable I'll have some rest."*

By 3:10 Vinogradov has the fourth cable connected. He is worried, though, because water seems to be seeping into the connection joints. He speculates that it is frost melted by the heat of their spotlight.

"Why do you think this is water?" Vladimir Solovyov asks.

"We tried it with our tongues," Vinogradov jokes. He asks for and receives permission to dry his work area. A few minutes later the station again passes back out of communications range. In the TsUP, Solovyov and his controllers push back from their consoles and stretch. They are behind schedule, but they should still have enough time to finish by six.

In the next hour Vinogradov manages to attach all seven remaining cables; the hardest part of the IVA is complete. Solovyov wriggles into the module, and by the time the station reestablishes ground contact, at 4:09, the two cosmonauts have managed to pack two bags full of Foale's personal supplies, including his laptop computer, pictures of his wife and children, and his toiletries.

"The Americans owe us a big debt," the commander jokes.

"You are taking all this garbage, will you still have enough space for yourselves?" Vladimir Solovyov asks. Everyone knows how cramped the node is.

"Don't worry, we'll fit in."

Vinogradov spends much of the next hour removing wall panels in search of the leak. Solovyov videotapes everything. The two men first open panel 208, then 211, 207, and 213. There is nothing inside any of the panels that remotely resembles an indentation or a hole. There are only pipes and drawers containing old NASA experiments.

"Listen, guys," Vladimir Solovyov says, *"by 17:45 you have to be in the process of locking [up] the module."*

By 5:30, while again out of communications range, both men have exited Spektr and are back in the node, having closed and resealed its hatch. A few minutes later the commander begins the repressurization procedure.

"How is everything?" Vladimir Solovyov asks when the station comes back into range at 5:47.

"We are back in the [node]," the commander replies. *"Pressure is 224 millimeters.'*

They have not found the leak, but on the TsUP floor a feeling of relief sweeps through the assembled controllers. By tomorrow they should have much of the station's power back, enough to turn on some of the other modules. "Nothing," a clearly relieved Vladimir Solovyov tells a press conference afterward, "can knock us out of the saddle."

Jubilation over the successful IVA is quickly tempered by the realization that Vinogradov failed to hook up one important cable—the cable that links Spektr's solar arrays to the station's main computer and allows the computer to reorient the arrays. Without this hookup, the arrays are unable to track the sun. Foale expected the IVA to allow the crew to turn on the power in Priroda and Kristall. Instead the two modules remain dark. The repair will now be accomplished on the station's outer hull, in the spacewalk Foale and Solovyov are to attempt on September 6.

The following week is largely quiet aboard the station, and several officials in both space programs, including Viktor Blagov, take the opportunity to grab fast vacations. The only glitch comes on Monday, when the NASA public affairs office in Houston issues a press release announcing that both of Mir's Elektron oxygen-generating systems have suddenly failed. Headlines in the American press trumpet yet another breakdown on the star-crossed Mir; the *New York Times* runs its story on page one. A day later chagrined NASA officials are forced to admit there is no emergency. The two Elektrons had only been shut down temporarily for still more repairs. The NASA ground team in Moscow, taking advantage of a slow day, had left work early and hadn't told Houston of the quick fix.

Tuesday, September 2
Moscow

That Tuesday, Energia officials announced results of their investigation into the collision. As expected, despite all of Culbertson's pleas, they blamed crew error. Though the Energia report was not released to the public—and remains secret to this day—it was, as Culbertson and other NASA officials long feared, a complete whitewash.

So what did happen that day that led to the collision?

In reality, the collision, as subsequent investigations by NASA, the Stafford Commission, and Energia itself would show, was the result of a confluence of events, none of which were under Tsibliyev's control. Far and away the most critical factor was the setup of the docking test itself, which was the TsUP's responsibility.

"Tsibliyev followed the procedures exactly as written," says Bob Castle, the veteran MOD flight director, who analyzed the collision for Mike Baker's NASA group, "and did everything that he was told to do." NASA analysts, in fact, were stunned by the slapdash nature of the experiment the TsUP asked Tsibliyev to perform. "The overwhelming thing for me," says Castle, "was that it looked like it hadn't really been thought through. I don't know why that is. I mean, there are a lot of smart people at Energia. But this is something I can safely say the Americans would never have done this way."

On the day of the collision, the NASA investigation showed, the initial approach of the Progress toward Mir had gone as the Russians planned. As in the early stages of most Russian dockings, it was what docking experts call a "hot approach"—that is, the Progress flew on a trajectory that sent it soaring toward Mir at a far greater rate of speed than NASA uses in its approaches. "The Russian approaches really use brute force," observes Jim Van Laak. "They come in fast and furious, using range rates that are ten to twenty times higher than we would use. Where they approach at meters per second, we would using inches per second. They come in faster, a lot faster, than we would. And then they brake as hard as they can." For the longest time NASA analysts couldn't understand why Russian ships approached at such high rates of speed. Eventually some at NASA, including Mike Foale, figured out the problem: It was the unreliability of the Kurs system. "They are afraid errors will build up in the Kurs system if they take too long in their approach," says Foale.

Already designed to approach the station at a high rate of speed, the Progress Tsibliyev controlled that afternoon in June came in even "hotter." The TsUP memo the commander had been given instructed him to rotate the incoming ship as it approached so that its camera would remain centered on the station. But rotating the Progress—adjusting its "pitch" downward—had the unforeseen result of increasing the ship's speed. This is a basic tenet of orbital dynamics: The sharper a spacecraft's downward angle, the faster it will fly. And so, at a time Tsibliyev was worried that the ship was approaching too slowly, it was actually approaching too fast, rocketing downward toward the station.

A commander using computerized range data would have instantly realized the ship was approaching too fast. But Tsibliyev had none, and this was the single most dangerous element of the TsUP's plan that day. Analyzing the collision later, NASA controllers like Bob Castle simply couldn't understand why the TsUP would allow Tsibliyev to fly with no telemetry data. If the TsUP wanted Tsibliyev to fly "blind," Castle and others suggested, there was an obvious, and a far less risky, way to do it. A strip of masking tape could have been placed over the incoming telemetry data on his screen, blocking the commander's view of the numbers; if he encountered an emergency, Tsibliyev could then have pulled up the tape and immediately read the speed and range data. Why hadn't the TsUP done something like that? The answer lies in the assumption underlying the test. The TsUP believed that the Kurs antenna had interfered with the camera aboard the Progress during the Near Miss in March; they couldn't risk turning it on again. As a result, "they took this huge, huge leap forward," notes Charlie Precourt, the American shuttle commander, "and they clearly were not ready for it."

Instead the TsUP directed Tsibliyev to deduce the incoming ship's range with the laser rangefinder and a stopwatch. But the rangefinder was all but useless. It could be employed only with the Progress in sight, and as TsUP controllers should have known, Mir's solar arrays blocked sight of the incoming Progress until the ship closed within several hundred yards of the station, at which point it was too late to halt the incoming ship. And so Tsibliyev was saddled with a ship approaching at a high rate of speed and no way to know it. By the time he realized his predicament, the ship was approaching too fast for him to bring to a full stop. "The guy," says Precourt, "was set up to fail."

The TsUP also erred by ignoring the repeated warnings of Star City psychologists and others, including the cosmonaut Sergei Krikalev, that Tsibliyev was too exhausted to perform the docking. The military commanders at Star City, who supervise cosmonaut training, must bear a small measure of responsibility as well. The test that day called for Tsibliyev to dock the Progress against a swirling backdrop of clouds, which, as Tom Stafford later pointed out, is an exceedingly difficult background to see into. None of the docking simulators at Star City, in which the commander had been trained, had clouds in the background; all were simply black. Even this, however, wouldn't have been a problem had Energia heeded the warnings of Aleksandr Serebrov and others to install flight-simulator software aboard the station so that commanders could refresh their docking skills. By the time Tsibliyev was asked to dock the Progress on June 25, he had not rehearsed a manual docking on a simulator since shortly before leaving Earth in February, a period of more than four months.

Saturday, September 6
4:07 A.M.
Aboard Mir

Crouching inside his Orlan space suit in the air lock at the end of Kvant 2, Foale nervously eyes the hatch that leads outside the station. Solovyov is right behind him, and at the commander's urging, he gives the rusty hatch a shove. Apparently some air remains in the air lock, because the hatch springs out forcefully, banging back onto its hinges. Hanging on to the hatch, Foale is jerked outside the station along with it.

"*Whoa!*" he says.

Regaining his composure, Foale quickly moves onto the ladder outside the station. Unlike Linenger, he feels no sensation of falling; Foale has walked in space aboard the shuttle and loves spacewalking. He and Solovyov have spent hours diagramming their work today. Their six-hour spacewalk is fairly straightforward. They are to make their way to Spektr's outer hull, check it for punctures, and construct handrails for later repair work. If they have time, they are to perform a minor repair on the Vozdukh system and retrieve a small radiation monitor.

Foale, ignoring Linenger's "razor-sharp" solar arrays, quickly makes his way down the length of Kvant 2 and crawls onto base block, where he straddles the crane at the base of the Strela arm. Behind him Solovyov muscles out a large bag containing the scaffolding they are to assemble on Spektr. It takes just over an hour for Solovyov to maneuver his way to the Strela arm, straddle its far end, and have Foale move him across open space to Spektr. At the Star City hydrolab, Foale had practiced using the Strela arm exactly once and wasn't at all sure he could do what was needed; as it turned out, the crane proved easy to operate.

Once at Spektr, Solovyov wastes no time. The TsUP has identified seven possible places the coin-size puncture might be located. Solovyov quickly takes out a knife and begins cutting through the foam insulation that covers portions of Spektr's outer hull. For ninety minutes, while Foale waits at the Strela's base, Solovyov methodically carves up the insulation near the impact area. The Mylar material keeps fluffing up as he cuts. *"I should have taken scissors but not a knife,"* he says at one point.

It is, in fact, an impossible job, akin to finding a lost coin in a junk heap. After two hours Solovyov gives up. There is no hole, at least none he can see. *"It is strange—to rumple this way and destroy nothing,"* he says.

A little before eight Foale shimmies down the Strela arm and joins the commander, who now moves farther out on Spektr's hull. He spends the last hour of the EVA manually rotating the three undamaged solar arrays to a position where they will more fully face the sun. In this way the TsUP hopes to regain more power from the damaged module.

By ten both men are back in the air lock. Solovyov's hunt for the puncture has gone on so long they have had no time to erect the scaffolding—which they leave tethered outside Spektr—or work on the Vozdukh system. It is Foale's job to close the outer hatch, and for some reason Vladimir Solovyov wants him to work faster.

"Hurry, Michael, hurry," he says.

But Foale senses something is wrong. The hatch doesn't feel right when it closes.

"Hurry," he hears someone say.

"Look, guys, you're rushing me," Foale says. *"This does not feel right. I have to reopen the hatch and do this again."*

Foale takes an extra minute to make sure the hatch closes tightly. His sixth sense will prove to be on target. It is the last time Mir's outer hatch will ever work correctly.

Tuesday, September 9
Washington/Houston

By the first week of September there were no fewer than four separate safety reviews of Mir under way at NASA. One of the most influential involved the Human Exploration of Deep Space committee, which consists of ranking NASA safety officials from across the country; although the HEDS committee, as it is called, had no official vote in the review process, it served as a key advisory panel for Dan Goldin. Blaine Hammond, who sat on the HEDS committee, arrived at the headquarters conference room that morning for the all-day meeting, determined to argue against sending Wolf aloft but expecting no one to listen to him. In two months of harping on Mir, he had yet to find a single NASA official or astronaut—other than Linenger, who had his own problems with the agency—who would stand by him and openly lobby against the launch. "I saw no chance of stopping the locomotive," Hammond remembers. "It was all a fait accompli. Dave Wolf was going to go up no matter what they said. I knew all along I was beating against a wall. I could have gone to Goldin, and he would have looked at me and laughed at me. I actually thought about doing it."

Teleconferenced in from Houston, the first speakers to address the committee that morning were John Blaha and Bob Castle, the senior flight director who had been asked by Abbey to analyze the collision. As Blaha spoke, there was no hint of his visceral anger toward the program. Instead he wore what might be called his NASA face, the civil expression and tone of the professional astronaut. He said he had never felt unsafe aboard Mir and could see no compelling reason to prevent Wolf from going as well.

Castle was there ostensibly to talk about NASA's investigation into the collision, but he spoke only in general terms. He said Abbey had directed him not to discuss details until the Stafford Commission had completed its own investigation; Stafford and several commission members were due to begin arriving in Russia the following week. This was too much for Hammond.

"I feel like it's impossible to analyze the Russian process without analyzing the information they have at hand to make their decisions," he said. But Castle wouldn't budge: There would be no freewheeling analysis of the collision, not today.

When both men finished, Fred Gregory, the headquarters safety chief who chaired the meeting, asked whether they felt it was safe for Wolf to replace Foale. Then, in a strange moment, Gregory said he would not tape-record their answers. Hammond, who had never seen anyone at NASA make a similar offer, thought this was done to encourage Castle and Blaha to be candid. Whatever the case, Blaha responded with a "qualified yes." He added: "The real answer should come from Mike Foale."

"It'd be okay to transfer Dave for Mike," Castle said, "but the commander should ask Mike when you get up there."

To Hammond this was classic NASA ass-covering. "What's Mike going to say?" Hammond recalls thinking. "He's not going to take that responsibility for the whole program coming apart. You'll never get an honest answer from him. It's like asking a crew on launch day if they want to go."

After Blaha and Castle signed off, the committee members engaged in a long debate about the merits of sending Wolf to Mir. Hammond put forward the argument he knew everyone was sick of hearing: The benefits of staying aboard Mir, he said, were now outweighed by the risks of having an astronaut killed. "By then it was clear I was just pissing in the wind," Hammond recalls. "They don't want to hear what I have to say. You know, it was just the ravings of ol' Blaine Hammond, who no one listens to anyway."

In fact, Hammond's argument did sway two members of the committee who had not heard it before. One was Axel "Skip" Larsen, head of the shuttle program's Payload Safety Review Committee. "In the morning session Blaine piqued my interest; I felt comfortable agreeing with him," recalls Larsen. "The morning session was kind of a wide-open discussion, considering all aspects and risks, not only to astronauts but to potential future space stations, if there was a problem. There was a concern that if we had a disastrous event on the station with a U.S. astronaut, it would cause the International Space Station program to go into a stand-down mode, and possibly it could have resulted in the Space Station program being canceled. We didn't want a new Challenger situation. It could cause the country to cancel habitable space flight altogether."

Before adjourning for lunch, Gregory informally polled the committee on whether to send Wolf to Mir. The vote was eleven to three, with Hammond, Larsen, and an ISS official named Kevin Klein dissenting. Larsen immediately noticed a subtle change in the air. "It was interesting," he recalls. "Up until the vote it was an open discussion. I actually felt somewhat estranged after that." The feeling deepened when Klein and Larsen found themselves lunching alone at the deli in the NASA lobby; everyone else, including Hammond, had gone to the headquarters cafeteria. "There was kind of an 'uh-oh' feeling, a sense maybe we had done something we shouldn't have," Larsen remembers. "I certainly perceived there was."

After lunch the committee reconvened to hear a presentation from Culbertson, Van Laak, and Gary Johnson, the Phase One safety chief, who were teleconferenced in from Houston. Culbertson, appearing haggard and worn, looked like he needed a shave. The strain of the last six weeks was etched into his face. In his mind, however, it was all worth it. Since the collision, the Russians, especially Ryumin, had grown noticeably freer with technical information than at any time in the three-year life of the program; funny, Van Laak had quipped, how the fear of imminent cancellation creates a climate of cooperation. For the first time since the collision, Culbertson felt NASA was now really Russia's partner aboard Mir. He had wrung from Ryumin a written pledge not to attempt any docking exercises or anything else remotely risky without first notifying NASA. The success of the August 22 IVA had capped Culbertson's budding confidence in the Russians. Not only had the TsUP shown it still had the ability to spring back from disaster; almost all the power generated by Spektr's solar arrays was again coursing through the station. Now, in answer to the critics who said there was nothing for an astronaut to do aboard Mir, they could be certain Dave Wolf would be able to perform his experiments in space. NASA science officials were hard at work finalizing his science program.

But first Culbertson had to persuade the rest of NASA, as well as Congress, that Phase One was worth continuing. He had been to Washington three times to brief White House staffers, including the Vice President's national security adviser Leon Fuerth, but had been unable to persuade

Fuerth or any other White House official to make a public statement backing the program. Culbertson knew the White House had a lot riding on the success of the Russian partnership; he also knew no politician was going out on a limb at a time like this. "There's a remarkable lack of fortitude inside the Beltway in terms of taking technical risks," Culbertson recalls.

Now, facing the HEDS panel, Culbertson went through the Russian systems one by one. In response to a question from Gregory, he acknowledged that the Russians had never engaged in "an American-style hazard analysis of [Mir] systems and controls. . . . As far as getting under the Mir and certifying their manufacturing process from the ground up, that was impossible at the beginning of the program. Nor could they get in and do that on the shuttle." After the fire, he said, "We did establish safety requirements to be met by each other. We really had to start from scratch." The only things to worry about aboard Mir now, Culbertson went on, were fire and decompression. "These are the two most serious situations that anyone can imagine aboard the Mir, and I think they were handled correctly, using judgment in addition to the rules. I believe we will see the same thing in the future."

Before leaving, Hammond asked about the fire investigation. "Have they worked that and come up with a cause for that, and if not, how can we justify letting them continue to let them burn these [things]?"

Gary Johnson took the microphone and explained the progress of the Russian fire investigation. The Russians, Johnson said, were now theorizing that a protective rubber seal on the burned cassette had been missing; they believed it was a manufacturing error. "I wouldn't say it's conclusive," he said, "but what they came up with sounds very plausible to me." Someone asked whether the Russians had succeeded in duplicating the fire in a ground test. "I understand they had not done that yet, but are pleased to do so," Johnson said.

"That sounds real plausible," Hammond agreed. "My only concern is we're [still] burning these candles. I don't know that NASA would have operated that way. I don't think we would have let that system [go on] until we knew what the case was."

"You're right, Blaine," Johnson said. "It's a case of being able to control it if it happened again, as opposed to being able to prevent it."

When Culbertson's group signed off, Gregory announced that he wanted to poll the committee members once more. "In the rediscussion, Fred pointed out that we were a safety committee, charged to make assessments only from a safety standpoint," recalls Skip Larsen. "[He] focused on the fact that our charter was a safety charter, not a broad-ranging charter that gave us, if you will, authority to include intangible aspects of this. Clearly that morning we had gone beyond our charter. We were charged only with looking at safety."

Based on safety considerations alone, and on Russian promises that the TsUP wouldn't convene any more docking exercises without telling NASA, both Larsen and Klein changed their minds and voted to send Wolf to Mir. Hammond, who had left for a prior appointment, did not vote. Larsen raised only one caveat: He wanted to know more about the collision investigation being done by the Stafford Commission. Gregory promised to reconvene a week later, when Stafford's finding would be known. At that follow-up meeting, only Hammond and a representative from Klein's office voted against sending Wolf.

Aboard Mir

Tensions aboard the station rise a bit as the various American investigations of Mir draw to a close. Members of the Stafford Commission begin arriving in Moscow on September 14 to conduct their own analysis of the station's safety, while Congressman Sensenbrenner has scheduled a hearing on the matter September 18, just a week before Wolf's mission is to launch. Foale, who unlike Linenger strongly backs a continuation of the American presence aboard the station, has been asked by Culbertson's office to videotape a quick, upbeat tour of Mir to show at the hearing. When the TsUP asks Foale to make his tape during a comm pass over one of the American radar sites, he objects, saying the quality won't be good enough. He wants a Russian radar site. Solovyov, who unlike Tsibliyev has few qualms about issuing orders to Americans, is immediately furious. It is the first and only time he blows up at Foale.

"You've got to do it!" Solovyov tells Foale as the U.S. comm pass nears. "You have an order!"

Foale just smiles. "Anatoli, I understand where you're coming from. But you Russians cannot order me to do this."

"You have to do it now!"

"No," says Foale, fighting back laughter. "No, I don't." Foale's bemused response wins the day; he makes the tape over a Russian site.

Foale understands why Solovyov is on edge. The commander has been deeply embarrassed by the ridicule heaped on the Russian program over the summer. More than anything, he wants to get the station back in fighting trim, make it something his countrymen can be proud of again. He and Vinogradov have spent long hours cleaning and straightening the modules, making sure everything is in its place. But their efforts have been undercut by Mir's continuing technical problems. Just two days after the EVA, on Monday, September 8, then again a week later, the central computer crashes, sending Mir once again into free drift. Both times the TsUP manages to regain attitude before losing all power, but the malfunctions leave everyone on edge—especially with the knowledge that NASA is poised to decide whether to continue the program by sending up Dave Wolf.

September 18
9:30 A.M.
Washington

For all the fears inside NASA that Congressman Sensenbrenner would find a way to cancel the Phase One program, the House Science Committee's two-month review of Mir safety proved anticlimactic. In the absence of any blockbuster new evidence that the program was technically unsound, Sensenbrenner and the other congressmen on his committee were left to make windy speeches and toss pointed questions at Culbertson, who sat squirming at a witness table in the hearing room beneath a giant model of Mir. "What will it take for Russia to decide that Mir has passed its prime or the United States to determine that it's not safe?" Sensenbrenner asked in his opening statement. "Does someone have to get killed?"

The criticisms, Culbertson noticed, tended to follow party lines. Phase One and the International Space Station were strongly backed by the

Democratic White House; before the hearing, two committee staffers had privately admitted to him that what Sensenbrenner most wanted to accomplish was to embarrass the President and Vice President. Not surprisingly, it was the Republicans who took most of the potshots. "The knowledge and experience we can gain from sending more U.S. astronauts to Mir no longer merit the risk involved," intoned Dave Weldon, a congressman from Florida. The Democrats, in turn, urged restraint. "I do not believe it is appropriate for us as Members of Congress to insert ourselves into the conduct of [NASA's] review process," said George Brown of California. "We cannot be NASA's safety engineers, and we should not pretend otherwise."

Culbertson gritted his teeth as the speeches droned on. Few if any of these legislators, he knew, understood the first thing about his program. They were here to make him look bad; if they were lucky, one of their needling questions might make the nightly news. Sitting with Shannon Lucid in the audience, Jim Van Laak quietly seethed. "This was just ludicrous to me," he recalls. "These people didn't know the first thing about space. You think *Frank* was pissed. *I* was coming apart at the seams. I was thinking about leaping over the table to throttle somebody. They were just twisting things, making these long soliloquies that were gross distortions of the truth. I was just beside myself."

Culbertson answered the legislators' questions calmly. He pointed out that his children went to school with Foale's. "I take the safety of my friends very seriously," he said. "I would not send anyone on something that I would not do myself."

The NASA inspector general, Roberta Gross, also testified that morning, laying out in dry verbiage her concerns about Mir's safety. To George Abbey's consternation, IG investigators had forced their way into the closed-door flight readiness meetings for Wolf's mission and, even as Gross spoke, were shadowing Stafford's people in Russia. Still, her brief preliminary report, submitted to Dan Goldin the week before, avoided making any conclusions about the continued viability of the station. Instead, in a thinly veiled jab at Abbey, it raised questions about how candid the safety debate at JSC really was and how independent Stafford's commission, packed as it was with former NASA officials, could be.

By and large, Gross's testimony was ignored by the press.

Near Moscow

That same afternoon, while Sensenbrenner's committee grilled Culbertson in Washington, a black Volga limousine was speeding north through the Russian countryside toward Moscow. In the backseat two old astronauts, Tom Stafford and Joe Engle, were doing their best to persuade the Russian scientist Vladimir Utkin that crew error was not the true cause of the collision. Utkin, a voluble academic with a head of swept-back gray hair who reminded Engle of Albert Einstein, chaired the Russian half of Stafford's commission. All week, at a remote conference center five hours south of Moscow, Stafford and his people had been pumping officials from Energia, the TsUP, and Star City for information about everything from the fire to the collision to ethylene glycol leaks.

To offset the inspector general's conflict-of-interest allegations, Stafford had named a "Red Team," consisting of four commission members with the loosest ties to NASA, to analyze the collision. Ronald Merrell, a Yale University physician who was one of two medical doctors on the Red Team, came away from the talks convinced that Mir's major systems—with the exception of the central computer, a replacement for which was arriving with the next shuttle—remained in robust health. "I thought Mir was as safe as it had ever been," Merrell recalls. "I really felt quite confident that the biggest danger to the crew had been human behavior and not the age of the craft." Merrell and his colleague Craig Fisher grilled the Russians on everything from the tranquilizers Tsibliyev had been taking—they were worried the medication was too strong—to the psychological training Star City administers to cosmonauts. "We wanted to make some recommendations about the fire, but we couldn't figure out exactly what happened, and so we couldn't," Merrell recalls. "They didn't know what happened. They had been unable to reproduce the [fire] on earth. So I didn't know what to recommend. We just didn't know anything about it."

The Russians tried to remain diplomatic, patiently answering the Americans' questions, but it was difficult. "If I were them, I would do the same thing," recalls Viktor Blagov. "It was no big deal. *We* didn't have any doubts, we proved it to them, and they left. That was all. We can understand their concern about safety. On the other hand, you can't just go and scare yourself like this all the time. We have an expression in Russia: 'You can't be like a bird scared of a bush.' Once a bird gets scared, he is afraid of everything. If you get scared, you should not be working in space. Space presents some danger. There is risk. If you are scared, you should go do something safe, like cleaning the streets."

For the most part, Stafford and Engle had avoided placing blame for the collision. Whenever the subject came up, all the Russians bridled. To a man, Russian officials seemed unwilling to even entertain the idea that anyone other than the commander was responsible. But now, in the back of the limousine, Stafford and Engle made their strongest arguments to Utkin: The docking test, the two men said, had been seriously flawed. Flying without range and distance data, against a dizzying backdrop of clouds, with no simulator training for four months, anyone—*anyone*—would have crashed that Progress. "Utkin was not a rendezvous and docking expert," Engle remembers. "[Until then] he didn't really appreciate all the variables and factors we were emphasizing. But that afternoon in the limo you could tell by his facial expressions, by his responses, he was in total 'receive' mode. By the time we got back to Moscow I think he realized that what we were saying was right."

Arriving at Utkin's offices the next morning for a final set of meetings, Stafford and Engle immediately noticed the scent of change in the air. From the first presentations by officials from Energia and Star City, the Russians now appeared willing to accept blame for the collision. To the Americans' amazement, the Energia men admitted that the docking test could have been set up better. Glaskov and the Star City generals admitted they could have done a better job training Tsibliyev. "It was like night and day from earlier in the week," remembers Engle. "Tom and I talked and realized obviously that Utkin had been on the phone the previous night."

By 8 P.M. that Friday the two delegations had sketched out a report enumerating the myriad factors that had led to the collision; Tsibliyev was absolved. The joint American-Russian report also urged NASA to continue sending Americans to Mir. This came as no surprise to anyone inside NASA. "We were not concerned about their recommendation at all," recalls Van Laak. "[Stafford] was there to validate whatever conclusion we arrived at, and he did. We were comfortable he would agree with us."

Houston

A few days before Dave Wolf was set to enter quarantine at The Cape, Blaine Hammond spied him walking through the JSC parking lot. Wolf had made it through the rest of his training in Star City without incident, returning to Houston just before Labor Day after sneaking off to attend a wedding in Florida. Hammond stopped his car, waved at Wolf, and got out. After some chitchat, Hammond brought the conversation around to Mir.

"Dave, now I'm your safety officer, I need to know," Hammond said. "Are you afraid to go do this? Because I have my doubts."

Wolf shook his head. "No doubts at all, Blaine. I have total confidence in my commander. He's a great guy. And I trust his judgment. He will do what he thinks is best, not what the ground says." Wolf talked for a few minutes about all he had learned about Mir. Foale had e-mailed words of encouragement and told him the station was safe. Linenger had lectured him about his own travails, but for the most part Wolf ignored him. The bottom line was, Wolf felt safe.

"Are you sure about that?" Hammond asked.

"Yeah, absolutely."

"Okay, I just needed to ask."

Hammond got back in his car, knowing it had been a useless exercise. "I knew Dave couldn't say anything," Hammond recalls. "If he said anything at all to me, he knew I could push it up the line and jeopardize his chance to fly. And he needs to fly. If he doesn't, his career is over. And we both knew it."

Saturday, September 20
Washington

All summer long the NASA safety-review process had been dogged by charges of conflict of interest. Thursday's Congressional hearing featuring the inspector general's allegations now moved Dan Goldin to create yet another independent committee, which he charged with reporting directly to him. That Friday, Goldin telephoned A. Thomas Young, a retired fifty-nine-year-old former president of Martin Marietta, and asked him to assemble his own committee. Young accepted the assignment, even though there were only six days until the shuttle launch; Goldin assured him he would delay the flight if Young's group needed more time. Afterward Goldin would cite the recommendation from Young's committee as perhaps the most influential on his decision whether or not to send Wolf to Mir.

The team Young assembled that Friday, though nominally "independent," hardly consisted of people who could be expected to stand up to NASA. Young himself chaired a NASA advisory committee on the International Space Station. Barbara Corn was a member of that and other aerospace-industry committees. Charles Bolden, a Marine general, was a former astronaut. Lawrence Adams was another former president of Martin Marietta. None entered the process believing that NASA was doing anything rash or unsafe. Corn, for one, says she took the assignment with little doubt that Mir was safe. "I was up on all this stuff, I really had no problem with it," she recalls. "I didn't think Dan Goldin was going to send someone up there into an unsafe situation."

On Saturday, Young's four-person committee gathered in a conference room at NASA headquarters in Washington and listened to yet another multi-hour video briefing from Culbertson and Van Laak. The two men went through every failure and every breakdown, and explained for the umpteenth time what had caused the collision and what they guessed had caused the fire. "By the time I finished listening to Culbertson and his people, I thought: Why are we here?" recalls Corn. "It seemed to me pretty obvious that everything was okay. I just think Frank Culbertson is wonderful."

Sunday morning at ten, Roberta Gross and two of her assistants briefed the team on their findings. Corn, who lives in Searcy, Arkansas, didn't think too much of the New York–bred Gross. "She was new, and I could understand where she was coming from," recalls Corn. "I did not share her concerns. A lot of what she was saying I attributed to her being from New York. She didn't know too much about this. She was the new kid on the block, and for whatever reason she had gotten upset about this."

That afternoon Young's team spoke by telephone with Linenger. The astronaut held nothing back, saying in no uncertain terms he felt Mir was unsafe. He gave a dramatic retelling of the fire and excoriated the Russians for a solid hour. Young, for one, was surprised. "Jerry set us back on our heels," he remembers. "Now we were seeing this through a different set of eyes. This was a guy who had a very traumatizing experience, a guy who fundamentally wouldn't go again. He was very good." Corn, however, was unimpressed. "It put Tom back on his heels more than it did me," she says. "The fire scared [Linenger] to death; a lot of what we heard was his reaction to that. He talked about all the trash, the litter, how their housekeeping hadn't been super. Well, who are we to judge? . . . I was really very concerned about Jerry, from the way he was expressing himself. Personally, I think he was a poor choice to send up."

After Linenger came Shannon Lucid, who had few safety concerns, and then Blaha, who despite his anger at NASA continued to believe Mir was safe. On Monday the team met with Tom Stafford, who had returned from Moscow that weekend. "That was wonderful. We actually went out to lunch with him," recalls Corn. "He told us how the accident happened. He actually drew a picture of it, on the back of a napkin. He told us this is never going to happen again. Look, the only two things that are going to kill you up there are fire and decompression. They have been through both of those, they had instituted procedures to handle a fire, and they were never going to do something as stupid as that maneuver again. [After that] I was feeling perfectly comfortable with sending Wolf up there. The four of us were unanimous."

On Tuesday, Young wrote a letter to Goldin, laying out the committee's findings. His only caveat was his concern whether astronauts were really free to stand up and defy NASA management; he didn't mention Abbey by name, but everyone inside the agency knew what he meant. "We did not feel there was a planned pattern of 'Don't listen to [internal critics],' but there was not a planned pattern of 'Let's listen to them,' " Young recalls. "One of our strong recommendations was indeed, that they be included in all Shuttle-Mir reviews. We thought that Linenger's input maybe had not been included the way it should have been."

Wednesday, September 24
Washington

Dan Goldin returned from a trip to Moscow that morning to make the decision on whether Dave Wolf would fly to Mir. He first held a videoconference with Wolf, who was at The Cape readying for his flight. The astronaut, to no one's surprise, was pumped and ready to go. At two o'clock Stafford and Tom Young entered Goldin's office and sat on the administrator's long couch. Both men gave a thumbs-up, then handed Goldin brief reports on their findings.

"All right," Goldin finally said. "I want you guys to leave the room."

Stafford and Young rose and left. Goldin closed the door behind them. Alone, he took out a pad and sketched out the pros and cons of sending Wolf to Mir. Eight minutes later he rose and walked to his outer office and told his secretaries what he had decided. That night he phoned The Cape and told Abbey and Culbertson. The next day, as the Space Shuttle Atlantis sat on the launch pad, waiting to take Wolf to Mir, Goldin, flanked by Stafford and Young, strode into a NASA briefing room and announced his decision.

"It is only after carefully reviewing the facts, thoroughly assessing the input from independent evaluators, and measuring the weighty responsibility that NASA bears for the lives of our astronauts that I approve the decision to continue with the next phase of the Shuttle-Mir program," Goldin told reporters. "This is a decision that all of us at NASA do not take lightly. We share our fellow Americans' deep concerns for our astronauts' safety and we have heard the calls of some who say it's time to abandon Mir. We at NASA, especially Michael Foale, are deeply, deeply touched by this outpouring of emotion. . . .

"In light of the increased scrutiny and heightened emotion, I can assure you this intensely rigorous internal and external review of the shuttle-Mir analyzed thoroughly risk, readiness, and, foremost, safety. I will not trivialize the risks associated with human space flight and exploration. Like all Americans, I know every time an astronaut travels into space there is risk. When we build the International Space Station, we will encounter similar problems and there will be danger. But NASA is ready. We are ready because the review assures us. But we're also ready because it's the right thing to do. We overcome the unexpected. We discover the unknown. That's been our history, and that's been America's destiny."

EPILOGUE

Dan Goldin's decision to continue the Phase One program was immediately and widely second-guessed. "It is my fervent hope that the safety evaluations submitted to the NASA Administrator are not a NASA whitewash of the many significant safety risks aboard Mir," Sensenbrenner said, in a clear dig at the Stafford Commission. "We have learned from the Challenger accident that ignoring safety warnings can lead to tragedy and setback of space exploration for years."

In comments for this book, two of the most respected figures in NASA history openly questioned the thoroughness of the agency's review of Mir safety. "We've got two different sets of standards we use, those (for) working with the Russians and those we follow in the U.S.," says Gene Kranz. "If we had these mistakes [in the United States], there would be a detailed investigation all through the chain of command. That didn't happen here."

"The Mir thing really surprised me," adds Dick Truly, the former NASA Administrator. "After the Challenger incident, one of the things we believed we had instilled in the system was, if you had a problem about safety, you could stand up and say that. It looked to me during Mir—well, it makes me wonder if that's being eroded." Another former NASA Administrator agrees. "I think it was the stupidest goddamn thing they ever did to continue flying [on Mir]. Christ, they had a fire, they had computer failures. Jesus Christ. I don't understand that. How could they fly?"

But Goldin's decision, however NASA may have stacked the deck in its favor, proved to be the right call. The night after the NASA Administrator's press conference, astronaut Dave Wolf and his shuttle crew departed for Mir. Wolf not only survived his four-month stay aboard the station, but he thrived, returning to Houston in January 1998 to be greeted as the Phase One program's unlikely savior.

It was an eventful mission. The centerpiece of Wolf's abbreviated science program—Spektr remained closed off—was the so-called bioreactor device the astronaut had helped design. During his mission Wolf was scheduled to use the machine in an attempt to grow human cancer cells. But during the changeover period, on Wolf's first weekend aboard the station, Anatoli Solovyov strongly objected to taking the machine aboard; apparently, Valery Ryumin explained to Culbertson, Solovyov was spooked by the idea of cancer cells somehow escaping the device and floating free inside the station. According to Culbertson and Van Laak, only their vocal resistance prevented Solovyov from tossing Wolf's pride and joy out the airlock into space.

The contast between the happy-go-lucky Wolf and his new Russian crewmates—not to mention Foale—was immediately and readily apparent. Over dinner one Sunday evening early in his mission, Wolf was pleasantly surprised when Solovyov offered him a rare treat: a packet of dehydrated "Blackcurrant Jelly with the Pulp." All Wolf had to do was snip one corner of the packet and add water to the purplish powder inside. Conscious that Solovyov and Vinogradov were watching him closely, he began to cut the packet and then realized to his irritation that he had cut the "drinking" end, not the "filling" end. He added water carefully, hoping that his two crewmates wouldn't notice his error, and in a minute he had a nice mouthful of jelly. Pleased with himself, Wolf put the packet aside. When he went to eat some more jelly a few minutes later, however, he instinctively squished the packet to mix its contents again and was amazed to see a huge cloud of purplish jelly spurting across the dinner table right toward Solovyov's head: He had forgotten to hold the packet closed when he squished it. Solovyov ducked, and the flying jelly shot past him, plastering the stereo, the rack of audio cassettes, the television set and most of the wall behind the commander, who rolled his eyes as Wolf furiously grabbed some towels and began cleaning up the mess.

It was the kind of mistake a child would make, and that is how the two Russians came to regard Wolf, as a kind of space teenager. "Dave is quite a peculiar man," says Vinogradov. "At his age, we were quite surprised by the problems he had with your FBI. The way he behaved, I don't know how to put it mildly, he was like a little child. His training, I wouldn't say it was insufficient, but it was quite fast."

Solovyov, as Foale had learned, was by far the most demanding and temperamental of the Russian commanders. Solovyov didn't ask Wolf to do things, he gave orders—harsh orders, often shouted. "Anatoli was extremely demanding, on himself and on the crew," Wolf recalls. "It was a matter of pride as a Russian that he was going to get this station back under control. He was hard on Pavel, and they were both hard on me. Thank you's were not a part of his routine. Just 'Do this!' " At one point early in his mission Wolf finally objected to the tone of Solovyov's voice, telling his commander, "That's an insult to me."

"No it's not," Solovyov insisted. "We're just trying to help you."

And slowly, Wolf realized he was right. He was being treated the way any Russian commander might treat a green recruit. "It would have been patronizing if they had done anything else," Wolf asserts. "I loved it when they yelled at me. It meant I was part of the team."

Solovyov's temper could rise out of nowhere. Perhaps the angriest Wolf saw his commander was one day early in the mission when Wolf was trying to figure out how to run a series of exercises the Russian doctors had mandated. The checklist Wolf was attempting to read was tattered and torn, so Wolf asked Vinogradov, who was busy repairing a system in base block, for help. Distracted, Vinogradov said, "That's not my job." Solovyov floated in a few minutes later, having missed the exchange. A few minutes after that, Vinogradov asked Wolf to give him a hand in his repairs.

"That's not my job," Wolf shot back.

Solovyov came flying at Wolf, angrily shouting, "I don't want to *ever* hear anything like that again! *Everything* is your job! If we say, 'Do it,' you will do it!"

Wolf tried to defend himself, but it was no use. "You know I was just saying that to make a point," he said.

Like Foale, Wolf instinctively took steps to integrate himself into the crew by volunteering for menial jobs. The day after the shuttle left, he noticed Vinogradov sticking his upper body behind one of the panels in Kvant 2 to wipe up pools of condensate that had collected there. Wolf spent several hours working with Vinogradov to mop up a basketball-size drop of water. "Pavel, you never have to do that again," Wolf said afterward, proudly volunteering to handle water-mopping chores in the future. He soon had reason to question his enthusiasm. "Frankly, I thought it was a one-shot deal," he remembers. "I didn't know exactly what I was asking for. I didn't realize I bought myself anywhere from two to six hours *per day* doing this for the rest of the mission." And so Wolf became Mir's new water boy. Nearly half his mission, he estimates, was spent mopping up water puddles.

Solovyov's regard for Wolf, in fact, was so low that the commander strenuously objected to performing an EVA with him. Between October and January, Solovyov and Vinogradov engaged in four rigorous EVAs to inspect the Spektr damage, including a second spacewalk into the module itself. Wolf was scheduled to replace Vinogradov for one of the EVAs. Then in November, Culbertson was hand-delivered a letter from Solovyov in his office. In the letter, according to NASA officials, Solovyev complained that Wolf did not possess either the "character" or the language skills to carry out an EVA. Culbertson read the letter in silence, then folded it and put it away. Afterward he and Van Laak walked up to George Abbey's office and gave what Van Laak considered a "very low-key" synopsis of Solovyov's concerns, playing down the problem. While Culbertson took up the matter with Valery Ryumin, Van Laak returned to Abbey's office several days later without telling Culbertson.

"I'm going to stick my neck out here a little," Van Laak said, explaining the letter's full contents.

"So why are we doing this EVA with them?" Abbey asked.

"That's a good question," replied Van Laak, who had never fully understood the need for Wolf to perform an EVA.

Ultimately Wolf and Solovyov did perform a single three-hour EVA together in January, but only after a strong push from Culbertson. Vinogradov confirms Solovyov's fears about working with Wolf. "First of all there was the language barrier," says Vinogradov. "Dave could speak Russian quite well, but in situations when we needed quick solutions, when we had to think fast, Dave was at a loss. It was dangerous to go with a person who couldn't understand your language. We weren't against him doing the EVA, we just asked management to postpone the EVA, so Dave could get used to the situation."

The EVA wasn't Solovyov's final dispute with Wolf. In January, during the last weeks of Wolf's mission, rumors spread through the TsUP that the commander had gotten several IBMP doctors fired, or at least suspended. The explanation that reached the Phase One office in Houston was that Solovyov had complained bitterly when the IBMP doctors had put on comm a woman friend of Wolf's from Moscow. During his training at Star City, Wolf had also befriended a Russian woman and her daughter; Solovyov—or so this version of events would have it—knew and liked the Star City woman and knew nothing of the Moscow woman. When the Moscow woman came on comm, Solovyov privately excoriated the IBMP doctors for putting "that woman" on comm with the station. Wolf says he knows nothing about the incident. Vinogradov says the commander's anger had nothing to do with the woman's identities. "We had nothing against Dave's girlfriends," says Vinogradov. "It was just that they put her on comm during one of Anatoli's private sessions. Anatoli didn't know that girl. Neither did I."

The last NASA astronaut to live aboard Mir, Andy Thomas, arrived at the station in January 1998. Thomas, an easygoing forty-five-year-old (and the astronaut corps' only Australian), had been Wolf's backup and hadn't expected to fly to Mir; the cancellation of Wendy Lawrence's mission that August changed all that. In fact, Thomas had only volunteered for Phase One in mid-1996 because he felt it would earn him points with the Astronaut Office. "I knew if I took the backup role they would have to take care of me," Thomas recalls. "I felt sure I could get a shuttle flight, plus I'd probably be a candidate for ISS."

Thomas's mission got off to an ominous start when, during the changeover period, he was unable to fit inside his Orlan spacesuit, a critical problem since he would need the suit in the event of an emergency evacuation in the Soyuz. Thomas had complained that the suit was too small in the weeks before his launch, but the Russian suit engineers insisted it would be fine; Thomas, to his regret, accepted their judgment. Once in orbit he insisted on cutting a pair of leg straps to lengthen the suit. The TsUP ordered him to try on Wolf's Orlan instead, but it proved far too big. Viktor Blagov became irritated with Thomas, and when news of the problem reached the newspapers, Blagov was quoted as saying Thomas was being "capricious." Thomas got mad. "I was not liking the way things were shaping up at all," he recalls. "Here the Russian flight director is shooting me down in the press. That was all my parents in Australia were hearing about. It was a bad time. It was ugly."

Thomas felt real pressure from the TsUP to don a space suit that he considered unsafe. "They expected me to make the sacrifice, to do it the Russian way," he recalls. "But that is entirely inappropriate. It's unsafe. It's a reflection of Russian culture, and the autocratic nature of their society. It's a very counterproductive approach and not the way we should do things in the twenty-first century. That kind of behavior is something we don't want on ISS. We just don't." Eventually Thomas got his way; his suit straps were cut and lengthened.

The most frightening moment of Thomas's mission came on February 26. That afternoon, one year and three days after the fire, Thomas had just finished running on the Kristall treadmill and was passing through the node on his way back to Priroda when he noticed that base block was full of light smoke. "It looked like someone had lit several cigars," he remembers. His new crewmate Nikolai Budarin was passing through the node at the same moment.

"What's going on?" Thomas asked.

"Oh, nothing," said Budarin. The Russian explained that a valve on the BMP, one of the station's atmospheric-cleansing systems, had somehow overheated and begun smoking. He and Commander Talgat Musabayev quickly fixed the problem, but not before carbon monoxide fumes reached levels so high that Thomas endured two days of vicious headaches.

With Andy Thomas's return to Earth in June, the Phase One program came to a close. It is easy to criticize NASA for its failures—and there were many—during the seven missions to Mir. But before doing so it is important to remember that the politicians who created the Phase One program, while acting with the best of intentions, put NASA in an untenable position. With minimal preparation the agency was asked to stage the longest mission in its history on a spacecraft it knew next to nothing about. It thus comes as no surprise that, from an organizational point of view, the early Phase One missions will go down as perhaps the ugliest in NASA's storied history. The ground teams that supported Norm Thagard, Shannon Lucid, and John Blaha were undermanned, underexperienced, overworked, and badly rushed. All three astronauts paid the price for NASA's failure to prepare adequately for their stays aboard Mir.

"They were pioneers," says Frank Culbertson. "They were doing all this for the first time. That's what people forget. It's been so long since we've done something completely new in this agency that people forget what it's like. We've had no new manned program since the shuttle in 1981. Well, it can be ugly. That's why Shannon Lucid and John Blaha had such a tough time. They were used to having MOD in place to do every little thing, and we didn't have that in Phase One. We didn't have *anything* in place. We had to learn how to do it as we went."

The nearly fatal flaw of the program was in its underlying premise that NASA astronauts would serve as guests aboard a Russian spacecraft. It was the first time in NASA's history that the agency had put astronauts in orbit without U.S. oversight. This decision placed American lives in the unsteady hands of a space program that NASA should have known was no longer prepared to guarantee their safety. At no time did NASA officials appear to confront the obvious question of what they would do in the event of an unforeseen disaster—a reality that blindsided them in the wake of the February 1997 fire. Only then did NASA attempt to do what it should have done long before: forge a real operating partnership with the Russians. Yet even then NASA proved asleep at the wheel. Amazingly the March 4 Near Miss alarmed no one. For all the talk of monitoring the Russians and their safety precautions, no one at NASA had the first clue what the TsUP was attempting on the morning of June 25.

Today Mike Foale is very lucky to be alive. An errant wiggle of Vasily Tsibliyev's left pinky could easily have sent him to a cold and lonely death in space. In the wake of the collision, the insistent refrain emanating from NASA headquarters held that Mir's troubles were actually *helpful*. Recovering from a disaster in space, this argument went, allowed the two programs to work together in a way they never could have otherwise. While this is no doubt true, it is also the worst kind of rationalization. A pilot who crawls from the wreckage of a 747 does not justify a plane crash by talking about how well it cemented his relationship with the control tower. NASA's lack of oversight was inexcusable and totally out of character for an agency that is legendarily obsessed with safety. Gene Kranz was right: NASA did develop two sets of standards, one for Americans on American spacecraft and one for Americans on other spacecraft.

This might be dismissed as an anomaly of history, a situation NASA will never encounter again. In fact, those two sets of standards could well live on aboard the International Space Station, the ISS. Phase Two of the ISS program, the assembly of the new station, will officially begin on November 20, 1998, with the launching from Baikonur of the Russian-built FGB, or space tug. Three weeks later, on December 4, the Americans will send aloft their first element. The ISS "service module"—its base block—will follow in mid-1999, and shortly thereafter will receive its first Russian-American crew, Commander Bill Shepherd and cosmonauts Sergei Krikalev and Yuri Gidzenko. As this book went to press, however, the schedule of all ISS missions remained in doubt. The Russians, gripped by a nationwide financial crisis, apparently have run low on funds needed to complete and launch the service module; in September 1998 NASA asked the White House for more than $600 million in emergency funding to help the Russians finish their ISS work. Until the arrival of the American science module in the fall of 1999, when NASA flight directors in Houston take over, responsibility for the infant station will rest in the hands of the TsUP.

The ISS service module is to contain a version of the solid-fuel oxygen generator that burned aboard Mir—even though nearly two years after the February 1997 Mir fire, neither the Russians nor NASA have solved the riddle of what caused it. Gary Johnson, who cochaired Phase One's safety group, pestered the Russians for months to produce the results of tests on the SFOG run by a Moscow criminological lab. "We've been insisting that we see the reports and testing results from the criminological institute, but we never have gotten that report," says Johnson. "Initially we heard Energia had not paid the institute, and they were refusing to give them the report. Then they paid them, and still did not get the report. So we really haven't gotten it."

What is clear is that Energia's initial conclusions were wrongheaded. The rubber ring that seals the SFOG's air filter—which Russian engineers suspected had been missing—had in fact been in place for the fire. Subsequent tests discovered residue of the ring in the wreckage; it had been burned during the fire. "So we're back to calling it an 'unexplained anomaly,' " says Johnson. "It's causing us concern, because they promised corrective action and a fix for that." To add to NASA's headache, the Russians had promised to redesign the system to replace the mysterious ignition system its engineers would never explain to Johnson. In its place they pledged to manufacture a conventional electric ignition. "Now," Johnson says, "because of their money problems, we're getting hints that the service module is going to launch with an unmodified unit, and *that's* got us concerned."

Nor are the TORU and Kurs systems to be a thing of the past. The Russian Progress ships will continue to dock with the ISS using the automatic Kurs system. The TORU system will also be onboard, ready to be used in the event of a Kurs failure. According to Frank Culbertson, who has been named the head of operations for ISS, the Russians have agreed that the TORU system will only be used for manual dockings in the immediate vicinity of the station. Culbertson promises that this time NASA will be paying attention to the Russians' actions.

What is Phase One's legacy? Throughout the Johnson Space Center, NASA officials spent much of the summer of 1998 pounding out lists of "lessons learned." To many of those involved in the Shuttle-Mir missions, the exercise was tinged with irony. They knew what the public and Congress didn't: that many if not most of the men and women involved in designing the ISS paid little attention to their work. "Phase One was sold as a learning experience for NASA, and the people in Phase Two never once paid attention," says Mark Bowman, the senior Krug Life Sciences engineer who worked throughout the Shuttle-Mir missions. "They didn't learn anything, they didn't give a shit about Phase One. I've talked over and over and over and over to people. They don't listen."

It's a common refrain but not entirely true. The ISS will include a number of significant design changes suggested by NASA's experiences aboard Mir, including added lights and sensors on the station's hull to aid rendezvous and docking missions; "quick-disconnect" cables that should come apart in seconds in the event of an emergency depressurization; rerouted cooling lines and electrical cables to prevent moisture buildups from causing the kind of leaks that plagued Mir in its old age; and, NASA officials promise, more flexible daily scheduling for astronauts. Some solid science was completed aboard Mir—NASA points especially to advances in crystallography and plant growth—but the real value of Mir to NASA's doctors and scientists was in learning new procedures for long-duration space flight. "We've learned a lot," says Mike Barratt, the young Phase One doctor who was named head of medicine for ISS, "but ironically, it's not that much new medically. It's how to do medicine and science on a station."

Even Culbertson, the program's head, downplays the science Mir astronauts performed. "Do we remember what scientific experiments were left on the moon?" Culbertson asks. "No, of course you don't. All you remember is that we were there. It was true of Christopher Columbus, it was true of Lewis and Clark, and it was true of Neil Armstrong. They went and they came back successfully. So did we."

Phase One also pointed up several nagging disagreements the Russians and Americans have yet to resolve. One is language. The contract NASA signed with the Russian Space Agency explicitly states that the language to be spoken aboard the ISS's English. To a man, the Russian cosmonauts and ground controllers will have none of it; they insist that Russians will speak Russian while speaking with Russians. "I don't see this as a problem," scoffs Viktor Blagov. "Whatever the Americans say, we are going to speak to Russian commanders in Russian. See? No problem. Well, the Americans say, their commander won't understand what is being said. Well, if we speak to a cosmonaut in English, we may have an emergency. No matter what the Americans say, we will never agree to speak in English."

Some of the disagreements are more subtle. Will underpaid cosmonauts, their flight bonuses always in mind, be prone to downplay safety hazards? The ISS will feature alternating Russian and American commanders. After debriefing Dave Wolf and Andy Thomas, the NASA psychologist Al Holland strongly urged NASA to apply pressure on the TsUP and Star City to somehow damp down the Russian commanders' autocratic tendencies. American and European astronauts, Holland argues, will not respond well to shouted commands and stern orders. Even Holland acknowledges it is probably a useless exercise. "We are a nation of badly brought-up children," sighs Viktor Blagov. "What can you do? We must wait for several generations to learn how to behave. It is just something the Americans will have to put up with."

For all their condescension toward the soft, egocentric Americans, the Russians are grudgingly willing to admit that they also learned some lessons from NASA. One was their obvious need for greater computerization. Jeffrey Manber, Energia's Washington representative, says an even greater lesson was "the need for decentralization of information. The Russians have never seen lower-ranking people with decision-making authority to the degree they saw working with NASA. That's had a significant impact on their thinking."

Frank Culbertson argues that the most important lesson NASA—and perhaps the Russians—learned during Phase One was humility. "We learned that we don't know everything about space flight," says Culbertson. "As good as we are, we learned things we did not know. Just setting the bit in your mind that you don't have all the answers, as an agency or a country—that somebody else has something to offer—that was a great leap forward." For the first time, Culbertson says, NASA developed an appreciation for just how different operating a space station is from the shuttle. Long-ignored areas such as astronaut psychology will now be seen as crucial. "We learned that long-duration flight on a space station is quite a bit different than short flights on the shuttle," he says. "The shuttle is an airplane, and the station is a ship, and it's going to be at sea for a long, long time. It may be a long way from land, but you still got to keep it afloat."

It is a lesson the American public, one suspects, will only learn the hard way. Schooled for years on NASA's assurances that space flight has become almost as safe as commercial air travel, it will take Americans and their media by surprise the first time the ISS spins out of attitude control or leaks antifreeze or suffers a fire. It will take time, maybe years, for Americans to learn what the Russians have long understood: Space stations break down. They can be dangerous. They have to be fixed. When an actual disaster strikes ISS—and it will happen—it will test the American public's appetite for space travel in a way the Challenger explosion never did. For NASA engineers there will be no disappearing into aircraft hangars and quiet offices to ponder ways to redesign the spacecraft. There will be repairs to be done—dangerous repairs—and astronauts will have to do them. In that sense the events aboard Mir during 1997 can be viewed as a glimpse of an uncertain future.

For all the technical, organizational, and psychological lessons learned, the most concrete legacy of the Phase One program is probably the simplest. For the first time astronauts, flight directors, doctors, and scientists from the two countries really did learn to work together. Friendships were formed, procedures adopted. At every level the men and women who will work aboard and support the ISS are no longer strangers to one another. One suspects the geopolitical and diplomatic implications of the program may be even more profound. Someday, perhaps a century or two from now, there will be no nations in space, no Russias or Americas or Japans. There will just be human beings. And when that day comes, schoolchildren may well look back on the seven missions of Americans to a rickety old space station called Mir as the first shaky step toward a unified tomorrow.

Most of the people featured in this book, including Dan Goldin and George Abbey, remain in their jobs. Blaine Hammond has retired from NASA and joined Gulfstream Aerospace in Southern California. Four of the seven Mir astronauts are still at NASA. Shannon Lucid has written extensively about her stay aboard Mir. Mike Foale serves as a special assistant to George Abbey. Dave Wolf was grand marshal of the 1998 Indianapolis 500. ("Highlight of my life," he jokes.) Andy Thomas has his eye on an ISS mission. Bonnie Dunbar is a NASA liaison to colleges and universities. Of the three Mir astronauts who have left NASA, Norm Thagard is an engineering professor at Florida State University in Tallahassee; he has served as a technical consultant on two recently completed movies, including *Armageddon*. John Blaha retired from NASA in late 1997. Today he is an executive with USAA, a financial services company in San Antonio. Jerry Linenger, whom *People* magazine named the "Sexiest Explorer" of 1997, retired from NASA at the end of that year. The following spring he moved with his family to a small town in northern Michigan, where there is speculation that he will eventually mount a run for Congress.

In a lighthearted ceremony in Moscow, Valeri Korzun and Sasha Kaleri were presented with fire helmets and and given Russia's highest firefighting award; both men remain cosmonauts. The ultimate fates of Vasily Tsibliyev and Sasha Lazutkin remain unclear. Though Russian officials announced that neither man would be fined for their actions aboard Mir, Lazutkin says that as of June 1998, they have yet to receive about a third of their flight bonuses. He declines to say how much money that involves; in all likelihood it is around ten thousand dollars. In the spring of 1998 Lazutkin went back to work at Energia after a months-long postflight vacation. He spends much of his time speaking to schoolchildren and other groups. In early 1998 he flew to New York to appear on an American home shopping network, hawking a set of Russian space memorabilia, including space suits. "We do not know what will happen to us," his wife Ludmilla said in June. "We hope he will return to space. We do not know."

Tsibliyev, who can still be seen smiling and smoking cigarettes with other cosmonauts around the TsUP, remains bitter. He says he has taken a job in Star City's training division. He, too, cannot say if he will ever return to space. In all probability he won't.

And what of Mir itself? Under pressure from NASA officials, who understandably don't want the TsUP's attention divided between the dowdy old station and their new ISS, the Kremlin had promised to deorbit Mir in December 1999. In the end the Russians simply had no money to keep it up even that long. If everything goes as planned, the station will begin to reenter the Earth's atmosphere early in the summer of 1999. In its final minutes on June 8 it will streak southeast across Japan, a fiery comet in the heavens, and plummet into the Pacific Ocean several hundred miles east of New Zealand.

ACKNOWLEDGMENTS

More than any other writing project I have attempted, this book was made possible by the kindness of strangers. For better or worse, I approached the Russian and American space programs as a rank outsider. I had not written about NASA since pitching in on the *Wall Street Journal*'s coverage of the Challenger disaster in 1986. In July 1997, when I first told NASA public relations officials I was writing this book, they encouraged me. In the ensuing year, I spent a week or two every month at the Johnson Space Center in Houston. As a native Texan, I found it good to come home again, to eat barbecue or Mexican food every night. But what I came to look forward to most was the kindness and openness of the NASA people I got to know. Jim Van Laak graciously allowed me to use empty offices and conference rooms for interviews and to review documents. Both he and Keith Zimmerman offered patient explanations of the arcana of both the Russian and American space programs. Charlie Precourt explained the TORU system. Al Holland helped me fathom the ins and outs of space psychology. Scores of other hardworking NASA officials dropped what they were doing to help me. I can count on two hands the people, most of them astronauts, who declined to be interviewed.

When I wasn't in Houston, I was in Moscow. I had worked in the city before, but over the course of the last year I fell in love with it and its people. For thirty years Western space enthusiasts like James Oberg and Dennis Newkirk, along with a handful of academic analysts, have written excellent histories of the Russian space program despite severe limitations, compiling their data from private sources and what little information the Russians chose to make available in news reports. With the transformation of the former Soviet Union into something approximating a Western-style democracy, it is now possible, if difficult, to employ traditional news-gathering techniques in writing about Russian space efforts. Generals and cosmonauts will sit for interviews, if sometimes impatiently. This book is not intended to be a history of the Russian space program, but the literature could use one, and the program's new openness provides an opportunity for the right observer to write it.

I had expected to find Russian officials less than enthusiastic about discussing Mir's problems, and at times I wasn't disappointed. Unaccustomed to Western press scrutiny, the men at Star City and the TsUP had been taken aback by the attention the station's problems drew during the long summer of 1997. In the event, the Russians ran hot and cold. Some officials, like General Yuri Glaskov, sat for repeated interviews with little complaint. Others, including psychologists at the Institute for Biomedical Problems, refused to be interviewed at all. Officials at the TsUP, while initially welcoming, ultimately cut off official participation in this project after unsuccessfully demanding cash payments as large as fifteen hundred dollars for a single interview.

To my surprise the cosmonauts themselves proved the most accessible and eager participants. Vasily Tsibliyev, while reluctant at first to rehash events aboard the station, eventually agreed to sit for a total of eighteen hours of interviews over the last year. At various times we talked alongside his family in a Prophylactorum conference room, over Russian champagne at a Star City restaurant and over tea in a tiny break room down from his office. Unfailingly courteous, he was at turns gregarious and defensive, morose and curious about what Americans were saying about his mission. Some incidents, including the collision itself, he described in vivid detail; other matters, chiefly his anger at the ground, he denied altogether. In many ways Sasha Lazutkin was the most eager of the Russian participants, sitting for twenty hours of interviews over the last year, all in his little apartment in northeast Moscow. Over tiny cups of wine and tea he smiled and laughed often as he recounted his misadventures in space.

Other Russian officials and cosmonauts were interviewed in their offices at Star City and the TsUP. Valeri Korzun sat for two lengthy interviews in the cosmonaut museum. Sasha Kaleri welcomed me into his room in a Moscow military hospital, where he was undergoing routine postflight tests. Gennadi Strekalov broke from a boisterous meeting of Russian waterskiing enthusiasts to share memories of his time aboard Mir with Norm Thagard. Viktor Blagov, Vladimir Solovyov, Oleg Tsygankov, and others sat for interviews in the break area outside their offices at the TsUP. I tracked down Valery Ryumin and Sergei Krikalev in Houston. I am grateful to them all.

With rare exceptions the American astronauts who lived aboard Mir and the NASA staffers who supported them recounted their memories with gusto and grace. Thagard sat for a marathon seven-hour interview in his office at the Florida State engineering school, and for regular telephone conversations after that. John Blaha was interviewed at his new home and office in San Antonio and in numerous telephone calls. Dave Wolf talked over barbecued salami at his Houston home and in his office. Mike Foale consented to twelve hours of interviews in his NASA office and on the phone. Jerry Linenger proved the most ambivalent of the Mir astronauts. He initially refused to discuss his time aboard Mir in detail, saying he was entertaining offers to write his own book and maybe a Hollywood screenplay. Eventually, toward the end of my research, he consented to several lengthy interviews, speaking forcefully and sometimes angrily over dinner at a waterside restaurant in Houston and, following his retirement from NASA, during a daylong tour of his new environs in northern Michigan.

Frank Culbertson and Jim Van Laak graciously made room in their schedules to talk during almost all of my visits to Houston. John McBrine was an invaluable guide in both Moscow and Texas. Mike Barratt provided summaries of his diaries. Dan Goldin and George Abbey sat for lengthy interviews in their NASA offices. Bonnie Dunbar, Shannon Lucid, Charlie Precourt, Bill Gerstenmaier, Phil Engelauf, Tommy Holloway, Rick Nygren, Al Holland, Bill Readdy, Richard Fullerton, Travis Brice, Caasi Moore, Keith Zimmerman, Gary Johnson, Tony Sang, Tom Marshburn, and many others were interviewed in their offices or in conference rooms provided by NASA.

For their guidance and help I am also indebted to NASA's many reliable public affairs officers, including Rob Navius, Barbara Zelon, Kyle Herring, and especially Laura Rochon in Houston; Kate Maliga in Moscow, and Debbie Rahn and Mike Braukus in Washington. At the Johnson Space Center, Stella Luna cheerfully and efficiently processed my many Freedom of Information Act requests. Mike Gentry and the folks in the Media Resource Center skillfully helped me navigate NASA's photo library.

Others who provided help and guidance include Bill Reeves, John Curry, Jessie Gilmore, Ken Cameron, Matt Muller, Ron Sega, Terry Taddeo, Dave Ward, and Seth Borenstein at the *Orlando Sentinel*, Scott Pelley at CBS News, Jim and Cookie Oberg, Carlos Fontenot, Charlie Stoegmuller, Mark Bowman, Lee Silviera, Christine Chiodo, Barbara Corn, Thomas Young, Ronald Merrell, Congressman James Sensenbrenner, and Katy McGregor and the staff of the House Science Committee, Arnaud Nicogossian, Caroline Huntoon, Igor Lissov and the staff of *Videocosmos* magazine in Moscow, Tamara Globa, Alexander Serebrov; Alexander Derechin, Jeffrey Manber and Chris Faranetta at RSC Energia in Alexandria, Virginia, Sam Keller, Arnold Aldrich, Bob Clarke, Brian Dailey, Mark Albrecht, Liz Prestridge, Bryan O'Connor, David Mobley, Porter Bridwell, Chuck Daniels, Leroy Chaio, Marcia Smith, Pia Pialorsi, Deedee Myers, Marla Romash and Jenny Turzano, Kathryn Linenger, Jonathon Spalter, Jim Keller, Richard Truly, Gene Kranz, Story Musgrave, Peggy Whitson, Rhea Seddon, Hoot Gibson, Drew Gaffney, Byron Lichtenberg, Susan Anderson, Charlie Brown, Linda Gavin, Tom Tate, Tom Short, Jim Nice, Geoff Perry in the United Kingdom, Chris Vandenberg and Bert Vis in the Netherlands, Reinhold Ewald in Germany, John Uri, John Charles, Robert Park, Charles Vick, Susan Flowers, John Fabian, and Steve Hopkins at Anser Corporation, Cindy Evans, Robert Murphy, Nick Kanas, Frances Linenger, Karen Brandenberg, Kristen Taraszewski, Byron Harris, Barbara McCoy, Joel Montalbano, James Medford, Mark Severance, Mark Ferring, Blaine Hammond, Melanie Jones at Southwest Airlines, Mark Hess, Guy Gardner, Patricia Santy, Dan Brandenstein, Alex McPherson, Harold Rosenbaum, James Beggs, Gaylen Johnson, Dave Larsen, Nick Fuhrmann, Marton Forkosh and Skip Larsen. A handful of people, including astronauts, consented to interviews on the condition they not be named.

The writings of several authors and newspaper reporters were also extremely helpful. Jim Oberg, in his book *Red Star in Orbit* and in numerous magazine articles, is a perceptive NASA critic. Kathy Sawyer in the *Washington Post,* Bill Broad in the *New York Times,* Seth Borenstein, formerly of the *Orlando Sentinel,* and Craig Covault of *Aviation Week* all covered elements of Phase One in detail.

I am immensely grateful to Graydon Carter, the editor of *Vanity Fair,* for funding my initial scouting trip to Moscow and then allowing me to turn the magazine article we envisioned into a book; if there is a more supportive magazine editor working today, I haven't met him. Andrew Wylie, Jeff Posternak, and everyone at the Wylie Agency remain the best agents in the business. At HarperCollins, I am indebted to Adrian Zackheim, Marilyn Allen, Paul Olsewski, Craig Herman, Steven Sorrentino, and many, many others. Doug Stumpf, John Helyar, Steve Swartz, and Marla Burrough read drafts of the manuscript and offered excellent tips on improving it. Nancy Cardwell expertly identified flaccid sections to be excised. Suzanne Regan was a whiz at gathering photos. In Moscow,

Inessa Rohm, Ludmilla Mekertycheva, and Lola Toptchieva were excellent guides and interpreters. In Boston my good friend Masha Pavlenko translated Russian-language documents. Through it all, Michelle Memran was the best research assistant an author could ask for.

For all their help, however, this book never would have been written without Marla, Dane, and Griffin. *Okay, guys, it's over now. . . .*

Bryan Burrough
Maplewood, New Jersey
September 1998

INDEX

Abbey, George, 35, 40, 42, 47, 245, 247, 249, 251, 266, 267, 278, 328, 346, 506
astronaut selection and, 21–27, 41, 261–262, 275–276, 348
background of, 19–20
Foale's mission and, 406, 418–419
Hammond's career and, 27–34
as Johnson Space Center director, 17–19, 24, 34–35, 513
safety concerns and, 34–35, 254, 425–426
STS–71 (1995) mission and, 24–27
Thagard's mission and, 261–262, 275–276, 317
Adams, Lawrence, 500
Advertising, on Russian missions, 59, 62, 213
Afanaseyev, Viktor, 66, 68, 70
Albrecht, Mark, 24, 239–244, 246
Aldrich, Arnold, 251, 257
Apollo program, 182, 185, 212, 253, 266, 421–422
Armstrong, Neil, 5, 7
Atlantis, 4, 63, 101, 322–323, 324, 334, 502
Augustine, Norm, 239, 241

Baker, Mike, 16–17, 39, 445, 463
Balandin, Aleksandr, 223–224
Barratt, Mike, 186, 268–270, 282–283, 289, 290, 297, 298, 299, 300–302, 303, 315, 316, 319–320, 322, 326
Bartholomew, Reginald, 248, 249
Bean, Alan, 185
Beggs, James, 241, 263
Berezhnov, Yevgeny, 13–14
Billica, Roger, 173
Blagov, Viktor, 64, 72, 100, 110, 168, 211, 278, 352, 360, 433–434, 485, 497–498, 507, 512
coolant systems problems and, 188, 308–309, 310–311
Linenger's mission and, 198, 200–201, 202–203
Progress docking near miss and, 160, 164, 165
reporting of Mir fire and, 146, 148, 154, 155
Thagard's mission and, 316
Tsibliyev's mission and, 391, 403–404, 408, 411, 416, 420–421, 437, 439, 454, 458, 460, 473
Blaha, Brenda, 94, 95, 96, 105, 111
Blaha, John E., 11, 44, 92–114, 180, 187, 202, 316*n*, 321, 514
anger at NASA of, 92, 98, 106–107, 114, 112
background of, 92–93
changeover period after arrival and, 95–98, 337
Culbertson and, 49–50, 92, 106–107, 112
depression experienced by, 111–112
experiments and, 98–100, 107–111, 167
Mir mission of, 63, 88, 107–114, 116, 141, 172, 325
personality of, 93–95
relations between cosmonauts and, 101, 105–106
relations between Lucid and, 94, 95, 107, 329, 330, 331, 336–337
safety concerns about Mir and, 490–491
Star City training of, 98–101, 105–106
working hours of, 108–109, 111
Bogdashevsky, Rostislav ("Steve"), 15, 198–200, 357
Borenstein, Seth, 51
Bowman, Mark, 93, 291, 292–293, 511
Brandenstein, Dan, 27, 33, 259, 260–261
Breeding, James, 293
Brice, Travis, 288, 440
Brown, Charlie, 105
Brown, George, 496
Budarin, Nikolai, 324, 387, 440
Buran space shuttle, 60
Bush, George, 239–240, 243, 246–247, 250, 253, 254, 265

Cabana, Bob, 21, 29, 32, 33, 40, 47–48, 49, 173, 300, 349, 430–431
Cameron, Ken, 28, 281–282, 287
Carr, Gerald M., 185–186
Carter, Sonny, 22, 24
Casper, John, 22, 424
Castle, Bob, 490, 491
Chadrin, Viktor, 190–191, 453–454
Chaffee, Roger, 20
Challenger, 18, 22, 23, 80, 212, 348, 431, 503
Chernomyrdin, Viktor, 70, 120, 266, 267, 273, 302, 304

Chiodo, Christine, 115, 118
Chretien, Jean-Loup, 143
Christopher, Warren, 267
Clarke, Bob, 252–253
Clifford, John, 44
Clifford, Michael R. ("Rich"), 21, 29, 30, 34, 227, 344, 441
Clinton, Bill, 35, 49, 243, 246, 262–263, 264, 266–268, 327
Coats, Mike, 29, 31
Cohen, Aaron, 24, 240
Collins, Eileen, 25, 307
Congress, 23, 119–120
Corn, Barbara, 500–501
Cosmonaut Museum, 55–59
Covey, Dick, 28
Crew On-orbit Support System (COSS), 184
Crippen, Bob, 22
Culbertson, Frank, 22, 28, 48, 80, 153, 262, 279, 305, 334, 359, 506, 508–509, 510
  background of, 40–41
  as Phase One director, 41–42, 142, 342, 511
  Blaha's mission and, 49–50, 95, 99, 100, 106–107, 112
  fire emergency aboard Mir and, 138, 147, 154–155, 179
  Foale's mission and, 391–392, 406–407, 408, 418–419, 420, 441–447, 455, 459
  Linenger's mission and, 40, 49–50, 169, 173, 177–179, 197, 232
  NASA's knowledge of Russian programs and, 138–139, 141, 512–513
  Progress docking near miss and, 163, 165
  safety concerns about Mir and, 210–212, 428–429, 440, 441–447, 464, 465–466, 492, 493–494, 496
  support for astronauts from, 49–50

Dailey, Brian, 246–249, 253, 254, 255–256
Daniels, Chuck, 272, 273
Darman, Richard, 240, 244
Dateline NBC (television program), 43, 45
Davis, Lynn, 267
Derechin, Alexander G., 66–67, 274*n*
Dezhurov, Vladimir, 88, 289, 312, 313, 318, 319, 320, 321, 323–324, 326–327, 344
Discovery, 4, 307–311
Dobrovolskiy, Georgiy, 57–58
Dronov, Vladimir, 290
Duce, Jack, 387
Dunbar, Bonnie, 349, 513
  background of, 277–278
  health issues of, 301–304
  Mir mission of, 324
  NASA medical experiments and, 297–301
  relations between Dunbar and, 284, 286, 287–288, 295–296
  Russian chauvinism and, 284–287
  selection of, as an astronaut, 276, 278–279, 280
Dunn, Marsha, 180
Dzhanibekov, Vladimir, 409

Earth Observation System (EOS), 244
Egan, John, 48
Elektron systems, 72, 91, 123, 166–167, 210–211, 334, 390, 485–486
Endeavour, 27
Energia, 56, 70, 73, 98, 140, 306, 445, 498, 510
  cosmonauts and, 63–64, 71–72
  docking unit design and, 66, 67, 87, 427, 486
  fire emergency on Mir and, 153, 154
  setting up Shuttle-Mir program and, 249–250, 252–253, 255–256
  space station construction and, 265, 266–267, 271–273, 292–293
Engelauf, Phil, 41, 142, 318, 328, 453–455
Engle, Joe, 462, 463–464, 497–498
Ermac, Sasha, 292
European Space Agency (ESA), 314–315
EVAs. *See* Spacewalks
Ewald, Reinhold, 85, 116, 121, 123, 148, 151, 181, 404
  Cosmonaut Museum ceremony and, 56, 58–59
  fire emergency and, 125, 127, 128, 131
  flight to the launch site by, 76–77
  launch preparations and, 82–85
  Soyuz's docking problems and, 86–87, 115
Extravehicular activity (EVAs). *See* Spacewalks
Eyharts, Leopold, 442–443

Fabian, John, 462
FBI, 43, 44–45, 46, 505
Ferring, Mark, 220*n*
Fisher, Craig, 463, 497
Fletcher, Jim, 241
Flynn, Chris, 468
Foale, C. Michael, 6, 85, 113, 122, 210, 307, 311, 487, 499, 509, 513
  arrival on Mir of, 233–235
  assessment of pressure problems and, 395, 396
  astronaut training and, 349–350
  background of, 347–349
  decompression problems and sealing off Spektr module and, 373–375, 377–378, 383, 385–386, 391, 392–394
  electrical power loss and, 399–406
  Linenger's mission and, 173–174, 219, 233, 350
  Mir mission of, 341, 347, 352–354, 360–396, 420, 448–450
  Mir training and, 350–352
  recovery strategy and, 411–416, 418, 421–422, 423, 426, 456, 459
  relations between Solovyov and, 470, 473–474, 475–486, 494–495
  repair work by, 435–436, 439

Russian language use by, 350–352
spacewalk by, 488–490
Star City training of, 9, 106
Tsibliyev's manual docking exercise and, 360–361, 362, 364, 367, 368, 371–373
Foale, Rhonda, 9, 349, 350
Fuerth, Leon, 267, 492–493
Fullerton, Richard, 220, 222
Fursow, Irene, 255–256, 257–258

Gaffney, Drew, 18, 22, 23
Gagarin, Yuri, 55, 56, 57, 58, 59, 69, 79, 80, 84, 212
Gemini program, 182, 266
Gerstenmaier, Bill, 101–102, 328, 332, 333, 334–336
Gibbons, Jack, 459
Gibson, Edward G., 185–186
Gibson, Robert L. ("Hoot"), 9, 21, 22, 23, 24–25, 26–27, 29, 34, 42, 45, 276, 277, 280, 296, 324, 325, 463
Gidzenko, Yuri, 510
Glaskov, Yuri, 80, 83, 217, 218, 284, 296, 458, 498
Glenn, John, 5, 40, 347
Globa, Tamara, 74–75, 174–175
Goddard, Robert, 79
Godwin, Linda, 227
Goldin, Daniel S., 18, 35, 119, 210, 218, 261, 274, 513
appointment as director of NASA, 24, 243–245
decision to continue Phase One program and, 462, 496, 500, 501, 503–504
Dunbar's mission and, 302–303, 324
Foale's mission and, 406–407, 408, 450
safety concerns about Mir and, 490
setting up the Phase One program and, 247–248, 253–254, 255–256, 262–263, 264, 265–268
Thagard's mission and, 276, 279–280, 283, 325
Wolf's mission and, 501–502
Goncharov, Boris, 308, 309, 310, 334–336
Goncharov, Igor, 137, 202, 439, 456–457, 459–460, 472
Gorbachev, Mikhail, 60, 246, 247, 265
Gore, Al, 70, 120, 245, 267, 273, 304
Gorshkov, Leonid, 272
Grechko, Georgi, 225
Gregory, Fred, 33, 210–211, 346–347, 423, 424, 425, 493–494
Grigorev, Eduard, 292
Grissom, Gus, 20
Gross, Roberta, 34, 463, 496, 500
Grunsfeld, John, 38–39, 51, 52
Gyrodynes, 92, 176, 422

Hale-Bopp comet, 172, 175
Hammond, Blaine, 262, 434–435, 444, 513
astronaut career of, 9, 27–34
Linenger's criticisms and, 345–347
safety concerns about Mir and, 419, 423–425, 430–431, 441, 490, 491, 493, 494, 499
Harris, Byron, 216–218
Hauck, Rick, 28
Hilmers, Dave, 258, 260–261
Holland, Al, 94–95, 100, 112, 181–187, 221, 315–316, 321, 356, 358–359
Holloway, Harry, 291
Holloway, Tommy, 41, 42, 274, 305, 317, 324, 325, 328
Horner, Connie, 242, 243
Horowitz, Scott, 31
House Science Committee, 119, 212, 218, 495–495
Human Exploration of Deep Space (HEDS), 490–494
Huntoon, Caroline, 278, 279, 295, 297, 298

Institute for Biomedical Problems (IBMP), 64, 185, 200, 283, 315, 435, 457, 507
International Space Station (ISS), 28, 35–36, 119–120, 142, 271–274, 351, 495–496, 509–510
Ivins, Marsha, 30, 39, 51, 52, 96

Jennings, Richard, 278–279
Jett, Brent, 39, 51
Johnson, Gary, 139, 144, 153–154, 492, 493, 510
Johnson, Gaylen, 334

Kaleri, Aleksandr ("Sasha"), 63, 87, 116, 118, 121, 123, 146, 148, 151, 154, 166, 514
fire emergency and, 125–127, 130, 131, 132, 133–134
relations between Blaha and, 96, 105–106, 108, 109, 111, 112–113
Kanas, Nick, 184
Kargopolov, Yuri, 301
Kato, Takao, 272–273
Keller, Sam, 249, 256, 257*n*
Kennedy, John F., 240, 263
Kerwin, Joe, 182
Khrushchev, Nikita, 79
Klimuk, Pyotr, 54, 62, 217, 281, 282, 350, 358, 458
Klein, Kevin, 492, 494
Knight, Jack, 142–143
Kobsev, Gene, 326
Komarov, Vladimir, 80
Kondakova, Elena, 284–285, 313
Koneev, Ekrim, 316, 452–453, 454
Koptev, Yuri, 59–60, 120, 147, 250, 255–258, 265, 267–268, 273, 279, 283, 302, 324, 450
Korolev, Sergei Pavlovich, 57, 59, 79
Korzun, Valery, 63, 116, 118, 121–122, 123, 148, 151, 154, 285, 289, 337, 514
Blaha's changeover period after arrival and, 95–98

Korzun, Valery *(cont.)*
docking problems of cosmonauts and, 86–87
fire emergency and, 125–132
fire precautions and, 134–135
fire reporting by, 133–134, 135–137, 147
as Mir commander, 116–117, 145–146
relations between Blaha and, 105, 106, 109, 110–111, 112–113
relations between Linenger and, 116–117, 145–146
salary and bonus of, 116–117
Soyuz's docking problems and, 86–87, 115
weighlessness effects on, 116, 123
working hours of, 167, 199, 200, 205–206
Kraft, Christopher, 20
Kranz, Gene, 20–21, 24, 34–35, 93, 183, 260, 274
Krikalev, Sergei, 62, 139, 268, 279, 351, 352, 359–360, 387, 391, 412–413, 414, 417, 510
Kristall module, 81, 88, 91
Kristol, Bill, 241
Krug Life Sciences, 99, 289, 291, 331
Kurs docking system, 65–66, 67, 86–88, 165, 359, 474, 510
Kvant modules, 66, 81, 90, 91, 188–189

Lake, Anthony, 264–265
Lane, Helen, 42
Larsen, Axel ("Skip"), 491–492, 494
Larsen, David, 316
Laviekin, Aleksandr, 81
Lawrence, Wendy, 48, 424, 430, 431, 462, 465–466
Lazutkin, Aleksandr ("Sasha"), 74, 118, 121, 166, 171, 172, 174, 345, 409–410, 435, 444, 448, 514
coolant systems problems and, 188–195, 215, 354–356
Cosmonaut Museum ceremony and, 56, 58–59
decompression problems and sealing off Spektr module and, 373–378, 383, 384, 385–386, 388–389, 391, 392–394
electrical power loss and, 399–404
family gathering before launch and, 75–76
fire emergency and, 124–127, 130, 134, 136, 147
flight to the launch site by, 76–77
Foale's mission and, 347, 352–353, 471
launch of mission of, 85
launch preparations and, 82–85
manual docking exercise and, 361, 364–373
Progress docking near miss and, 157–159, 161, 162, 165
recovery strategy and, 411–416, 417–419, 426–428, 432, 455–460
relations between Linenger and, 221, 232–233, 234
repair work by, 166–167, 435–436, 437, 451–453, 454
return to Earth by, 471–473, 475
safety concerns and, 73, 211
salary of, 64
Lebedev, Olga, 294
Lebedev, Valentin, 284
Lee, Mark, 258–259, 260
Leestma, David, 26–27, 31, 33, 155, 173, 274, 276, 277, 278, 288, 296, 430, 433
Legostaev, Viktor, 271–272
Linenger, Fran, 6, 7, 37, 38, 51, 52–53
Linenger, Jerry, 1–15, 123, 167, 170, 173–174, 277, 514
arrival on Mir by, 88, 95
concerns about attitude of, 14–15, 177–180, 181, 187, 198–299, 200
coolant systems problems and, 190, 191, 192–193, 194, 195–196, 198
departure from Mir by, 234
exercise by, 191, 196–197, 205–206
family conferences from Mir, 121
family gathering before launch and, 37–38
fire emergency and, 128–130, 131–132, 345, 460, 509
fire precautions and, 134–135
fire reports and, 146, 147, 155–156, 179, 180–181
health concerns of, after fire, 134, 135, 144–146, 148, 149, 148, 149–150, 171–172
launch of mission of, 52–53
launch preparations and, 49–52
letters to his son from, 117–118, 128, 174, 175–176, 230
medical tests and, 11–15, 172–173, 174, 203–205
moods of, during Mir voyage, 118–119
opinions of NASA held by, 342–343, 345, 419–420, 441
personality of, 14–15
Progress docking near miss and, 156–159, 161, 162, 163–164, 345
relations between Korzun and, 116–117, 145–146
relations between Lazutkin and, 221, 232–233, 234
relations between Tsibliyev and, 11, 166, 190–192, 198–199, 205–206, 209–210, 214, 232–233
return to Earth by, 342–346
safety concerns about Mir and, 464, 500–501
Soyuz's docking problems and, 86–87, 115
space shuttle experiments and, 115, 126, 167–168, 177–179, 207, 208–209
spacewalk by, 201–203, 215–216, 219–232, 343
voice communications with, 152, 168–170, 171, 190, 316*n*
Linenger, John, 6, 38, 117–118, 128, 174, 175–176, 230
Linenger, Kathryn, 7–8, 9–10, 115, 121, 187, 201

Lockheed Martin, 104, 239, 314–315, 331
Lovell, Jim, 422
Lucid, Shannon W., 10, 42, 44, 98, 126, 261, 298, 419, 430–431
  background of, 329–330
  media and, 49, 284, 327
  Mir mission of, 62, 99, 101–102, 104, 105, 141, 189, 259, 334–336, 513
  relations between Blaha and, 94, 95, 107, 329, 330, 331, 336–337
  safety concerns about Mir and, 423, 424–425
  space station experiments of, 331–333, 334
  training of, 329, 330–334
  voice communications with, 316*n*
Lunney, Glynn, 423–425

McBrine, John, 44, 46, 47, 48, 99, 103, 114, 289–290, 297, 314, 315, 317, 332, 465–466
McCulley, Mike, 30, 31
McGinnis, Pat, 104–105, 109–110
McPherson, Alex, 291
Maliga, Kate, 341, 342, 382
Malyshev, Mike, 192, 196, 381–382, 387
Manakov, Gennadi, 105, 269
Manarov, Musa, 66
Manber, Jeffrey, 71, 249, 512
Marshburn, Tom, 10, 40, 42, 104, 117, 122, 168, 170–171, 172, 176, 189–190, 191, 192
  Linenger's attitude and, 118–119, 178, 180, 187, 203–204
  Linenger's health concerns and, 11, 12, 149–150, 155, 214
  Progress docking near miss and, 160–161, 162–163
  spacewalk and, 215–216, 221, 232
  voice communications between Linenger and, 115, 168–169, 196–197
Mars probes, 79, 80, 240, 241
MEAT (Mir Environmental Assessment Team), 140
Medford, James, 341, 381–382, 387, 391–392
Media, 27, 74
  communications with Mir and, 145
  cosmonauts' launch and, 83, 85
  Discovery problems and, 308–309
  Foale's mission and, 234–235, 408, 410–411
  Linenger's mission and, 50, 51, 180, 216, 345, 460
  Lucid's mission and, 49, 284, 327
  reporting of Mir fire in, 146, 460
  safety concerns about Mir reported in, 434
  Sensenbrenner criticisms in, 119–120, 147
  Shuttle-Mir program in, 248–249
  Thagard's mission and, 326
  Tsibliyev's return and, 474–475
Melroy, Pam, 51–52
Merbold, Ulf, 314
Mercury program, 4–5, 182, 266
Merrell, Ronald, 497
Mir, meaning of, 80
Mir Environmental Assessment Team (MEAT), 140
Mission Control Center. *See* Russian Mission Control Center
Missions Operations Directorate (MOD)
  attitude toward astronauts in, 41
  Kranz's supervision of, 21
  Mir fire reporting and, 142–143
  Phase One program and, 141–142, 328–329
  Thagard's mission and, 317–318
Mobley, David, 271–273
Modules of Mir space station, 80–81, 88–92
Moon probes, 79, 80, 240
Moore, Isaac ("Caasi"), 101, 104, 107, 108, 109, 110, 111–112, 113, 328
Morgun, Valery, 14–15, 283, 297, 300, 302, 303–304
Muller, Matt, 11, 14, 99, 105, 219, 220, 289–290, 332–333
Musabayev, Talgat, 467, 508
Musgrave, Story, 34, 311
Mussara, Gerald, 257*n*

Nagel, Steven R., 26
National Public Radio, 174, 175–176
National Space Council, 24, 239
Navius, Rob, 420
Nedelin, Mitrofan, 59, 79
Nikiforov, Nikolai, 110, 316, 379, 380–381, 383, 387
Node, 90
N-1 moon rockets, 78
Nygren, Rick, 99–100

Oberg, James E., 78, 79, 144, 210, 216, 217–218
O'Connor, Bryan D., 23, 251–257, 278
Onufriyenko, Yuri, 63, 334
Operation Lightning Strike, 45
Ostroumov, Boris, 252, 257

Parazynski, Scott, 173, 349
Patsayev, Viktor, 57–58
Phase One program (Shuttle-Mir program), 36
  astronaut selection for, 21–27
  bureaucratic snafus in, 122
  completion of, 508–509
  Culbertson as director of, 41–42, 142, 342
  Goldin's decision to continue, 501, 503–504
  legacy of, 511–513
  Linenger's criticisms of, 342–347
  NASA's view of, 42
  negotiations to establish, 247–258
  reaction to Mir fire in, 138, 142–143
  safety reviews and, 34–35
Pickering, Thomas, 280
Pogue, William R., 185–186
Polyakov, Valery, 143
Portree, David, 139–140

Precourt, Charlie, 69, 211, 233–235, 342, 392, 446, 487
Press. *See* Media
Prestridge, Liz, 245
Priroda module, 73, 81, 90–91, 291
Progress cargo ships, 63, 65, 66, 67, 81, 90, 154, 156, 202, 207–208
  docking near miss during Linenger's mission involving, 156–159, 161, 162, 163–164
  Tsibliyev's manual docking exercise with, 359–362, 363–373
Proton rockets, 80, 85, 267
Psychological support. 14–15, 181–187, 315–316

Quayle, Dan, 24, 239, 241, 242, 243, 245, 248–249, 250

Radiopribor, 66–67
Rahn, Debbie, 331
Rea, Chris, 85
Readdy, Bill, 25–26, 276, 277, 280, 296, 299–300
Reagan, Ronald, 263
Reeves, Bill, 307–311
Resnick, Judy, 41
Richards, Dick, 32
Romanenko, Yuri, 81, 225
Ross, Brian, 45
Ross, Jerry, 45
Russian Mission Control Center (TsUP)
  cosmonaut's compensation program and, 64
  docking system design and, 65–67, 86
  downplaying of problems by, 145, 146–147, 165
  fire emergency on Mir and, 133–134, 145, 146–147, 153–154, 180–181
  physical state of, 61
  Progress docking near miss and, 161–162, 164–166
  U.S. space shuttle experiments and, 100, 168
Russian Space Agency
  Energia's power and, 71–72
  partnership with NASA, 81–82, 511–512
  role of cosmonauts and, 63–64
Russian space program
  funding of, 59, 61–63, 65, 66–67, 71, 81, 120
  historical overview of, 79–82
  NASA's knowledge of, 138–143, 163–164
  state of, 59–64
  tensions between NASA and, 120
Ryumin, Valery, 41, 70–71, 73, 212, 284–285, 327, 334, 465, 504
  Discovery problems and, 309, 310–311
  on support for the Mir program, 81–82
  safety concerns over Mir and, 304–305, 306, 324, 428–429, 442
  setting up the Shuttle-Mir program and, 252, 253, 257
  Tsibliyev's mission and, 449, 450, 455, 457
Sagan, Carl, 239, 241
Salyut space station, 57–58, 80, 143, 225
Sang, Tony, 140, 170–171, 176–177, 192, 328, 464
  fire emergency aboard Mir and, 137–138, 145, 148, 181
  Linenger's support and, 10, 114, 116, 173, 180, 201–203, 208, 214, 215, 219
  Progress docking near miss and, 160–161, 162–163
  voice communications between Linenger and, 115, 168–169, 171–172, 176–177
Santy, Patricia A., 18, 182, 183
Savinykh, Viktor, 409
Savitskaya, Svetlana, 284
Sawyer, Kathy, 248–249
Schirra, Wally, 185
Sega, Ron, 288, 296, 301–303
Semenov, Yuri, 71, 212, 249, 255–256, 265, 273, 324, 422–423, 433–434, 457, 465
Sensenbrenner, James, 119–120, 147, 212, 407, 494, 495–495, 503
Serebrov, Aleksandr, 69, 70–71, 72–73, 226, 473, 488
Sevastyanov, Vasily, 63, 64
Severance, Mark, 151–152, 358, 386, 387, 404, 410–411, 437, 438–439
Shepherd, Bill, 5, 27–28, 120, 262, 275–276, 510
Short, Tom, 462
Shuttle-Mir program. *See* Phase One program
Skinner, Samuel, 242
Skylab, 185–186
Solovyov, Anatoli, 120, 166, 301, 324
  Mir mission of, 457, 469–471, 473–474, 475–486, 494–495, 504–507
  relations between Blaha and, 106
  spacewalks by, 223–224, 489–490, 506–507
Solovyov, Vladimir, 62, 198, 209–210, 218, 232, 309, 335, 358
  Anatoli Solovyov's mission and, 477, 483–484, 485
  fire emergency aboard Mir and, 137, 138
  Progress docking near miss and, 162–163, 164–165
  Tsibliyev's mission and, 279–281, 383–385, 386, 387–391, 394–398, 401, 405–406, 408, 411, 422, 437, 438, 456, 458, 472
Sotnikov, Boris, 153–154
Soyuz spaceships, 59, 65, 69, 70–71, 79–80, 86
Space Cooperation Working Group, 265
Space News, 27, 51, 144
Spacesickness, 116, 123
Space suits, 12, 39–40, 82, 84, 221–222, 226, 349, 436
Space Station Freedom, 239, 250, 263
Spacewalks, 91
  Dezhurov and Strekalov, 322–323
  Solovyov and Balandin, 223–224
  Solovyov and Foale, 488–490

Solovyov and Vinograd, 506–507
Tsibliyev and Linenger, 201–203, 215–216, 219–232, 343
Wolf and, 467–469
Spektr module, 73, 81, 91, 291
Sputnik satellite, 79
Stafford, Thomas, 25, 210, 253, 462–463, 491, 497–498, 501, 502
Stafford Commission, 462, 486–488, 491, 494
Stone, Randy, 142
Strekalov, Gennadi, 70–71, 312–313, 314, 315, 318, 319, 320, 321–322, 324, 326–327
Sununu, John, 240, 242
Synthin, 65

Taddeo, Terry, 122, 341, 359, 376, 396, 404, 434, 436
Tate, Tom, 22
Television. *See* Media
Thagard, Danny, 184
Thagard, Kirby, 260, 275, 282, 288–289, 315, 316, 320, 326
Thagard, Norman E., 23, 107, 285–286, 297, 298, 302, 304, 513–514
background of, 259–260
first Mir mission of, 8, 10, 25, 26, 41, 88, 89*n*, 91, 183–184, 312, 313–24
medical emergencies on Mir and, 321–322
relations between Dunbar and, 278–279, 284, 286, 287–288, 295–296
return to Earth by, 324–326
selection of, as an astronaut, 258–259, 260–261, 275, 280
space shuttle experiments of, 98–99, 313–315, 317–319
training of, 280–284
weight loss of, 318–320
Thomas, Andy, 440, 507–508, 512, 513
Titov, Vladimir, 25, 59, 70–71, 268, 307
Topol, Igor, 416
TORU system, 66, 67, 69–70, 156, 360, 510
Truly, Richard, 19, 21, 24, 240, 241, 242, 245, 503
Tsibliyev, Larissa, 68, 90, 121, 415, 429, 473
Tsibliyev, Vasily ("Vasya"), 68–82, 106, 116, 121, 123, 151, 172, 204–205, 429–430, 470, 514
assessment of pressure problems and, 394–398
background of, 68–69
backup oxygen canisters and, 170–171
coolant systems problems and, 188–195, 198, 212–213, 215, 354–356
Cosmonaut Museum ceremony and, 55–59
decompression problems and sealing off Spektr module and, 373–375, 383–385, 386, 387–394
electrical power loss and, 398–406
family gathering before launch and, 68
fire emergency and, 125–130
fire reporting and, 134, 136, 147
first flight to Mir by, 69–70
flight to the launch site by, 76–77
Foale's mission and, 347, 352, 353–354, 471, 509
launch of mission of, 85
launch preparations and, 82–85
manual docking exercise and, 359–362, 363–373, 379–381
mental health of, 358–359, 360, 434–435
physical health of, 199–200, 421, 444, 447–450, 459
premonition of disaster felt by, 73–75, 174–175
Progress docking near miss and, 156–159, 160–162, 165, 166, 445, 446, 465, 486–488, 498
proposed internal spacewalk of, 420–421, 435, 450–451
recovery strategy and, 411, 414–419, 420–422, 426–428, 432, 455–461
relations between Linenger and, 11, 166, 190–192, 198–199, 205–206, 209–210, 214, 232–233, 234
repair work by, 436, 437, 438–439, 452–453
return to Earth by, 471–473, 474–475
safety concerns and, 73, 211
salary of, 64, 88
sleep study and, 357–357, 361
Soyuz's docking problems and, 86–88, 115
spacewalk by, 201–203, 219–222, 225, 226–232
working hours of, 167, 199, 200–201, 205–206
Tsiolkovsky, Konstantin Edvardovich, 79
TsUP. *See* Russian Mission Control
Tsygankov, Oleg, 427–428, 429

Uri, John, 251–252
Usachov, Yuri, 63, 334
Utkin, Vladimir, 462, 497–498

Van Laak, Jim, 80, 100, 153, 173, 317, 328, 334, 341, 360, 506
fire emergency aboard Mir and, 138, 139, 141, 142–143
Foale's mission and, 392, 406–407, 408, 418–419, 455, 459
Linenger's mission and, 343, 344–345
Progress docking near miss and, 164, 165, 486–487
safety concerns about Mir and, 210–212, 429, 440, 442, 443, 445, 464, 492, 496, 498
spacewalks and, 221, 225
Venus probes, 79
Versekis, Albertas, 390
Viktorenko, Aleksandr, 311
Vinogradov, Pavel ("Pasha"), 105, 457, 469–471, 476, 477–486, 495, 504–507
Volkov, Vladislav, 57–58

Vorobiev, Pavel, 272
Vozdukh system, 92, 188, 193–194, 212, 390

Ward, David, 12–14, 104, 278, 281, 282, 283–284, 287, 312, 315, 316, 319, 326, 330
Watson, Cathy, 341, 382
Weldon, Dave, 496
Wetherbee, Jim, 22, 25, 28, 33, 307, 309, 311, 352
White, Ed, 20, 43
Whitson, Peggy, 89*n*, 291–292, 293, 294–295, 331
  Dunbar's mission and, 297, 304–306
  Thagard's mission and, 317, 320, 324–325, 327
Williams, Don, 30–31
Wolf, David A., 12, 42–48, 466–467, 490–492, 494, 499, 501–502, 504–507, 512, 513

Yeager, Chuck, 4, 5
Yeltsin, Boris, 35, 59, 60, 70, 72, 246, 247, 250, 253, 265, 266, 267, 274, 279, 472–473
Young, A. Thomas, 500–501, 502

Zimmerman, Keith, 113, 140, 328, 341, 358, 375–376, 380, 386, 387, 391, 396, 404, 413–414, 416, 417, 420